Combined Proceedings, International Plant Propagators' Society 2019

IPPS Volume 69

2019

Editors:

Robert Geneve
Clive Larkman
Charlexs W. Heuser Jr.
Tim Lawrance-Owen
Masanori Tomita
Hayden Foulds
Elsa du Toit
Fred T. Davies Jr.
Ryan Stewart
Rakhi Chaturvedi

ISBN: 978-1-7138-1369-9

Printed from e-media with permission by:

Curran Associates, Inc.
57 Morehouse Lane
Red Hook, NY 12571

Some format issues inherent in the e-media version may also appear in this print version.

International Plant Propagators' Society (IPPS)
174 Crestview Drive
Bellefonte, PA 16823-8516
USA

Phone: +01 814-355-2467

secretary@ipps.org

Additional copies of this publication are available from:

Curran Associates, Inc.
57 Morehouse Lane
Red Hook, NY 12571 USA
Phone: 845-758-0400
Fax: 845-758-2633
Email: curran@proceedings.com
Web: www.proceedings.com

THE INTERNATIONAL PLANT PROPAGATORS' SOCIETY

Combined Proceedings

Volume 69

2019

IPPS

Sharing Plant Production Knowledge Globally

Published 2019

Combined Proceedings International Plant Propagators' Society (ISSN 0538-9143) is a yearly publication of the International Plant Propagators' Society (IPPS) headquartered in Pennsylvania, U.S.A.
Email: Secretary@ipps.org
Website: www.ipps.org

Business Office. Memberships are processed at the regional level, with dues being allocated to both international and regional expenses. Contact the regional secretary / treasurer in your area of the world for membership information and applications.

Table of Contents

Japan Region

Western Region – North America

Eastern Region – North America

Southern Region – North America

European Region

Singapore Symposium

NGPPBUT Conference Abstracts – Assam, India

Rakhi Chaturvedi - Convener

PROCEEDING'S PAPERS

SOUTHERN AFRICA REGION

Dr. Elsa du Toit, Regional Editor

Twenty-second Annual Meeting - 2019

Stellenbosch, South Africa

Combined Proceedings IPPS 69:1. 2019.

Horticulture on the Wild Side

David Hancock – Perth Plant Propagation Pty. Ltd. Western Australia 6068, Australia

david@naturalarea.com.au

Keywords: Oilfield water, constructed wetland, Oman wetland species, seed, propagation, nursery construction, nursery operating procedures

Abstract

The design and construction of a new nursery in the Omani Desert and the growing of 800,000 plants for a constructed wetland to purify oil field wastewater. The challenges, risks and environmental benefits to deliver all from scratch in a 12month timeframe.

INTRODUCTION

In early 2018, I was approached by the Project Manager for German environmental company Bauer Resources to provide consulting services for a new nursery development and propagation project for the Nimr oilfield in Oman. The introduction came via an IPPS connection.

Nimr is home to the largest constructed wetland in the world, used to purify oil laden water. To facilitate a 250-hectare extension of the wetland, I was engaged to consult on the project in a wide brief to advise on seed collection and processing, propagation, nursery design & construction, nursery operating procedures and risk management.

Support was provided by Dermot Molloy, Horticulturist and propagator at Royal Botanic Gardens Victoria as a subconsultant.

Four species of wetland plants were proposed to achieve the 800,000 in total required and the presentation discussed the methodologies recommended to assist the project team, as well as recommendations for all requirements of the consulting brief.

A comprehensive written report was compiled and submitted within 14 days of the site visit. The demanding brief, short timetable, remoteness of the location and the growing conditions made for an intensive and demanding project.

Combined Proceedings IPPS 69:2-3. 2019.

The Future of Landscaping: Understanding and Embracing Resilience

Marijke Honig, Beyer Honig Landscapes

Cape Town, South Africa

landscapes@beyerhonig.com

Keywords: Genetic diversity, water, conservation

Abstract

With climate change and dwindling water resources, **resilience** is the new criterion for plants and landscapes. Resilience not only to physical events (drought, storm winds, flooding), but also to human disturbances such as bark stripping, trampling.

We need to identify suitable plants that can adapt to a changing climate and bounce back after an extreme event. They may be indigenous or exotic. Need to find suitable combinations of plants – plant mixes – that require minimal maintenance.

INTRODUCTION

What changes can we expect with climate? Weather patterns will become more unpredictable and extreme.

How can we respond?

By creating resilient landscapes which can adapt to changing conditions and bounce back after an extreme event.

Resilience to what?

What are some of the general attributes of resilient plants and landscapes?

- Adaptable, can tolerate a wide range of conditions (not fussy about soil type, moisture, temperature).
- Can 'bounce back' (regrow) after disturbance or an extreme event.
- Resprouters (plants that can survive and regrow from an underground rootstock after fire), and/or plants that can be heavily pruned and will regrow.

IPPS Vol. 69 – 2019

- Requiring little or no human input to survive (e.g. no watering or fertilising).
- Genetic diversity: plants grown from seed have different genotypes and therefore slightly different attributes. When there is an extreme event there is a greater chance that one / some individuals may survive than when there is a single genotype (i.e. a plant cultivar or variety which is clonally regenerated).
- Biodiverse landscapes tend to be more adaptable than those with low species diversity - more likely to include some species that will survive extreme events.

In addition to these general resilience traits, one can consider specific potential threats or future conditions that may affect landscapes.

- A site with a borehole in a region affected by power cuts is likely to have an interrupted water supply, resulting in temporary DROUGHT.
- A site on the urban edge and adjacent to flammable vegetation may be exposed to FIRE.
- Plantings in public areas or school grounds may be subject to TRAMPLING.
- In some area plants may be dug up (e.g. Porcupines, dune mole rats) or eaten by wild animals (HERBIVORY).

PROCEEDING'S PAPERS

AUSTRALIA and NEW ZEALAND REGION
Joint meeting

Clive Larkman, Australian Regional Editor

Hayden Foulds, New Zealand Regional Editor

Joint Annual Meeting - 2019

Brisbane, Australia

Variations and Peculiarities in Plant Breeding

H. William Barnes

Barnes Horticultural Services, Warrington, PA

bill@barnhortservices.com

Keywords: Hybrid crosses, *Hibiscus*, *Capsicum*, pollen

INTRODUCTION

Plant breeding can be approached via a variety of avenues. Time tested methods of hand pollination either by direct transfer by fingers, popsicle sticks or artists brushes are still quite valid and useful. Some new techniques include caging plants with bees, and using mechanical devices such as electric tooth brushes to dislodge pollen. Still other methods can be as simple as interplanting two different cultivars or plants in a given area and let natural forces do the job.

Simple crosses such as corn hybrids can be done by planting two types adjacent to one another and emasculating one clone so that it can only be pollinated by its neighbor. Similar techniques are used for the crossing of *Castanea dentata* hybrids. But these applications work only when standard Mendelian Genetics apply. In the case of annuals specific clones are back bred to F6 status. This eliminates most if not all extraneous genes so that the crossing of two F6 clones result in a specific given hybrid. It could be safely assumed that a cross of a specific red petunia with a specific white petunia will give rise to a chosen pink petunia. However, when it comes to woody plants and some perennials it is not always possible to go to the F6 status. As a result, the offspring of crossing two distinct but variable individuals will result in significant variation in the resulting seedlings. Of course, this in turn requires the rigors of planting out and selection both visually and environmentally to achieve a desired goal. Somewhere, somehow, the bulk of the offspring under testing have to be dispatched and eliminated in preference to the best possible choices.

Many of today's roses are the product of such crosses where for instance Mr. Lincoln is crossed with Chrysler Imperial. The results of such a cross will be quite varied and will at times give some unexpected results. In work at Barnes Horticultural Services a cross of *Hibiscus syriacus* hybrid 'Blue Bird' as the female parent and *Hibiscus* x*tosca* (*H. syriacus* x *H. paramutabilis*) as a male, gave rise to 15 different individuals based on form and flower color. Variations include those that looked like the parents but also a vivid pink, a dwarf white and a fastigiated purple amongst the 15 individuals. Some are worth keeping and others not so. Some had unique leaves but non-descript flowers. Again, selection pressure and evaluation were applied to discover those with the best merits. It should be noted that this is with plants that are chromosomally compatible. But when disjunct and poorly coordinated chromosome pairings are made problems immediately jump to the forefront.

In *Capsicum*, there are two separate lines based upon chromosome numbers (12,14), one line is 2x= 24 and the other is 2x=26. Crosses between these two lines is not possible and stymies attempts at breeding for acquiring novel traits. However, if these lines are doubled to tetraploid then crosses could be made but the resulting offspring would be most likely sterile. Researchers (personal communications) at the University of Florida have told me that they cannot get large fruited hot peppers to cross with small fruited hot peppers even if the chromosome numbers are compatible and they are mystified about why this situation exists. Possible answers to this which will be discussed further along in this paper.

In my personal work I have found that crossing *Hippeastrum x Clivia* will work but only in that direction. *Clivia x Hippeastrum* will not take. The late Luther Burbank (1914) suggested that reciprocal crosses will be just as effective as the opposite crosses but sad to say he was wrong.

Explanations as to why this might be the case are complex and varied. But sometimes things can be elucidated. In the Genus *Hibiscus* timing of pollination seems to be a consistent pattern. In Zone 6, *H. syriacus* stigma are not generally receptive until 10 AM and pollen generally is not available prior to 10 AM and sometimes matures much later than 10 AM. However, this is temperature related and things can speed up with warmer ambient temperatures, although it is not clear if the stigmatic surface receptivity functions on the same pace as pollen maturing. Cooler temperatures result in significantly delayed pollen maturity. These mechanisms seem to be a device to limit self-pollination and serve to foster outcrossing. It should be noted that many of the cultivars of *Hibiscus syriacus* are self-sterile and out crossing is a must. However, *H. syriacus* goes one step further and generally shuts down the whole operation at 2 PM. Few if any pollinations will occur in the *Hibiscus* after 2 pm. Like the morning phase the afternoon phase can be altered by temperatures being hotter or cooler on any given day. Things are speeded up with warm temps and delayed with cooler ambient air temps. I work with about 40 species of *Hibiscus*, *Abutilon* and *Pavonia* and they all follow similar timing regimens. This of course makes me wonder if other plants have similar built in "rules" on outcrossing. Perhaps this is the source of the reluctance of the *Capsicum* at the University of Florida not following a predictable pattern. It may be predictable, but it simply is not known what it is.

Several factors that govern pollination effectiveness include:

a. Pollen is not ripe at the time of pollination.
b. Stigma surface is not receptive due to poor timing.
c. Stigma surface is not receptive due to a hostility to incoming pollen.
d. Temperature and humidity are unsuitable for effective pollen germination.

Pollination involves two processes - pollen germination upon sticking to a conducive and receptive stigma followed by pollen tube growth allowing for a pathway for sperm to move down into the ovary. Another aspect of the Malvaceae, is that the pollen often varies with a certain degree of stickiness. Some pollen when fully mature will be extremely sticky and will adhere to just about anything, at other times the pollen will not adhere to brushes, or other implements or for that matter to insect pollinators. Unless conditions are exactly in sequence and in complete agreement an effective pollination will not occur.

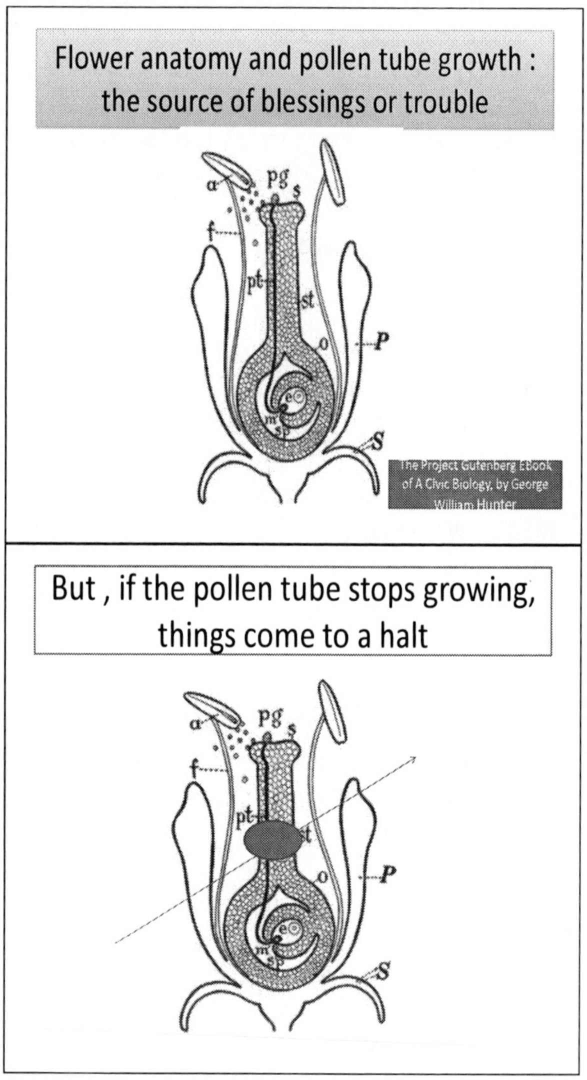

Figure 1. Pollen tube growth impacts successful fertilization.

The female/mother plant controls all of the interactions with pollen and has complete control over whether a successful pollination comes about. The process is not fully understood but somehow the mother plant can detect unsuitable pollen. Most likely this is chemical in nature, but the exact mechanisms are not known. Undoubtedly each species of pollen has unique signatures that the female plant can interpret with rejection appearing to be the norm, primarily to prevent inbreeding and self-pollination. This system also applies to pollen where the chromosome numbers are out of phase or if it is a foreign pollen that has no merit for the female. Think *Quercus* pollen landing on a *Hibiscus* stigma. The chemical barriers will prevent the pollen from germinating and if by chance it does then there are other agents in service to her majesty that come into play. However, this runs contrary to the role of a plant breeder who is trying to circumvent the processes and get two plants to form a union.

Hibiscus rosa-sinensis offers an interesting look into some of the nuances found in the plant world. Of the multitude of cultivars known most are sterile, of the fertile clones breeding research at BHS shows some 25% of them are yellow flowered. Having the fertility of yellow flowers being so high in comparison to the rest is a bit of an enigma. Do the genes regulating fertility have a link to yellow pigmentation? It should be noted that in the genus *Hibiscus*, many of the flowers especially of the species found in Africa are yellow. The highly colored forms of *H. rosa-sinensis* in comparison are out of phase with the genus as a whole. The late Barbara McClintock (Creighton and McClintock, 1931; McClintock, 1950) showed that plants do have linked genes and this situation is not an anomaly. But the key question is why is yellow the chosen color for fertility? This would be an interesting research project should someone somewhere have the time and interest to pursue it.

What about the "Controlling Hand"? Can we as plant breeders bring our expertise to bear on fooling the system and getting successful pollinations that otherwise would not occur. A number of chemicals are known to affect flowering, pollination and seed formation. (Addicot, 1943; Brewbaker and Kwack, 1963). These include auxins, gibberellins, cytokinin, ethylene as well as sugar.

This is not unexpected but what is somewhat of a mystery is how the relationships of the various hormones and chemicals play a role in pollination and seed formation. The question that comes to mind is, can we overcome barriers to pollination. The answer seems to be yes, but the systems are not necessarily straight forward. We can apply auxins, gibberellins, and cytokinin, we can use ethylene either as a gas or a liquid application with products such as Florel (Chacko et al., 1976; Costa, 2012; Rajasekharana and Ganeshan, 1994) We can alter both temperature and humidity to be optimal. As demonstrated with the *Hibiscus* genera we can alter receptiveness and the timing of pollination. These things are possible, but the application is more of a jigsaw puzzle than something more straight forward.

There are other chemicals when applied to flowers are known to influence pollination. Calcium salts, and boron compounds, are two definite agents (Brewbaker and Kwack, 1636). Other possibilities are manganese and zinc salts as well as napthyl acetamide (a synthetic auxin derivative) is well known to influence the process (Addicott, 1943). Sugars such as fructose and sucrose (glucose+fructose) seem to have an influence as well.

There are other tricks that can be applied to fool the system. One is to mix an acceptable pollen with that of a pollen that is not readily accepted. Another is to mix a killed acceptable pollen with that of a pollen that is not readily accepted (Jones, 1920).

One of the biggest dilemmas that breeders have is when two plants bloom out of phase. *Hibiscus* from Africa such as *H. platyphyllos* (yellow flowered) always wants to bloom in fall and winter in the Northern Hemisphere. Of course, potential Northern Hemisphere natives want to boom in spring and early summer. Trying to get the two trains to meet in the dark is challenging. Or consider the yellow flowers of *Cornus officinalis* and *Cornus mas* which bloom early in the spring but a good recipient such as *C. kousa* blooms several months later. There are examples of manipulation of environment to get two strangers to connect but it is tedious and not always reliable. However, pollen storage might be a useful key to solve the problem. There are problems. Many pollens do not store well and are short lived. Storage varies from species to species and they are not created equal. Liquid nitrogen storage of pollen is a given (Gaseshan, 1986; Pozzobon et al., 2006; Rajasekharana and Ganeshan, 1994). But it is hard to handle, it is expensive, and most operations are not readily equipped to deal with liquid nitrogen storage. Are there alternatives?

Using the techniques outlined by Jain et al. (1988), Kopp et al. (2002) and Parfitt and Ali (1983) indicate that a standard home or commercial freezer at -20C /-30C will do the trick.

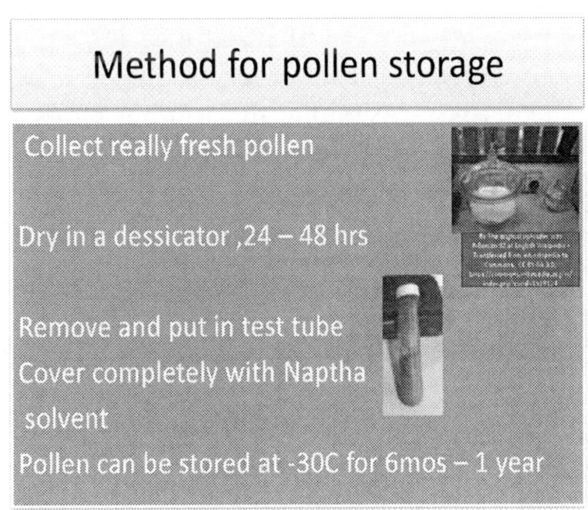

Figure 2. Methods for storing pollen.

Other Issues in Breeding

A common problem in breeding is that the fruit does form but the seed is hollow. Enough fertilization occurred to get the system going but then most likely the pollen tube failed, or the gametes were killed once they entered the ovary. A bigger problem is for seed to be normal and filled but fails to germinate. While embryo rescue will work it is restricted in application and out of reach of some breeders. A third issue is when normal seed germination can be offset, and a new regimen installed. For example, seed that normally would take one season to germinate will shift to a two-year cycle. Seed can also exhibit very poor germination, but something does get through the process. However, many of the resultant seedlings can be sickly and will die, leaving but a few examples of the hybridization. Those that remain could be well worth the wait by hanging on to for further evaluation.

Eureka, all systems are go and the result is a bonafide hybrid seedling. But and this is a big but, the resultant plant will not bloom. Sometimes nothing applied, altered or manipulated will change this and the plant simply will not bloom. This situation can occur especially when a wide cross is accomplished by manipulation. Chances are there is such misalignment of genes and chromosomes that some systems are permanently turned off. Grafting onto a more freely flowering root stock might be a help and this is a technique that Luther Burbank employed to great use. (Burbank, 1914) But it is not always a cure all.

Can we be sure a cross pollination has been successful? Sophisticated labs make use of flow cytometry to ascertain if a hybrid has been accomplished, but many small-scale breeders do not have access to such expensive and technical hardware. Can a specific hybrid be determined with less complex technologies? In general, the answer is yes, although with the caveat that such methods are not truly definitive but do point to a particular outcome. Bear in mind that a negative result tells nothing, but a positive result indicates something did occur.

One of the first indicators that a successful cross has been made is that resultant fruit or seed capsule is misshapen, or irregular compared to the norm for the mother plant. Along these lines is the degree of fill of seeds in a given fruit, if there are gaps and chambers are sparse or not filled the likely hood is that some form of cross pollination did take place. It should be remembered that the mother plant exercises a great many controls on pollination and seed development and some seed will be absent due to maternal influence. For instance, in a wide cross of *Hibiscus palustris* 772 with *H. coccinea* 45, the fruit is smaller, and the seed count is greatly reduced, 105 for the norm verses 43 for the cross. This is almost definitive that a successful pollination did take place.

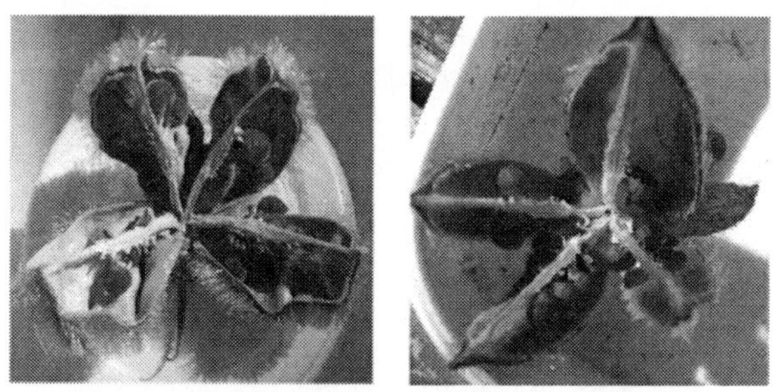

Figure 3. Partially filled *Hibiscus* fruit.

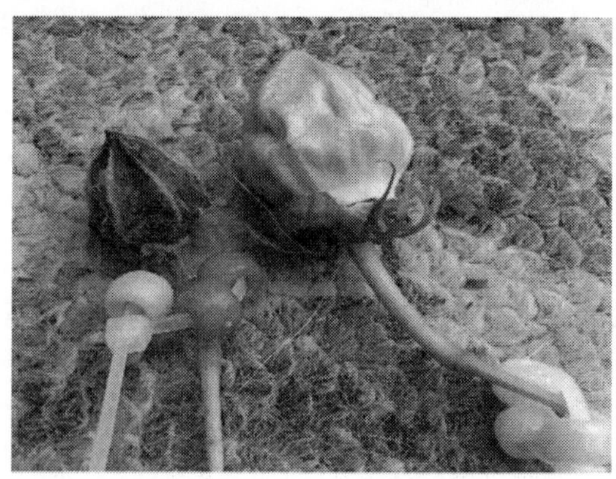

Hybrid fruit Normal fruit

Figure 4. Variable fruit size in normal and hybrid *Hibiscus* crosses.

Another indicator of successful hybridization is relative seed size. In the *Hibiscus* hybrid seed is often smaller and, in some cases, misshapen. The exact cause of this is not known but a good assumption is that the hybrid seed is under stress by the mother plant and development is slower than that of seed which is more normal. Fig. 8 shows the disparity with hybrids of *Hibiscus aculentus* with a wide cross of another *Hibiscus* species.

Figure 5. Seed size as a clue indicating a successful hybrid pollination in *Hibiscus*.

Other morphological and physiological changes to detect hybrids

- 1. Seed is good but refuses to germinate
- 2. Seed is good but germination stratification and time is altered
- 3. very low seed germination but a clear sign of a hybrid
- 4. seedlings appear to be weak or deformed and slow to begin growing well

Figure 6. Morphological and physiological indicators of successful hybridization.

There are other indicators of a successful hybrid being formed. Sometimes juvenile leaf morphology is changed from the norm but might revert upon more mature leaves forming. In some instances, the seedlings will not grow straight away compared to those of a normal pollination. Cold hardiness in the case of cold tolerant plants could well be affected and those with a dramatic new genetic constitution might not be as hardy as before. It is reasonable to assume the cross of a non-hardy plant with a hardy plant will result in less hardiness. This is not necessarily the case and often depends which was the mother plant. In many cases the mother plants genes have a dominant effect on the seedlings and seedlings will commonly behave like the mother plant.

Conclusion

The variations that are present in hybrid seedlings can be extensive and not always easily seen. Part of the process of selection and testing is to uncover the hidden traits. Sometimes this is a rather quick process, leaf color or morphology can quickly tell which is which. For instance, *Cercis canadensis alba* seedlings can be quickly discerned from those of the pink forms by a difference in the colors of the leaf petioles. Telling one pink from another is not such an easy task. Leaf morphology and seed germination times also might indicate fundamental differences. Cold, heat or salt tolerance can be tested reasonably easily, and the results can be dramatic, a large block of seedlings can quickly be reduced to a mere few. With the advent of new and much superior genetic testing a high value plant might be easily and quickly selected over those of less stature.

There is hope for the small breeder regardless. Small breeders can concentrate on specific plants not in the main stream. Small breeders can also look for traits that might be lost in the fog for a large breeder. Small breeders might not have the resources to go after genetically modified plants that seem to incur a great deal of ire from certain segments of the buying public. Finally, small breeders might be more adept at producing specific clones and new hybrids that are regionally adapted. At present there is room for all with enough initiative to tackle new projects and species not found in the main stream.

Literature Cited

Addicott, F.H. (1943). Pollen Germination and pollen tube growth as influenced by pure growth substances. Plant Physiology *8*:270-279.

Brewbaker, J.L. and Kwack, B.H. (1963). The essential role of calcium ion on pollen germination and pollen tube growth. American Journal of Botany *50*: 859-865.

Burbank, L. (1914). Luther Burbank: His methods and discoveries and their practical applications. http://digital.library.wisc.edu/1711.dl//HistSciTech.LutherBurbank.

Chacko, E.K, Kohli, R.R., Swany, R.D. and Randhawn, G.S. (1976). Growth regulators and flowering in juvenile mango (*Mangifera indica* L.) seedlings. Acta Horticulture *56*:173-176.

Costa, A. P. (2012). Effects of florel and promalin application on branching, flowering and fruiting in Jatropha. (*Jatropha curcas*. L.) Master's Thesis. University of Florida. http://ufdc.ufl.edu/UFE0044685/00001

Creighton, H.B. and McClintock, B. (1931). A correlation of cytological and genetical crossing over in *Zea mays*. Proceedings of National Academy of Sciences *17*:492-497. https://doi.org/10/1073/pnas.17.8.492

Gaseshan, S. (1986). Abstract. Cryogenic preservation of *Papaya* pollen. Scientia Horticulture *28*: 65

Jones, D. F. (1920). Selective fertilization in pollen mixtures. Biological Bulletin *38*: 251-289.

Jain, A., Shivanna, K.R. (1988). Abstract: Storage of pollen grains in organic solvents: effects of solvents on pollen viability and membrane integrity. Journal of Plant Physiology *132*:499

Kopp, R.F. Maynard, C..A., de Niella, P.R., Smart, L.B., and Abrahamson, L.P. (2002). Collection and storage of pollen from *Salix* (Salicaceae). American Journal of Botany *89*:248-252

McClintock, B. (1950). The origin and behavior of mutable loci in maize. Proceedings National Academy of Sciences *36* :344-355.

Parfitt, D.E., and Almendi, A.A. (1983). Abstract. Cryogenic storage of grape pollen. Am. Journal of Enol. Vitic. *34*: 227.

Pozzobon, M.T., Wittmann, M.T.S., and Bianchetti, L.D.B. (2006). Chromosome numbers in wild and semi-domesticated Brazilian *Capsicum* L. (Solanaceae) species: do x=12 and x=13 represent two evoluntionary lines. Botanical Journal of the Linnean Society *11*: 259-269

Rajasekharana, P.E. and Ganeshan, S.S. (1994). Abstract: Freeze preservation of rose pollen in liquid nitrogen: Feasibility, viability and fertility status after long term storage. Journal of Horticultural Sciences *69*:3.

Improving Pest Management for the Nursery Industry

Chris O'Connor

BHort MSustAgric

Chris.oconnor@mq.edu.au

Keywords: IPM, sustainable systems, pesticides, Biosecurity

INTRODUCTION

Pressure on the industry both locally and globally to use pesticides in an environmentally and socially responsible manner is increasing. Such pressures have the potential to cause damage to the industry's generally positive reputation and negatively impact our social license to operate, and some industry actors are already responding.

A recent example of a supply chain decision by a major green life retailer, and the subsequent flow-on effect from other retailers, will see growers restricted in utilizing a key pesticide group (neonicotinoids) after 2020. This has been an important part of the suite of tools available for industry insect pest management over the past two decades. This decision is the first market-driven restriction on a registered pesticide in Australia that has been made without supporting evidence, nor a general regulatory backing, all driven primarily through international influences.

The implementation of this decision has highlighted the degree of exposure Australian production nurseries have to decision drivers within the supply chain, and the real issue that they have limited opportunity to combat. This is however not the only pressure that the industry is facing concerning pesticides. The maintenance of pesticide efficacy is becoming an increasing concern. Each year more pests adapt and become resistant to the chemistries which have stood the industry in good stead in the past rendering these chemistries as ineffective. Likewise, the funnel of new chemistries entering the market has reduced as the high costs of R&D and registrations of new chemistries make the process difficult. Additionally, new release pesticides may not exhibit the same level of utility over the breadth of pests which previous pesticides have done.

Combineed with rising farm labour costs and other production input costs as well as the changes to pest pressures influenced through climate change and it is easy to see the burden that these issues place upon industry.

Consequently, the industry must increase its awareness of international activities and the potential it has, to influence the industry supply chains as well as investigate new pest management options such as synthetic pesticides, bio-pesticides, and bio-controls. Additionally, it is essential that the industry understands and implements actions that preserve the access, efficacy, and cost of current synthetic pesticides through informed decision making, careful application including targeted use, improved application technology and through rotation and pesticide resistance management strategies.

Referring to one of the conference themes, "if we don't change the direction we are going, then we likely end up where we are headed", which leads us to the concept of IPM.

Integrated Pest Management (IPM) has been an established philosophy of pest management since the 1960's evolving from the early work of Stern et al. (1959). Over time the definitions of what IPM is have evolved and consequently, there are several interpretations. A current definition however is that IPM is "a comprehensive approach to managing host stress that is economically and ecologically sustainable" (Peterson et al., 2018).

There are numerous models describing how this works such as the one suggested at farmbiosecurity.com.au. At its core, however, IPM is a decision support system which is used by a grower to inform the selection and application of appropriate pest management options (practices & control) in a cropping system to keep plant pest populations below economic thresholds.

In nursery production, these IPM options include preventative practices that minimize pest entry, establishment, and spread, hygiene strategies such as traffic control and disinfestation, access to and storage practices of raw materials, crop production strategies plus beneficial biological control and targeted synthetic pesticide use. All successful IPM programs are supported by a well-structured crop monitoring, surveillance, import/despatch inspections and recording system that drives the decision-making process and enhances knowledge capture, which in turn achieves the most from each IPM option employed including synthetic pesticide application.

Industry has invested in a range of systems and resources targeted at on-farm adoption of high health practices, designed to assist growers in building a holistic integrated cropping system including the IPM components. Some examples of tools and resources available include the AgriLink published Integrated Pest Management in Ornamentals: Information Guide (Goodwin et al., 2000) and the Pest ID Tool Levy-funded projects addressing pest management such as NY11001 Plant Health Risk Management, Planning & Capacity Building Project, and more recently the NY15002 Building the resilience and on-farm biosecurity capacity of the Australian production nursery industry project. Resources from the NY15002 Building the resilience and on-farm biosecurity capacity of the Australian production nursery industry such as pest factsheets, management plans and biology and pest images for the Pest ID Tool are key supporting documents and resources for growers when implementing an IPM program.

Production nurseries in Australia also have access to the Australian Plant Producers Standard programs that lay the foundations for growers implementing a nursery production IPM program. The Guidelines of NIASA and particularly BioSecure HACCP offer a robust process for pest prevention, crop monitoring, inspection, surveillance and recording which are the key drivers for IPM success.

Whilst, the field of IPM in nursery production has been researched, promoted, extended and in cases 'adopted' over the past, consistent and structured uptake of IPM strategies across the nursery production sector has been low. Poor rates of adoption have been attributed to a lack of understanding of IPM, practical guides and resources, skills shortage and a reliance on current methods, namely pesticides that are effective, legal and familiar.

In 2018 CSIRO surveyed elements of the production nursery industry focusing upon the grower conducted pest surveillance activity of the industry (Loechel et al, 2018). This survey identified several findings.

- While most growers appear motivated to monitor it is unclear how systematic and comprehensive these efforts are.
- Site surveillance was identified as an area for improvement.
- There was a sizeable proportion of the survey group who never inspected imported stock.
- Dispatched plants were almost universally inspected.
- Record keeping did not appear to be common practice

This indicates that industry has some opportunities to improve its monitoring activity which is itself an essential foundation of IPM.

Anecdotal data, captured in 2017 under the levy project NY15004 National Nursery Industry Biosecurity Program, suggests a 2.5 - 3% increase in annual productivity for businesses implementing BioSecure HACCP pest management procedures. Beyond this anecdotal data not a great deal of information is available on the economic impacts of IPM, in particular monitoring, for the production nursery industry.

A report conducted by Newman et al. (1999) from the university of California in 1999 indicated a pest management cost reduction of 50% per acre for a petunia grower adopting an IPM scouting approach compared to their standard practice. Likewise, the same report highlighted a 37% reduction in pest management costs for a protected cropping rose grower. In 2013, Xu and Khachatryan (2013) undertook an analysis of US production nursery sales and IPM adoption and noted a positive sales increase correlation for businesses conducting IPM activity including monitoring.

Given that both the anecdotal information and previous studies indicate a positive correlation with a business's economic and monitoring activity it is prudent to investigate this further. Providing economic drivers to support the crop monitoring, site surveillance, dispatch inspection and recording components of BioSecure HACCP and IPM will give growers greater confidence in their adoption of the program elements and assist in driving change across the industry.

For this reason, NGIA with support from the industry levy is undertaking a research project examining a regionally diverse cropping system based economic analysis - (before & after) of adopting IPM practices (crop monitoring, site surveillance, consignment inspections, and activity recording) consistent with the BioSecure HACCP procedures.

The Production nursery industry is highly diverse with a large range of crops produced across several primary supply chains. It is also geographically widespread, therefore, to value the overall investment in assessing the economic benefits of IPM procedures and provide growers with access to educational activities the project will undertake the following crop monitoring program/cropping system/regional location over a 24-month period.

Table 1. Cropping system monitoring frequency.

Number	State	Cropping System	Monitoring Frequency	Total Monitoring Activities
1	QLD	Propagation	Fortnightly	26
2	QLD	Potted colour perennial	Fortnightly	26
3	NSW	Fruit Tree (Citrus)	Fortnightly	26
4	VIC	Tree and Shrub	Fortnightly	26
5	VIC	Indoor house plant	Fortnightly (6 months) Monthly (6 months)	19
6	SA	Potted colour annual	Weekly	52
7	WA	Farm forestry seedlings	Weekly (9 months)	36

Each of the 7 participating production nurseries has been visited by an economist contracted to the project, before crop monitoring begins, to undertake a financial assessment of business costs related to the pest, disease and weed management and implement the appropriate structures to capture pest management related costs and financial data. These metrics including costs associated with stock losses, stock turnover, picking/despatch, pesticide use, labour, etc. will be used for financial benchmarking of data collected before and after crop monitoring activity is implemented. The metrics will be consistent across each of the participants and related to the pest management influenced costs of the business. For the first year of the project, the data captured will be based upon the business' normal operations.

Data capture will be based upon the records being generated by the business such as pesticide spray records as well as additional forms such as a form to record crop losses and throw outs. However, to introduce better utility of these records, a process of electronic data capture has been implemented. The digitisation of the data captured by the business will enable the project team and the businesses involved a greater ability to analyse trends and interrogate the data captured to a higher level. To facilitate this data capture, a suite of google forms have been developed for the project participants. These forms enable data to be entered at the point of collection, for example, a pesticide spray record on a mobile device. Once the data has been entered it can be easily permit aggregated and exported into a usable spreadsheet format.

The advantage of using google forms is that while they may lack the level of sophistication which a dedicated program may have, they are adaptable and are an extremely low-cost option. Some businesses are already utilizing the suite of Google tools in their business and for these operations the adoption of the forms has resulted in few technical issues. Some businesses are however, still exploring the opportunities presented by this new data capture technique, and so require additional assistance, however paper-based forms are still being used as an alternate measure.

After the first year of baseline data collection (business operations as normal – year 1) a structured crop monitoring, site surveillance and inspection program will be introduced.

A professional crop monitoring service will be used to provide all the professional crop monitoring and decision support services (advice/recommendations on managing pests detected utilising all elements of IPM including pesticide recommendations that consider impacts on beneficial organisms) across the seven businesses over 12 months (Year 2). The monitoring activity will be conducted in accordance with the procedures documented in the BioSecure HACCP Guidelines. This will provide a consistent approach to the monitoring process and assist in further validating the documented processes. The production nurseries engaged in the project will undertake the structured site surveillance and dispatch inspection activities according to the procedures in the BioSecure HACCP guidelines. Again, using the procedures developed in BioSecure HACCP allows for a consistent approach and provides further validation to the method. Data on the business costs related to the pest, disease and weed management as well as the business finances in general, will continue to be captured during this period.

Following the second year, the economic data captured will be analysed by the project economist. This will allow him to determine if adopting IPM principles, in particular, crop monitoring, site surveillance, consignment inspections and activity recording consistent with the BioSecure HACCP procedures, is of economic benefit to the production nurseries.

It is intended that the information gained through the project will then be presented and discussed at several field days across the country. Practical guided exercises will be undertaken at these field days focusing on introducing structured crop monitoring, site surveillance, consignment inspections and activity recording processes. Furthermore, exercises will be developed on how to apply IPM principles and developing tailored site-specific strategies for pests of prime concern.

The validated economic benefits achieved on-site by adopting IPM techniques will also be presented including commentary by the production nursery owners and/or key staff on their experiences including the challenges they faced, how they overcame these challenges, and the benefits they have realised within their business through the adoption of IPM techniques.

To ensure that information from the project and field days is communicated as broadly as possible videos will be developed and hosted on digital platforms. Alternate communication activities such as this is important given the vast geographic spread of the industry.

Considering the broad issues which face our industry and the imperative being placed upon it to change, it is prudent to reflect upon another of the conference's themes – It is not the strongest or the most intelligent who will survive, but those who can best manage change. Hopefully this project will see a positive outcome for the industry, providing the tools and evidence to encourage change and further embrace IPM approaches. Doing so will protect our industries reputation, make us more resilient and steward the tools and resources we have, to manage our pest pressures, now and into the future.

Literature Cited

Farmbiosecurity.accessed 2018. https://www.farmbiosecurity.com.au/what-is-integrated-pest-management/

Goodwin., S., Steiner, M., Parker, R., Tesoriero, L., Connellan, G., Keskula, E., Cowper, B., Medhurst, A. and Rodriguez, C. (2000) Integrated Pest Management in Ornamentals: Information Guide. Horticulture Series: Agrilink,your growing guide to better farming. Manual. Agrilink Series QAL0004, Queensland Horticulture Institute. Brisbane, Queensland.

Loechel, B., Hobman, E.V., Collins, K.; Robinson, C., Singh, J., Pavey, C., and Chilcott, C. (2018). Improving grower-led surveillance: Case-study of the Nursery Production Industry. Dutton Park: CSIRO.

Newman, J. Robb, K. and Tjosvold, S. (1999) Pest Management Grants Final Report: Training Scouts and Developing Demonstration Sites to Promote Floriculture IPM Programs (Contract number 97-0224) University of California.

Peterson, R.K.D., Higley, L.G., and Pedigo L.P. (2018). Whatever Happened to IPM? American Entomologist *64*:146–150.

Stern, V., Smith, R., van den Bosch, R., and Hagen, K. (1959). The integration of chemical and biological control of the spotted alfalfa aphid: The integrated control concept. Hilgardia *29*:81-101.

Xu, W. and Khachatryan, H. (2013). The impact of integrated pest management practices on U.S. national nursery industry annual sales revenue: an application of smooth transition spatial autoregressive models. Southern Agricultural Economics Association Annual Meeting Orlando, Florida.

Role of Nutrients in Plant-Disease Interaction

Reyhaneh Pordel

Garden City Plastics, Victoria Australia

Reyhaneh.pordel@gardencityplastics.com

Keywords: Fertilizer, pathogens, minerals, smoke water.

Abstract

In recent years, the importance of sustainable horticulture has risen to become one of the most critical issues in the plant industry. Besides, plant diseases continue to play a significant limiting role in horticulture production. The control of plant diseases using classical pesticides raises serious concerns about food safety, environmental quality, and pesticide resistance, which have dictated the need for alternative pest management techniques. Nutrients could affect the disease tolerance or resistance of plants to pathogens. However, there are contradictory reports about the effect of nutrients on plant diseases and many factors that influence this response are not well understood. There is a difference in the response of obligate parasites to Nitrogen (N) supply, as when there is a high N level, there is an increase in the severity of the infection. In contrast, in facultative parasites at high N supply, there is a decrease in the severity of the infection. Potassium (K) decreases the susceptibility of host plants up to the optimal level for growth, and beyond this point, there is no further increase in resistance. In contrast to K, the role of Phosphorus (P) in resistance is variable and seemingly inconsistent. Among the micronutrients, Manganese (Mn) can control several diseases as Mn has an essential role in lignin biosynthesis, phenol biosynthesis, photosynthesis and several other functions. Zinc (Zn) was found to have several different effects as in some cases, it decreased, in others increased, and in a few, it did not affect plant susceptibility to diseases. Boron (B) was found to reduce the severity of many diseases because of the function that B has on the cell wall structure, plant membranes and plant metabolism. Calcium (Ca) application can enhance host plants' resistance to diseases. Silicon (Si) has been shown to control several diseases, and it is believed that Si creates a physical barrier which can restrict fungal hyphae penetration, or it may induce accumulation of antifungal compounds. Integrative plant nutrition is an essential component in sustainable horticulture. It can be a cost-effective and environmentally friendly way to control plant diseases with no pesticides. Nutrients can reduce disease to an acceptable level, or at least to a level at which further control by other cultural practices or conventional organic pesticides are more successful and less expensive.

INTRODUCTION

Sustainable agriculture is the management and utilization of the agricultural ecosystem in a way that maintains its biological diversity, productivity, regeneration capacity, vitality and ability to function, so that it can fulfil - today and in the future - significant ecological, economic and social functions at the local, national and global levels while does not harm other ecosystems. Therefore, it is crucial to find alternative measures to control plant diseases which do not harm the environment and at the same time, increase yield and improve product quality. It is certain that the use of pesticides has stabilized our food supply and permitted millions of people to live longer lives. However, Heavy use of pesticides has been associated with degradation of the environmental resources and quality of water. There is also widespread public concern about the use of pesticides and their potential effect on our food. Therefore, planting resistant varieties in the first place and then growing healthy, strong plants will help to eliminate the usage of pesticide when it comes to plant-pest interaction.

Role of Nutrients in Plant-Disease Interaction

Nitrogen (N)

Nitrogen key role in driving growth, development and yield is well known. N has an important role in plant defense against pathogens. Lack of nitrogen shows up as overall yellow-green leaves instead of dark green. Nitrogen taken up by plants is used in the formation of amino acids which is the building block for proteins. Nitrogen is a structural component of chlorophyll.

Ammonium nutrition enhances the content of sugar and amino acids, thereby increasing the availability of nutrients to the invading pathogen Conversely, under NO–3 nutrition, increased resistance to the pathogen is observed. One of the reasons is polyamine production which is known to increase plant resistance.

Phosphorus (P)

Phosphorus improves resistance and tolerance to fungal diseases. The protection that P provides is often related to its role in plant development. P is vital in early root development. The role that P plays in promoting rapid root development in young plants is well established. Especially under adverse or stressful conditions, early root development is essential.

One of the best defenses against root diseases is a robust and well-developed plant root system. P is the most beneficial when it is applied to control seedlings fungal diseases where vigorous root development permits plants to escape the disease condition.

It has been shown that phosphate fertilization of wheat can have a significant effect and almost eliminate economic losses from Pythium root rot. Foliar application of P can induce local and systemic protection against powdery mildew in cucumber, roses, wine grapes, mango and nectarines.

Potassium (K)

When there is a lack of enough potassium in plants, low molecular weight compounds begin to accumulate. This accumulation of soluble nitrogen-containing compounds (such as amino acids and asparagine) and sugars (such as sucrose) make a particularly favourable environment for numerous pathogens and insects. For example, aphids are severely nitrogen-limited, making potassium stressed plants an attractive host as an abundant nitrogen source.

The presence of enough potassium also promotes the production of defensive compounds (such as phenols) which are an important component in plant pest resistance.

This results in more resources available for hardening cell walls and tissues to better resist the penetration of pathogens and insect pests and to repair any damaged tissues. Airborne pathogens are more rapidly shut out from stomatal invasion when adequate K is present.

Iron (Fe)

Iron is involved in the synthesis of chlorophyll, and it is essential for the maintenance of chloroplast structure and function. Fe is involved in chlorophyll synthesis, and it is essential for the maintenance of chloroplast structure and function. Fe works as a cofactor for enzymes involved in a wide variety of oxidation-reduction reactions (i.e., photosynthesis, respiration, hormone synthesis, DNA synthesis). This function makes Fe an essential nutrient, and its deficiency causes iron chlorosis, which significantly constrains normal plant development.

Boron (B)

Boron plays a direct role in the cell wall structure, and, thus, likely directly affects pathogen susceptibility. Besides, B plays an essential role in carbohydrate transport through the phloem. Low B can cause phloem to collapse and leaf veins to become "corky". In addition to its role in the cell wall structure and plant metabolism, B is known to be toxic to pathogenic fungi.

Sulphur (S)

In summary, Sulfur-containing metabolites that are supposed to be involved in pathogen resistance are glutathione, glucosinolates, the gaseous release of volatile S, phytoalexins, S-rich proteins, and the formation of elemental S.

The S status of the crop is affecting many different plant features such as colour and scent of flowers, pigments in leaves, metabolite concentrations and the release of gaseous S compounds which are directly influencing the desirability of a crop for a variety of different organisms from microorganisms, over insects.

Manganese (Mn)

Manganese is probably the most studied micronutrient about its effects on disease. It is essential in the development of resistance in plants to both root and foliar diseases Manganese is used in plants as a significant contributor to various biological systems including photosynthesis, respiration, and nitrogen assimilation. Manganese is also involved in pollen germination, pollen tube growth, root cell elongation and resistance to root pathogens. Manganese fertilization can control several pathogenic diseases such as powdery mildew, downy mildew, take-all, tan spot, and several others. Manganese controls lignin and suberin biosynthesis which Both are significant biochemical barriers to fungal since they are phenolic polymers resistant to enzymatic degradation.

Fertilizers and Soil conditioners

The following is technical information regarding products that improve the nutrition level of the plants and structure of the media. Application of these products results in a vigorous seedling which has excellent resistance to and abiotic, biotic stresses.

ICL Fertilizers Products

Start&Gro

Start&Gro is the starter fertilizer offered by ICL Specialty Fertilizers for potted, bedding and all container nursery plants. Start&Gro is designed to be premixed in the substrate to fertilize crops for the first weeks of the cultivation. Useful features include Balanced NPK analysis, Free-flowing, Good mixability, 100% water-soluble, Not dusty, High chelated trace elements level.

Osmocote Start

Osmocote Start offers a good fertilizer for annual plants with a short growth cycle, such as violets, primulas, impatiens and marigolds. It is also ideal for salt-sensitive vegetable crops and cuttings that have difficulty with rooting. Osmocote Start offers a good fertilizer program for annual plants with a short growth cycle, such as violets, primulas, impatiens and marigolds. Osmocote Start is also ideal for more salt-sensitive vegetable crops and cuttings that have more difficulty with rooting. In contrast with traditional fertilizers, Osmocote Start avoids the risks of an excessive salt level; your crop will always receive the right nutrients. The high potassium level of Osmocote Start ensures compact and even growth. Osmocote Start can be mixed with the growing media. Topdressing is possible as long as you ensure that no granules remain on the leaves. Safe to use, no salt stress, Better development of roots and crop colour. Features include Uniform release: 100% coated, very efficient due to greatly reduced leaching, no de-mixing of nutrients; every granule contains all the required elements, Easy to apply, dust-free.

Osmoform NXT

Osmoform NXT is based on the new granulation technology SILK (silica-based), which includes slow-release nitrogen and potassium. The slow-release K is embedded in a 3-dimensional matrix, which slowly dissolves during the time. The product sticks to the surface of the growing medium, Quick plant reaction, strong greening up effect, very efficient nitrogen release, contains magnesium and trace elements with high iron content, Includes silicium for stronger plants, Long lasting effect up to 8-10 weeks, Easy application. Osmoform contains NPK, magnesium and a package of trace elements. The nitrogen composition is balanced to have a quick plant reaction after application and a long-lasting effect afterwards. Osmoform NXT contains 1.2 - 2.8 mm granules. And is developed for top-dressing applications in container nursery stock crops but can also be used for other purposes where a long-lasting effect of nitrogen and potassium is required. The nitrogen and potassium in the product are releasing over approximately 8-10 weeks after application—slow release of nitrogen and potassium, no loss of fertilizer when plants fall over. In the case of top dressing, it is essential to spread the amount of fertilizer evenly over the surface of the growing media. In case of the application over the plants, it is recommended to remove the granules from the leaves. Store under dry conditions. Partly used or damaged bags should be closed well.

Grayson Smoke-based Products

Smoke master

It triggers natural germination, through direct effects of heat in breaking seed-coat imposed dormancy. A relatively small number of investigations has indicated that products of fire rather than direct effects of heat may stimulate germination. Thus, water-soluble products of charred are promotive to germination of the native.

Smoke water

Liquid smoke extracts are an important tool in the germination of many Australian native species. While the mechanism is unclear, the chemical signature of smoke (aqueous or otherwise) interacts with the seed structure to stimulate germination, possibly by signal molecules or by hormones. Some plant varieties remain dormant until exposed to smoke-based stimuli; a study indicated that smoke signatures are the most critical trigger to the germination rate and cumulative germination of several varieties of *Grevillea* seeds.

Regen Shield

Liquid Smoke Condensate plus Polysorbate 80, Acetic Acid and Water (Acidic Smoke Extracts).

Smoke extracts may produce a natural barrier to the liverwort growth while acting to promote plant growth of desirable plant species. Spray applications of acidic smoke extracts (Regen Shield) seemed to demonstrate a benefit to the tested plants. However, the underlying processes that cause this benefit have not been discussed in the literature. Likewise, the mode of action on the liverwort is unknown.

CONCLUSIONS

Nutrient management through amendment improved genetic efficiency, and modification of the environment is an essential cultural control for plant disease. Disease resistance is genetically controlled but mediated through physiological and biochemical processes interrelated with the nutritional status of the plant or pathogen. The nutritional status of a plant determines its histological or morphological structure and properties, and the function of tissues to hasten or slow penetration and pathogenesis. Various nutrients also condition pathogen virulence and their ability to survive; however, most nutrients influence disease potential more than inoculum potential.

The intricate relationship of the plant's nutritional status with plant pathogens, the abiotic environment and organisms in the environment is dynamic, and the severity of most diseases can be significantly decreased by proper nutrient management. Also, there are still some plant diseases where the efficiency of chemical fungicides is limited. For example, currently, no fungicides are available to control Verticillium wilt. Therefore, fertilizer strategies which improve the plants potential and resistance against fungal diseases are still of high importance. Knowledge of the relationship of plant nutrition to disease provides a basis for reducing disease severity in intense as well as integrated crop production systems.

Literature Cited

Agrios, N.G. (2005). Plant Pathology, 5th ed., Elsevier-Academic Press.

Batish, D.R., Singh, H.P., Setia, N., Kohli, R.K., Kaur, S., and Yadav, S.S. (2007). Alternative control of littleseed canary grass using eucalypt oil. Agron. Sustain. Dev. *27*:171–177.

Bloem, E., Haneklaus, S. and Schnug, E. (2015. Milestones in plant sulfur research on sulfur-induced-resistance (SIR) in Europe. Front Plant Sci. *5*:779. doi:10.3389/fpls.2014.00779

Dordas C. (2008). Role of nutrients in controlling plant diseases in sustainable agriculture. A review. Agronomy for Sustainable Development *28*:33-46.

Grunert, O., Hernandez-Sanabria, E., Vilchez-Vargas, R., Jauregui, R., Pieper, D.H., Perneel, M., Van Labeke, M.C., Reheul, D. and Boon, N. (2016). Mineral and organic growing media have distinct community structure, stability and functionality in soilless culture systems. Scientific Reports *6*:18837.

Heckman, J.R., Clarke, B.B., and Murphy, J.A. (2003). Optimizing manganese fertilization for the suppression of take-all patch disease on creeping bentgrass. Crop Sci., 43:1395-1398.

Huber, D.M., and Haneklaus, S. (2007). Managing nutrition to control plant disease. Landbauforschung Volkenrode *57*:313.

Lewandowski, I., Hardtlein, M., and Kaltschmitt, M. (1999) Sustainable crop production: Definition and methodological approach for assessing and implementing sustainability, Crop Sci. *39*:184–193.

Mur, L.A., Simpson, C., Kumari, A., Gupta, A.K., and Gupta, K.J. (2017). Moving nitrogen to the centre of plant defence against pathogens. Annals of Botany *119*:703-709.

Simoglou K., and Dordas, C. (2006). Effect of foliar applied boron, manganese and zinc on tan spot in winter durum wheat. Crop Prot. *25*:657–663.

Spann, T.M., and Schumann, A.W. (2009). The role of plant nutrients in disease development with emphasis on citrus and Huanglongbing. Proc. Fla. State Hort. Soc. *122*:169-171).

Wang,M., Gao, L., Dong, S., Sun, Y., Shen, Q., and Guo, S. (2017). Role of silicon on plant–pathogen interactions. Frontiers in Plant Science *8*:701.

Trading Gray Cubicles for Rainbow Skies: How I Changed Careers and Ended up Growing Plants in Hawaii

Emily Teng

University of Hawaii, Dept of Tropical Plants and Soil Sciences, 3190 Maile Way, Honolulu, HI 96822

eteng@hawaii.edu

Keywords: Careers, poinsettia, tropical plants

INTRODUCTION

Most of you reading this pay a power bill, right? Here is a question: have any of you ever wondered how your account information with the power company gets transferred over when the power company changes over to a new software system? Anybody? Neither did I. At least not until my job was to get that information moved over by writing COBOL computer code. If this sounds boring, then I have to agree. But twenty years ago, I was freshly graduated from college with a bachelor's degree in Chemistry from the University of South Carolina Honors College, and I was sitting in front of a computer screen for eight to twelve hours a day typing code like this.

```
PROGRAM ID - CUSTOMER ACCOUNT

MOVE CUSTOMER NAME         TO     ACCOUNT NAME
MOVE CUSTOMER ADDRESS      TO     ACCOUNT ADDRESS
MOVE CUSTOMER ACCOUNT      TO     ACCOUNT NUMBER
MOVE CURRENT BILL          TO     ACCOUNT BALANCE
MOVE PAST BALANCE          TO     PAST BALANCE
MOVE AMOUNT PAID           TO     ACCOUNT PAID

ADD PAST BALANCE           TO     ACCOUNT BALANCE GIVING     CUSTOMER BILL
SUBTRACT ACCOUNT PAID      FROM   CUSTOMER BILL   GIVING     TOTAL DUE
```

Figure 1. Writing computer code.

During my undergraduate years, I thought I was going to become a veterinarian. I love animals! I worked at a veterinarian's office part-time. I studied for the VCAT, the veterinary school entrance exam. I visited veterinary schools. But sometime during my senior year, I realized I was too faint-hearted. Or maybe I was too soft-hearted? I don't know the right term, but I couldn't stand to see sick animals who were suffering. I get attached to animals much too easily and it flat out made me too sad to see sick animals. Even though I knew I would be helping them, I decided becoming a veterinarian was not the right path for me.

There I was, a college graduate with a degree in Chemistry, unsure of what I was going to do with my life. I interviewed for a few jobs and ended up working for Accenture, one of the largest consulting firms in the world.

I ended up in the Utilities sector, working on projects at power and water companies. I converted these companies to new software systems. I tested new software and wrote computer programming code to customize the software. I would fly out from my base in Charlotte, North Carolina every Monday morning to my project sites in Florida, Delaware, Missouri and other places. I generally stayed on a project for a few weeks or for many months. I'd fly home every weekend and spend a few days living my real life before flying out early Monday morning again.

Now don't get me wrong, the job was not bad. I met great friends along the way. I was paid well. It was a great firm to work for. But for me, this was just a holding period while I figured out what I really wanted to do with my life. I was living in corporate apartments during the week. I had two of everything; one for the week and one for the weekend at home. On some projects we sat in cubicles. On other projects we would all sit in a room at conference tables hammering away at our laptops, usually for at least ten hours a day.

By 2001, I had been at this for about two years. I was on a project at a water company in Cherry Hill, New Jersey. My roommate decided to germinate an avocado seed in the corporate apartment we shared. I thought to myself, "Well, that is fun." She named him Oscar.

Figure 2. Growing avocado on a window sill.

We grew more avocados. Then we started growing other plants in our apartment. Pretty soon, we had quite the collection.

Figure 3. Houseplants -One of those avocado trees is Oscar.... My friend will be very upset that I don't know which one he is!

At home, I started some more growing projects. I renovated the backyard of my parents' house and built a gazebo and garden. I took a vacation to Hawaii (using all those airline miles I had racked up flying every week). Of course, I had to start trying to grow tropical plants inside my apartment in South Carolina. I grew all sorts of potted plants outside. I even had a Chia Turtle!

Figure 4. Backyard tropical plants.

I don't remember the exact thought process I went through. But at some point, it dawned on me that I could make a career out of this new hobby of mine. It was a true light bulb moment!

So I committed. There was a bit of a recession with the e-commerce bubble popping, and Accenture offered employees a voluntary Flex Leave program where I could basically do whatever I wanted and still get paid 20% of my salary. I took full advantage of this opportunity and enrolled in a couple plant sciences courses at the University and got a job at a landscaping company.

I planted and maintained bedding gardens for shopping centers, office complexes and suburban neighborhood entrances. It was the most physically demanding job I had ever had but I learned to till soil, drove a dump truck, and built up enough strength to haul around fifty-pound bags of mulch. Most importantly, I learned how to grow things like beautiful beds full of pansies, petunias, marigolds, and snapdragons that created masses of color! I loved it.

Since I had no education in this field, I decided I would get a master's degree in horticulture. I applied for master's programs and took this chance to move to Hawaii. So off I went on a road trip across the USA to Hawaii 5000 miles away - but not without some hiccups along the way…

And I did it. I studied and received my master's degree in Tropical Plant and Soil Sciences at the University of Hawaii.

Instead of having no idea what to do this time around when I finished my degree, I was faced with several possible directions. I could go work for one of the major seed companies like Pioneer or Monsanto. They paid well but working there made no sense to me. I didn't quit my high paying corporate job at Accenture, move to Hawaii, and get a master's degree in horticulture to work on crops like corn and soybeans at a major corporation. I pursued horticulture because I loved growing beautiful flowers. I wanted to work in ornamental horticulture. So, I looked in the newspaper (yes, the newspaper - newspaper listings were still useful even in 2007!) for part-time jobs to earn some money while looking for a full-time position. I found an opening for an hourly nursery worker at a local family-owned nursery growing potted ornamentals and various other plants. At the interview, the owner told me their propagation manager had just quit and I would probably be a good fit for that job.

The owner and I hit it off and she quickly became one of my best friends. I learned to grow a wide variety of plants. I learned how to run a small wholesale nursery. I had gone from sitting at a computer all day in a gray cubicle to being outdoors in beautiful Hawaii growing all kinds of flowering plants. And there was a bonus as we have many animal friends at the nursery!

Figure 5. Seasonal plants at the wholesale nursery.

The owner worried that I would get bored because I had a master's degree and was overqualified for the job and wanted me to promise I would stay at least a year. Well, she didn't need to worry. I've been there for 12 years now.

One of the themes of the IPPS conference this year was to "CHANGE DIRECTION NOW." Although I didn't consciously think too much of it back then, I did just that. I'm not a bold person. In fact, I was always the quietest, most reserved student in school growing up. I was even voted "Most Likely NOT to Be Heard" in high school. But I suppose you gain confidence as you get older. Or maybe it isn't so much confidence, but that you start worrying less about what others think. In any case, I somehow succeeded in doing that thing people say so much so that it has become cliché: I followed my passion. I am sometimes jaded by these touchy-feely inspirational themes, but looking back, I must admit that my life has sometimes embraced these themes quite well.

That is how I "Traded Gray Cubicles for Rainbow Skies." It truly changed the trajectory of my life. I can't imagine where I would be or what my life would be like if I hadn't made those decisions to change direction. It would have been pretty easy to stay in that cushy consulting job. I could have climbed the corporate ladder and become a very highly paid manager. I certainly would have been better off financially by now. But I don't regret the decisions that got me to where I am in my career today. I wouldn't have twelve years of growing experience under my belt. I wouldn't have decided to go back to school (AGAIN!) to pursue a Ph.D. I wouldn't be the poinsettia growing expert in our department at the University. And I certainly wouldn't have gone to Australia to tell everyone about my career and taken a once in a lifetime chance to tour New Zealand with IPPS members.

Figure 6. Traded Gray Cubicles for Rainbow Skies.

Current Trends in Propagation and Commercialization of Aquatic/Aquarium Plants

Dr. S. K. Unnikrishnan

Partner (Technical Head) Thirth Aquatics, Cochin, Kerala, India 683503

unni_sk@hotmail.com

Keywords: Tissue culture, micropropagation, aeroponics, hydroponics

INTRODUCTION

Aquarium plant/aquatic plant production is estimated to be 400-million-dollar worth industry with annual growth of 2-3%. As popularity of aquarium plants is increasing through the introductions of dynamic Planted Aquariums, Vivariums, Paludariums,etc, the aquatic plant growers are embracing innovative methods by adopting modern cultivation practices and packing methods in order for minimizing mortality and damage during transportation. The modern agricultural practices such as Hydroponics, Aeroponics, Plant Tissue Culture, etc, have certainly helped deliver premium quality aquarium plants to end users. This review highlights not only its adaptation to growth conditions depending upon the seasonal variations but the modern techniques for commercial production by employing innovative packing methods also.

Aquarium Plants in Natural Habitat

Still there is a wrong notion prevailing among the majority of hobbyists that these plants are growing always in submerged conditions and not able to survive in relatively dry conditions. To be frank, majority of the aquatic plants are amphibious in nature and can adapt to both growing conditions viz, underwater (emersed) and above water (submersed) growth conditions. As the habitat is seasonally inundated, majority of them acquired the ability to survive in both wet and relatively dry conditions through the course of evolution. Only a few such as water lilies, Apo*nogetons*, *Nuphar*, etc are considered true aquatic plants, as they are not able to display its vigorous vegetative growth during dry season, but can survive in the form of seeds, rhizomes and tubers during dry spells.

Amphibious plants exhibit strikingly different leaf morphology when they survive in different growth conditions. When they grow above water or relatively dry conditions the leaf colour and its leaf shape vary. This striking leaf morphology differences is evident from the illustrated examples (Fig. 1).

Figure 1. Comparison of leaf growth in immerged vs. submerged *Rotlala* sp 'HRA' and *Ludwigia arcuata*.

Even though submerged plants (submersed plants) exhibit attractive leaf colouration and shape, they are more fragile and vulnerable to damage during transportation. As emersed plants (above water grown amphibious aquatic plants) are much stronger and have less mortality and prolonged shelf life, modern propagation systems being exploited for promoting emersed growth. True aquatic plants always pose a challenge, but tissue culture technique coupled with innovative packing methods has enabled its transportation long distance without losing its quality.

Modern Propagation Systems Hydroponics

Employing hydroponic techniques to promote the emersed growth of amphibious aquarium plants in climatically controlled greenhouses has tremendously improved the health and shelf life. All parameters such as light, temperature, humidity are optimized through automation, and nutrient concentrations, water quality and pH of nutrient solution are maintained consistently uniform throughout the growth period (Fig. 2). As it is an enclosed propagation system, pest infestation and disease outbreak are very minimal. Control of disease and pest can be achieved by applying biological pesticides and fungicides. Using strong synthetic pesticides and fungicides should be completely avoided as the residue can pose serious threat to the life of fauna (fishes and invertebrates} in aquariums.

Figure 2. Modern hydroponic production systems under protected cover.

Hydroponic Cultivation Aeroponics

Nutrient fog instead of nutrient solution is used to cultivate aquatic mosses and certain amphibious plants such as many species of *Anubias*, *Bucephalandra* and ferns. Conventionally aeroponic system is employed to provide nutrient mist to the root system in an enclosed chamber to cultivating economically important crops, but for growing aquatic mosses and selected amphibious plants, these plants are wholly exposed to nutrient fog (Fig. 3).

Generally, aquarium plants are equally capable of absorbing nutrients trough leaves and roots, hence this growing technique is well suited for its large-scale propagation. Moreover, the soggy and over wet conditions can be avoided in this system which will certainly minimize the disease outbreak.

Figure 3. Aeroponic system for growing aquatic mosses.

Plant Tissue Culture

Tissue cultured aquatic plants in plastic containers for sale are disease free, pest free and being embedded in nutrient rich sterilized medium (Fig. 4 and 5). Now professional aquatic plant growers use this technique to provide top quality plant with prolonged shelf life to end users. However, they are more expensive but, quality is assured. As they are in sealed plastic containers being anchored in nutrient rich medium, many varieties such as *Anubias* spp, Alternanthera spp and aquatic mosses can survive for many months, provided being kept in cool area with sufficient light. The retail market prefers aquarium plants in this format, as shelf life is really good and shop owners can keep these plants for longer period before being sold out. Another advantage, as they are in its prime health, its survival rate is very high in newly set tanks which are not fully cycled.

Figure 4. Tissue cultured plants for sale in plastic containers.

Figure 5. Tissue cultured plants in growth room.

This technology is used for the commercial propagation of true aquatic plants such *Aponogetons*. The plants are sold in plastic pouches anchored in sterilized gelled nutrient medium and layered with nutrient solution to keep the leaves and crown submerged (Fig 6.).

Aponogeton longiplumosis

Aponogeton medagascerensis

Figure 6. Tissue culture production systems for *Aponogeton*.

Another advantage of plant tissue culture is bulking up of aquarium plants in lesser place through bioreactor production. Such bioreactors can act as mother stock in order for facilitating the production of containerized aquarium plants for sale later (Fig. 7).

Figure 7. Well maintained planted tanks in the author's gallery in India

Figure 8. Bioreactor production.

The Result of Using Growth Regulators Paclobutrazol and Chlormequat Chloride on Reducing the Height in Pots of The Sunflower (*Helianthus annuus*) 'Vincent'

Ou Wenjin (Owen) Australian Botanical Garden

Qi'ao Island, Tangjia Town, Zhuhai City, GD, People's Republic of China

owen@birkdaleinternational.com; 245091664@QQ.com

Keywords: Vincent sunflower, PGR, height reduction

Abstract

The result shows that all concentrations of the Paclobutrazol (applied at rates of 7.5, 15, 30 mg a.i. / 2 litre pot) can reduce the height of the sunflower variety 'Vincent' in varying degrees. For this experiment, the rate of 30 mg a.i. / 2 litre pot produced the most desirable result for reducing their height. But all concentrations of the Chlormequat Chloride (applied at rates of 0.025%, 0.05% & 0.1% / 2 litre pot) had little effect or no effect on reducing plant height of the Sunflower variety 'Vincent'.

INTRODUCTION

Helianthus annuus (Sunflower) is a highly economic global plant species known for its seed production for food & edible oil production. It is also a popular cut flower species for floristry & is becoming 'very fashionable' as a flowering pot plant for interior & exterior horticultural displays.

To meet a client request for a 12-month continuous supply of 400 – 800 mm tall, flowering Sunflowers for an interior themed display, a production template had to be developed within the company nursery facilities on Qi'ao Island to satisfy this request. From a very limited seed range of Sunflower

The Future of Landscaping: Understanding and Embracing Resilience

Marijke Honig, Beyer Honig Landscapes

Cape Town, South Africa

landscapes@beyerhonig.com

Keywords: Genetic diversity, water, conservation

Abstract

With climate change and dwindling water resources, **resilience** is the new criterion for plants and landscapes. Resilience not only to physical events (drought, storm winds, flooding), but also to human disturbances such as bark stripping, trampling.

We need to identify suitable plants that can adapt to a changing climate and bounce back after an extreme event. They may be indigenous or exotic. Need to find suitable combinations of plants – plant mixes – that require minimal maintenance.

INTRODUCTION

What changes can we expect with climate? Weather patterns will become more unpredictable and extreme.

How can we respond?

By creating resilient landscapes which can adapt to changing conditions and bounce back after an extreme event.

Resilience to what?

What are some of the general attributes of resilient plants and landscapes?

- Adaptable, can tolerate a wide range of conditions (not fussy about soil type, moisture, temperature).
- Can 'bounce back' (regrow) after disturbance or an extreme event.
- Resprouters (plants that can survive and regrow from an underground rootstock after fire), and/or plants that can be heavily pruned and will regrow.

From previous initial research experimentation undertaken in May – June 2018 when the air temperatures in this Qi'ao Island location are 20-26°C(16 - 22°C) at night and 24-32°C(19-27°C) through the day time (Fig 2a). Due to higher temperatures at this time of the year, all the treated plants grew significantly taller than the Feb – March 2019 trial. Observations from that trial showed that the dwarfing effect of 7.5mg a.i./pot treatment was not as effective. The plant height was taller at approx. 900mm. The dwarfing height of the 30 mg a.i./pot treatment was perfect but at this rate we observed a visually unacceptable, leaf deformation problem at the higher temperatures.

It was observed that all concentrations of the Chlormequat Chloride (apply as 100ml drench / pot at rates of 0.025%, 0.05% & 0.1%.) had a visually insignificant effect of reducing the height of this sunflower variety (Fig 2b).

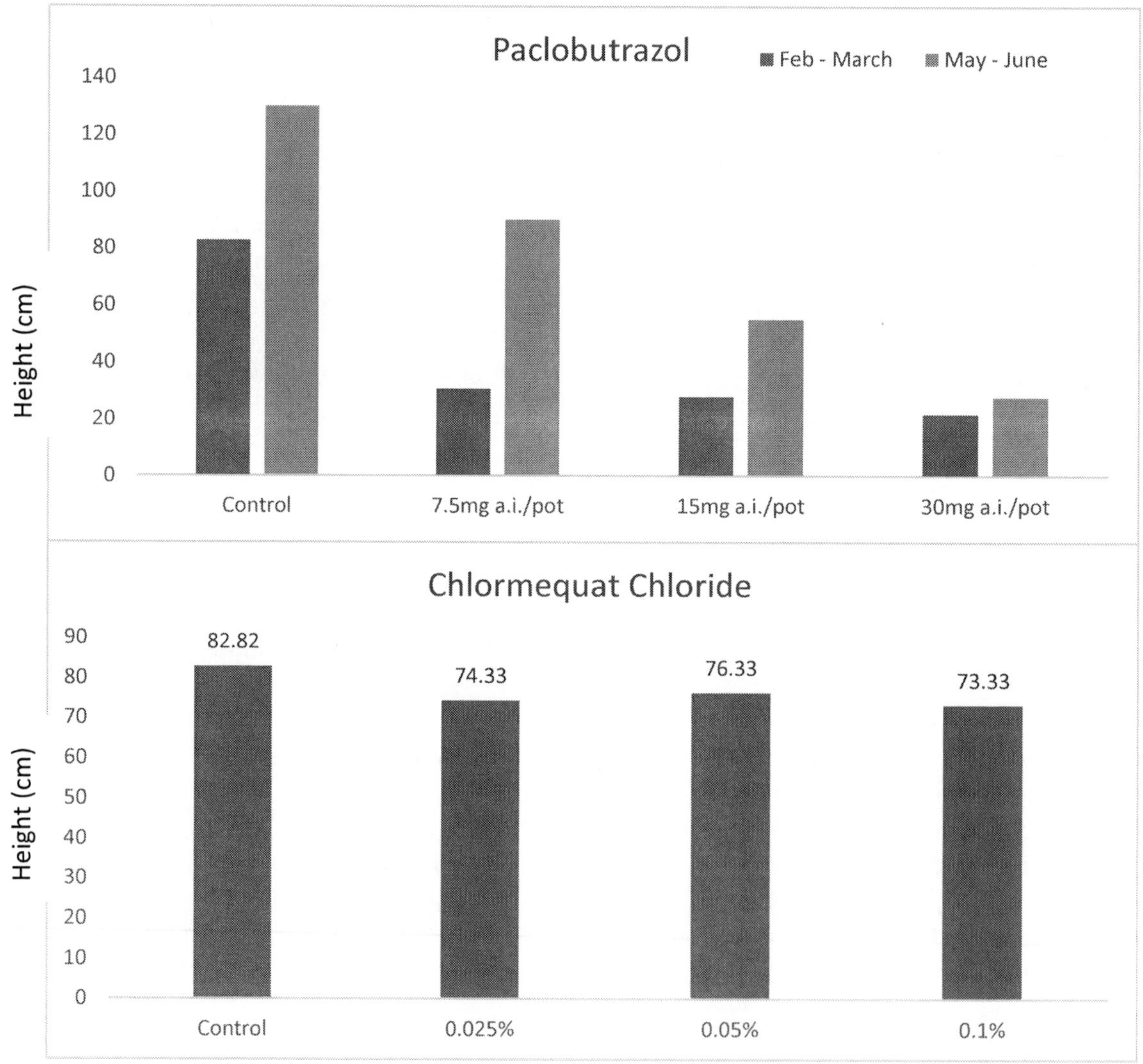

Figure 2. The reduction in Vincent sunflower after Paclobutrazol and Chlormequat Chloride.

The production time from sowing to a flowering product ready for installation into an interior themed display was 59 days in the February – March 2019 period.

Control | 7.5 15 30 (mg a.i./pot) Paclobutrazol | 0.025% 0.05% 0.1% Chlormequat Chloride

Figure 3. Comparison for height reduction in Vincent sunflower after Paclobutrazol and Chlormequat Chloride.

Literature Cited

Dasoju, S., Evans, M.R. and Whipker. B.E.. (1998). Paclobutrazol drenches control growth of potted sunflowers. HortTechnology *8*:235-237.

Spitzer, T., Bilovsky, J., Kazda, J. (2018). Effect of using selected growth regulators to reduce sunflower stand height. Plant Soil Environ. *64*:324-329.

Koutroubas, S.D. and Damalas, C.A. (2015). Sunflower response to repeated foliar applications of Paclobutrazol. Planta Daninha *33*:129-135.

Biosecurity: Who Wants to Be A Millionaire?

Lisa Burton

Otago Polytechnic, Private Bag 1910, Dunedin 9054, New Zealand.

Lisa.Burton@op.ac.nz

Keywords: Education, weeds, pests, diseases, New Zealand

INTRODUCTION

Part of what I do as a Principal Lecturer in Horticulture is to teach Plant Health. I really enjoy teaching adult students and apprentices because I know I can make a difference to their learning. When students start, most can identify a ladybird, an aphid, and a dandelion, but beyond that, it's a bit of a void. That's not too surprising as it's a big wide world of pests, diseases, and disorders out there.

Training people in Plant Health is vital no matter what sector of horticulture you're in. if a business doesn't have knowledgeable staff, who can confidently identify and then apply the right control methods, at the right time, you will be ditching a lot of sub-standard plants or worse, your reputation as a quality grower.

Beyond any individual business's concern about pest control, is the global issue of the huge negative impact that pests have on society's wellbeing and wealth.

So let's talk money. Put your hand up if you want to be a millionaire. Good, lots of people keen on that. You know the people that didn't put their hand up, that's because they are already millionaires, which is awesome because I've heard that Australians are generous and I'd like to ask them for a donation to cover some of New Zealand's pest control costs!

Counting the costs of weeds

Professor Philip Hulme (2018) reported that weeds cost the New Zealand economy billions of dollars per year. "Recent estimates suggest about $2 billion is lost every year in lost productivity and management costs simply from our pastoral, arable and forestry sectors.

The Royal Society Te Aparangi (RSTA) noted that, "Weeds pose a threat to one-third of all New Zealand nationally threatened plant species, and could potentially degrade 7% of the conservation estate within a decade, corresponding to a loss of native biodiversity equivalent to $1.3 billion." (RSTA, 2014:2).

Counting the cost of animal pests

Recent data (Ferguson et al., 2019) found that invertebrate pasture pests, most commonly targeting New Zealand's pastures, are costing the economy up to $2.3 billion per year.

In terms of vertebrate plant pests e.g. rabbits, possums, birds, the total direct economic cost of these pests to the primary sector, is estimated to be about $1 billion per year, but with multipliers included could be as high as $3.3 billion (RSTA, 2014).

New Zealand is known for its fairness, so I'm prepared to take donations of $10 billion, partly in Bitcoin if that helps.

So why are there so many pests to control in New Zealand?

"New Zealand remains under intense pressure from pests which threaten our economy despite investing heavily in biosecurity and pest management systems", according to the Royal Society Te Aparangi (2014:1).

Part of the reason we have so many pests is that scientists believe we became an isolated island about 85 million years ago and therefore much of our flora and fauna are unique. The pests that have established here have no natural enemies to keep them in check and so they can spread relatively quickly.

New Zealand's economy is reliant on primary industry exports and tourism. In other words, we rely heavily on our natural resources such as the land, sea, fresh water, native biodiversity and the ecological functions that support their sustainability.

However, we are no longer isolated. Pests and diseases have found their way to our country by chance, by design and by human ignorance of ecological systems. New Zealand is one of the weediest places in the world and *74%* of our weeds started as ornamental garden plants that then made a break for it in their new home (Landcare Research).

Australia in the past, generously, sent us some cute wildlife – the common brushtail possum (*Trichosurus vulpecula*) in 1837 to be exact. We have whittled the possum population down to a mere 30 million thanks to extensive government and volunteer efforts (Landcare Research, 2011).

A more recent arrival from Australia has been myrtle rust (*Austropuccinia psidii*). OK, myrtle rust sent *itself* on the wind currents, but we are all devastated that it has arrived and worry what the consequences of this will be for our native Myrtaceae such as the iconic pohutukawa (Metrosiderous sp.) as well as food crops.

Figure 1. Myrtle rust arrived in New Zealand in 2017 and has now spread as far south as the top of the South Island. (Source: L. Burton).

Why is biosecurity important?

The first line of defence is a biosecurity system to prevent unwanted organisms entering our countries in the first place. Both New Zealand and Australia's biosecurity systems monitor and assess risks to minimise outbreaks of pests that already exist in our countries and to prioritise the resources available.

To test your knowledge of Australia's biosecurity priorities I have a question for you.

What is the common name of the most threatening organism for Australia that destroys plants according to the Department of Agriculture and Water Resources? Here are some clues:

This pest is not in Australia or NZ (yet) The term pest covers animals, insects and diseases. Sorry there is no 50:50 or phone a friend on this one.

The answer is *Xylella fastidiosa* – a bacterial disease that originated in the Americas and is spreading to other countries. It causes browning and loss of leaves, stunting, reduced fruit size, dieback and death of the plant. The Department of Agriculture and Water Resources (DAWR, 2019) reports that over 350 native, commercial and ornamental plant species are at risk from *Xylella.*

Xylella is just one of Australia's top 40 exotic and unwanted plant pests. It is estimated that the cost to the industries at risk if these organisms enter Australia are (in Australian dollars) $27.9 billion to the broadacre and horticulture crop industries and $2 billion to the forestry industry (DAWR, 2018).

The Department of Agriculture and Water Resources has a useful website, with a short video clip and information sheet on each of the top 40 plant pest risks to Australia.

In New Zealand, it's been hard recently not to feel depressed by the ongoing and significant pest issues that we are facing – be it *Mycoplasma bovis*, myrtle rust, the continuing spread of guava moth or the heart breaking kauri dieback. This disease is killing thousands of our giants - *Agathis australis* grow up to 50 metres and can live for more than one thousand years. The disease was identified in 2015 and named *Phytophthora agathidicida* – the kauri killing *Phytophthora* (Landcare Research, 2019). Unfortunately, kauri dieback continues to spread despite efforts to control it.

I know that many of you are working hard on protocols in the nursery industry and in the revegetation industry to prevent or control the spread of *Phytophthora species.*

Do we understand biosecurity?

It was encouraging in 2018 to hear that New Zealand's Ministry for Primary Industries (known as MPI), had formed a new business unit called Biosecurity NZ to provide a stronger focus on biosecurity. We have scientific experts, trained staff in Biosecurity NZ, Regional Councils and other community groups and professional networks and I thought "*we had it under control".* And then two things in particular made me think that maybe *we didn't...*

Last year the Government's 'One Billion Trees' programme was launched with the aim of planting One Billion trees in a decade: good news in itself. The bad news was that the Ministry released a list of trees as a guide to councils and the public, which included a number of well-known tree weeds such as hawthorn (*Crataegus* sp), sycamore (*Acer pseudoplatanus)* and Taiwan cherry (*Prunus campanulata*). It took others such as Forest & Bird (2018) to point out the mistake which included one species that was illegal to sell or propagate and others that ratepayers are spending millions on trying to eradicate.

How could anyone working for the Ministry for Primary Industries fail to notice, or understand the difference, between suitable trees and invasive tree weeds. Then, I read an article with the heading, 'Biosecurity Not Well Understood'. It related to a government report that included the scary statistic that only 2% of New Zealanders see biosecurity as relevant to them and 40% do not know, or have limited knowledge of, what the word even means (Biosecurity 2025, 2018).

These two events combined, made me think, what is going on here? Have we run out of people with even the most basic knowledge of biosecurity and plant health? Given that a substantial amount of New Zealand's wealth is due to our primary industry exports, and that these industries rely on quality biosecurity systems, it was shocking to read that so few people understood how any failure in our biosecurity would directly affect them economically. For example, about 60% of New Zealand's total production of fruit and vegetables is exported and this trade was recently valued at $3.62 billion by the Horticulture Export Authority (2018).

The millions that some people dream of, via the lotto draw, or the television quiz game, might be their only hope if the plant equivalent of 'foot and mouth' disease, should ever reach our shores.

I'm worried that we are not training enough entomologists, plant pathologists, horticulturalists, researchers and IT specialists to deliver a robust biosecurity system to manage these complex challenges. Unsurprisingly given my role, I believe knowledge is power, and an accessible, quality education is the building block of a resilient and innovative knowledge-based economy.

I believe that both our governments need to wake up and look at the numbers of those training in plant sciences and horticulture, and significantly increase funding to *attract* and *train* people at every level - from degrees to certificates, *whether it's on campus, on job or online*. The number of people studying horticultural science in New Zealand at degree level has undergone a massive decline since the 1980's – I should know because I was there.

Experienced horticulture industry members are already playing their part in biosecurity vigilance, as are many organizations such as Predator Free New Zealand and Weedbusters but is this enough?

Who's responsible for biosecurity?

I took a closer look at Biosecurity New Zealand and was pleased to see that it had taken stock of its essential role, and with partner organisations, called 'Biosecurity 2025' were producing strategies, setting targets and conducting surveys to measure their progress.

Some of the results of their public survey (Biosecurity 2025, 2018a) were encouraging. For example, nearly all New Zealand adults (96%) recognised the importance of protecting New Zealand from unwanted pests, weeds, and diseases. However, only 29% agreed that they could identify the main pests, weeds and diseases that pose a threat to New Zealand's environment. The survey results showed there was also a need for greater public awareness of biosecurity, why it mattered in our everyday lives and how Kiwis could help.

Biosecurity New Zealand also correctly identified that relying on their own staff and resources was never going to be enough to win against pests; it was going to take all of us. They have set out to achieve a biosecurity team of 4.7 million people. In other words, they are trying to encourage every New Zealander to contribute to minimizing our biosecurity risk.

Figure 2. A Biosecurity Team of 4.7 Million. Source: With the permission of Biosecurity NZ, Biosecurity 2025.

One of their key messages is that we all suffer the consequences if we don't have a strong and resilient biosecurity system (Biosecurity 2025, 2018b). Their aim that ''biosecurity education and training is building the necessary knowledge and skills for all ages across the system" is heartening but I have yet to see the details of how they plan to achieve this (Biosecurity 2025, 2018c).

Without governments seriously valuing the educators and science staff that they already have, and the crucial experience and knowledge from industry people like yourselves, plant health skills will not easily transfer to the next generation of horticulturalists, let alone the general public.

One of the themes of this conference is 'Manage Change', and one of the biggest issues in biosecurity, is how to predict and assess the effects of a warming climate on the spread of unwanted organisms. There should be a sense of urgency about this as a warming climate makes an already difficult biosecurity situation, even worse.

In general, the warmer the climate, the faster pests reproduce. New Zealand has a temperate climate but there is an increased risk of tropical and subtropical pests entering our country, surviving and spreading due to global warming. Increases in storms, and changes to wind patterns and ocean currents will also contribute to new distribution patterns of unwanted pests.

On this theme, I'd like to play a two-minute clip about climate change produced by NIWA which is New Zealand's National Institute of Water and Atmospheric Research. The video is called 'Our Climate Is Changing' and was produced in June 2018.
https://www.niwa.co.nz/videos/our-climate-is-changing

It asks the question, are we ready for change?

CONCLUSION

We all have a part to play to ensure our biosecurity systems are as robust as possible, so we can continue to enjoy the financial benefits from our primary sector exports, built on our natural resources. Invasive pests, diseases, and weeds erode that wealth and the cultural taonga (treasure) that is our responsibility to sustain and protect.

Knowledge of plant health requires continued investment in the education of those training in horticulture, plant sciences and associated sectors. Funding is also needed to increase the public's awareness of why biosecurity matters and how each of us can together contribute to this challenging task.
Kia ora and Ko Tatou: This is Us.

Literature cited

Biosecurity 2025. (2018a). 'Strategic Direction 1: A Biosecurity Team of 4.7 million Public Survey'. Biosecurity NZ, Ministry for Primary Industries.

Biosecurity 2025. (2018b). 'Strategic Direction 1: Engagement Plan'. Biosecurity NZ, Ministry for Primary Industries.

Biosecurity 2025. (2018c). 'A project to develop biosecurity system health monitoring and reporting'. Biosecurity NZ, Ministry for Primary Industries.

Department of Agriculture and Water Resources. (2019).Retrieved from http://www.agriculture.gov.au/pests-diseases-weeds/plant/xylella

Department of Agriculture and Water Resources. (2018). http://www.agriculture.gov.au/pests-diseases-weeds/plant

Ferguson, Colin M. et al. (2019). Quantifying the economic cost of invertebrate pests to New Zealand's pastoral industry. New Zealand Journal of Agricultural Research *62:* 255-315.

Forest & Bird. (2018). 'MPI includes pest weeds in Billion Trees list'. Press release 4 July 2018.
Retrieved from https://www.forestandbird.org.nz/resources/mpi-includes-pest-weeds-billion-trees-list

Horticulture Export Authority (2018). 'Horticulture growth retains momentum'. Press release 3 December 2018. Retrieved from http://www.scoop.co.nz/stories/BU1812/S00036/horticulture-growth-retains-momentum.htm

Hulme, P. (2018). 'Philip Hulme: 'NZ is one of the weediest places on earth''. Radio New Zealand, 6 May 2018. Retrieved from https://www.rnz.co.nz/national/programmes/sunday/audio/2018643664/philip-hulme-nz-is-one-of-the-weediest-places-on-earth

Landcare Research. (2019). 'Our history with kauri dieback'. Retrieved from https://www.landcareresearch.co.nz/about/news/snippets/our-history-with-kauri-dieback

Landcare Research. (2011). 'Forecasts and 'nowcasts' of possum distribution in New Zealand'. Retrieved from https://www.landcareresearch.co.nz/publications/newsletters/kararehe-kino/kararehe-kino-23/possum-distribution

Royal Society of New Zealand Te Aparangi. (2014). 'Challenges for pest management in New Zealand. Retrieved from https://royalsociety.org.nz/assets/documents/Challenges-for-pest-management-in-NZ.pdf

IPPS Western Region Exchange 2018

Megan Robinson

Headford Propagators Ltd, 24 Horsnells Road, Morven, Waimate

meganrobinson901@gmail.com

Keywords: Hawaii, United States, scholarship

INTRODUCTION

When I first got offered the application to apply for the IPPS Western Region Exchange, I actually thought Grant Hayman my employer was having me on. Little did I know that applying for this and getting accepted was going to be one of the most mind - blowing, eye-opening and memorable experiences of my life.

I had never experienced travelling overseas before from my home country, so it was a very nervous but exciting feeling and buzz that kicked in when I first flew out of Auckland on Saturday 28th July. I'll try to summarise in this write-up as much as I can about the whole overall exchange.

I attended the IPPS Western Region Conference in Hawaii where I was picked up by Phil and Gail Gredler in Honolulu. The first day there I got to try the local beer and got to go to Pearl Harbor and learn about the history of the place. The rest of the tours included visiting the Punchbowl Memorial Cemetery of the Pacific, the Foster Botanical Garden, visiting a Hibiscus breeder and even made it to the waterfall at the Waimea Arboretum. From swimming in the gorgeous beaches of Hawaii to taking a whole lot of pictures, the start of my exchange it was an absolute breath-taking experience.

Figure 1. Corpse flower (*Amorphophallus titanium*) at Foster Botanic Garden.

On Sunday 29th July, I flew over to beautiful Kona, also known as 'Big Island'. We visited the USDA National Germplasm Repository, Hawaiian Sunshine Nursery, Nani May Gardens, Mauna Loa Macadamia Nut Factory and shop, Akastsuka Orchid Gardens (which were amazing, I never knew an orchid could smell so good!) and then we made our way back around the Southern end of the Kona Island.

On the last day in Kona I presented my speech to everyone at the conference. I think I did my employers Grant and Robynne Hayman and the New Zealand IPPS Region very proud and touched a few souls at that. After speeches, we went on our last tour for conference which included a visit to the Mountain Thunder Coffee Plantation and the Kona Cloud Forest Sanctuary (that was so incredible!). We ended at Dean's Gardens and then headed back to our motel to get ready for our Hawaiian Theme Dinner!

I flew out of Kailua and flew into San Jose on Wednesday 4th August 2018. This is where the IPPS Western Region Exchange began. I got picked up by Don Dillon and spent my first night in Corralitos with Don and his wife Donna at their beautiful home.

The next morning in Corralitos, Don decided to show me around. We started with going and visiting a place called Roses of Yesterday and Today which was a fenced off section filled with roses, some alive and some dead. The roses were budwood plants. Our next stop was Don's nursery, Four Winds Growers which is a wholesale citrus nursery. The nursery is mainly run by Lexa, Don's daughter as a shared partnership with another guy. Four Winds got its name when alcohol was illegal during prohibition = speak easy, and the orange behind the logo is from Don's grandad - Minneola tangelo. This was a very well set out nursery. Everything stays shut up to protect plants from ACP (Asian Citrus Psyllid) and HIB (also known as Citrus Greening or Yellow Dragon Disease).

Four Winds grow 60 different cultivars of citrus plants including Buddha Hand (the original dwarf citrus since 1950), Seedless Midnight Valencia Orange (semi dwarf citrus), Lamb Hass (PAT), Lemon Allen Eureka and Kaffir Lime. The shade house has bottom heating to help with rooting. All houses are kept very hot. The nursery has joined rails in houses to help with the transport of aluminum benches (massive trays) which seemed like a very good solution for transportation of plants to benches. 1 acre holds 125,000 plants which is 4x4 by 9 inches which then gets transported to a 3-gallon pot. They mostly sell to wholesale nurseries in California, Canada and are going to start selling to the East Coast. The glass house has insect proof vents with insect fabric on sides. For my first nursery to visit in California it seemed huge compared to Headford Propagators back home, I could not wait to see what was next. That about sums the tour of the Don's Nursery.

Figure 2. Citrus plants at Four Winds Nursery

We went down what the Californian's call a rough road which in New Zealand is just known as bumpy tar seal. I'd hate to think what their opinions of the some of the New Zealand roads would be. Along our journey, Don pointed out some of the Californian natives and told me a few reasons why they're known worldwide. Redwoods are the tallest tree in the world at over 116 metres. Giant Redwoods are the largest trees in the world on the terms of trunk volume. I also learnt a cool little fact that 'casa' means house in Spanish. Over in the states, pretty much everywhere that is a working environment you have to be able to speak two languages: Spanish and English.

Later on that day I got picked up by Nicky Hughes. I stayed with Nicky and her partner for two days in Soquel. Over the two days we visited a number of places. We started at the Arboretum at the University of California, Santa Cruz. I was amazed by the number of gardens at this place from succulent gardens to a butterfly garden. They even had a New Zealand native garden and a place for the plants that attracted the bees. It was so bewildering seeing so many new, but different plant species and some that had so many similarities to the plants back home like the Red Fountain Cordyline. They even had White sage (*Salvia apiana*).

Next morning was filled with tours and sight-seeing. We visited Yamagami's Garden Centre and Nursery. There I learnt that a plant called *Asarum caudatum* is pollinated by ants, butterflies have tongues and Gazania 'Daybreak White' is from the sunflower family. We also visited the Elizabeth F. Gamble Garden in Palo Alto CA, alongside an absolute beautiful garden placed in the centre was a big old historic home. While driving along checking out the ocean and beaches off Highway 1, CA, we pulled over to go check out the Pigeon Point Lighthouse, which was just in the process of being restored. It's always windy there so a good place to lose your hat.

Soquel Nursery Growers is a nursery covers 9 acres and is right next door to where Nicky and her partner lived. The nursery has a crew of only 11 people - two propagators, a crew of six on potting and orders, a full-time manager and two part time sales people. They buy in liners and do propagation by cuttings and seed. For transportation they have one pick-up truck and a box truck. Electric carts can hook on a trailer. They grow their plants in pots from one gallon at the smallest to 15 gallons roughly being the biggest. Potting mix is from Sunland and has Redwood bark, peat, coir and lava rock instead of pumice. This is used as a four-month slow release formulation. We also visited Gurdsneidner Nursery where they use whitewash to stop the sunlight coming through their roofs.

Figure 3. Part of the large selection of plants at Yamagami's Garden Centre and Nursery.

Nicky works at a community college called Cabrillo College in Aptos, CA. We went for a tour and she explained to me what they do at their college. They were currently doing an Aquaponic lettuce crop, which involved a big tub filled with water and wee lettuce plugs were placed individually into polystyrene and the roots were just growing freely into the water.

Later on, that day we went out for tea in a gorgeous town called Capitola. This had to be one of my favourite places with my stay with Nicky. We went out to Margaritaville - what better place to have a margarita and some seafood. I really loved my stay with Nicky and her partner, not only did she give me awesome tours but she left me with a whole head filled of new knowledge and was never short of an answer for every question I asked.

The next day I met up with Danny Taiko, my next host and his family in Fresno. This is where I first got a glimpse from a distance of just how bad the bush fires and droughts over in California are. Danny is the owner of Taiko Nursery which is a propagation nursery and his daughter Lisa mainly runs the business now. Danny's nursery is a very well-established nursery and I was shown a demonstration of his Elle pot machine. Elle pots seems to be very popular overseas.

Now for some more of the fun stuff, I got the opportunity to go to Disneyland from morning to night! What a blast that was so much fun with Lisa and her partner Darren.
The next morning, we drove to San Jacinto, California. This was my next drop off to my hosts Jim and Andi Conner. I had already met Jim and Andi in Hawaii, so it was a pleasure to see them again. I also wanted to spend more time with Andi so that she understood my kiwi lingo before I left the country and flew back home.
Jim had no time to waste with tours starting right away. First up I got the pleasure of a tour round their nursery, Alta Nurseries and wow, what a beautiful portrait of scenery you get with the nursery and the San Jacinto mountain ranges in the background. I visited so many different nurseries it was hard to write them all down. Jim gave me so much of not only an open mind, but also taught me so much about horticulture in California.

Before our first nursery visit, we stopped in at a winery and although it was too early for wine testing, I was able to shake hands with the owner of the winery who just happened to be a billionaire. That will probably be the first and last time I will ever get to do that.

Figure 4. Alta Nurseries with the San Jacinto mountains in the background

We visited Olive Hill Greenhouses, which produce hardier indoor plants, and also Western Cactus where I never knew you could grow so many different varieties of cactuses and succulents! Jim also took me to a flower market, which is a real talent the way people can design bunches of flowers.

Figure 5. Indoor plant production at Olive Hill Nurseries.

We visited Shearman Nursery which is owned by a German guy called Jim Shearman. Some plants he grows that I took a major interest in were *Eucalyptus deglupta, Ficus dammaropsis* from New Guinea and the Grape tree (trunk gets huge). Here I also got the opportunity of holding my first ever snake!

I had the pleasure of staying in Jim and Andi's condo in Oceanside. Oceanside has a gorgeous waterfront with a pier that reaches 400 metres out to sea. They also have a big surfing competition on one side of the pier. Andi and I even got in some time to do some local shopping at the Oceanside Street Faire. They were lovely hosts and made me feel so at home, Jim couldn't have showed me around more in the small time I stayed with them.

Next Jim dropped me off to Glenn and we drove through Malibu and the majority of the cruise was just beautiful by the coastline. It was also different to see how close the big oil rigs were to the coastline which being a kiwi we don't have that over here. We finally made it too San Luis where I got dropped back off to Lexa for my last couple of nights before flying back to New Zealand.

The next day I got the opportunity of going for a visit with Don (Lexa's Dad) and her two children Max and Cora to go see the Monterey Bay Aquarium, which was just awesome! We saw everything from Otter's being fed, heaps of different kinds of fish, eel, Sand dollar, Sting rays, birds, jelly fish (Moon Jelly had to be my favourite) and plenty more, I would go back there in a heartbeat.

I spent my last day at Four Wind Growers with Lexa while she completed her work in the office for the morning before driving to San Francisco. To fill in some time before my flight, we visited the famous San Francisco Golden Gate Bridge which honestly TV and movies do it no justice until you see it in person. Lexa also showed me the Palace of Fine Arts, a beautifully designed building with so much texture and art that was incredible to see.

A massive thank you to the IPPS New Zealand Region and IPPS Western Region Exchange for giving me this amazing opportunity and making this trip possible. It has given me an insight to a bigger world of horticulture. It has also made me realize how lucky we are as New Zealanders that we don't get droughts, nor do we have a high risk of bush fires like over in the United States. Thank you to all my hosts, IPPS members and to my employers Grant and Robynne Hayman for getting me out of my comfort zone. This has been a life-changing experience and has made me have so much more knowledge that I can put to use.

PROCEEDING'S PAPERS

JAPAN REGION

Dr. Masanori Tomita, Regional Editor

Twenty-sixth Annual Meeting - 2019

Mie, Japan

Effects of Medium Constituents on the In Vitro Flowering Response in *Celosia argentea* L.

Wakanori Amaki, Yoko Harada, Yuko Yamamoto, Hiroko Nakatsuka and Arisa Noguchi

Department of Agriculture, Tokyo University of Agriculture, 1737 Funako, Atsugi, Kanagawa 243-0034, Japan.

amaki@nodai.ac.jp

Keywords: Tissue culture, nutrient solution

Abstract

The effects of medium constituents on the growth and flowering in *Celosia argentea* L. were investigated. At first, the influence of the difference in medium strength was examined. The flowering rate was almost 100 % from 0 MS to 1/8 MS, but no flowering was observed at concentrations higher than 1/4 MS. As the medium strength was lowered, the number of true leaves formed until flowering decreased gradually, and on the 0 MS medium, inflorescence was formed without the formation of true leaf after cotyledons unfolding. Inflorescence formation without true leaf formation was observed only when macro-nutrients in the MS medium constituents were not added. Furthermore, it became clear that inflorescences formed without true leaf formation occurred only when NH_4NO_3 and KNO_3 in the macro-nutrients were not added. All of 16 cultivars tested formed inflorescence without true leaf formation on the 0 MS medium. When the time of transplantation from 0 MS medium to 1/30 MS medium was changed, we confirmed the NH_4^+ and NO_3^--free period required for the phenomenon of inflorescence formation without true leaf formation was at least 3 weeks

INTRODUCTION

Tissue culture techniques have been used to precisely analyze the effects of environment such as light and temperature and/or the effects of chemical substances such as plant growth regulators on the growth and flowering in plants. Many studies using tissue culture techniques reporting the in vitro flowering response of plants and proposed several in vitro flowering systems (Van Staden and Dickens, 1991). For example, tissue culture systems using stem thin cell layer (TCL) explant of *Nicotiana tabacum* L. (Tran Than Van, 1973, 1999), internodal stem explant of *Torenia fournieri* Linden ex E. Fourn. (Tanimoto and Harada, 1981a, 1981b; Tanimoto et al., 1985) and shoot apex explant of *Chrysanthemum morifolium* Ramat., *Ipomoea nil* (L.) Roth (Harada, 1967; Ishioka et al., 1990) and *Kalanchoe blossfediana* Poelln. (Dickens, 1987; Yang et al., 1999), have revealed the influences of medium constituents, culture environment conditions and physiological conditions of mother plants for promoting in vitro flowering (Van Staden and Dickens, 1991). We have accidentally discovered that *Celosia argentea* L. easily flowered in vitro through the series of experiments which the growing of sterile plants in many plant species (Yamada et al., 1997). In this report, we have examined in detail the effects of medium constituents on the in vitro flowering in *Celosia argentea.*

MATERIALS AND METHODS

The following experiments were performed using commercial cultivar seeds of *Celosia argentea.* The seeds were immersed in 70% ethanol and 3% sodium hypochlorite solution containing 0.01% Tween 20 for 30 seconds and for 10 minutes respectively, and then washed three times with sterilized pure water. The culture vessel was a glass flat-bottom test tube (40 × 150 mm). Murashige and Skoog (1962) medium (MS) added 30 gL^{-1} sucrose and 8 gL^{-1} agar and the pH was adjusted to 5.8 as a basal medium. Each of test tube dispensed 30 mL of respective media. The test tubes were closed with aluminum foil, and autoclaved (120°C, 15 minutes) before use.

After sowing 3 seeds in each test tube, placed them in the dark at 24°C for the first 3 days for dark germination. From the 4th day, they were illuminated with white fluorescent lamps (FL40SS-N/37, Toshiba Lighting & Technology Co., Ltd., Yokosuka city, Kanagawa prefecture) for 16 hours (30 μmol m^{-2} s^{-1} PPFD) / 8 hours in the dark at 24°C for 12 weeks after sowing. Two weeks after sowing, thinning was performed to obtain one seedling in a test tube, and 10 seedlings were used in each treatment plot.

Effects of medium constituents on flowering reaction

Seeds of *Celosia argentea* 'Castle Yellow' were used. At first, effect of medium strength on flowering was examined. All constituents of MS were diluted to 0 (means no addition), 1/1000, 1/500, 1/100, 1/50, 1/15, 1/8, 1/4, 1/2 and 1/1 concentrations. Secondly, using 1/15 MS, we divided the MS constituents into 3 groups, macro-nutrients, micro-nutrients and organic components such as vitamins, and set up 8 treatment plots combined with and without them to examine the effect on flowering (Table 1).

Thirdly, using 1/15 MS, 8 treatment plots combined with or without the addition of N (NH_4NO_3, KNO_3), P (KH_2PO_4) and K (KNO_3, KH_2PO_4) in the macro-nutrients were set up, and the effect of each macro-nutrient on flowering was examined (Table 2). When N and P were not added, the amounts of potassium equivalent to KNO_3 and KH_2PO_4 were supplemented with KCl. When K was not added, the amounts of nitrogen and phosphoric acid equivalent to KNO_3 and KH_2PO_4 were supplemented with NH_4NO_3 and $NaH_2PO_4 \cdot 2H_2O$, respectively.

Cultivar differences in flowering response on 0 MS medium

Since there are many strains and cultivars in *Celosia argentea* (The Royal Horticultural Society, 1992), 16 cultivars were tested for the flowering response on the 0 MS medium. The strains used were 10 cultivars of Plumosa group, 5 cultivars of Kurume group and one cultivar of Cristata group (Table 3).

Table 1. Effects of MS medium strength on growth and flowering in *Celosia argentea* 'Castle Yellow'.

1/15 MS medium			Flowering	No. of	Days to	Plant height
Macro-nutrients*	Micro-nutrients*	Vitamins*	(%)	true leaves	flowering	(mm)
−	−	−	88.8	0c	49.0	7.2d
−	+	−	100	0c	45.0	7.5d
−	−	+	100	0c	35.8	6.9d
−	+	+	100	0c	55.3	7.0d
+	−	−	100	10.8a	46.7	42.1b
+	+	−	100	9.8b	42.9	44.1b
+	−	+	100	10.4a	52.9	55.8a
+	+	+	100	9.0b	56.0	37.5c

* Macro-elements: NH_4NO_3, KNO_3, $MgSO_4 \cdot 7H_2O$, $CaCl_2 \cdot 2H_2O$, KH_2PO_4, Micro-elements: Fe-Na-EDTA, H_3BO_3, $MnSO_4 \cdot 4H_2O$, $ZnSO_4 \cdot 7H_2O$, KI, $NaMoO4 \cdot 2H_2O$, $CuSO4 \cdot 5H_2O$, $CoCl_2 \cdot 6H_2O$, Vitamins: myo-inositol, glycine, pyridoxine-HCl, thiamine-HCl, nicotinic acid

Table 2. Effects of N, P, K addition or absence on the growth and flowering in *Celosia argentea* 'Castle Yellow'.

1/15 MS medium			Flowering	No. of	Days to	Plant height
N*	P*	K*	(%)	true leaves	flowering	(mm)
−	−	−	100	0d	49.7	6.8d
−	+	−	87.5	0d	52.0	6.9d
−	−	+	100	0d	52.1	6.7d
−	+	+	88.8	0d	49.9	6.9d
+	−	−	33.3	2.0c	46.7	10.0b
+	+	−	12.5	3.0c	35.0	8.8c
+	−	+	100	4.3b	47.8	37.5a
+	+	+	100	6.1a	49.0	41.4a

* N: NH_4NO_3, KNO_3, P: KH_2PO_4 , K: KNO_3, KH_2PO_4. When N and P were not added, the amounts of potassium equivalent to KNO_3 and KH_2PO_4 were supplemented with KCl. When K was not added, the amounts of nitrogen and phosphoric acid equivalent to KNO_3 and KH_2PO_4 were supplemented with NH_4NO_3 and $NaH_2PO_4 \cdot 2H_2O$, respectively.

Effect of N application time on flowering reaction

In order to reveal the relationship between the N (NH_4^+ and NO_3^-) containing period in the medium and the phenomenon of flowering without the formation of true leaf, the following experiment was performed. Seeds of *Celosia argentea* 'Castle Yellow' were used. The seeds were sowed in 0 MS medium, and then transplanted to 1/30 MS medium at 1, 2 and 3 weeks after sowing. In addition, a plot without transplantation (all culture period used 0 MS) and another plot which was sowed to 1/30 MS from the beginning were set up (Table 4).

Table 3. Cultivar differences in the growth and flowering responses of *Celosia argentea* on the 0 MS medium.

Cultivar name (Strain)	Plant form	Earliness	Flowering (%)	No. of true leaves	Days to flowering	Plant height (mm)
Kimono Red(P)	Very dwarf	Very early	100	0	32.2d	9.0b
Kimono Yellow(P)	Very dwarf	Very early	100	0	36.4d	12.0b
Scarlet Charm(P)	Very dwarf	Early	100	0	43.4cd	13.0b
Castle Scalet (P)	Very dwarf	Early	66	0	49.0c	13.3b
Castle Pink (P)	Very dwarf	Early	100	0	46.2c	12.5b
Castle Yellow (P)	Very dwarf	Early	100	0	42.0c	11.9b
Red Kewpie (P)	Dwarf	Early	100	0	43.4cd	20.5a
Yellow Kewpie (P)	Dwarf	Early	80	0	49.0c	18.0a
Century Red (P)	Medium tall	Early	100	0	60.2ab	8.0b
Century Yellow (P)	Medium tall	Early	100	0	64.8a	6.9c
Fire Glow (K)	Medium tall	Very early	100	0	52.5b	10.0b
Golden Glow (K)	Tall	Very early	60	0	60.7ab	8.0b
Early Rose (K)	Tall	Early	40	0	42.0cd	8.5b
Kurume Kagayaki (K)	Tall	Early	100	0	40.6cd	8.5b
Kurume Gold (K)	Tall	Late	80	0	71.8a	9.0b
Jewel Box (C)	Very dwarf	Very early	100	0	30.8d	8.0b

Table 4. Effects of nitrogen supply period on the growth and flowering in *Celosia argentea* 'Castle Yellow'.

Media		Transplant timing	Flowering (%)	No. of true leaves	Days to flowering	Plant height (mm)
First	Second					
0 MS		No	100	0c	56.0	9.0c
0 MS	→ 1/30 MS	After 1 week	100	6.0a	49.3	36.0a
0 MS	→ 1/30 MS	After 2 weeks	100	2.1b	49.0	17.4b
0 MS	→1/30 MS	After 3 weeks	100	0c	48.5	10.4c
1/30 MS		No	90	4.8a	45.4	33.8a

RESULTS AND DISCUSSION

The flowering rates were almost 100% from 0 MS to 1/8 MS, but no flowering was seen at concentrations higher than 1/4 MS (Fig. 1). As the medium strength was lowered, the number of true leaves formed until flowering decreased linearly (Fig. 1), and on the 0 MS medium, inflorescence was formed without the formation of true leaf after cotyledons unfolding (Fig. 2). Inflorescence formation without true leaf formation was observed only when macro-nutrients in the MS medium constituents were not added (Table 1 and Fig. 3). Furthermore, it became clear that inflorescences formed without true leaf formation occurred only when N (NH_4NO_3 and KNO_3) in the macro-nutrients was not added (Table 2 and Fig. 4).

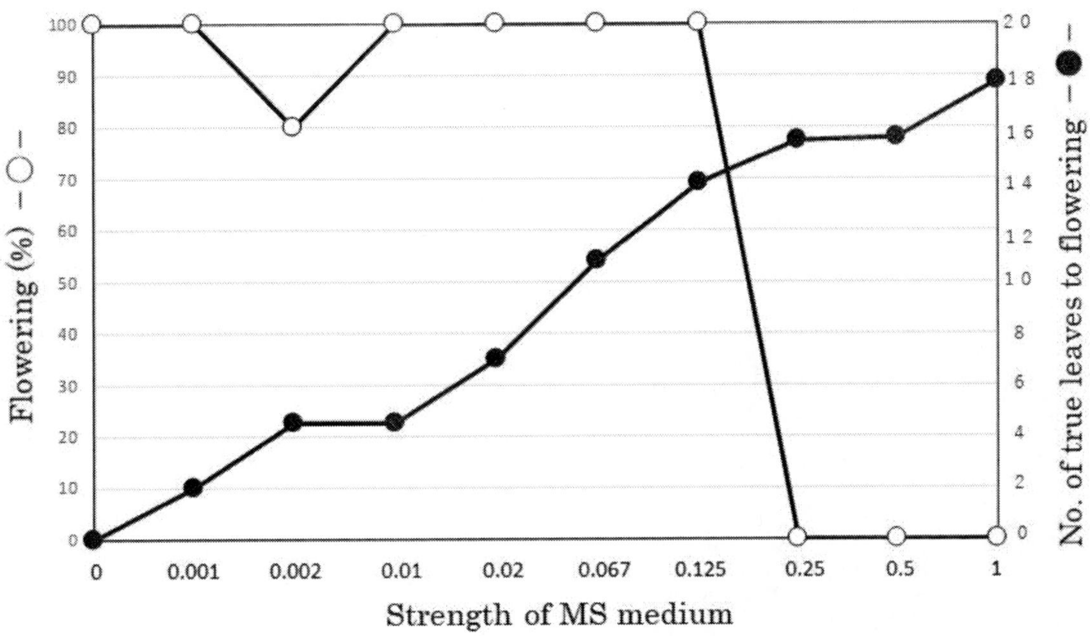

Figure 1. Effects of MS medium strength on the in vitro flowering rate and number of true leaves to flowering in *Celosia argentea* 'Castle Yellow'. All seedlings were cultured for 12 weeks.

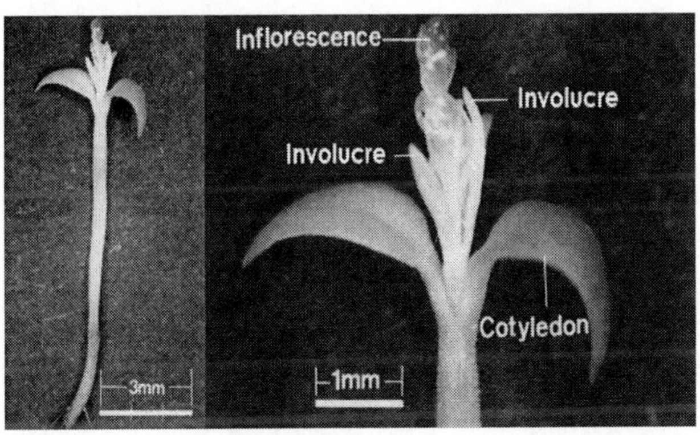

Figure 2. Photographs of seedlings of *Celosia argentea* 'Castle Yellow' cultured in vitro on the 0 MS (only added 30 gL^{-1} sucrose and 8 gL^{-1} agar) medium for 12 weeks. Left: Whole seedling. Right: Enlarged view around the inflorescence.

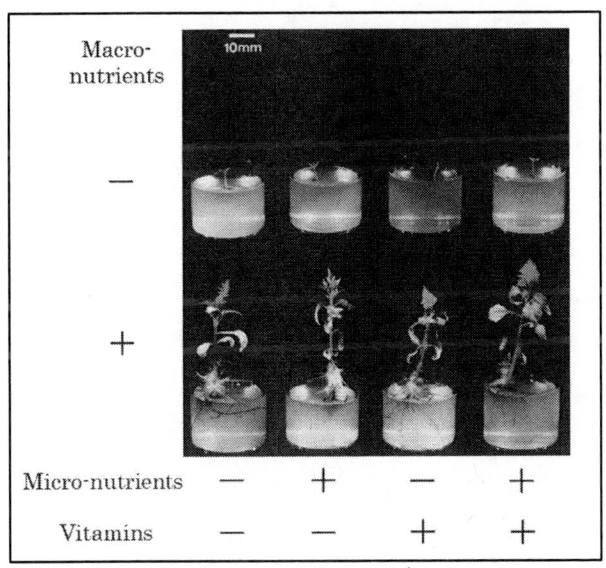

Figure 3. Effects of deletion of macro-nutrients, micro-nutrients and/or vitamins from 1/15 MS medium on the growth and flowering in *Celosia argentea* 'Castle Yellow' (12 weeks after sowing).

When K (KNO_3 and KH_2PO_4) was not added, the number of true leaves decreased to 2.0 - 3.0, but at the same time the flowering rate was significantly reduced, that is, potassium deficiency markedly inhibited the both of vegetative and reproductive growth of *Celosia argentea* (Table 2 and Fig. 4).

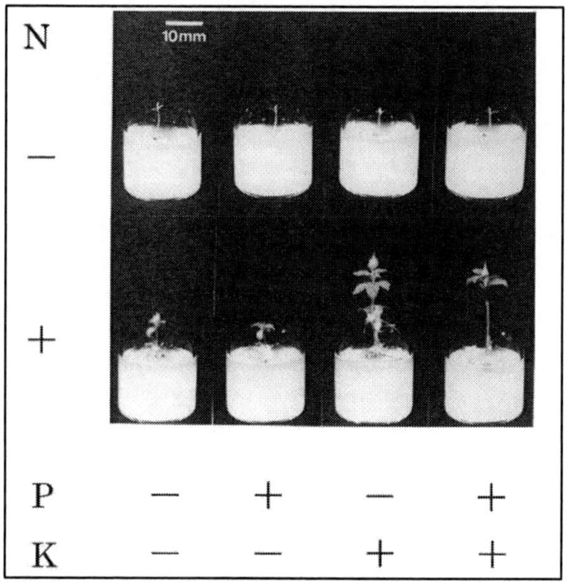

Figure 4. Effects of deletion of N (NH_4NO_3, KNO_3), P (KH_2PO_4) and/or K (KNO_3, KH_2PO_4) in the macro-nutrients from 1/15 MS medium on the growth and flowering in *Celosia argentea* 'Castle Yellow' (12 weeks after sowing).

When N and P were not added, the amounts of potassium equivalent to KNO_3 and KH_2PO_4 were supplemented with KCl. When K was not added, the amounts of nitrogen and phosphoric acid equivalent to KNO_3 and KH_2PO_4 were supplemented with NH_4NO_3 and $NaH_2PO_4 \cdot 2H_2O$, respectively.

From the above results, it has been clarified that the phenomenon of inflorescence formation without true leaf formation was caused by the no addition of inorganic nitrogen salts (N) to the medium. As a result of investigating the differences between cultivars in the phenomenon, all of 16 cultivars tested formed inflorescence without true leaf formation on the 0 MS medium (Table 3). The number of days from sowing until the visible inflorescence (Days to flowering) varied greatly between cultivars (from 30.8 to 71.8 days), but about 30 days for 'Kimono Red', 'Kimono Yellow' and 'Jewel Box' (Table 3). It has long been known as the Carbon to Nitrogen ratio (C/N) theory that the flowering response is promoted by reducing the inorganic nitrogen salts application rate (Corbesier et al., 2002). Past studies on the in vitro flowering using *Torenia fournieri* (Tanimoto and Harada, 1981a) and *Ipomoea nil* (Ishioka et al., 1991; Wada and Shinozaki, 1985) also reported that in vitro flowering is promoted by lowering the NH_4^+ and NO_3^- concentration and increasing the sugar concentration in media. However, *Celosia argentea* is unique compared to other plants reported in the past that it reacts very sensitively to the NH_4^+ and NO_3^- concentration in the medium and does not form true leaves without the addition of NH_4^+ and NO_3^-.

When the time of transplantation from 0 MS medium to 1/30 MS medium was changed, the NH_4^+ and NO_3^--free period required for the phenomenon of inflorescence formation without true leaf formation was examined. The NH_4^+ and NO_3^--free period of at least 3 weeks was required (Table 4). Therefore, it is considered that *Celosia argentea* was physiologically converted to reproductive growth phase without vegetative growth phase within 3 weeks after sowing on the NH_4^+ and NO_3^--free medium. If 'Kimono Red', 'Kimono Yellow' and 'Jewel Box' which are have about 30 days to flowering would be used, there is a possibility that the nitrogen-ions content in soil, media and freshwater such as rivers, ponds, lakes, etc. can be evaluated by the number of formed true leaves. In other words, there is a possibility that *Celosia argentea* can be used as an indicator plant for evaluating eutrophication. In addition, since the number of leaves can be controlled by the amount of nitrogen applied, the plant size also can be regulated. In fact, an item named "Celosia candle cake" of pot-plants that have been cultivated like a colorful decoration cake have been commercialized by seeding of dozens mixed flower- color seeds on potting media with reduced fertilizer and applying growth retardant in Japan.

Literature Cited

Corbesier, L., Bernier, G. and Pĕrilleux, C. (2002). C:N ratio increases in the phloem sap during floral transition of the long-day plants *Sinapis alba* and *Arabidopsis thaliana*. Plant and Cell Physiol. *43*:684-688.

Dickens, C. W. S. (1987). The physiology of flowering with contributions by in vitro techniques. Ph. D Thesis, University of Natal.

Murashige,T. and Skoog, F. (1962). A revised medium for rapid growth and bioassays with tobacco tissue cultures. Physiol. Plant. *15*:473-497.

Harada, H. (1967). Flower induction in excised shoot apices of *Pharbitis* and *Chrysanthemum* cultured in vitro. Nature *214*:1027-1028.

Ishioka, N., Tanimoto, S. and Harada, H. (1990). Flower-inducing activity of phloem exudate in cultured apices from *Pharbitis nil* seedlings. Plant and Cell Physiol. *31*:705-709.

Ishioka, N., Tanimoto, S. and Harada, H. (1991). Roles of nitrogen and carbohydrate in floral-bud formation in *Pharbitis* apex cultures. J. Plant Physiol. *138*:573-576.

Tanimoto, S. and Harada, H. (1981a). Chemical factors controlling floral bud formation of *Torenia* stem segments cultured in vitro. I. Effects of mineral nutrients and sugars. Plant and Cell Physiol. *22*:533-541.

Tanimoto, S. and Harada, H. (1981b). Effects of IAA, zeatin, ammonium nitrate and sucrose on the initiation and development of floral buds in *Torenia* stem segments cultured in vitro. Plant and Cell Physiol. *22*:1533-1560.

Tanimoto, S., Miyazaki, A. and Harada, H. (1985). Regulation of abscisic acid of in vitro flower formation in *Torenia* stem segments. Plant and Cell Physiol. *26*:675-682.

The Royal Horticultural Society. (1992). The New Royal Horticultural Society Dictionary of Gardening. Vo. 4, p.560. The MacMillan Press Ltd., London.

Tran than Van, K. (1973). *In vitro* and *de novo* flower, bud, root and callus differentiation from excised epidermal tissue. Nature *246*:44-45.

Tran than Van, K. (1999). Floral and vegetative differentiation *in vitro* and *in vivo*. pp. 215-233. In: So, W-Y. and S.S. Bhojwani (eds.). Morphogenesis in plant tissue cultures. Kluwer Academic Publishers, Dordrecht.

Van Staden, J. and Dicken, C.W.S. (1991). In vitro induction of flowering and its relevance to micropropagation. pp. 85-115. In: Bajaj, Y. P. S. (ed.). Biotechnology in agriculture and forestry. Vol 17. High-tech and micropropagation I. Springer-Verlag, Berlin Heidelberg.

Wada, K. and Shinozaki, Y. (1985). Flowering response in relation to C and N contents of *Pharbitis nil* plants cultured in nitrogen-poor media. Plant and Cell Physiol. *26*:525-535.

Yamada, J., Nakata, K., Amaki, W. and Higuchi, H. (1997). In vitro flowering in *Celosia argentea* var. *cristata*. Bull. NODAI Res. Inst. *8*:22-24. (Japanese with English summary)

Yang, S.-J., Amaki, W. and Higuchi, H. (1999). Effects of cultivars and ambient environments on the in vitro flowering in *Kalanchoe blossfeldiana* Poellniz. J. Japan. Soc. Hort. Sci. *68*:1170-1177. (Japanese with English summary)

Effect of Plant Enhancement Liquid "FFC-Vegemake©" on Growth of Edible Flower Cultivated by Semi-Hydroponic Method

Sachiko Hasegawa and Tadao Fujimori

Institute of Biological Process Research, AkatsukaGarden Co., Ltd.,
1868-3 Takanoo-cho, Tsu, Mie 514-2293, Japan

s.hasegawa@akatsuka.gr.jp

Keywords: Edible crop, vegetables, hydroponics, pansy, torenia, nasturtium, snapdragon

INTRODUCTION

Since 1984, Akatsuka Garden Company has focused on the behavior of certain ions, especially iron ions in water and interactions of water molecules with them. We have continued research on various solutions to not only accelerate plant growth, but also activate physiological functions of plants.

Based on this research, we have developed FFC materials such as "FFC-Ace" for soil improvement, "FFC-Vegemake" for plant enhancement, and others. In addition, many agricultural producers in Japan have been utilizing FFC materials to rejuvenate plants and increase profits. Those producers have also explored many other possible methods of using FFC materials and consequently found good ways that benefit their actual production sites.

On the other hand, our company has been cultivating and selling garden plants for many years. In addition, we started trial cultivation of edible flowers as a new business in 2015. There is no special species as an edible flower. Among horticultural plants, (1) non-toxic flowers, (2) flowers cultivated with no pesticides or using pesticides within the specified range are distributed as edible flowers. It is said that there are about 100 kinds of edible flowers, such as Pansy (*Viola × wittrockiana*), Torenia (*Torenia fournieri*), Nasturtium (*Tropaeolum majus* L.), and Snapdragon (*Antirrhinum majus* L.). In recent years, it has been widely used in Japan as a decoration for cooking and confectionery. In this paper, we will introduce the trial cultivation of edible flowers using FFC-Vegemake and two edible flowers that are currently being considered for sale.

MATERIALS AND METHODS

The effects of FFC-Vegemake on the growth of semi-hydroponic viola was studied by sowing 40 seeds of three kinds of viola were sown on a flat and thin sponge soaked with tap water, and cultured in the dark at 20° C. After 5-6 days, the seeds with roots larger than 0.5 cm were selected, and the roots were inserted into a cube sponge notch. The seedling transplanted to the sponge was put in a tray containing fertilizer solution for hydroponics (Hyponica liquid fertilizer, Kyowa Co., Ltd.) diluted 750 times with tap water, and grown for about one month at 20° C. under artificial light (Toshiba Plantlux FL40S / BRN, Toshiba Corporation).

Ten seedlings of average size were selected from each kind and transplanted to the pot containing coconut chips. Liquid fertilizer (control) or liquid fertilizer in which FFC-Vegemake was diluted 1000 times (FFC) was poured into each tray to a height of about 1 cm and soaked 5 pots each transplanted seedlings of each kind. These seedlings were cultivated for about 2 months in a glass greenhouse with an average temperature of 15-20° C (winter). Three seedlings of average size were selected from each kind in each treatment and transplanted to planters. They were cultivated for about 5 months and the number of each flower was investigated.

RESULTS AND DISCUSSION

The cultivated seedlings for about 2 months in the pot containing coconut chips promoted the growth of the above ground parts and roots in the FFC treatment compared to the control in kinds of all. The average number of flowers about 5 months after transplanting in the planter was about 100 more in the FFC treatment than in the control. Although there were individual differences, it can be said that the growth of seedlings and the number of flowers increased by FFC-Vegemake treatment.

CONCLUSION

We tried the cultivation test of about 10 kinds of edible flowers by applying the semi-hydroponic cultivation test of viola above. As a result, good growth and flowering were observed in torenia, nasturtium, basil (*Ocimum basilicum*), and begonia (*Begonia × semperflorens-cultorum*). Among them, begonia showed stable growth and flowers (Fig. 1), and it has become possible to harvest all year round by heating in November to April when the temperature is below 15° C.

Figure 1. Pink begonia in flower

Begonia, which started cultivation in September 2017, is growing well without any problems such as disease even now (August 2019). Sometimes thrips and aphids occur, but they are not severely damaging to growth.

Currently, we are conducting trial and error on cultivation, harvesting, packing and transportation methods in order to sell the original cultivar 'Titanbicus' (*Hibiscus moscheutos × Hibiscus coccineus*) as an edible flower (Fig. 2).

'Titanbicus' has a texture like a soft lettuce and has a little stickiness. The taste is light and has a little sweetness that goes well with meat and fish. 'Titanbicus' is very large compared to other edible flowers with many small flowers. 'Titanbicus' will attract many people because of its impact, beauty and taste.

Figure 2. 'Titanbicus' (Rhea - Registration number 27541)

Development of F1-hybrid Strawberry of Seed Propagation Type Named 'Yotsuboshi' by Collaborative Breeding among Institutes

H. Kitamura[1], T. Mori[1], J. Kohori[1], T. Inokuchi[2], I. Kato[2], K. Sone[3], M. Ishikawa[4] and F. Maeda[4]

[1]Mie Prefecture Agricultural Research Institute, Mie, Japan; [2]Kagawa Prefecture Agricultural Experiment Station, Kagawa, Japan; [3]Kyusyu Agricultural Research Center, National Agriculture and Food Research Organization, Fukuoka, Japan; [4]Chiba Prefectural Agriculture and Forestry Research Center, Chiba, Japan

kitamh03@pref.mie.lg.jp

Keywords: Everbearing strawberry, *Fragaria*, breeding.

INTRODUCTION

'Yotsuboshi', a seed propagating type of strawberry (*Fragaria ×ananassa* Duchesne) was developed from the collaborative breeding program including four institutes of Mie, Kagawa and Chiba prefectures and National Agriculture and Food Research Organization (NARO) and registered as a new variety in 2017 in japan (No.25605). It is an F1-hybrid, who's maternal (ovary) and paternal (pollen) parents are 'Miebohon 1 gou' developed in Mie prefecture and 'A8S4-147' developed in Kagawa prefecture, respectively (Fig.1).

Figure 1. Pedigree for 'Yotsuboshi'

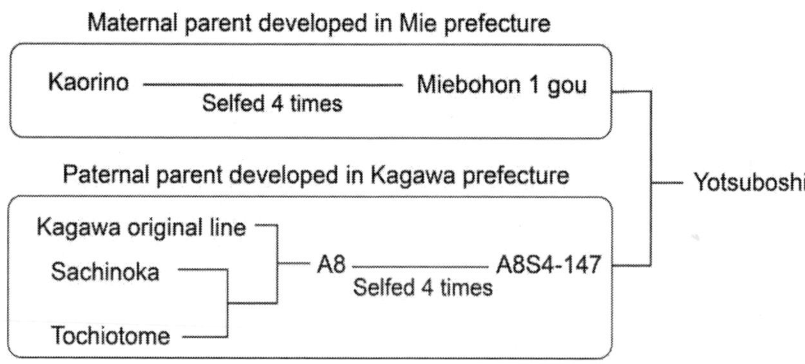

The behavior of flower initiation is quite unique. 'Yotsuboshi' exhibits stable extra earliness under natural conditions with a lower temperature and shorter day length, while it possesses the ever-bearing gene which results in the induction of flower initiation under long-day condition. 'Yotsuboshi' has several attractive traits, such as a clear red fruit color, excellent fruit shape and size, high yielding ability, and superior taste with a high content of soluble solids and moderate acidity. The seeds and/or plug tray seedlings were commercially provided from 2016 for the member of the 'Research Society for Seed Propagating Type Strawberries', 'Yotsuboshi' became widely cultivated across Japan in 2019.

Introduction and Possibility of Kumano Cherry: The First New Cerasus-Type Cherry Species to Be Discovered in 100 Years

Masayuki Nakamura

Kiraku-Nakamura, Tomiya 1-7-8, Suzuka, Mie Pref. 513-0037, Japan.

Kirakuna@mecha.ne.jp

Keywords: Cherry, *Prunus kumanoensis*, *Cerasus*.

INTRODUCTION

Kumano cherry (*Prunus kumanoensis* T. Katsuki) is a *cerasus*-type wild cherry species that was first described in 2018 by Dr. Katsuki, who studies the classification of flowering cherry at the Forestry and Forest Products Research Institute. The species represents the first newly discovered wild species since the discovery of Ohshima cherry (*Prunus speciosa*) in 1915. The species' distribution spans 90 km, from north to south, and 70 km, from east to west, and overlaps with the Mie, Nara, and Wakayama prefectures in the southern part of the Kii Peninsula of Honshu, Japan (Fig.1).

Fortunately, I was able to witness the first discovery of Kumano cherry at Wakayama Pref. in March 2016 and investigated with Dr. Katsuki to the distribution at Mie Pref. As a result of the investigation, the species was also confirmed to occur in Kumano City (Mie Pref.) in July 2016 and has since been confirmed at other locations in Mie Pref., including Kiho, Mihama, Owase, and Kihoku. Many individuals have been identified in Kiwa-cho (Kumano City), and the species' beautiful flowers can be viewed there.

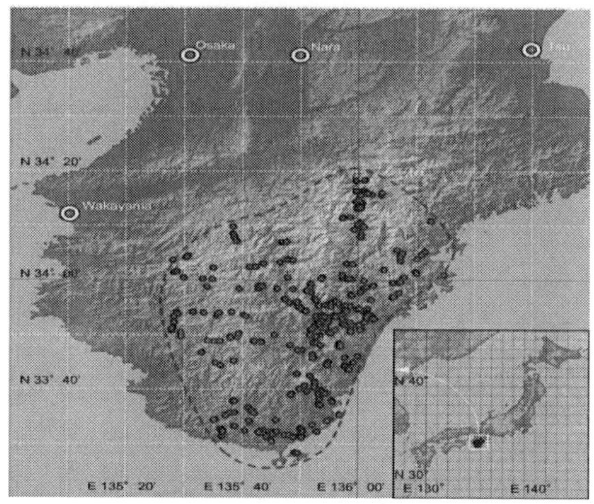

Figure 1. Distribution of Kumano cherry. Red dots and dashed lines indicate confirmed localities and estimated distribution, respectively. (Provided by Dr. Katsuki,).

Kumano cherry blooms earlier than Japanese mountain cherry (*P. jamasakura*) or Tokyo cherry (*Prunus* ×*yedoensis* 'Somei-yoshino'). In Mie Pref., the species begins to bloom between March, in the lowlands of Kiho and Mihama Town, and early April, at the high altitudes of Kumano City. Because

Kumano cherry is a wild species, individual trees differ in flower size, color, and blooming time. However, the flowers generally vary from white to pink. In contrast to Japanese mountain cherry, which blooms after the emergence of leaves, Kumano cherry blooms before leaf emergence and is considered to have high ornamental value. Kumano cherry is also expected to be used as an ornamental tree for parks and gardens because it is smaller and more elegant than Japanese mountain cherry.

Kumano cherry has attracted attention as a tourism resource in the future plan of township in natural habitat, and many activities of protection, preservation and propagation as a new regional treasure have begun. As a first step, the first large-scale planting of Kumano cherry was installed in Kiho and Mihama Town in February 2019. In Kumano City, tree planting has been carried out in the mountainous area of Kiwa-cho as a "Scenery project for Kumano cherry", and the scenery making for Kumano cherry has begun (Fig. 2).

Figure 2. Viewing event at Kiwa-cho for Kumano citizens in March 2019.

The seedlings planted in Mie Pref. were propagated from seed, but in the seedling propagation, a serious problem for the conservation of Kumano cherry may arise. Cherries are self-incompatible, and fertilization requires outcrossing. As such, crossing occurs among individual Kumano cherry trees but rarely occurs between Kumano cherry and other species. Many cultivars, such as Tokyo cherry, have been introduced to the natural habitats of Kumano cherry, and natural hybrids have been confirmed. This so-called "gene contamination" has contributed to the decline of Kumano cherry. Owing to the high possibility of hybridization in habitats near Tokyo cherry trees, the collection and propagation of seeds from such habitats should be avoided.

Finally, in regard to horticultural appeal, Kumano cherry exhibits variation in floral characteristics (Fig. 3). The species' petals, for example, may be white, pink, or white to pink. Because Kumano cherry produces numerous flower buds along its branches, it is possible to grow and display the plants in pots, like bonsai. I am trying to grow the excellent individuals found during the survey by grafting and confirming their characteristics with the help of Dr. Katsuki. Collection of seeds and branches was accomplished with permission from the owner of trees. In the near future, I would like not to offer Kumano cherry from Mie Pref. to various towns in Japan and to introduce it to the world as a beautiful flowering cherry.

Figure 3. Kumano cherry with pink and white petals. It appears beautiful and light red when viewed from a distance.

Effects of Intensity and Quality of Light on the Coloring of Leaves on a Succulent Plant, *Graptopetalum paraguayense* 'Bronze'

Arisa Noguchi, Hiroko Nakatsuka and Wakanori Amaki

Department of Agriculture, Tokyo University of Agriculture
1737 Funako, Atsugi, Kanagawa 243-0034, Japan

a3noguch@nodai.ac.jp

Keywords: Light quality, succulent plant, anthocyanin, blue light

Abstract

Graptopetalum paraguayense 'Bronze' is a small succulent plant with reddish-bronze colored leaves. This plant contains anthocyanins as pigments. Therefore, the effects of light intensity and quality of light on the color and anthocyanin content of leaves in 'Bronze' were investigated. The leaf color became more reddish as the light intensity increased. The a* values of the leaves under the green and red light decreased with increasing of treatment days, but the values hardly changed under white and blue light. The anthocyanin content in the leaf epidermis was highest under white light containing blue light component, but the total polyphenol content was highest under blue light. From the above results, it is considered that anthocyanin and polyphenol synthesis in 'Bronze' leaves are promoted by blue light.

INTRODUCTION

Plant groups called succulent plants have been attracting attention from especially young peoples in Japan in recent years due to their distinctive form and color of leaf and stem. Recently, *Graptopetalum paraguayense* (N.E.Br.) E. Walther (Crassulaceae) plants had gained popularity due to its ease of leaf cutting propagation and easy cultivating management. Among them, one of the cultivars, named 'Bronze' is particularly popular because the green leaves turn beautifully red color at the beginning of autumn in outdoor cultivation (Matsui, 1988). It has the characteristic that leaves spread in a rosette shape at the top of the stem. Already, the authors have revealed that the red pigment in 'Bronze' leaves was anthocyanin (Noguchi, 2019).

Anthocyanins are thought to play a role in plant sunscreen and scavenging reactive oxygen species (Chalker-Scott, 2002; Karageorgou and Manetas, 2006; Hatier and Gould, 2009; Lev-Yadun and Gould, 2009; Llorens et al., 2015). The production of anthocyanin pigment is expected to be affected by cultivation conditions, particularly light. Therefore, the effects of light intensity and light quality during cultivation on the surface (epidermis) color and anthocyanin contents of the 'Bronze' leaf were investigated.

MATERIALS AND METHODS

Potted plants of *Graptopetalum paraguayense* 'Bronze' with about 25 leaves of uniform size were selected and used in the following experiments.

Experiment 1. Effect of light intensity on the leaf color and anthocyanin content

White light emitting diode (LED) lamps (LLM0172A, Stanley Electric Co., Ltd., Yokohama, Japan: 460nm peak blue LED covered with Yttrium Aluminum Garnet (YAG) phosphor) were installed at the top of the plant so that the photosynthetic photon flux density (PPFD) was 60 or 600μmol·m^{-2}·s^{-1}, and three plants were arranged for each light intensity. All plants were subjected to 24-hr photoperiod and room temperature kept at 23±2 °C.

Experiment 2. Effect of light quality on the leaf color and pigment contents

Light quality treatment was performed in a temperature-controlled room at 24±2°C. Three monochromatic LED lights (blue, green and red LEDs had peak wavelengths of 470, 530 and 630 nm respectively) and a combined white LED light (blue: green: red = 1: 1: 1) put on 24-hr photoperiod. Each of treatments was screened with black cloth to preclude extraneous light. The PPFD of each LED was adjusted to 100 μmol·m^{-2}·s^{-1} at the top of plants. Three plants were placed in each light quality treatment area for 7 or 14 days.

Analysis of leaf color and pigment content in leaf epidermis

For the analysis of leaf color, three leaves from the fifth to the seventh counted from the top were sampled. The measurement points were the apex part (within 5mm from the tip) and the center of leaf lamina. L*, a*, and b* values of those leaves were measured with a color meter (ZE6000; Nippon Denshoku Industries Co., Ltd., Tokyo, Japan).

For the measurement of anthocyanin or total polyphenol contents, epidermis of whole three leaves from the fifth to the seventh counted from the top were sampled. The epidermis samples were dipped in 1% HCl-methanol for 24 hours for anthocyanin extraction. The absorbance of each extract was measured at 530 nm by a spectrophotometer (U-2000, Hitachi, Ltd., Tokyo, Japan). Anthocyanin content was expressed as mg of cyanidin chloride (Wako Pure Chemical Industries, Ltd. Tokyo, Japan) equivalent per gram fresh weight of epidermis. Using other epidermis samples, total polyphenol content was analyzed by the Folin–Ciocalteu reagent method (Singleton et al., 1999). Gallic acid (Acros organics, Thermo Fisher Scientific Inc., USA) was used as the standard for the calibration curve, and the total polyphenol contents were expressed as mg gallic acid equivalent per gram fresh weight of epidermis. The means values were compared using ANOVA followed by Tukey's multiple range tests at the 1 % level.

RESULTS AND DISCUSSION

Experiment 1. Effect of light intensity on the leaf color and anthocyanin content

The leaf color became reddish as the light intensity increased. The a* value of the leaves and anthocyanin content of epidermis after 14 days of treatment were 3.16 ± 0.9 and 0.17 ± 0.02 mg/g for 600μmol·m^{-2}·s^{-1} PPFD, and -3.20 ± 1.4 and 0.04 ± 0.00 mg/g for 60μmol·m^{-2}·s^{-1} PPFD respectively. There was a statistically significant difference between the values of 600 and 60μmol·m^{-2}·s^{-1} PPFD at the 1 % level (Table1). On the seventh days from the start of treatment, this difference had already been made visible. It was assumed that the leaves of 'Bronze' were very sensitive to light intensity.

Table 1. Effect of light intensity on the a* value and anthocyanin content of *Graptopetalum paraguayense* 'Bronze' after 14 days from the start of treatment.

Treatment period (Days)	PPFD (μmol·m^{-2}·s^{-1})	a* value	Anthocyanin content (mg/gFW epidermis)
0^z	-	3.02a	-
14	60	-3.20b	0.04b
	600	3.16a	0.17a

z: Anthocyanin content was not measured.
n=3, Different letters indicate a significant difference at the 1% level by Tukey's test.

Experiment 2. Effect of light quality on the leaf color and pigment contents

The a* values of the leaves under the green and red light decreased with increasing of treatment days. On the other hand, the values hardly changed under white and blue light (Figure 1).

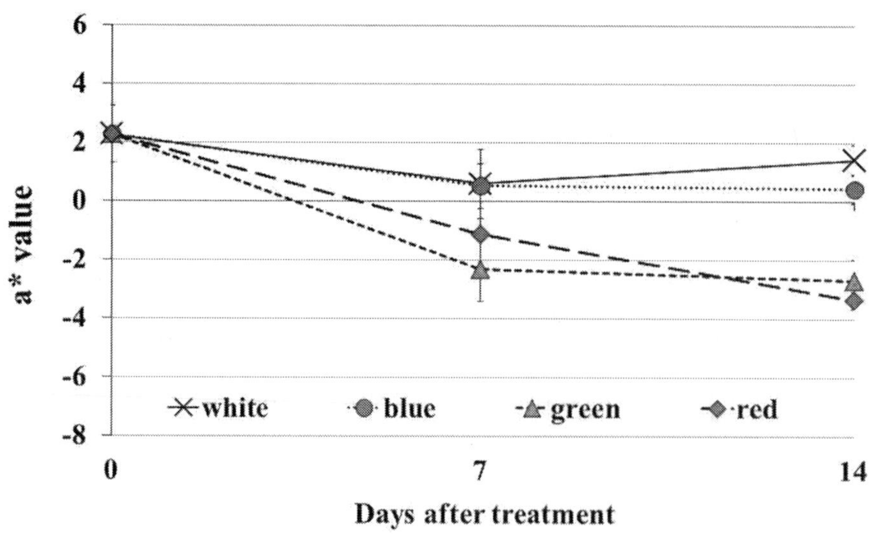

Figure 1. Effect of light quality on the a* value of *Graptopetalum paraguayense* 'Bronze' leaf lamina.

The a* values of the leaves tended to be higher at the apex of the leaves than at the center of the leaf lamina. However, the green light treatment showed the lowest value at both parts (Figure 2).

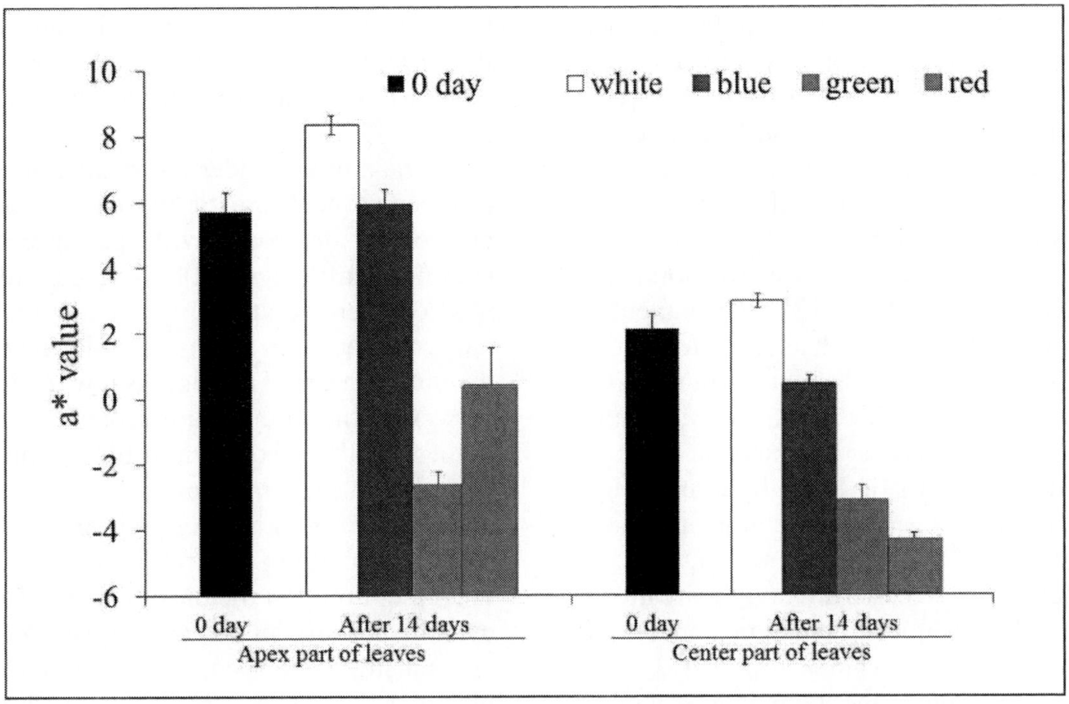

Figure 2. Effect of light quality and the position of leaf lamina on the a* value of *Graptopetalum paraguayense* 'Bronze' leaf lamina after 14 days from the start of treatment.

The content of anthocyanins in the epidermis after 14 days from the start of treatment was about 2 times higher under the white light than under the green and red light (Table 2), and there was a very high correlation (r = 0.93) between the anthocyanin contents and the a* values (data not shown).

Table 2. Effect of light quality on the content of anthocyanin and total polyphenol of *Graptopetalum paraguayense* 'Bronze' leaf epidermis after 14 days.

Treatment period (Days)	Light quality	Anthocyanin content (mg/gFW epidermis)	Total polyphenol content (mg/gFW epidermis)
0	-	0.42 ab	30.55 b
14	white	0.47 a	25.17 c
	blue	0.32 abc	34.88 a
	green	0.24 bc	28.78 b
	red	0.17 c	18.67 d

n=3, Different letters indicate a significant difference at the 1% level by Tukey's test.

Besides, the value of total polyphenol content was the highest under the blue light, and significantly increased from the start of treatment. There are some reports that the coloration of strawberry and apple fruits is improved by blue light irradiation (Yunting et al., 2018; Kokalj et al., 2019). In the leaves of 'Bronze' as well, it is considered that the blue light irradiation promotes anthocyanin and polyphenol synthesis and maintains the anthocyanin content in the leaf epidermis.

Anthocyanin biosynthesis was found to be differentially modulated by environmental and biological factors such as light, temperature, sugar content, and plant hormones. In light, it has been clarified that UV, blue, and red lights are signals for anthocyanin synthesis (Gu et al., 2019). High light intensity stimulates anthocyanin production in many plant species (Maier and Hoecker, 2015). Also, leaves that receive direct light are more intense anthocyanin pigmentation than leaves that shaded (Mazzucato et al, 2013). In 'Bronze', the higher the light intensity, the higher the a* value, especially in the leaf apex parts that are more likely to receive light (Table 1 and Figure 2), indicating that anthocyanin biosynthesis is controlled by light as in previous reports. On the other hand, the degree of change in the a* value was larger for the mixed white light of 100 $\mu mol \cdot m^{-2} \cdot s^{-1}$ PPFD (Figure 1 and 2) than for the YAG white LED of 600 $\mu mol \cdot m^{-2} \cdot s^{-1}$ PPFD (Table 1).

In addition, under the irradiation of red or green monochromatic light, the anthocyanin content decreased with the lapse of the treatment time (Figure 2). From these results, it is considered that the light quality that promotes anthocyanin synthesis is blue light, and the involvement of red light is low in 'Bronze'. However, the anthocyanin content was the highest in the mixed white light containing blue rather than blue monochromatic light (Table 2). From this result, it is expected that the light quality of the irradiation light is involved not only in the biosynthesis of anthocyanins but also in the metabolism of sugar as a substrate, the accumulation of anthocyanins in cells and the subsequent metabolism.

Literature Cited

Chalker-Scott, L. (2002). Do anthocyanins function as osmoregulators in leaf tissues? Advances in Botanical Research *37*:103-127. https://doi.org/10.1016/S0065-2296(02)37046-0

Gu, Kai-D., Wang, C-K., Hu, D-G., and Hao, Y-J. (2019). How do anthocyanins paint our horticultural products? Scientia Horticulturae. *249*:257-262. https://doi.org/10.1016/j.scienta.2019.01.034

Hatier J.H.B., and Gould K.S. (2009). Anthocyanin function in vegetative organs. p.1-19. In: K. S. Gould, K. M. Davies, C. Winefield (eds). Anthocyanins. Springer, New York, NY.

Karageorgou, P. and Manetas, Y. (2006). The importance of being red when young: anthocyanins and the protection of young leaves of *Quercus coccifera* from insect herbivory and excess light. Tree Physiology *26*:613–621. https://doi.org/10.1093/treephys/26.5.613

Kokalj, D., Zlatić, E., Cigić, B., Kobav, M.B. and Vidrih, R. (2019). Postharvest flavonol and anthocyanin accumulation in three apple cultivars in response to blue-light-emitting diode light. Scientia Horticulturae *257*:17. https://doi.org/10.1016/j.scienta.2019.108711

Lev-Yadun S., and Gould, K.S. (2009). Role of anthocyanins in plant defence. p.22-28. In: K. S. Gould, K. M. Davies, C. Winefield (eds.), Anthocyanins. Springer, New York, NY.

Llorens L., Badenes, F. R. P., Julkunen., R. T., Zidornd, C., Fereres, A. and. Jansen, M. A. K. (2015). The role of UV-B radiation in plant sexual reproduction. Perspectives in Plant Ecology, Evolution and Systematics *17*:243-254. https://doi.org/10.1016/j.ppees.2015.03.001

Maier, A. and Hoecker, U. (2015). COP1/SPA ubiquitin ligase complexes repress anthocyanin accumulation under low light and high light conditions. Plant Signal Behav. *10*(1):e970440. https://doi.org/10.4161/15592316.2014.970440.

Matsui, K. (1988). *Graptopetalum* Rose. p.143-144. In: Aiga, T. (ed.) The grand dictionary of horticulture Vol. 2. Shogakukan, Tokyo. (in Japanese)

Mazzucato A., Willems, D., Bernini, R., Picarella, M., Santangelo, E.E., Ruiu, F., Tilesi, F. and Soressia, G. P. (2013). Novel phenotypes related to the breeding of purple-fruited tomatoes and effect of peel extracts on human cancer cell proliferation. Plant Physiol. Biochem. *72*:125–133. https://doi.org/10.1016/j.plaphy.2013.05.012

Noguchi, A. (2019). Effect of light quality on coloring of *Graptopetalum paraguayense* 'Bronze'. Horticultural Research 18 (supplement.1) :*423*. (in Japanese)

Singleton, V. L., Orthofer, R. and Lamuela-Raventós, R.M. (1999). Analysis of total phenols and other oxidation substrates and antioxidants by means of folin-ciocalteu reagent. Methods in Enzymology *299*:152-178. https://doi.org/10.1016/S0076-6879(99)99017-1

Yunting, Z., Leiyu, J., Yal, L., Qing, C., Yuntian, Y., Yong, Z., Ya, L., Bo, S., Xiaorong, W. Xiaorong W. and Haoru. T. (2018). Effect of red and blue light on anthocyanin accumulation and differential gene expression in strawberry (*Fragaria* × *ananassa*). Molecules *23*:820. https://doi.org/10.3390/molecules23040820

Research and Development at Hakusan Company

Yoshikuni Suzuki

Hakusan co., ltd. Research and Development Division.
321-1, Sanbanwari, Iwafuji, Nisshin, Aichi,470-0401, Japan.

suzuki@hakusan1.co.jp

Keywords: Tissue culture, micropropagation, import, export.

INTRODUCTION

Hakusan co., Ltd. is an import-export company specialized in flower seeds, seedlings and horticultural goods. We have made a great effort to understand various market conditions and have unique opportunity to market our outstanding products to fulfill the needs especially for professionals and consumers. We are pleased to assist you build your successful horticultural fields through our marketing techniques (Fig. 1).

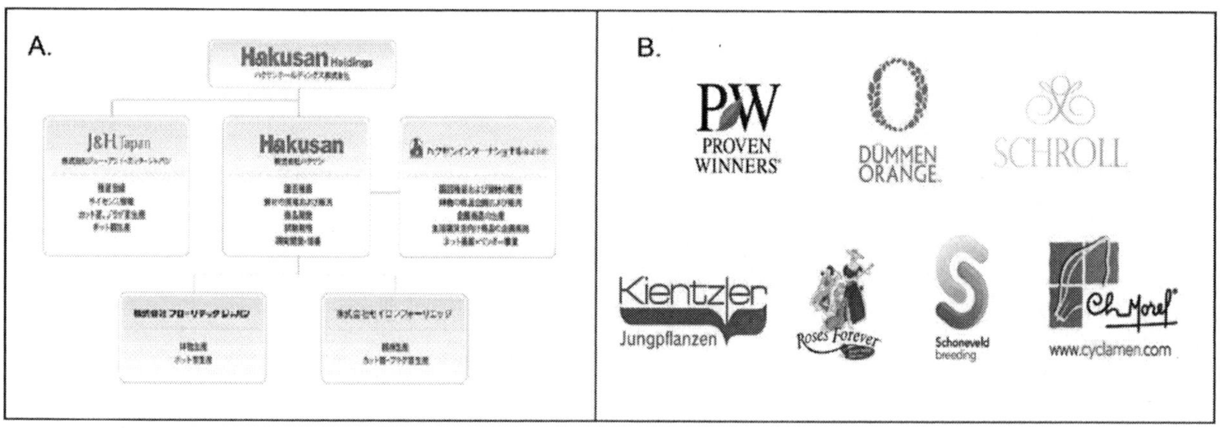

Figure 1. (A) Hakusan group. (B) Brands handled by the Hakusan group.

Research and Development Division

Our Research and Development Division applies the knowledge of biotechnology to ensure a supply of clean virus-free plants, the development of a mass propagation system, and new and improved breeding techniques (Fig. 2).

Figure 2. Research and development at Hakusan. (A) Elisa test; (B) Clean tissue culture room; (C) Clean stock plants in tissue culture; (D) TC growth room; (E) Multiplication stage cultures; (F) individual microcuttings.

Tissue culture lab in Ceylon Foliage in Sri Lanka

We are constructing TC laboratory close to Ceylon foliage nursery at the present. We will try to provide various propagation and supply station in Ceylon Foliage, in addition to cuttings and plug propagation for Japan and Europe (Fig. 3).

Figure 3. (A) Hakusan group; (B) Brands handled by the Hakusan group.

PROCEEDING'S PAPERS

WESTERN REGION OF NORTH AMERICA

Dr. Ryan Stewart, Regional Editor

Sixtieth Annual Meeting - 2019

Santa Cruz, California USA

Avoiding Diseases in Propagation

Ann R. Chase

Chase Agricultural Consulting, LLC, Cottonwood, AZ 86326

archase@chaseresearch.net

Keywords: Fungi, bacteria, *Botrytis*, *Pythium*, RootShield

INTRODUCTION

The characteristics of propagation that promote disease are excess water, high humidity, poor air movement, and handling. In addition, there are generally open wounds on unrooted cuttings for obvious reasons. Water trumps fungicides! That means poor water management cannot be repaired by application of fungicides and bactericides. I've heard really good growers say that some diseases never occur unless the water is not managed correctly. Getting plants out of misting as soon as possible is always the best idea.

Combating shipping damage with rehydration under cool/cold conditions before sticking can really help. In addition, ethylene monitoring during shipping can alert you to possible problems. Dipping is not the best idea for many herbaceous unrooted cuttings as it can lead to water-soaking and cutting loss.

Unfortunately, pathogens on cuttings and seeds are more common than I would like. The fact that our propagators are often overseas makes the possibility of introducing a "new" disease more likely than previously. There do appear to be fewer seed-borne diseases than 20-30 years ago, but I would watch out for "heirloom" varieties. These special crops are not usually handled the same way as the more common, newer varieties.

It is critical that you learn to recognize common diseases or at least when something is wrong with your cuttings. This will allow you to do something therapeutic to stop losses and damage. Some fungi and bacteria are on cuttings, including downy mildew, bacteria, *Sclerotinia*, and even *Pythium* (Table 1). Keep track of all treatments you do, including dates and rates. It never helps to guess what worked or did not work after the crop is done.

Table 1. Pathogens common to widely cultivated ornamental species and their primary source of infection on propagative material.

Plant species	Source(s)	Pathogen
Basil	Seed	*Fusarium,* downy mildew
Begonia	Leaf cuttings	*Xanthomonas*
Calla lily	Bulbs	*Pythium, Erwinia, Rhizoctonia*
Coleus	Seeds, cuttings	Downy mildew
Geranium	Cuttings	*Botrytis, Pythium, Sclerotinia,*
Hydrangea	Cuttings	*Rhizoctonia*, powdery mildew
Impatiens	Seed	*Acidovorax*
Lavender	Cuttings	*Pseudomonas, Xanthomonas*
Molucella	Seed	*Cercospora*
New Guinea impatiens	Cuttings (appears in shipping)	*Myrothecium*
Petunia	Cuttings	Tobacco mosaic virus
Poinsettia	Cuttings, flats, benches	*Rhizoctonia, Erwinia*
Purslane	Cuttings	Alternanthera mosaic virus
Ranunculus	"Bulbs"	*Pythium, Xanthomonas*
Rosemary	Cuttings	*Pseudomonas, Xanthomonas*
Salvia	Cuttings	*Pseudomonas, Xanthomonas*
Spathiphyllum	Liners (appears in shipping)	*Erwinia, Myrothecium*
Zinnia	Seed	*Alternaria, Xanthomonas*

It is also important to track the exact source and how they root for each set of unrooted cuttings. If you do not, you cannot determine which source is most reliable. If you "clean" up bad leaves or cuttings, do not just pile them in the middle of a bench where they can act as a source of more infections. Botrytis moves really well from debris piles.

You also must choose the right products and apply them when the cuttings are stuck instead of 3-5 days later when the bench or entire greenhouse is full. Delaying treatment for *Botrytis* effectively means no control. Use a product that stays on the unrooted cutting. For *Botrytis* we found that Daconil WeatherStik was most effective under average propagation conditions.

Do not use plugs within the immediate area of where others have died or rotted out. They are probably infected, too. Scout frequently and quickly discard infected propagation flats. A little problem can quickly turn into a big one. If you use infected cuttings or plugs, you will simply put more dollars into something that will end up in the trash.

What kind of water treatments are safe and effective? ZeroTol is very good on algae prevention, but not very safe on small plants. X3 is not as effective on algae but is safer on small plants and can control some diseases. KleenGrow can be very effective on algae for prevention, but should be checked for safety on sensitive crops, such as basil. Be sure to regularly clean trimmers. KleenGrow is effective and safe on plant material and machinery.

Do not use fungicides on everything. They are costly, and some of them can delay or even stop rooting. This has been shown with a few active ingredients (e.g., fludioxonil and triflumizole) on a few plants. It is also common to see copper used under conditions where it will not dry quickly, such as in propagation facilities. The result is often phytotoxicity, which may not be apparent to the naked eye. If you wound a plug or unrooted cutting, that gives *Botrytis* an entry point, and can lead you to assume that the fungus is resistant to copper. It is not clear that this is the case since copper is not the best control for *Botrytis* and can cause phytotoxicity as mentioned above.

Get plants out of the propagation environment as quickly as possible. This can mean using under-bench heating in the winter to speed up rooting, using fans to move stagnant air around, and careful monitoring for signs of rooting and/or signs of disease.

Some products even help seeds germinate and unrooted cuttings root. Strobilurins are a good example of this. A single spray with Pageant Intrinsic (4 oz/100 gal) or RootShield Plus (4 oz/100 gal) has improved germination and growth of *Ranunculus* seed, rooting of *Solidago* cuttings, and rooting of myrtle cuttings. We have also seen Pageant Intrinsic help with drought stress. Similar benefits in rooting on poinsettias (and other crops) using either Heritage or Mural has been demonstrated.

Most recently we have seen benefits in doing a bulb dip employing KleenGrow. The product has been shown to reduce losses due to *Fusarium* on infected bulbs. Our trials started by checking out the possibility that the dip solution might be contaminated with plant pathogens, leading to the spread of disease. We found that treatment with either 1- or 2-mL L^{-1} actually killed bacteria and fungi in the dip water, making disease spread unlikely. We also saw improved growth after a 5-minute soak in 1 mL L^{-1} for several bulb crops, which was probably due to improved wetting of the bulbs given that KleenGrow is a wetting agent. These trials included calla lilies (three cultivars), caladiums, tulips, Asiatic lily, hyacinth, and gladiolus.

CONCLUSIONS

Do not use contaminated seeds or cuttings. Do not try to cure trash. Be careful about using fungicides or certain water treatments on unrooted cuttings. Do not use recycled water in propagation at all! Manage your water, both the timing and amount. Get plants out of propagation as soon as possible. Use RootShield Plus or a strobilurin, such as Heritage, Mural, or Pageant.

Characteristics of propagation that lead to disease include high humidity and poor air movement. Handling spreads pathogens. Combat shipping damage with rehydration and ethylene monitoring. Dipping is not the best idea for many herbaceous unrooted cuttings given that it can cause water-soaking and cutting loss.

Lean Flow in the Green Industry

Gerson "Gary" Cortés

FlowVision, LLC, Parker, CO 80134

cortes@flowvision.com

Keywords: Work efficiency, progressive work, shipping, labor

Abstract

Labor is typically the biggest slice of the cost pie in the Green Industry. Whether we are in a strong or weak economy, labor efficiency is a paramount issue. FlowVision's Lean Flow methodology works to make your employees more efficient. We work together with your team to come up with the most appropriate solutions for your business, and then we will guide you to bring those concepts and visions to reality. We have developed several concepts that have increased productivity, improved quality, reduced credits/claims, and drastically reduced overtime during the busy shipping season. Two of the biggest areas of benefits are in Progressive Work and Dock Shipping Supermarket.

INTRODUCTION

FlowVision was founded in 1998 with expertise in lean flow business practices and supply chain management. By offering a complete suite of lean services from consulting to patented software, we have demonstrated that we are able to solve any supply-chain or process challenges that our customers have. We have hundreds of global customers in a variety of verticals, ranging from small startups to Fortune 500 companies.

In 2003, we worked with our first Green Industry customer in Homestead, Florida, an orchid and bromeliad grower. Since that first successful implementation, we have worked with more than 200 growers. In 2013, we introduced our Rack Item Optimizer (RIO) software solution. RIO allows our Green Industry customers to increase productivity at the loading dock by optimizing racks to increase payload while enabling sales teams to dynamically upsell their customers who are receiving less than full racks. Our customers are now seeing an increase in payload and their trucks are going out at 97% full.

Master Pull report

Master Pull #: 2
Farm: FV
Zone: A
Load Date: 9/1/2017

Location	Item Number	Description	ContainerID	Height ID	Qty To Ship	Rack Req	Rack Seq	Comments
A01	100020	2.25-GAL ALMOND FLOWERING	3G	H07	270	4.23	1,2,3,4,5	
A03	100125	2.25-GAL EUONYMUS BRNING BUSH	3G	H05	10	0.11	5	
A05	100151	SYRINGA LILAC 3G ASST BFN	3G	H04	10	0.08	5	
A04	100112	2.25-GAL FORSYTHEA ASSORTED	3G	H06	160	2.12	6,7,8	
A06	100051	2.25-GAL PUSSYWILLOW ASSORTED	3G	H11	180	4.55	9,10,11,12,13	
A09	100021	2.25-GAL BOXWOOD WINTERGREEN	3G	H04	360	3.04	14,15,16,17	
A08	100097	2.25-GAL QUINCE ASSORTED	3G	H03	20	0.12	17	
A11	100022	2.50-QT NANDINA FIREPOWER DWAR	1G	H04	224	0.59	17	
A10	100126	2.50-QT BOXWOOD WINTERGREEN	1G	H04	128	0.34	18	
A13	100024	2.25-GAL YEW UPRIGHT	3G	H05	270	2.93	19,20,21	
A12	100023	2.25-GAL YEW SPREADING	3G	H04	100	0.84	22	
A15	100052	2.50-QT SPRUCE COLORADO BLUE	1G	H04	160	0.42	23	
A17	100025	2.5 QT ARBORVITAE EMERALD GREE	1G	H03	352	0.86	24	
A16	100053	2.5 QT SPRUCE ALBERTA DWARF	1G	H03	160	0.36	24	
A18	100026	2.50-QT BOXWOOD WINTERGEM	1G	H04	672	1.77	25,26	
A21	100055	3.25-GAL BOXWOOD GREEN MOUNTAI	5G	H07	160	3.13	27,28,29,30	
A20	100054	2.50-QT JUNIPER OLD GOLD	1G	H02	224	0.25	30	
A23	100002	ILEX DWF BURFORD HOLLY 3G PRO	3G	H06	310	8.22	31,32,33,34,35,36,37,38,39	
A22	100001	ILEX DWF YAUPON HOLLY 3G PRO	3G	H04	100	0.84	40	
A25	100056	2.25-GAL VIBURNUM ASSORTED	3G	H04	80	0.67	41	
A24	100027	2.25-GAL PHOTINIA RED TIP	3G	H06	10	0.13	41	
A24	51169018	BOX GRN VELVET	18" BB	H06	6	0.00	41	
A24	65300036	EUONYMUS ALATUS COMP	36" BB	H06	6	0.00	41	
A26	100174	ALTHEA APHRODITE PK N03	3G	H05	10	0.11	41	
A27	100028	2.50-QT LILAC ASSORTED	1G	H02	23	0.03	41	
A28	100057	2.25-GAL BOXWOOD WINTERGEM	3G	H04	80	0.67	42	
A30	100029	2.5 QT JUNIPER BLUE STAR	1G	H02	864	0.98	43	
A32	100058	2.50-QT EUONYMUS GOLDEN	1G	H03	64	0.12	44	
A50	100002	ILEX DWF BURFORD HOLLY 3G PRO	3G	H06	30	0.80	44	

Thursday, September 14, 2017 Page 1 of 7

Figure 3. Example of master pull report used in Dock Shipping Supermarket.

There is then another crew that picks or shops the orders from a smaller footprint, which is the "Supermarket". The pickers are then given a rack sheet/sticker, which tells them what to put on the rack (Fig. 4). The rack sheet used is based on FlowVision's patented algorithms. The racks are optimized to have as little air as possible.

Load Seq: 1 of 17
Order Seq: 1 of 3
Load #OHIO **Drop #6**

Lowes **Order #SI-12222**

SM Rack	Shelf	Item	Description	Qty	Height
5	3	11878	ROSEMARY, Barbecue	28	34.5
5	3	15807	ROSEMARY, Barbecue	10	34.5
5	3	11882	ROSEMARY, Common	19	22.5
12	2	20711	SENECIO crassissimus	80	46.5
22	1	60147	ROSEMARY, Barbecue	160	6.5

Load Date: 2/2/2018
Printed Date: 2/9/2018
SI-12222

Fig. 4. Example of rack sheet/sticker used in Dock Shipping Supermarket.

Once the order is picked, if required, the rack is then taken to a processing table to be cleaned; tagged and labeled; and then staged.

Regardless of whether you ship to big-box retailers, wholesalers, retail garden centers, or landscapers, FlowVision's Dock Shipping Supermarket can improve your shipping process. Typically, our customers see improvements of 15-40% in productivity, reduced overtime, and fewer credit/claims.

Lean Flow has been around since the creation of Henry Ford's assembly line. It has been implemented successfully around the world in every industry from high tech products, tractors, motors, leather, and every widget you can imagine. Many growers in the Green Industry are now turning to Lean Flow to help them solve their labor issues.

Learn more at www.flowvision.com.

Plant Collecting: Reason and Process

Dan Hinkley

Heronswood Garden, Kingston, WA, 98346

info@danieljhinkley.com

Keywords: Germplasm, plant introductions, genetic resources

INTRODUCTION

Over the past thirty years, I have had the profoundly rewarding opportunity to look at, collect seed of, evaluate, and introduce to cultivation, if warranted, plants from throughout the world. My interest in this regard has been catholic – encompassing plants considered native to my region and those thought of as exotic in both northern and southern hemispheres.

My fascination with looking at plants in the wild and the compulsion to grow them began at a young age. It has propelled me into a vocation that I love and for that alone I remain grateful to have discovered my passion at such an early age.

The reasons for what I have undertaken in my life's work I believe to be inherently justifiable. Among others, these include the following.

First, having collections with known provenance make them inherently valuable to the taxonomist and ecologist. The data recorded during the time of collection provides a snapshot of what was growing in any region at any point in time. In a methodical and fast paced destruction of ecosystems worldwide, having this record provides a reasonable assessment of the degree of alteration of landscapes worldwide.

Second, collections by seed increase the genotypic reserve of any taxa, especially important in terms of plants already in cultivation with just a few clones. Having additional genetic material provides a deeper vault of possible resistance to the rigors of climate change, insect predation and disease.

Third, collections by seed and/or divisions of plants unknown to cultivation increases the genetic resources for breeding work in both food and ornamental crops.

Fourth, observing the light, water, and soil requirements of any plant species in its native environment provides insight into its proper cultivation. Such an approach leads to better communication to the gardening public in regard to proper cultivation of known as well as unknown plant species.

Fifth, a reserve of genetic material of known provenance provides at least some opportunity in terms of conservation and possible re-introduction to the wild. This can be overstated, but it is applicable more, at least initially, to flora than fauna.

The process of collection varies considerably from country to country. However, the mechanics remain the same. Correct latitudes and altitudes corresponding to the desired climate to introduction must be identified. Rainfall patterns, i.e. summer vs. winter, must be identified for long term success. Proper import permits must be in place with a need to keep abreast of an ever-changing subset of requirements by APHIS prior to every importation. Permission by host countries and/or institutions, where required, must be obtained. Detailed collection data should be recorded including date of collection; GPS coordinates; altitude; habitat and common plant species of the same general vicinity; description of the taxa including height, leaf size and shape; leaf type; infructesence type; fruit color or characteristics of dry fruit; and number of seed per fruit.

In addition, seed collections must be thoroughly cleaned of flesh or dried material. Seed sufficiently large enough to host parasitic larvae should be provided adequate time after cleaning to mature and emerge from the seed embryo or endosperm. Seed must be thoroughly dried before final packaging to prevent rotting and damage to the embryo. Seed must be sent directly to APHIS from the country of origin. Cuttings or divisions must be pre-inspected and provided a phytosanitary certificate in the country of origin before sending directly to an APHIS facility.

Identification of the taxa collected must take place before sending to the any APHIS facility, including plant family, genus and species. Taxa restricted from import per APHIS or CITES rules must be abided by. A protocol for sowing upon receiving the inspected material after release by APHIS should be well considered before collection. Sharing of genetic material with as many institutions as possible, pre or post sowing, should be considered to increase likelihood of success in germination and preservation of the seed collected. Labeling and/or distribution of plants with appropriate collection numbers should be considered in the event of nomenclatural changes and/or tracking purposes.

With the best intentions and astute attention to the threat of bio-invasion, the introduction of unknown quantities into cultivation remains a possibility. These concerns can be mitigated to a degree by being acutely aware of red-flagged genera and/or families that have already proven to be invasive somewhere in the world. Also, the model of Reichard to predict invasiveness of collected plant material should be used. In addition, efforts should be made to evaluate and monitor collections for possible invasiveness before widespread distribution. Collectors and horticulturists should maintain the ability to destroy any collection and alert those who have received the collection if bio-invasiveness is noted.

Through proper procedure and etiquette, the introduction of new plants into cultivation and/or new clones of taxa already existing in cultivation remains a worthwhile endeavor, making our gardens more opulent while providing a greater appreciation of the floral richness of our planet.

Overview of the 2021 International IPPS Tour in Southwest British Columbia, Canada and Northwest Washington State, USA

Douglas Justice

University of British Columbia Botanical Garden, Vancouver, British Columbia V6T 1Z1

douglas.justice@ubc.ca

Keywords: International tour preview, Pacific northwest

Abstract

The 2021 meeting of the IPPS International begins with the International Tour, which starts in Vancouver, British Columbia and finishes in Bellingham, Washington, the site of the IPPS Western Region Annual Meeting. On the tour, we will see magnificent scenery, excellent parks and gardens, and a wide array of first-rate propagation and production nurseries. Our itinerary includes stops in and around Vancouver and picturesque southern Vancouver Island; mild, protected Puget Sound; the dry interior of British Columbia; and the coastal mountains and the transition zones between. We will be travelling through several biogeoclimatic zones as we see many distinctive landscapes. The following is a taste of what to expect.

Metro Vancouver (pop. 2 million) is the largest city in western Canada, and one of the largest seaports in North America. Vancouver was named and incorporated in 1886 and is now considered one of the most livable cities in the world.

Victoria is located at the southern tip of Vancouver Island. Southern Vancouver Island is known for its many gardens and forests with remnant giant conifers.

Sea to Sky Corridor. From Howe Sound, north of Vancouver to the Pemberton Valley, this area is noted for spectacular mountain scenery and all kinds of outdoor recreation opportunities.

Okanagan Valley. The dry, rain-shadow climate area east of the Coast and Cascade Mountains is British Columbia's prime fruit-growing region.

Similkameen Country. Settled during the Interior gold rushes of the 1880s and 90s, this area is known primarily for ranching and outdoor recreation.

The **Fraser River** is one of the world's major salmon rivers and the Fraser Valley lowlands immediately east of Vancouver has the largest concentration of berry farms and nurseries in British Columbia.

North Puget Sound. The protected waters of the Sound create a cool, maritime climate, ideal for berries, bulbs, potatoes, and nursery stock production. The protected islands and peninsulas are also home to world-famous American gardens.

Research Roadmap Created for Environmental Horticulture

James S. Owen, Jr.[1], Anthony V. LeBude[2], Jill M. Calabro[3], Jennifer K. Boldt[4], Jennifer Gray[5], and James E. Altland[6]

[1]Assoc. Prof., Virginia Tech, School of Plant and Environ. Sci., Hampton Roads Agric. Res. Ext. Cntr., 1444 Diamond Springs Rd., Virginia Beach, VA 23455, jsowen@vt.edu. [2]Assoc. Prof., NC State Univ., Dept. of Horticultural Sci., Mntn. Hort. Crops, Res. & Ext. Cntr., 455 Research Dr., Mills River, NC 28759, avlebude@ncsu.edu. [3]Sci. & Res. Prgms. Dir., AmericanHort & Horticultural Res. Inst., 525 9th St NW, Washington, DC 20004, jillc@americanhort.org. [4]Res. Horticulturist, USDA-ARS, Application Tech. Res. Unit, Greenhouse Production Res. Group, 2801 West Bancroft St., Mail Stop 604, Toledo, OH 43606; jennifer.boldt@usda.gov. [5]Admin., Horticultural Res. Inst., AmericanHort, 2130 Stella Crt., Columbus, OH 43215, OH, JenniferG@americanhort.org. [6]Res. Horticulturist & Res. Leader, USDA-ARS, Application Tech. Res. Unit, 212 Food, Agric., and Biol. Sci. Bldg., 1680 Madison Ave., Wooster, OH 44691, james.altland@usda.gov

Keywords: Horticultural Research Institute, Specialty Crop Research Initiative, AmericanHort, consumer preferences

INTRODUCTION

Environmental horticulture, or the green industry, is an integral component of specialty crops and agriculture. In fact, our industry generates one third of both all specialty-crop revenue (over $19 billion each year!) and its workforce. Yet, our industry receives only 12% of federal funds earmarked for specialty crops from USDA Agricultural Research Service and USDA's Specialty Crop Research Initiative. The Horticultural Research Institute (HRI) recognized this disparity and sought a means to bring federal funding levels more in line with our economic contributions.

HRI realized that the industry lacked a unified, strategic vision of research needs to provide a roadmap for future priorities that can be used to better leverage federal funds. With this new mandate, HRI set to work. The result was a professionally moderated, 2-day stakeholder meeting, where attendees shared their collective understanding of industry challenges, trends, and opportunities. Through these efforts, HRI identified research priorities to tackle the challenges, capitalize on the trends and opportunities, better direct our research investments, and to leverage federal funding.

METHODS

About 45 delegates, representing all segments of environmental horticulture and regions of the country, gathered for a face-to-face summit to listen, learn, and share their insights with each other. Participants were encouraged to expand their reach by engaging in conversations with their peers in advance of the meeting to widen their perspectives. During the summit, the moderator led the group through a series of discussions that culminated in a consensus on four key research priorities.

RESULTS

Four research priorities were identified by the group. They are detailed below.

Quantifying Plant Benefits

Research that quantifies and validates the benefits of plants on ecosystems, human health, and society can be used to craft improved value propositions. Ultimately, this will boost industry sales and services and increase interest in industry careers.

Our industry benefits when society understands how plants contribute to the health and well-being of individuals and their ecosystems, and the undervalued resources plants produce, which are often overlooked or taken for granted. New research regarding plant benefits needs to be aligned with industry priorities to maintain environmental horticulture at the forefront of providing sustainable green solutions for the world.

Creating Innovative Solutions

Research that creates or adapts biological, mechanical, and technological systems makes practices and processes more efficient and productive for horticultural businesses of all segments and sizes. Ultimately, this will increase efficiencies and profits.

Our industry continually needs improved systems to produce new or improved crops with less labor, water, nutrients, time, and/or pesticides in a safe work environment while adding value to quality plants that thrive during shipping, marketing, and consumer use. Whether in the supply chain, current inventory, or on the road to end-users, crops and inputs need to be traced, evaluated, ordered, managed and/or improved upon to continually provide cost-effective solutions for producers to integrate into existing production practices. This would include (but not limited to) advances in plant breeding, crop production and protection, software, automation, mechanization, and logistics. Recognizing and addressing barriers to adoption will be crucial.

Gathering Consumer Insights

Research that evaluates consumer behavior, preferences, and trends empowers horticultural businesses to optimize products and services. Ultimately, this will lead to industry-wide profitability and growth.

Consumers are responsible for the health and prosperity of our industry. Therefore, producers need to understand generational shifts in consumer demographics, as well as how those shifts affect consumer purchasing behaviors. Examples include emerging market preferences, relative purchasing power, and general gardening confidence. Markets, consumers, and the products they desire interact and change over time. To adapt, industry producers need information that considers all this, yet is easily understandable and crafted for various segments of the industry. Research on consumer preferences, attitudes, needs, motivations, and purchasing behaviors for our industry's products and services will help companies make better business decisions by capturing what consumers want – not what the industry thinks they want.

Producing Practical and Actionable Solutions

Research that tackles ongoing and emergent industry challenges in production, resource management, and pest and disease management provides practical and actionable advice for horticultural businesses to improve sustainability and profitability.

Disruptive, ongoing, emergent issues that challenge short-term profitability and success of environmental horticulture will continue to rise. Therefore, providing solutions to these challenges must remain a research priority for funding agencies.

Moving forward, HRI has adopted these four research priorities and will use them to guide future HRI funding and leveraging decisions.

HRI, the foundation of AmericanHort, supports scientific research and students for the advancement of the environmental horticulture industry. HRI was established by industry leaders on the premise that no one could better direct needed research to advance environmental horticulture than the very people who work in it. We adhere to that same vision today: we fund and guide environmental horticulture research efforts with direct input from industry professionals. It is the strong foundation upon which to build the industry.

What Do Plant Breeders Do?

Chuck Pavlich,

Terra Nova Nurseries, Canby, OR 97013,

chuck@terranovanurseries.com

Keywords: Germplasm, plant introductions, hybrids, breeding, coleus, begonia, *Heuchera, Bergenia, Mukgenia*

INTRODUCTION

It all starts with an idea – or a curiosity and the question; "What if?" Figure 1 illustrates in a simple way what a plant breeder goes through to get a product to market. Because horticulture is simply a reflection of fashion or fad, breeders have to be one step ahead of the game or be lucky enough to cash in on serendipity. Sometimes we're the windshield, sometimes we're the bug.

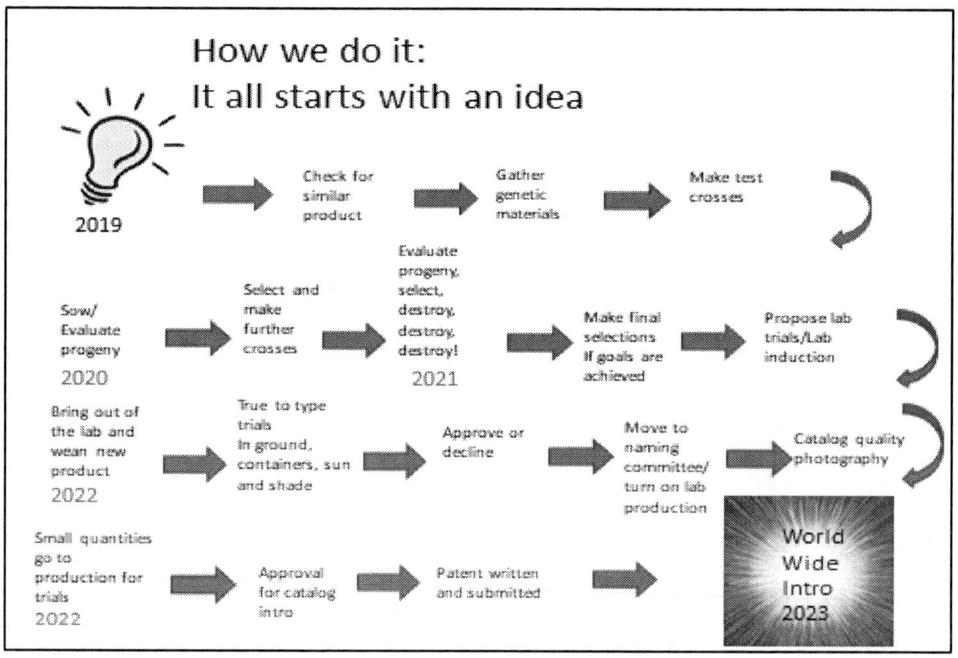

Figure 1. Conceptualization of the process involved in getting a new horticultural product to market.

Terra Nova relies on traditional breeding methods and the creative mind to bring out new products. However, but we're not afraid to use and have used most genetic manipulation tools to increase our chances of developing novel products.

With ploidy manipulation, we are able to restore fertility in some infertile hybrids. We are also able to increase flower or plant size or make a plant more stiffly upright in order to support the weight of relatively larger flowers. We have also made a few bi-generic crosses possible by raising the ploidy level of a distantly related plant.

Our most famous bi-generic plant; *xHeucherella*, a cross between Heuchera and Tiarella. Another up-and-comer is *xMukgenia*, a hybrid between *Bergenia* and *Mukdenia*. *Mukgenia* came about by reading phylogenies and plant family theories formulated by Dr. Doug Soltis, formerly of Washington State University.

Ploidy manipulation is one of my favorite methods of conversion of plant material. Doubling the amount of genes material leads to both interesting and not-so-interesting novel plant material. It is our job, as breeders, to weed through the plant material and destroy it. If you are not throwing out plants, then you are not a breeder, you are a hoarder - plain and simple.

Irradiation is a total roll of the dice. Deciding the Gy rate of gamma radiation to supply is sometime difficult. We work with genera that have little to nothing published about them, and so we have to make educated guesses about them based what are they related to, where they sit on a phylogeny tree, etc.

Chemical mutagens, such as E.M.S. (ethyl methanesulfonate), are useful for producing a wide range of results. Point mutations, insertions, and deletions are possible with this and several other chemical mutagens. EMS has also earned the name "Ethyl Methane-nasty-stuff" because of its extreme toxicity. Fortunately, it has a very short half-life, and can reach it within hours.

Embryo rescue is a method that can be used to intervene in the life of a seed. We often make crosses with divergent chromosome numbers or crossing two genera together. Pollen grains may germinate and grow down the style to the ovary. Ovules may form but are aborted due to some incompatibility between mother and embryo. Sometimes, only hours after making a cross, we will harvest the ovary, excise the ovule, and, rescue the resulting embryo – all done painstakingly under a microscope. After harvest, we will sterilize the embryo, and plate it on replication media in the lab. Weeks later, the embryo grows and begins to form a plant. To achieve the colors in our *Echinacea* plants, we had to rescue thousands of embryos. After several more generations of *Echinacea* breeding, incompatibility was not much of an issue, which allowed for the resumption of traditional breeding.

And then, there is the internet. It has been an incredible tool for research. However, one does have to sift through cubic yards of debris to find a nugget of information, and a lot of what used to be free information is now sold by universities and publishing houses.

And, then there is CRISPR, which stands for Clustered Regularly Interspaced Short Palindromic Repeats. I do not have an opinion on this yet, but the genie is out of the bottle.

Mutations! We love mutations! Some of our best plants have come into commercial culture due to spontaneous mutations. 'Obsidian', the world's best-selling black foliage *Heuchera* mutated to 'Midnight Rose'. (We pay tens of thousands of dollars annually to the finder, Behnke Nurseries.)

'Georgia Peach' begat 'Georgia Plum'. The unique thing about 'Georgia Plum' is that the mutation is an *extra* layer of pigment on the exterior of the plant. One mutation gave a purple coat to the entire plant. Foliage, petioles, stems, and even the flowers are purple. This mutation proved very valuable in breeding.

Heuchera 'Marmalade' lost a leaf layer in a mutation and became 'Lime Marmalade'. *Heuchera* 'Amber Waves' may be the most influential mutation at Terra Nova. A chimera mutation of the variety 'Whirl Wind', 'Amber Waves' was nothing more than a streak of color in the petiole of one leaf. We did a tissue pull from the petiole, cultured it, and turned it into a plant. From this plant, all pink-, coral- and red-colored selections originated. This plant made it possible.

Heuchera 'Snow Storm' provided the seed money (no pun intended) for Terra Nova. This *Heuchera* sold by the millions in Great Britain. One of my *Begonia* hybrids, COCOA™ 'Enchanted Evening', sported 'Enchanted Moonlight' *Heucherella* 'Solar Power' sported 'Solar Eclipse'. Heucherella 'Buttered Rum' sported 'Mojito' and 'Mojito' sported HAPPY HOUR™ 'Lime'. 'Buttered Rum' also gave us 'Firecracker'.

Discovery, research, and trial and error are our best friends and worst enemies. Whenever possible, we start from straight species, make selections, make crosses, and then throw out thousands and thousands and thousands of plants annually.

Our *Echinacea* selections started from almost nothing. Today, they are regarded as some of the world's best.

Our *Heuchera* selections are the world's best. That's why almost every other "breeder" in the world uses our genetics. Primrose Path Nursery in Pennsylvania may be the only one not to use ours. And we do not use theirs. The only two named cultivars in our breeding are 'Montrose Ruby' and 'Palace Purple'. The rest are species accessions from North America, and lots of careful breeding. With that said, we really had drab stuff to begin with.

Terra Nova is one of the most prolific coleus breeders around and perhaps the only breeder of genetic dwarf, non-flowering *Coleus* (or *Solenostemon* or *Plectranthus* (as they are now known)).

Our *Begonia* selections are some of the world's most exciting and the hardiest. Creating all new hybrids from the best and hardiest species give our plants excellent heterosis. We find they have great vigor, color, and ease of growth.

PROCEEDING'S PAPERS

EASTERN REGION OF NORTH AMERICA

Dr. Charles W. Heuser, Jr., Regional editor

Sixty-ninth Annual Meeting - 2019

Madison, Wisconsin USA

Paperbark Maple (*Acer griseum*) Conservation Project

Kris R. Bachtell[1], Anthony S. Aiello[2], Michael S. Dosman[3], and Kang Wang[4]

[1]Vice-President of Collections and Facilities, The Morton Arboretum
Lisle, Illinois 60532 U.S.A.; [2]The Morris Arboretum of the University of Pennsylvania, Philadelphia, Pennsylvania 19118 U.S.A.; [3]Arnold Arboretum of Harvard University, Boston, Massachusetts 02130 U.S.A.; [4]Beijing Botanic Garden, Beijing, People's Republic of China, China

kbachtell@mortonarb.org

Keywords: *Acer griseum*, conservation, plant introductions, Asian plants

INTRODUCTION

Despite being a well-known highly ornamental and popular garden plant, paperbark maple, *Acer griseum*, is listed as endangered in its native habitat in central China (Fig. 1).

Figure 1. *Acer griseum* form, fall color, and bark.

As far as can be determined, there have been a limited number of introductions from the wild, with five into the United States, one into the United Kingdom, and one into Finland. The first collection was made by British plant collector and explorer Ernest H. "Chinese" Wilson (E.H. Wilson) for James Veitch and Sons Nurseries in 1901. One other collection of two seedlings was collected by Wilson in 1907 for the Arnold Arboretum. Remarkably, there was good germination from the seed collected in 1901, something that is highly unusual for this species. By 1912, Veitch and Sons Nurseries was offering *A. griseum* for sale in its catalogue. These two collections became the basis for cultivated material in the West.

Besides the two known introductions of *A. griseum* by Wilson, seed was received in the early 1990s from China by Heritage Seedlings, (Salem, Oregon) but it is not known if the seed actually germinated, seedlings collected by the North American-China Plant Exploration Consortium (NACPEC) on its 1994 expedition to Hubei Province, and seed collected in China for Arboretum Mustila (Elimaki, Finland) in 2010. It is likely that plants grown from seed from the Wilson collections and Heritage Seedlings is the source of all plants currently in cultivation. With this in mind, the Morris, Morton, and Arnold Arboreta, and the Beijing Botanical Garden initiated the *Acer griseum* Conservation Project to determine whether the diversity of cultivated plants in the U.S.A. and U.K. accurately reflects the genetic diversity of plants in the wild, or if further efforts are needed to conserve this species.

MATERIALS AND METHODS

This project was divided into the following phases:

- In the summer and fall of 2013 and fall of 2014, Aiello, Bachtell, and Dosmann sampled trees from throughout the U.S.A.
- In July 2014, Aiello and Bachtell visited mature trees throughout the U.K., visiting a total of nine locations and sampling 24 *A. griseum*.
- In September 2015, Aiello, Bachtell, Dosmann, and Wang visited and sampled native populations of *A. griseum* in five provinces throughout its native range in central China.
- Additional collections from China were made by Dosmann in Hubei Province (2014), by staff from the Hunan Forest Botanical Garden in Hunan Province (2015), and from Hangzhou Botanical Garden (2016).
- Samples were sent from Arboretum Mutila, Elimaki, Finland (2016).
- In the fall of 2015 and summer 2016, Aiello sampled trees from other sites on the East Coast of the U.S.A.
- In the summer of 2016, Aiello sampled trees at three locations in the Pacific Northwest.

As far as can be determined, no seed was collected in China by Western botanists for most of the 20th century. Since the 1990s Heritage Seedlings has been the largest source of *A. griseum* seedlings sold throughout the U.S.A., using a seed orchard that was grown from seed collected from a tree at North Willamette Research & Extension Center, Aurora, Oregon, from the Highland Park trees, and possibly from the Henan collection.

Subsequently, the next wild collection occurred on the 1994 North America-China Plant Exploration Consortium (NACPEC) expedition. On the expedition, 16 seedlings were collected, and 10 of these are still alive: four at the Morris, three at the Arnold, and three at the U.S. National Arboretum in Washington, D.C. One other tree of wild origin was included in this study — in the garden of plant explorer Dan Hinkley (Seattle, Washington) — although in the summer of 2015 unfortunately, this tree suddenly died.

In July 2014 Aiello and Bachtell visited venerable *A. griseum* specimens throughout the U.K. (Fig. 2) with a goal of visiting as many trees known or purported to have been collected by E.H. Wilson in 1901.

Figure 2. Example of an old specimen Aiello and Bachtell visited of *A. griseum* in the U.K.

We visited a total of nine locations and sampled 24 *A. griseum* — some of these were known to be grown from the Wilson collection, others were suspected to be from Wilson, and others originated are from subsequent generations of seedlings. The gardens that were visited were as widespread as the Royal Botanic Garden Edinburgh (Scotland), Newby Hall and Gardens (North Yorkshire, England), Dyffryn Gardens (Vale of Glamorgan, Wales) and Highdown Gardens (West Sussex, England). Many of these collections include original introductions from China, and together they provide an intriguing insight into the world of collecting 100 years ago.

In September of 2015, we completed the third phase of this project, when a NACPEC expedition sampled wild populations of *A. griseum* across its native range in central China. Aiello, Bachtell, Dosmann, and Wang travelled within an approximately 500-mile radius of capital city of Xi'an in Shaaxni Province. The locations stretched from Gansu south to Sichuan and Chongqing, into Shaanxi, Hubei and Hunan, and north into Henan and Shanxi.

Even more helpful was recent work on the distribution, conservation status, and genetic diversity of natural populations of *A. griseum* that had been conducted by a group of researchers at the Chinese Academy of Forestry, Beijing (Chen et al. 2013; Sun et al. 2014). Among these provinces the populations are disjunct from each other, and plants are often scarce within a given area. Despite this, on this expedition we visited nine locations of *A. griseum* in five provinces, resulting in 66 trees sampled, along with two seeds collections.

In late 2015 and then in the summer of 2016, samples were collected from various locations in the eastern U.S.A. A final phase of the project was in late July and early August 2016 when Aiello travelled to Oregon and Washington to sample trees at Heritage Seedlings, North Willamette Research & Extension Center, and Washington Park Arboretum.

In total, the U.S.A. trees that were sampled included:

- Twenty American trees of wild-origin (nine form E.H. Wilson, ten from NACPEC, and one from Hinkley) — 2013 and 2014.
- Six plants of nursery origin from the Morton Arboretum, Lisle, Illinois — 2013.
- One tree at the Glenn Dale, Maryland research station — 2015.

- Two trees at the New York Botanical Garden, Bronx, New York — 2015.
- Two trees from Mount Auburn Cemetery, Cambridge, Massachusetts — 2016.
- Five trees of cultivated origin at the former site of Princeton Nurseries, Allentown, New Jersey — 2016.
- Five trees from Washington Park Arboretum, Seattle, Washington — 2016.
- Two trees at the North Willamette Research & Extension Center, Aurora, Oregon — 2016.
- Forty trees from Heritage Seedlings, Salem, Oregon — 2016.

Leaf samples from all of the sites were collected for RAD-sequence analysis to help answer the question of the degree of genetic diversity represented in cultivation compared to that in the wild. Dr. Andrew Hipp, Systematist and Director of the Herbarium at The Morton Arboretum conducted the analysis of the DNA of the wild and cultivated trees to answer our question of how much genetic diversity is represented in cultivation. The results support the idea that Wilson's 1901 collection provides the basis for all trees in the U.K. and most of what had been in the U.S.A. in the 20th century (Fig. 3).

PROPAGATING TREES FROM TWO UNPROTECTED POPULATIONS

In September 2015 during the third phase of the project, most of the populations that we sampled were located in areas with some level of forest preserve and protection status. But we encountered two populations of paperbark maple that were growing in remote hillsides managed by local farmers. These two populations were located in Chengkou, Chonqing Municipality and near Xixia, Henan Province. The population in Chengkou was growing in an area that is periodically clear cut for fire and construction wood. These trees ranged from 4 to 15 m in height, but some had large basal diameters (up to 45 cm; 17.7 inches) which demonstrated that they had been coppiced several times over a period of many years. The population in Henan was more scattered, and we observed six trees scattered over an area of a few miles.

Wang exchanged contact information with the farmers in both locations so that they could collect dormant scion wood from these trees at a later date for us to use for propagation. The farmers were each paid a stipend, and instructed not to cut down the trees and to cut away vegetation near the trees in order to open up the nearby canopy in the hope to initiate stronger annual growth, and thus better scions for grafting in the future.

In February 2018 we conducted a grafting trial to determine the best understock to use and to hone our grafting skills with *A. griseum*. Two different bare-root rootstocks, sycamore maple (*A. pseudoplatanus*) and sugar maple (*A. saccharum*) were used in the trial. The scions for this trial were collected from trees growing at the Morris Arboretum and Arnold Arboretum from plants of the 1994 NACPEC expedition. Depending on the diameter of the scions either a side veneer or cleft grafting technique was used. Following graftage, the graft wound was wrapped with 0.004 mil 1-in. wide × 2¾ in. long pre-cut pieces of clear poly grafting tape.

Then the grafts were placed in a hot pipe calluser tube in a heavily shaded and vented greenhouse (Fig. 4). The tube's temperature was set to a temperature between 20°C – 22°C (70°F – 74°F) and the grafts were kept in place for 21 days. The grafts were rotated 180 degrees after 10 days to allow for more even heat exposure to both sides of the graft union. After this, the grafts were placed in a sealed box and held in a cooler at 4.4°C (40°F) for 14 days before being potted up. In total 99 grafts were made (49 using *A. pseudoplatanus* rootstock and 50 using *A. saccharum* rootstock).

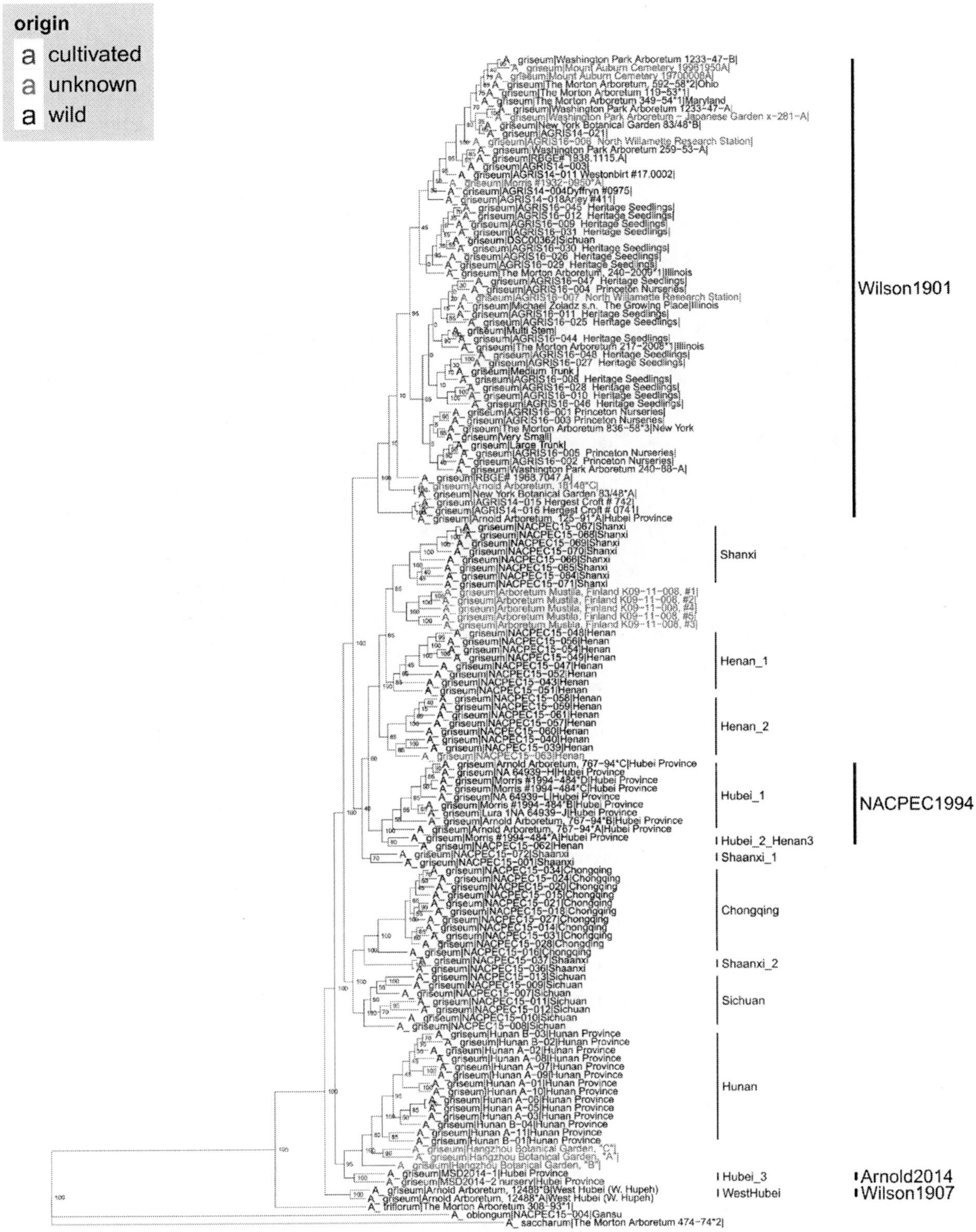

Figure 3. Phylogenetic relationship between different *Acer griseum* accessions.

A total of 71.4% of the grafts were successfully using the *A. pseudoplatanus* rootstock — this species has the remarkable trait of being a "universal rootstock" for many maples, including *A. griseum*. None of the grafts using the *A. saccharum* rootstock were successful. Due to the lack of visual evidence of callus formation when the grafts were in the hot-pipe calluser the grafts were allowed to remain there for an additional 10 days, but to no avail. A noticeable flow of sap arose from the rootstock (and eventually a rather foul odor). It appeared that the sap flow from the rootstock somehow impeded callus formation on both the scion and rootstock.

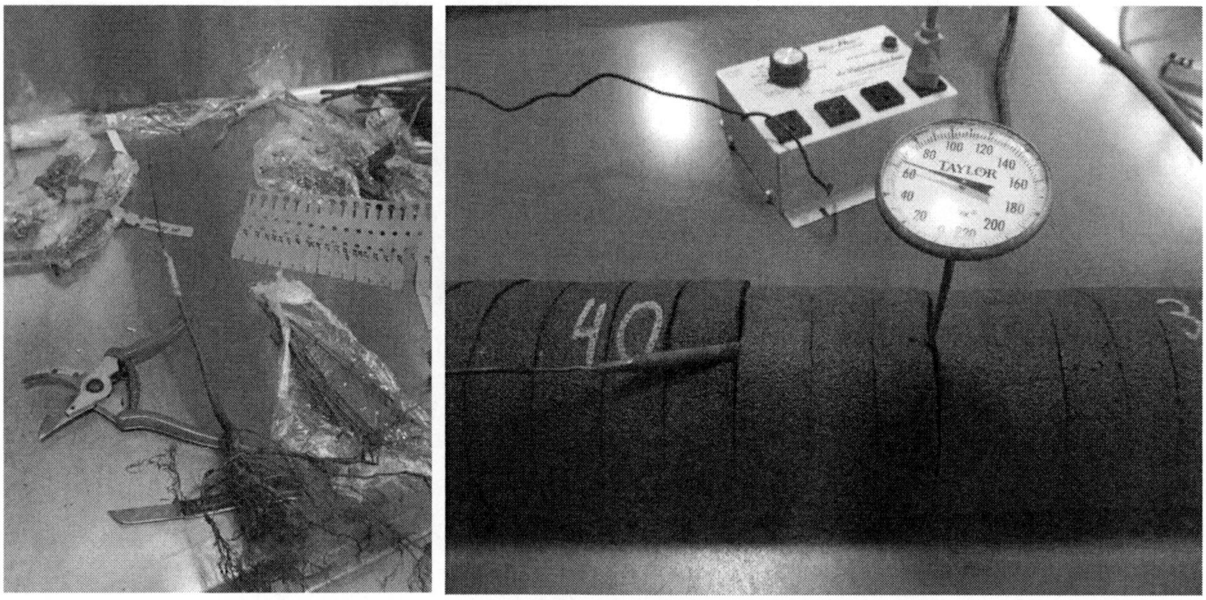

Figure 4. Example of grafted scion (left) and hot pipe calluser tube (right).

Scions from the two unprotected populations from China (Henan and Chongqing) were collected in February 2019. Due to various bureaucratic delays, it was only in late March of 2019 they reached the USDA National Plant Germplasm Plant Quarantine Center located in Beltsville, Maryland. Grafting was performed in early April 2019. Based on the success experienced in the 2018 grafting trial only *A. pseudoplatanus* rootstock was used. As in the earlier propagation trial two graft cuttage techniques were used, either side veneer or cleft, and this was determined largely by the diameter of the scions. Many of the scions were exceedingly thin likely due to the non-vigorous growth of the parent trees. This made the graftage difficult to perform.

Due to the shipping delay and the typically warm ambient temperature during April in Beltsville, Maryland the hot pipe calluser was actually placed in a walk-in cooler that was maintained at 3.3°C (38°F) so the buds on the scions would remain dormant. Following graftage the grafts were treated similarly as the graft in the earlier trial. After 28 days in the heat calluser the grafts were potted up and placed in a shaded greenhouse. After another 45 days the grafts were inspected for success. The percentage of success was lower than the earlier trails and this was believed to be due to the thin diameter of the scions and the substantial delay in time involved with their collection and transport to the U.S.A. from China. There were 54 total grafts attempted with 21 being judged as successful: a 38.9% success rate.

These plants are required to remaining under quarantine until they are released which is currently scheduled for 2022. Once released they will be distributed to members of NACPEC for further observation, cultivation, and propagation. An additional set of grafts was taken successfully in China. Both groups of young plants will provide the basis for ex situ conservation of the two unprotected populations. It will be interesting to observe their growth performance and to see if this new genetic material improves the viability of the seed set by other *A. griseum* already well established in these NACPEC member gardens public gardens.

Literature Cited

Aiello, A.S. 2016. *Acer griseum* in cultivation and in the wild. The Plantsman (New Series) *15* (4):250–255.

Aiello, A.S., Bachtell, K.R., Dosmann, M.S., Olshefski, P.M., Goff, E.I. (n.d.) *Acer griseum* conservation project. n.p.

Chen, P., Yu, X., Zhang, C., Zhang, Y., Sun, S., Cheng, B., and Zhu, C. 2013. Natural regeneration of *Acer griseum*, an endemic species in China. Scientia Silvae Sinicae. *49* (3):159–164.

Davies, Jr., F.T., Geneve, R.L., Wilson, S.B. 2018, 2011, 2002. Hartmann and Kesters: Plant propagation: principles and practices (9th Edition) (What's new in trades and technology). New York, New York. Pearson Education, Inc.

Franchet, A. 1894. Plantes Nouvelles de la Chine Occidentale. Jardin de Botanique *8* (17):290–297.

Gibbs, D. and Chen, Y. 2009. The red list of maples. BGCI, Richmond, England.

Hipp, A.L., Hahn, M., Aiello, A.S., Dosmann, M.S., and Bachtell, K. In prep. Hunting down the horticultural history of paperbark maple, *Acer griseum*.

Macdonald, B. 1986. Practical woody plant propagation for nursery growers. Portland, Oregon. Timber Press.

Sun, S., Yu, X., Zheng, Y., Chen, P., and Zhang, C. 2014. Isolation and characterization of twenty-seven polymorphic microsatellite loci in endangered Chinese paperbark maple, *Acer griseum* (Dicotyledoneae: Aceraceae). Biochem Syst Ecol. *56*:99–103.

Wilson, E.H. (1925). *Acer griseum*. The Garden. *89* (2773): 20.

Growing Natives: Lessons Learned

Kyle Banas

Pizzo Native Plant Nursery, LLC, 10729 Pine Rd, Leland, Illinois 60531 U.S.A.

kyleb@pizzonursery.com

Keywords: Native plants, genetic diversity, propagation, herbaceous plants, production, ecosystem restoration

INTRODUCTION

At Pizzo Native Plant Nursery (PNPN) we grow over 450 species of plants native to the Midwest and Eastern United States. We are a propagation and liner producer of herbaceous perennials focusing mostly on wetland, prairie and woodland species. Native plants can be a difficult term to define and not the goal of this presentation but for simplicity let's just say they are species that evolved naturally within an ecosystem. PNPN is located on 40 acres in Dekalb County Illinois. Illinois is a state in which less than 0.1% of the land area was left unaffected by agriculture or urbanization. Fortunately, for us and those in horticulture, one result of this massive loss of native ecosystem and habitat has spurred a renewed interest placed on these endangered landscapes. This has resulted in the Midwest becoming a hotspot for ecosystem restoration and the use of native plants in the landscape. It makes sense that one of the most disturbed landscapes in the country would hunger the most for restoration or at least some sense of place to be put back into the landscape.

It's for these reasons, among others, that a few of the native plant nurseries in the Midwest have been growing and expanding over the last decade. At PNPN we have increased production from 200 thousand units annually in 2009 to over 1 million units in 2019.

I want to share three main lessons I learned since I began growing natives. The first is a lesson in what fundamentally sets a native plant nursery apart from traditional ornamental growers as it applies to production and the markets they serve. The second is a lesson in what working with native plants has taught me in the way I approach plant production challenges. The last lesson is how working in ecosystem restoration and with native plants has changed the way I view employee management and labor in general. My hope in this presentation is that you leave with a better understanding of what differentiates a native plant nursery from a traditional growing operation and perhaps a few new ideas you can implement in your operation.

LESSONS

Lesson One: Dealing with Genetic Diversity

The largest idea that sets a native plant nursery apart from a traditional grower is that a native plant nursery works exclusively with genetically diverse crops (Fig. 1).

Figure 1. Genetic diversity within an ecotype in *Symphyotrichum novae-angliae* (New England aster).

Native plant nurseries grow wild origin material that has not been selected or further cultivated to meet any certain requirements. This was one of the most difficult things for me to grasp as an ornamental grower. I was used to observing relatively similar attributes among individuals across a crop such as growth rate, heights, flower times, etc. and this is not the case with a genetically diverse crop. This diversity in a crop is apparent at all stages of growth from the size of the seed though the actual finish time of the crop. This makes mechanization and automation more difficult but not impossible. With natives we end up looking more for trends and averages among the whole crop rather than at each individual when making decisions. For instance, when cutting plants back, we have to look at the whole crop and make a determination on what the best height would be for the crop as a whole and not each individual. By cutting all the plants at the same height you can begin to achieve some uniformity. Many of our decisions are driven by averages in an effort to create a more manageable and uniform crop. This method of production increases the uniformity in crops the further along the crop cycle you go. By the time we have a finished liner, the product will lend itself to most mechanical and automated processes.

At PNPN the markets that we sell into are almost as diverse as the species we grow. We service the landscape industry including landscapers, ornamental growers, and park districts as well as the ecological industry including restoration contractors, forest preserve districts, and the USACE. We ship plants across most of the continental United States and into Canada. Some of these restoration type projects require crops of a specific genetic origin, meaning seed stock must have been collected from remnant landscape within a certain region or distance from a particular site that is being restored. Sometimes this is 200 miles, sometimes it is only 20 miles from the site. This is achieved by sustainably collecting small amounts of seed from wild populations that are then used to establish a seed bed at the nursery. This stock bed is then collected from for many years. This adds another layer of complexity to crop management. A native plant nursery may have crops of the same species that are from different genetic populations. For instance, this year we had three distinct lots of *Schizachyrium scoparium* (little bluestem) at the same time; one from Dekalb Co, Illinois; one from Cook Co, Illinois; and one from Berrien Co, Michigan.

In my experience it has been a good practice to not rely on one specific seed lot or origin for a species due to the wide variance and unpredictability in wild genetics. To overcome this, we routinely stock multiple different genotypes for each species. We have developed a numerical system to track each crop. When seed is collected or purchased, we assign a 4-digit lot number, along with the year it was collected and a code regarding its source. This allows us to track origin and seed treatment methods throughout production and provide this information to the customer if needed. So, in addition to tracking the species, location and ready date we are also tracking genetic origin of all of our crops. Example SCHSCO50-4521-18-LP = *Schizachyrium scoparium* that is our 4,521 seed lot collected, in the year of 2018 that came from our Leland prairie. If I were to go back in our records, I would be able to tell exactly when and how much of this seed was collected, how it was treated, when it was sown, when it was transplanted into the next tray size as well as how many propagules resulted from each ounce of seed sown. We use Microsoft Excel for all pre-transplant information and SBI Software for all finished liner production that consists of deep 50-count plug trays or 3-in. pots in a 32-count tray.

One reason we strive to maintain genetic diversity within our crops is an effort to build resiliency in the landscape, especially when it comes to ecosystem restoration projects. The more genetic diversity that is initially provided increases the likelihood that the correct traits will be present to reproduce and sustain a population. For instance, say there is a really early frost in the fall and there was a wide variance of flower time among a population. Those individuals that flowered and set seed earlier in the year would be more likely to reproduce and survive than those that set seed later.

While we do try to maintain as much genetic diversity as possible there is some genetic selection during propagation and production that is inevitable. For instance, *Baptisia alba* (white indigo) seed usually germinates over a very long window and those seeds that germinate very late in that window are usually not used, so we are inadvertently selecting for the quicker germinating individuals. When growing *Monarda fistulosa* (wild bergamot) from seed there tends to be certain individuals that are far more susceptible to powdery mildew and are removed from the crop, thus we are selecting for more tolerant individuals. For these reasons we do not collect seed from our crops but instead always go back to our seed blocks or second generation wild collected stock at the most. This unintentional selection process does help create a more uniform crop and the further from propagation in the production line you go the more uniformity that is achieved.

Lesson Two: Nature Knows Best

The second lesson is the importance of observing plants in their natural setting. When it comes to making plants grow, IPPS is arguably the most knowledgeable group in the world, the only possible exception would be the plants themselves. I am beginning to learn the importance of observing plants growing in the wild and applying some of those ideas to our production practices, sort of a reverse engineering of nature or biomimicry. Some of the species we grow such as *Phyla lanceolate* (northern frog fruit) or *Lysimachia thyrsiflora* (tufted loosestrife) are rare to see in production at other nurseries so there is not much if any information to start with. I am learning that a good starting point is with natural observations. The following are a few examples of how observing plants in a natural setting has helped us overcome challenges in a production setting. These are not entirely my own observations or conclusions but more so how these practices came to be used in our nursery.

At PNPN we grow a range of wetland species used in retention basins and wetland restorations. For years we had problems overwintering many of the wetland plants for several reasons. One was that the large fleshy roots were a favorite of mice and voles. Wetland plants are also more susceptible to cold damage than many of the upland plants so extra protection was always required for us to successfully overwinter these plants. As a result, we were finding ourselves over sold in the spring for wetland plantings due to overwinter loss and struggling to fulfill early orders. I remember discussing the issue with our grower while standing at the back of our nursery that overlooks a wetland. The area is a restored wetland ecosystem that also serves as a seed collection area for these species that we grow. We were looking over the wetland and discussing how they seem to overwinter so well in the wild and it dawned on us that in the fall the water levels typically rise and the plants are frozen in a few inches of water. So that fall we built some trial wetland beds. We leveled an area with a little sand and put down ground cloth. On top of that we built two rectangles out of railroad ties and dropped in a pond liner. In the fall the plants were cut back and packed into the beds. We flooded the beds and let winter take care of the rest. In the spring the water levels were dropped, and we were happy to see zero rodent predation and that our overwinter success rate was significantly higher than in previous years with the typical hoop house method (Fig. 2).

Figure 2. Wetland beds used for growing and overwintering wetland perennials. Biomimicry in action — natural solution in background.

Another example of this type of reverse engineering is the use of rice hulls on our ephemeral crops (Fig. 3). We grow a handful of spring ephemerals such as *Mertensia virginica* (Virginia blue bells) and *Claytonia virginica* (spring beauty). These plants sell very well for us and are in demand in the spring but typically take 2 years from seed so we were left with essentially an empty pot for the remainder of the year. Having the potting mix exposed for the majority of the growing season was a prime spot for weeds, liverwort and algae to grow and colonize. We spent significant amounts of time weeding and spraying to keep the stock clean. We collect the seed for most ephemerals in early summer as the plants begin to go dormant. One year as I was collecting seed and shuffling though the leaf litter from the previous year, I was thinking about it as a natural mulch and possible ways to replicate that in a container.

Figure 3. An example of reverse engineering is the use of rice hulls on our ephemeral crops for weed control, crop is *Mertensia virginica* (Virginia blue bells).

So that summer we waited for the ephemeral crops to go dormant and applied about ¾ in. of rice hulls to the pots. This reduced our labor and spray needs by almost 70% and yielded a much cleaner looking plant the following spring. Since then we have taken to using rice hulls on most of our crops as it seems to lessen weed and liverwort pressure, as well as help retain moisture and reduce some irrigation needs. One interesting observation I have made about the rice hulls is that it lowers the temperature of the potting mix. This can be used to our advantage when growing cool season crops, but it does delay warm season grasses from breaking dormancy in the spring as well as increasing finish times in the summer.

A third natural observation is the dependency that some crops have with other species (Fig. 4). A good example is *Asarum canadensis* (wild ginger), which in the wild I usually compete with the ants to collect the seed. It is a species that we have struggled with to achieve high germination rates, even after trying several stratification methods. The observation that has been documented is that the ants carry the ginger seed into their colony where they ingest only the elaiosome, discarding the rest of the seed in a special chamber where it eventually germinates and establishes its own colony.

Figure 4. *Viola pedatifida* (praire violet) seed showing the elaiosome that attract ants.

A deeper thought here leads me to ask - Who is farming who in this situation. I could make the argument that plants are using the ants in this situation to survive. As a way to try and crack the germination protocol for a given species, I sometimes leave it up to interns to try and figure it out. This season one intern chose to try and crack germination protocol for *Asarum canadensis*. The student set up a trial and tried several chemical and mechanical means inspired by the ants to break dormancy on the seed and improve germination results. It is still an ongoing experiment so let me know if you are interested in the outcome.

Lesson Three: The Importance of Meaningful Work

Since I was 13 years old sweeping floors in the greenhouse after school and asking why sanitation prevents pests, it has been my impression that if you understand the importance of your work you will find meaning in the task. When you have meaning for your tasks you will be more satisfied with your job. That thought was recently vindicated by a study done by Payscale.com and referenced by the ILCA contractor magazine. The article discusses how important job satisfaction and the feeling of making a difference are when considering employee retention. At PNPN we have worked hard to develop a strong company culture and as a result we have developed a team that is engaged, empowered, and motivated (Fig. 5).

Figure 5. Part of the production team at Pizzo Native Plant Nursery celebrating hitting the production goal for the year.

I think these are traits that are valued by potential employees and make recruitment and employee retention much easier. In an increasingly competitive setting to find labor I believe these traits, along with a good salary will give companies the edge. I believe strongly in having a mission and vision statement that are not only applicable to our production and sales teams' everyday decisions, but also gives them reason to show up every day.

Our vision is:

> "That native plants are used in all planting applications because their value in enhancing and restoring our environment is widely understood and accepted."

Our mission is:

> "To build ecologically balanced communities through education, promotion, cultivation and trade of top-quality native plants."

We regularly revisit these statements and ask for all full-time employees' feedback and input. Unlike many other nurseries today, employee fulfillment is not our nursery's largest hurdle, and we are not exactly located in a heavily populated area. I have had people show up asking to volunteer their time because their interests align with our mission. While this is not a realistic way to staff a wholesale production operation, it does point out that developing good company culture is important when it comes to recruitment.

Finding people to hire is only half the battle, employee retention is the other half and I have found in most cases the longer we keep someone employed the more efficient and effective they become. I have found that getting our entire staff out for tours of other facilities and job sites that our plants are sent to has helped tremendously. It's a team building exercise and it brings meaning and closure to plant production and completes the cycle of what can seem to some as routine and monotonous production tasks. I think as an industry we tend to get caught up on efficiencies and cutting costs, but it's important to remind ourselves and new prospects that there is a lot of passion and career satisfaction in the horticulture industry.

CONCLUSION

I believe the increased use of native plants in our landscape is not a trend but is more of an evolution in the way we view the landscape. It is the result of more people connecting to the larger systems around them and the realization that we are part of a fragile ecosystem. The use of native plants acknowledges that we are not the only species that needs plants to survive and we are capable of creating landscapes that serve functions beyond our own desires.

It is interesting to think that only by removing parts of the puzzle we realized the complexity of the larger picture, and through this lesson I feel a responsibility to help the pieces fall back in place. In the last 10 years, natives have taught me a lot of lessons about growing. These lessons will help guide my decisions in the next decade as we continue to adapt and innovate to evolve our industry.

Systems Approach to Nursery Certification

Tom Buechel

Buechel Horticulture Solutions, Waterloo, Wisconsin U.S.A.

tjbuechel@gmail.com

Keywords: Nursery certification, non-invasive plants, sustainability, pests, disease

INTRODUCTION

It takes a team to make positive impact and extraordinary improvements in a nursery operation. The evolution of systems approaches in the green industry is facilitating this change to better workforces and their growing practices that promote healthy plants. Where did these systems approaches evolve from?

The Hazard Analysis Critical Control Point (HAACP) program is really the originator of the systems approaches that are discussed, allowing you to move forward producing healthy green industry products. This program was developed for the food industry to prevent contamination in all products that could cause serious human health issues. Systems approaches is essentially total quality management on steroids. Food industry examples include:

1) Consumption safety from bacteria, fungus and virus.
2) Create multiple checks, visual and documented, to ensure product safety.
3) Set a standard of integrity to promote the food industry image.

Let's take a look at what a systems approach actually is. It is individual measures that are integrated into a system that has a cumulative effect. It looks at situation analysis in nursery areas and determines risk by identifying critical control points. After this is achieved, growers can establish BMPs (best management practices) that will establish pest prevention, rapid detection, elimination or containment, and recordkeeping using multiple methods for each acting as check points to be sure issues do not arise.

WHY SYSTEMS APPROACH?

We can always digress into the past to find clues to better processes. The nursery industry was highly reactive, putting out plant problem fires instead of preventing them (control, control, control). This was a very typical mindset of nurserymen for a number of reasons. Two of the most glaring were the lack of technology and mechanization in production that limited growers' ability to spend more time on scouting. Physical labor ruled past decades. Tired growers equal drop in awareness — leading to plant production by volume to generate sales inventory. There was also a lack of regulation, especially for invasive plants, and discrepancies with use of plant names among operations.

Presently we have many laws and regulations and no shortage of pest, disease, and invasive plant problems. The nursery industry is scrambling to produce non-invasive plants to minimize future threats. The regulatory environment is also using mapping and web-based crawlers to help slow the spread of pests, diseases, and invasive species. There is a higher sense of environmental awareness about issues that cause harm to our surroundings. Social media has enhanced this and is often used as an accountability tactic to keep the green industry honest. Unfortunately, the negative use of social media can be destructive and drive the price of products down. In this era, many green industry operations are still operating reactively to minimizing the spread of pests, disease, and invasive species because of labor constraints.

The future brings hope. Systems approaches are true solutions to ensuring that the green industry moves forward doing things the right way. Self-auditing and policing will be the answer to proactively solve growing problems. This will include the use of sound teamwork among industry, states, federal government, organizations, and universities — hence utilizing the Systems Approach Model. Within systems approaches data becomes "king" and ever more important to be efficient and optimize plant health to keep our industry image at a very high level.

TWO VIABLE SYSTEMS APPROACH SOLUTIONS THAT CAN BE USED IN UNISON TO KEEP AN OPERATION SUSTAINABLE

Two viable systems approach solutions that can be used in unison to keep an operation sustainable for years are possible. What are some reasons we are racing down this path?

Pressure continues to mount from invasive organisms. They are hitchhiking with nursery and greenhouse plants. The public is much more conscientious. They want a premium plant. Any defects sound the panic alarm, and worse, cause the general consumer to demand products for less or even free (margin compression). There is also regulatory pressure to control and slow the spread so agronomic and general commodities are protected. There are also financial constraints by nursery operations and in the regulatory environment. Less staff is common in most operations and at the state/federal regulatory level, leading to more work and less time to do the diligent items it takes to prevent problems.

Systems Approach to Nursery Certification

The first solution was developed by industry professionals, The National Plant Board and AmericanHort / Horticulture Research Institute.

Systems Approach to Nursery Certification (SANC) is a voluntary program for production nurseries and greenhouses to achieve state plant regulatory certification to be able to ship plants domestically (interstate) based on a systems approach. The program is in pilot phase in 19+ growing operations from coast to coast. The fundamental program elements of SANC can be seen in below (Fig. 1).

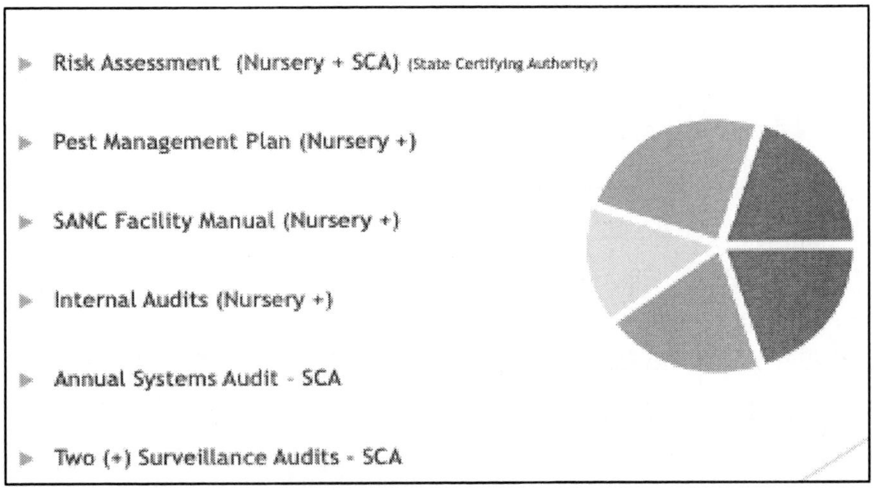

Figure 1. Systems Approach to Nursery Certification program elements.

SANC was developed to improve current practices that revolved around a one-time inspection. It helps alleviate the plant load by load inspection. If one thinks about it, scouting is not a once in a while type of thing. Producing great plants depends on watching them all the time with more sets of eyes. Annual inspections are just a one-day snapshot. Do we really want to rely on that? My bet is no. SANC focuses on the processes of producing the best plants. It trains growers to think of tough questions to ask in the important processes they do (the who, what, when, how and why). These questions lead to verification at multiple check points to prevent plant issues from starting in the first place. SANC habits help the workforce identify where a system is failing, forces them to review the process and put forth viable solutions to improve plant production going forward. SANC relies heavily on scouting and documentation. These two duties alone if done correctly can prove that healthy plants are being produced. In SANC a grower must do surveillance and systems-based audits (scouting/documentation).

The systems approach to nursery certification provides:

- Multiple ways to look at and solve risk-based growing issues.
- Employee engagement (more eyes on plants).
- Extended employee learning (certifying job done right).

Growing is all about risk management. No grower wants crop issues and they do their best to prevent shipment of pest/pathogens. SANC aids in preventing problems in a nursery setting by monitoring plants using effective practices. It also helps build an operations team to help accurately diagnose pest, diseases and plant related problems. When issues are seen it allows for growers/teams to address problems and document them, so they are not forgotten about. This documentation can then be used to make massive improvements within an operation typically leading to greater profitability. Below are examples of critical control points and how an operation could prevent them by clean growing examples (Table 1).

Table 1. Examples of critical control points related to *Phytophthra ramorum* on many nursery plant species.

Critical Control Point	**Best Management Practice**
Placement of container plants on contaminated ground	Do not place containers on contaminated ground
Use of contaminated irrigation water	Treat irrigation water before use
Use of contaminated pots	Use new pots or properly disinfected used pots
Buy-ins of infected pots	Buy only from certified suppliers; quarantine plants for 90 days
Poor drainage	Prevent standing water
Accumulation of infested leafy debris	Prevent accumulation of infested leafy debris

SANC allows for process development that effectively hits all areas within an operation like shipping/receiving, scouting, irrigation, media, container usage, equipment, and disposal of plant material. Cleanliness is often a key to success like discussed in the HAACP program to prevent food-borne illness. SANC develops superb learning/training programs with a bit of help from professionals that help write the program and put it into action. We discussed that data is "king." Systems approaches put what is in the brain into digital action where it can be used to analyze efficiencies and get better health and success of key production process. In a nutshell, SANC engages workforces and helps them adapt to start winning the battle against harmful pests and diseases.

Plant Sentry™

A key piece of the puzzle was missing within the SANC program that growers consistently struggle with — compliance. Intra/interstate regulations are numerous and can change, making it nearly impossible for growers to navigate based on past methods. Most regulations/summaries end up on PDF documents so trying to identify the correct ones and pair to revolving inventory has not proven to be totally reliable until now. To prove this point, take 500 nursery plants and multiply it by 16 ship states. The total equals 8000 checks that would have to take place to identify potential restriction matches based on pests, diseases, and invasive species alone, not to mention others.

The solution has arrived to make shipping compliant plants super easy — Plant Sentry™ compliance tool (Fig. 2). Plant Sentry™ evolved out of the systems approach for nursery certification to help eliminate purchase of invasive and non-compliant plants. Plant Sentry™ was developed with organizational partnership and critical feedback from industry and the regulatory environment.

Figure 2. Plant Sentry™ seal.

This systems approach tool is the first ever user-friendly regulatory approach for industries, ensuring shipment of compliant plants intra and interstate. It helps growers deal with challenging regulations that are very difficult to keep up with and revolving and changing statutes. It offers huge relief for those who enter into online sales and have to know all the regulations in the 50 states.

Plant Sentry™ gives growers the necessary options to work with sales vendors, state/federal officials, or a team of industry professionals confidentially who understand both best growing practices and regulation. The detail and advanced capabilities of this systems approach reaches further than growers, creating a winning situation for all to prevent, minimize and eliminate pests, diseases, and invasive species.

Growers are very conscientious of costs and worry about image. Plant Sentry™ is the solution to save countless hours of labor and costs when it comes to verifying compliance. It maintains clear organization and is up to date, a problem that growers had no good way to do. This systems approach tool also gives growers a fighting chance to ensure that plant shipments are done accurately and legally without the hassle of having to spend oodles of time figuring out complex regulation or treatment protocols.

What are some Plant Sentry™ highlights? It locates the source of regulatory information which in its own right can prove to be really challenging. Invasive species are not handled by the same organization in every state. Plant Sentry maintains state/federal regulatory changes the entire year for all 50 states. In fact, this systems approach usually knows when a regulation is coming down the pipeline well ahead of the actual law. Plant Sentry™ makes lists of regulatory items for multistate shipments with revolving plant inventory, identifying matches between plant name and regulatory information. This model eliminates extra steps in the already complicated area of shipping a good plant.

Lists of regulatory items for multistate shipments:

- Plant care prior to shipment.
- Multiple plant inspections prior to staging.
- Greater focus on final plant inspection and secondary treatments.
- Attentive detail to plant loading and packing.
- Placing correct regulatory documents with the plant article.
- Final shipment check- seal packaging and close transport doors.

Plant Sentry™ generates an easy list to be taken to state or federal officials creating a winning situation for both grower and inspector. Besides, it is impossible for a regulatory official to know a grower's inventory. It is also the responsibility of the grower to know the regulations. Other Plant Sentry™ advantages exist to help improve plant health and quality. Some examples may include directing growers to areas of need in order to treat plants timely, or provide them with the capability to eliminate non-compliant and invasive plant shipments before product purchase. Plant Sentry™ also creates a positive atmosphere between growers, officials, third parties, and consumers because all would know that due diligence was taken to prevent problems from starting. It also archives critical records that prove a grower is verifying.

Plant Sentry™ process is simple. The systems approach model receives a grower's inventory, plant sizes, and types (states of origin/destination) and does a thorough botanic nomenclature clean-up to ensure that the plants in question can be verified. This is a huge benefit to growers because it double checks that they are representing plants correctly, especially when dealing with branded items.

After this is complete, we send back any differences to be sure we are talking about the same plant. Once this is complete, we are ready to cross this information with the 50 states regulations maintained within the Plant Sentry™ database. A beautiful restrictions report summary is generated that a grower can review and take to the regulatory official to be sure of obtaining the necessary compliance to ship to point A or B. The third basic step is use of the portal tool that can be used by marketing, sales forces, production, receiving, and shipping areas to maximize correct sales and shipment, keeping consumer confidence high. The Plant Sentry™ tool can be used manually and also has API capabilities that can tie into inventory systems and send verification upon request. This is another beauty of systems approach — multiple methods to create success in all areas of an operation! The program also offers emergency response in the accidental shipment of a pest, disease, or invasive plant. The timely and organized approach that is used puts fires out before they can get out of control and damage a grower's image.

Since Plant Sentry™ eases compliance for an operation, it allows for use of other predictor models, aiding in production of non-invasive, pest- and disease-resistant plants. It allows focus that can improve practices and opens up other time to complete other certification programs like SANC/MPS and view efforts by invasive species organizations like Plant Right® Pre-Tool. The verification seal is also a signal that Plant Sentry™ and its users are here to end bad plant shipments (Fig. 3).

Figure 3. Plant Sentry™ verified shipment seal.

QUESTIONS COMMONLY ASKED ABOUT SYSTEMS APPROACH

Three questions commonly are asked about systems approach.

- How will it make you more profitable?
- Why should you adopt a systems approach philosophy?
- What steps does it take to "certify" profitability?

The How Profitability?

- It helps grow and ship a healthy quality product leading to return customers = greater profits.
    - SANC trains staff eyes for constant vigilance through scouting, and audits the processes necessary to keep product quality high.
    - Problems avoided or caught early mean money saved.
- Helps develop your workforce into exceptional leaders.
    - Your product can only be great if you have exceptional people growing it and have them focus on what they do best.
- It places a solid team around your company name and product in a tough competitive environment (inside personnel and outside-TEAM concept).
- Plant Sentry™ eliminates the hassle of figuring out what you can ship where, allowing your staff to focus on SANC, Best Practices and your #1 asset, employees.
- Plant Sentry™/SANC will help you eliminate unnecessary crop inputs and labor, redirecting it into positive gains.

The Why to Adopt?

Systems approach programs will be recognized by all states and most likely our North American neighbors.

The What Certifying Results?

Systems approaches hold you accountable by multiple avenues. The auditing that is complete using this model ensures that processes have been completed. Multiple methods that include internal verifying power (learning and training), external certifying power (optional oversight) and external certifying authority (national and state" Help" line) have been time tested to generate extraordinary results. The military has used similar models to ensure success in the field. This success is documented to improve upon in subsequent operations. The systems approach way of certifying boils down to simple documentation. Are you doing what you say you are doing? Both systems approach models discussed are an unbeatable combination if you want to stay at the top of your game.

Propagation of Chautauqua Oaks at Niagara College

Mary Jane Clark, Melissa Drake, Daynan Lepore, Rain Avila, and Elias Abraham

School of Environmental and Horticultural Studies, Niagara College, Niagara-on-the-Lake, Ontario, Canada

marclark@niagaracollege.ca

Keywords: White oak seed propagation, *Quercus alba*, root system development, RootSmart™ propagation trays

INTRODUCTION

The Chautauqua Oak Propagation Research Project aims to evaluate nursery propagation containers and production practices for growing oak trees with healthy root systems. During the propagation phase of growth in the Niagara College Greenhouse, RootSmart™ propagation trays were evaluated in combination with fertilizer and watering strategies to optimize white oak seedling root growth and survival.

MATERIALS AND METHODS

White oak (*Quercus alba*) were collected by Niagara College students from oak trees in the Chautauqua community in September 2018. Acorns were kept moist and planted shortly after collection into Ellepot-filled RootSmart™ trays in the Niagara College Greenhouse and grown on the propagation bench. To all plants in each tray, one fertilizer rate and one watering method was applied, totaling 6 treatment combinations as follows: Fertilizer 1 (F1): Osmocote Bloom 12–7–18, 5–6 month, 2 $g \cdot L^{-1}$; Fertilizer 2 (F2): Osmocote Bloom 12–7–18, 5–6 month, 4 $g \cdot L^{-1}$; Water 1 (W1): Overhead mist watering, 30 sec every 2 h.; Water 2 (W2): Water applied by hand, ½ capacity of an Ellepot; Water 3 (W3): Water applied by hand, full capacity of an Ellepot.

Plant height was measured at the completion of the propagation phase of growth (17 Jan 2019) and five representative plants per treatment had the growing media washed from the roots to evaluate root dry weight and architecture. Plant survival (n=50) was evaluated after a 3-month (January — April, 2019) (5 °C) cold period in a climate-controlled chamber.

RESULTS AND DISCUSSION

After the propagation phase of growth, on 17 January 2019, white oak seedling shoot height showed differences among treatments. The average height of F2 W3 was the tallest, followed by F1 W3 and F2 W2 (Fig. 1). Plant height responded primarily to watering treatments, followed by fertilizer rate. Seedlings preferred full capacity watering (W3), compared to watering to half capacity (W2) or overhead mist watering (W1).

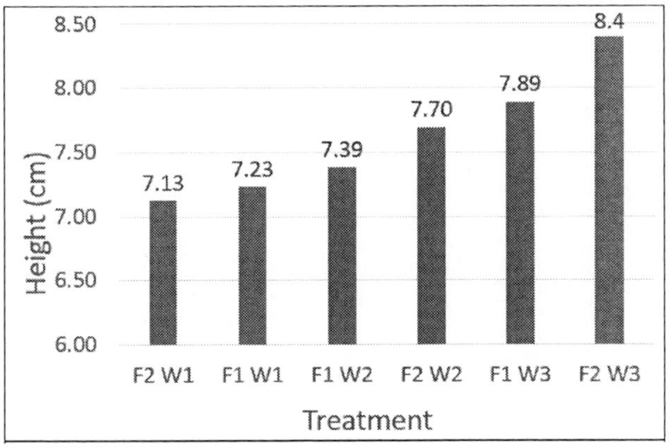

Figure 1. Shoot height among fertilizer and watering method treatments, evaluated on 17 January 2019.

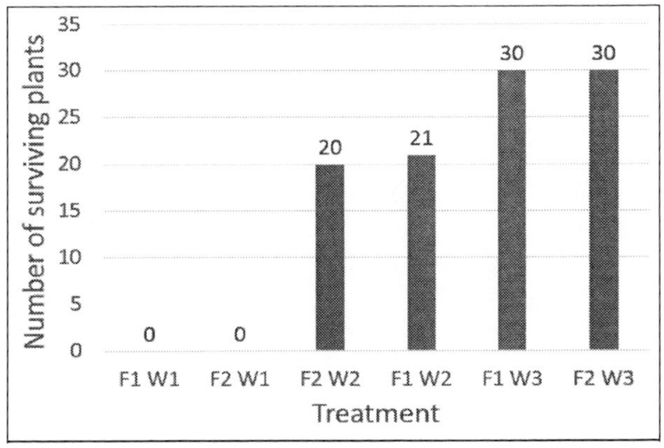

Figure 2. Number of surviving white oak plants among fertilizer and watering method treatments following a 3-month cold period.

Survival rate of seedlings clearly showed the impact of watering method during propagation (Fig. 2). Seedlings in the W3 watering method had 60% survival, while lower survival (40%–42%) occurred for seedlings in the W2 watering method. Overhead mist watering (W1) was insufficient to ensure survival of any white oak seedlings, evaluated after the 3-month cold period. Fertilizer rate did not impact survival of white oak seedlings.

Fertilizer rates had minimal impact on root dry weight; therefore, plants were combined per watering method for evaluation. Root dry weight of seedlings in both watering method W3 (0.928 g) and W2 (0.849 g) were significantly greater than W1 (0.627 g).

Figure 3 clearly shows the difference in root size and architecture among treatments for one representative white oak seedling per treatment.

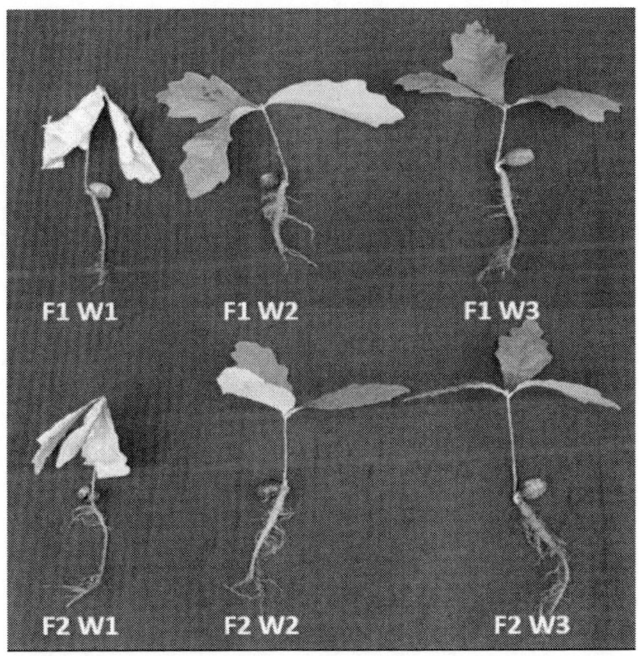

Figure 3. Comparison of root architecture and shoot growth among fertilizer and watering method treatments.

Acknowledgements Thank you to our research partners: AMA Horticulture, ICL Specialty Fertilizers and the community of Chautauqua.

Are You Prepared for The Future? Succession Planning Panel Discussion Points

Kathy McGinty

North Creek Nurseries, North Creek Road, Landenberg, Pennsylvania 19350 U.S.A.

kathy@northcreeknurseries.com

Keywords: Succession planning, nursery industry, business planning, strategic planning

PANEL DISCUSSION OUTLINE:

I. Company Organization

A. Legal Document Review — Mitigate Risk
   1. Articles of Incorporation, Charter, Bylaws
   2. Governing Body (Executive Committee, Board of Directors)
   3. Insurance Policies (D&O — are right people covered)
   4. Non-Compete v Non-Solicitation Agreements

B. Office Operating Procedure Manual — Need Important Tasks Documented
   1. Examples:
      a) Import Permit Application/Renewal/Amendment
      b) Office Alarm Codes & Error Messages
      c) Credit Card Processing
      d) Payroll Processing
      e) Sales Tax Remittance Procedure

C. Financial Systems/Assets
   1. Bank Accounts and Passwords, including contact information
      a) Will also be included in succession plan
   2. Credit References
   3. Asset Inventory (also useful for Insurance Company)
   4. Contractor Records & Processing Procedures, including W-9s

D. Employee Manual
   1. Ensure Annual Review of Policies (keep up to date)

E. Emergency Procedure Manual

1. Key Personnel Designation
2. Utilities Information (heat, electric, gas, water, phones)
3. How to run company systems (boilers, irrigation, temp alarms, computers)
4. First Responders & First Aid/CPR Trained Employees
5. Disaster Preparedness (e.g., assembly points)
6. Emergency Phone Tree

II. Succession Plan

A. What is the Plan?

1. Take control of inevitable! Every business owner will leave the business. If no planning is done — lawyers and government will control the process. Do you want that?
2. Options
    a) Turn Over to Next Generation
    b) Take in Partners/Sell to Outsider with Buy Out Plan
    c) Sell to Employee Stock Ownership Plan (ESOP) (or partial sell)
    d) Management Buy-out (or part management/part ESOP)
    e) Liquidation
3. Considerations in Decision
    a) A good plan MUST begin with clearly defined goals of the owner — the question should be "What do I want for my future, my family and my business?" rather than "How can I avoid paying taxes?"
    b) If owner plans for an orderly transfer, can reduce taxes paid, get maximum value out of business, leave business in hands of chosen successors and avoid family/business crisis (a quick, unplanned transfer is crisis management and not good for any one)
    c) There are many alternatives/tools to use to achieve a successful transition and the desired results for the owner
4. Procedures and Conditions
    a) When succession plan takes effect, communication to leadership (Board, Officers, Executive Committee/Key Personnel, etc.)
    b) Communications Plan
    c) Approval Process (if required)
5. Key Information in One Location
    a) Bank Accounts and Passwords (including contact information)
    b) Tax ID #s; Electronic Payment System Information; Records
    c) All Insurances (including policy numbers, renewal dates, contact information, who and/or what covered)
    d) Supplier and Vendor Contacts (e.g., website, IT/data service, PayPal, telephone/cell phone, credit card companies)

B. Leadership

1. Owner(s) need to make succession planning a priority — shows management and key employees the company/their future is secure

2. Owner's goal/succession plan needs to be shared with management/key employees who have been identified as leaders — provide data-driven direction and understanding
3. Groom/develop and engage your outstanding leaders for succession, which may include new positions/responsibilities; include them in the process; reward employees with stock ownership; determine if outside employees needed

C. Knowledge

1. How do you share your outgoing staff's knowledge with others?
    a) Standard Operating Procedures or Standards Documents
    b) Mentoring/Coaching
    c) Artificial Intelligence (AI)
    d) Written Manual
    e) Training Tools

III. Strategic Planning

1. For strategy to work, need critical support of top leadership, and any Board or Executive Committee support, when formulating, executing and adjusting succession plans
   NOTE: particularly for HR who typically formulate succession plan, as well as other top management who will be remaining at the company
2. Determine What Positions Included in the Succession Plan
    a) how and when to communicate if a position is involved in the plan
    b) how to gain buy-in from top leadership and any Board or Executive Committee throughout the process
3. Communication of Succession Plan (to staff, customers, vendors, etc.)
4. How/When to Begin the Transfer of Duties — when to take the back seat

Succession-Planning Panel Discussion

Brian M. Decker

Decker's Nursery, 6239 Rager Road, Groveport, Ohio 43215 U.S.A.

bdecker@deckersnursery.com

Keywords: Succession planning, nursery industry, business planning, strategic planning

INTRODUCTION

In advance of this panel discussion, we were presented with an outline for "Are You Prepared for The Future? Succession Planning Panel Discussion Points".

We realized that we were lacking in some of the emergency programs / policies / files that might be needed. We had our Special Projects Director, Pam Dukes, review the IPPS outline and develop a Master Company Organization File on a shared Network Drive that addresses these issues.

After significant thought and review of the options for succession at Decker Nursery, I decided the best option was for the company to continue in the event I am no longer available. This is facilitated by the fact that a solid management team is in place, 100% of the stock is owned within the family, and the real estate is owned separately with LLC corporations. Continuing operations would avoid a panic sale and asset loss.

In order to prepare for this transition a CEO had to be designated. In addition, it was determined to develop an Advisory Board to meet two times per year to review the status of the company. The Advisory Board would consist of:

- Owners/family members.
- Nursery and/or business advisors.
- Banking/accounting professionals.
- Company CEO.

The tasks of this Advisory Board are to review budgets, forecasts, and proposed capital expenditures for the upcoming year presented by the CEO. In addition, the Advisory Board would review company long and short-term debt, loans to asset ratios, and working capital/cash flow availability. Without this Advisory Board and the banking representative member, banks could become fearful in the event of a change of leadership and critical operating line of credit could be reduced or removed. The Advisory Board would also advise on Stockholder macro decisions.

In addition, a personal estate plan for the owner should be developed as it relates to the business. This would address future company stock ownership and next generation distribution. It would also address Real Estate ownership and leasing.

Succession Planning Creek Hill Style

Ron Strasko

Creek Hill Nursery, 17 W Main St, Leola, Pennsylvania 17540 U.S.A.

ron@creekhillnursery.com

Keywords: Succession planning, nursery industry, business planning, strategic planning

INTRODUCTION

In 1992, after working in the horticultural industry for 15 years I decided to start my own business, a small perennial plant liner production company. Creek Hill Nursery has grown to be a corporation with five locations and 60 employees. I'm now engaged in the process of stepping back and passing the baton of ownership to a new entity.

As you look at your handout you can see that there are many options to sell your business. I am choosing to pass my business to my son. Ross has been working at Creek Hill Nursery for the past 7 years. He started as a grower in the greenhouses and has worked his way up to vice-president. He has learned the ins and outs of the business from the daily workings to future financial planning.

How do we make a smooth transition that is fair to all concerned parties? In our case my wife and I, Ross, his sister and the Creek Hill Nursery employees?

We decided to enlist the help of a financial advisory group that handles succession planning. We interviewed four different companies and then selected the one that we felt fit best with our goals and personalities.

They in turn gave us a selection of businesses that could evaluate our worth. We picked the one that we felt would give us a fair value. Don't look for the cheapest evaluation — they are not all equal. Do ask if they have had experience in evaluating businesses in the horticultural industry.

There are two factors to look at — what is the value of the business itself and what is the value of the real estate. Once that is established my wife and I will have to determine how much per month we would need to live a comfortable life. This raises many questions, and the answers will be different for every owner; what does that really mean? It can include travel, dining out, giving to charities, education funding for grandchildren, etc.? What legacy do I want to leave? And also, how far out do you plan for? 95? 100? As we age will there be health problems? Long term care?

On the other hand, what can Creek Hill Nursery afford to pay and still be a thriving business? We want to live well but not drain the life of the business. It's important while making these decisions that everyone involved is aware of your decisions and how and why they were made. So that, in the long run there are no surprises to anyone concerned.

We have been working on this for a few months now, but we aren't there yet. I expect to slowly relinquish my responsibilities to Ross and others, so that in another 2 years I can be out of the business. Of course, I will be available in an advisory capacity.

It is hard to give up control but as I see Ross bringing in younger people with new ideas, I have confidence in his ability to continue and improve Creek Hill Nursery. I know that he will make mistakes, but I've made plenty and Creek Hill Nursery is still thriving.

Are You Prepared for The Future?

David Thompson

Foxborough Nursery, Inc, 3611 Miller Road, Street, Maryland 21154 U.S.A.

dthompson@foxboroughnursery.com

Keywords: Succession planning, nursery industry, business planning, strategic planning

INTRODUCTION

My part of this panel is to talk about the Succession Planning for a family business. I started the company, Foxborough Nursery along with my wife, Marilyn in the Fall of 1977 with the thought of eventually changing our business model within 10 years. My primary goal was to start a propagation production company selling liners and continue my landscaping business to help finance this goal. Once the nursery was up and running, we would phase out the landscaping division. At this time, this was a major undertaking and it took and incredible amount of my time and energy. I had developed a business plan while at college, University of Maryland, with the help of my advisor, Dr. Robert Baker. Succession planning really starts when an individual decides to make these career choices.

It wasn't until years later in the late 1980s that the true development of our succession plan really began. During this time, I was fortunate to be a member of the Advisory Board in the Department of Horticulture at the University of Maryland. A fellow board member, an older gentleman, ask me one day what plans I had for the future and who was carrying on the legacy of Foxborough Nursery. He suggested that I find some way to get my two very young sons interested, he recommended starting a small plant business just for them. This was accomplished by starting a cut branch business for the floral market. It would be their company with the help from us till they were old enough to manage it on their own. They would see the reward of money given to them when they gave the customer the product they grew and cut. At this point, Foxborough Nursery's true succession plan really began. It was important to let my son's see this side of a business that requires making business decisions, working long and hard hours and realizing the income that you will make when you are committed.

As years passed by, my sons grew up, graduated from different colleges in Horticulture, came back to the Nursery to work. Our business model changed, for the better, which changed our future. The oldest son wanted to grow and sell plants on the wholesale market, the youngest wanted to run and expand the landscape business. It was at this time we split off Foxborough Nursery Inc. (Wholesale Division) and formed Foxborough Inc. (Landscape Division). Now we have two companies that will require a succession plan for each.

With any type of family business or businesses, you can run into many obstacles especially if you start your business where you live. Unfortunately, that is what we did, we built these companies around our home on the family farm we live on. Knowing this now and not when we started, I would greatly emphasize that any serious business plan should avoid this potential problem. A true commitment from the next generation must be made for these Corporations to move into the future. Company evaluations must be made continuously, yearly meetings with accountants, lawyers and financial planners are critical.

Many components for a successful transition or succession plan for a family business are involved and they require insight and input from all of those that are family partners. Those components consist of what does the next generation want to do, what do we think their children will want to do, how can we keep the family farms in the family, how can we grow these companies and be profitable and productive year to year.

Succession planning for a family business is more than having a successful plan, it is knowing where you are at this point in your career, where you see the next generation going and how will they achieve this and still have a family business. All these factors must be decided on by the family and by outside consultants that can help direct the future of what we have started.

Using Soil Water Sensors to Evaluate Plant Available Water in Engineered Landscape Soils

Kevin Donnelly

Midwest Trading Horticultural Supplies Inc., 48W805 Illinois-64, Maple Park, Illinois 60151 U.S.A.

kdonnelly@midwest-trading.com

Keywords: Water availability, engineered soils, water sensors

INTRODUCTION

The performance of plants is subject to many variables. One key aspect is the availability of water in soil. The team at Midwest Trading's Center for Horticultural Soils Testing and Research have begun a long-term research program investigating the soil water interactions in horticultural soil, specifically for use in container production and landscapes. In soil science, plant available water can be characterized through a moisture release curve that plots moisture content with matric potential. Volumetric moisture content is a measurement of the total volume of water held in a soil. Matric potential is a measurement of how tightly the water is held to the soil surface. Unlike naturally occurring mineral soils, container substrates and engineered landscape soils can vary drastically from one project to another due to site conditions, raw material inputs and recipe.

There is limited understanding of plant available water in these systems, however, there is an opportunity with engineered soils to tailor soil blends to meet project requirements.

This research is the first step in an extensive research program to better understand site conditions and plant available water characteristics of engineered soils. The objectives were:

- Evaluate use of moisture and matric potential sensors.
- Determine how sensors can be used as a tool to automate irrigation systems.
- Use plant available water data to inform engineered soil design.

MATERIALS AND METHODS

Site Selection

This research was conducted at the Gardens at Ball Horticultural Company in West Chicago IL. This is a display garden used to showcase varieties offered through Ball Seed. Two sites were selected that contained the same engineered landscape soil but were under different site conditions. Both are in raised beds with drainage tile and are on drip irrigation.

Site 1 is a perennial display bed in the shade near a creek that is prone to flooding. Site 2 is a vegetable display bed at higher ground and in full sun (Fig. 1).

Figure 1. Site 1 (left) is a perennial garden plot; Site 2 (right) is vegetable garden plot.

Soil and Sensors

Soil: The soil for each site was PM-35 provided by Midwest Trading. It is a blend of coarse pine back, loam topsoil, sand, and compost.

***Sensors*:** Sensors for the sites were sourced from Meter Group, Inc of Pullman, Washington. For each site, two Teros 21 and Teros 12 sensors were installed together at depths of 4 in. and 8 in. The T8 sensor was placed at 6-in. depth. Cables were run through garden hose and PVC to protect against damage. The five sensors were connected to a ZL6 data logger. Measurements were taken at 5 min intervals and uploaded to a Zentra Cloud platform.

The system was installed May 24 and data was collected through Oct 1st. Teros 21 reports matric potential by measuring the moisture content in a ceramic disk with a known moisture release curve to convert volumetric moisture to matric potential.

The T8 sensor measures matric potential through a sealed column of water with a permeable ceramic cap that allows water to be pulled through creating a tension measurement. Teros 12 measures volumetric moisture content, temperature, and EC.

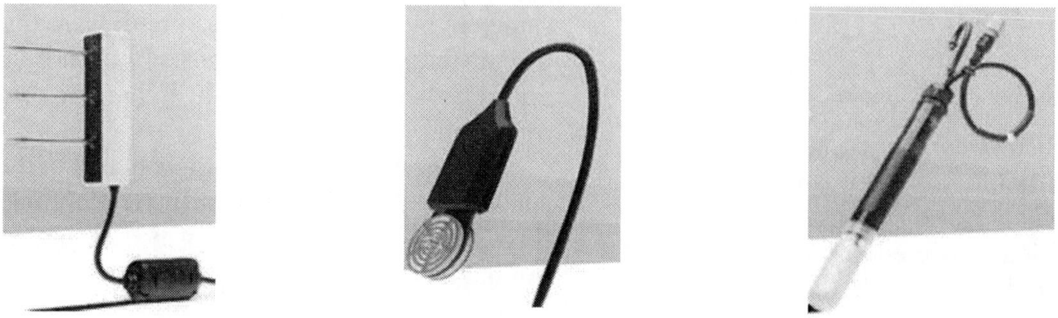

Teros 12. Moisture, Temperature, EC

Teros 21. Matric Potential, Temperature

T8 Matric Potential

Figure 2. Sensors from Meter Group, Inc of Pullman, Washington (www.metergroup.com).

RESULTS

At time of publication, preliminary analysis was conducted on data sets collected during the course of trial. Figure 3 represents data from Teros 21 matric potential sensor and T12 volumetric moisture sensor. Data is displayed in a scatter plot of the volumetric moisture compared to matric potential. This displays one of the limitations of the Teros 21 showing full saturation at 10 kPa, where the T8 tensiometer will measure at or near 0kPa. The data set for the scatter plot is from both sites and shows water being abundantly available from 50% to 35% moisture content. The crops would start to see some water stress below 35% moisture content when the matric potential increase past 40-60kPa.

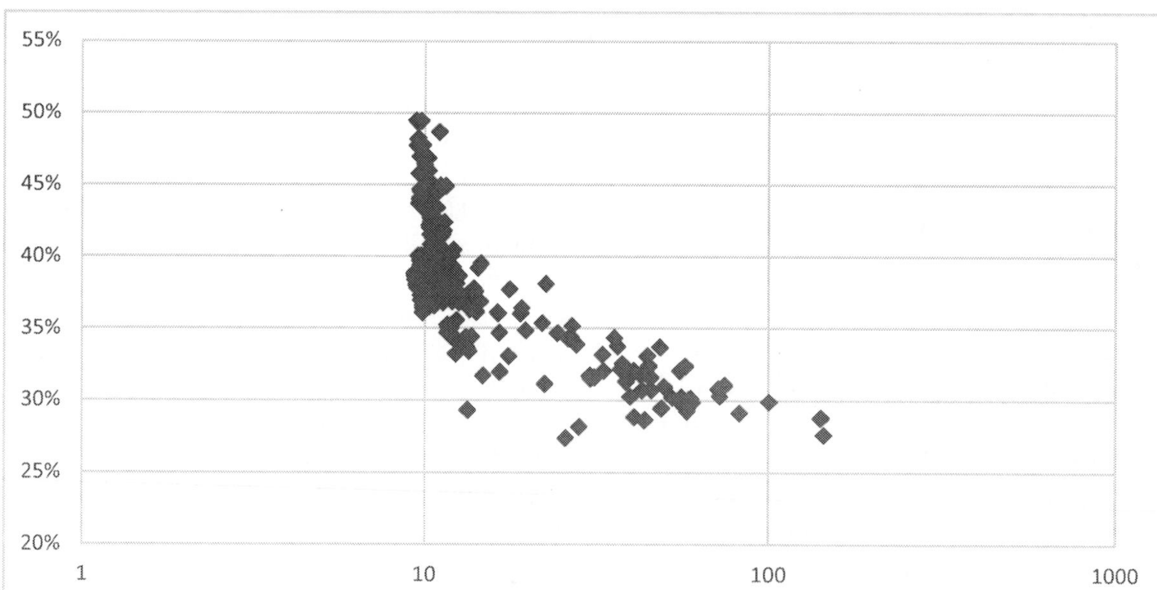

Figure 3. PM35 moisture release curve.

With two sites under different exposure conditions we were able to see differences in soil temperature between sites and depths. Plotted in Figure 4 is a 4-day span showing the temperatures in °F on a 24 h cycle. The shallow depth in the sun had dramatic swings in temperature of 10–14 °F during the day. Deeper soils and those in the shade had much more moderate soil temperatures. The sun site is open and more exposed to heat loss during the night resulting in lower night temps for 4 in. vs 8 in. depths.

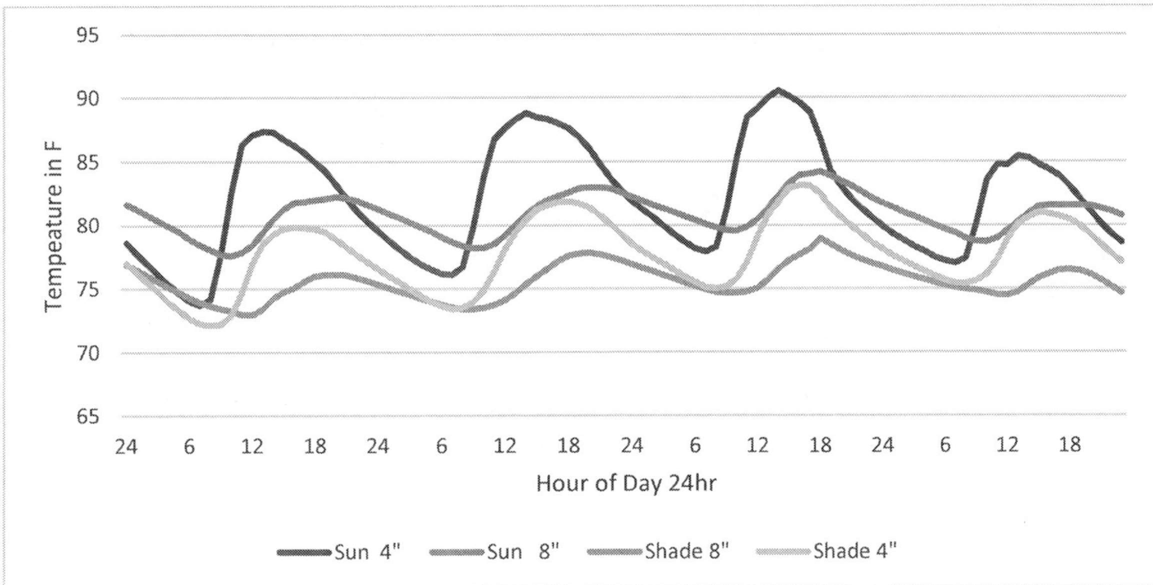

Figure 4. Soil temperature in shade and sun.

First Season Feedback and Next Steps

There were a lot of lessons learned through this first season of field research using these sensors.

- 2019 was a record setting year for rain fall and the soils did not have much time to dry down; this limited the high kPa measurements that could help interpretation of the systems and substrate. Future research will be needed at these sites and in more controlled environments where dry down can be properly tested.
- Usability of sensors for integrating into irrigation controls was one of the main objectives of the study. For a display garden where irrigation is frequent, it is likely a moisture sensor would provide better reactivity compared to a matric potential sensor. Matric potential sensors could be better applied to cropping systems where the soil or the crops will experience more water stress.
- The EC values fluctuated (data not presented) quite a bit from rain events and fertigation through the drip line. This measurement is of interest for the site to better monitor fertility of crops to ensure performance over the summer.
- The sensors were placed at different depths and different places in the beds. With the beds being on drip irrigation, it is possible that there was localized var-

iation based on the proximity of the sensors to the drip line. This would need to be considered in future seasons

- The measurement interval for this research was set at 5min. This interval provided 600,000 data points which can be a challenge to run analysis on. The sensitivity of the measurements does not need to be as detailed. An interval of 15-30 min should be sufficient in the future.
- More analysis is required of the data set and we will be looking into:
  - Correlation of soil EC readings, fertigation of beds and performance of plant material.
  - Identifying conditions that caused water stress.
  - Mapping the impact of temperature on dry down patterns.

CONCLUSION

This research has promising use for the industry, specifically engineered soil applications. We will continue to work with this sensor technology to evaluate different sites and different types of soils. This will allow us to look at how different raw ingredients impact the functionality of the soil and will influence recipe design to optimize water use.

Effects of Ecological Restoration Decisions on Native Plant Horticulture

Elliott Duemler and Corrine Daniels

Applied Ecological Services &Taylor Creek Restoration Nurseries, 17921 Smith Road, Brodhead, Wisconsin 53520 U.S.A.

Elliott.duemler@appliedeco.com

Keywords: Ecological restoration, native plants, ecology

INTRODUCTION

Applied Ecological Services, Inc. (AES) is a 40-year-old ecological consulting firm whose vision is one of bringing the science of ecology to all land use decisions. We bring to this mission a diverse consulting staff of ecologists, botanists, ornithologists, engineers, landscape architects, GIS specialists and other science specialists. Our experienced field contracting services personnel install and manage ecological restorations in a variety of ecosystems, adapting to challenges. in addition, since the firm's inception, success of these efforts has relied on the growth of our robust native plant nursery, Taylor Creek Restoration Nurseries (TCRN).

We learned, developed and refined our restoration process in Wisconsin and Illinois. And we are continually translating this experience to other locations. Although cultures vary, the ecological restoration and native plant horticulture principles we have developed can be applied throughout the world as demonstrated by work AES has done in such places as South America, Romania, Alaska, and China.

It is this synergy of consulting, contracting field services, and native plant nursery that has uniquely positioned AES to examine the ways ecological restoration decisions act as driving forces for innovations in native plant horticulture. In response to ecological challenges presented by our consultants and contractors, TCRN has continually solved the emerging challenges inherent in acquiring genetically appropriate propagule stock, producing and managing adequate volume native seeds and plants, matching species' traits to intended field uses, and storing and transporting viable products in a timely manner to far-flung project locations.

AES takes a long ecological view with all its ecological restorations. Our mission commits us to the use of wild genotypes (rather than cultivars) that can persist and adapt over time, knitting together intact, sustainable restored ecosystems. This commitment — not only to native species but also to regional genotypes — has allowed us to push the envelope of native plant horticulture at all stages of production.

METHODS AND DISCUSSION

In this section we describe seminal ecological land use and restoration projects AES has conducted in diverse landscapes (e.g., agricultural, urban, natural areas, restored lands). Without the 40 years of widely varied AES projects and the demands each presented, the innovations and improvements in native plant horticulture experienced by Taylor Creek Restoration Nurseries would not have occurred. Truly, we have seen how "necessity is the mother of invention."

We describe some of our "game-changing" projects that have translated to innovations in the techniques of both ecological restoration and native plant horticulture. We begin by summarizing challenges and then, in the context of the projects, describe innovations that produced successful project implementation.

Challenges of native plant horticulture derived from our experience in ecological restoration include:

1) Baseline assessments to accurately characterize restoration actions and goals and identify target species needed for successful restoration.
2) Creation of a restoration palette with sufficient species diversity to form viable, sustainable ecosystems.
3) Selection of plant species adapted to specific conditions (e.g., toxins and heavy metals, droughty soils, fire management, competition with invasive species).
4) Nursery propagation of rare species with unknown growth specifications.
5) Identification of geographic propagule sources for species that are genetically representative of target restoration location.
6) Identification of most appropriate propagules (e.g., seed, tuber, rhizome, plant plug) for successful installation and growth of each species.
7) Design of propagation schedules that complement projected planting schedules.
8) Need for remote nursery production for large-scale projects distant from Taylor Creek Restoration Nurseries.
9) Propagation, processing and handling of seeds and plants in sufficient volume in a timely fashion to meet scheduled demands.
10) Concerns about nursery production costs as well as client procurement costs.
11) Capacity for storage as needed to accommodate staggered production and scheduling and potential delays
12) Secure transport of propagules to restoration installation site and viable staging on site.

The following projects were instrumental in the creation and evolution of Taylor Creek Restoration Nurseries.

Flambeau Copper Mine Restoration (Wisconsin)

AES participated on a team that designed the backfill of an open pit copper mine in northwestern Wisconsin. The main ecological challenge was one of identifying plant species that could tolerate and grow on a substrate of ground-up rock that originated from the mine along with high concentrations of heavy metals. Droughty conditions and low nutrient concentrations due to limited organic soil were also prime limitations for species selection. Once suitable species were identified, seed sources were identified and propagules produced and installed to create a self-sustaining, soil-producing restored ecosystem.

Prairie Crossing Conservation Development (Grays Lake, Illinois)

Acclaimed in National Geographic, New York Times, Wall Street Journal, and honored with more than a dozen conservation awards over the past two decades, Prairie Crossing is truly a national model of conservation development (Fig. 1). The project was designed with a vision of maintaining agrarian, ecological and historical connections to the landscape.

Figure 1. Prairie Crossing, Grayslake, Illinois with homes over wetlands.

One of the pioneering, and much-studied, strategies of the native landscape at Prairie Crossing is the treatment of stormwater runoff via the first major installation of the AES Stormwater Treatment Train^{TM}, a system composed of open swale stormwater conveyance, upland prairie biofiltration, wetlands and a man-made lake. Working in combination, these native landscapes and features are highly efficient at pollutant removal through biological and mechanical means while significantly reducing the volume and flow rate of stormwater runoff.

Project plant species were selected for considerations of aesthetics, biodiversity, ongoing maintenance costs, and resiliency for stormwater management. AES restored over 200 acres of prairie, wetland and agricultural hedgerow communities. AES continues to maintain the health of ecological communities on an annual basis.

Rare Ecosystem Restoration on Albany, New York Landfill Site

AES worked collaboratively with government and community stakeholders on the design, permitting, construction and monitoring phases of a habitat restoration and wetland mitigation to restore globally imperiled pine bush community. Furthermore, the project was designed to help connect fragmented areas of the Albany Pine Bush Preserve, a significant natural area. The mitigation plan also included two stream relocations, creation of a biofilter wetland complex and protection of a pine barrens vernal pond community.

The entire restored site includes 55 acres of hardwood forest converted to high quality oak savanna and enhancement of 35 acres of forested wetland. The project also involved sand placement and planting on closed portions of the landfill leading to final landfill closure.

Particularly noteworthy was the construction a 5-acre test plot on a closed portion of the landfill to test adaptability of selected plant species for closure conditions, and the creation of a 3-acre native plant nursery used to supply plant products for the project.

Parkland Development on Former Airbase (Northerly Island)

Applied Ecological Services was retained to assist the City of Chicago to design a beneficial re-use plan for the old Meigs Field Airport, on the coast of Lake Michigan in downtown Chicago. AES partnered with several other Chicago-area and regional firms for the planning, public process, technical design and engineering.

Shoreline stabilization of the lake fill zone combined with stormwater management via biofilters and dune restoration placed new demands on native plant production by TCRN. AES specifically focused on shoreline stabilization of the lake fill zone — recontouring needs for the lake fill to create native restored ecosystems that used to occupy parts of the historic island, which had been expanded to create the airfield. For the new park, we designed the stormwater management system, conceptual grading plans, restoration plans, phasing plans, and prepared a regulatory environmental assessment and permit documents.

The AES team prepared technical strategies for stabilizing the island perimeter and lake fill. The decommissioning and removal of other costly airport infrastructure were also a significant challenge. Water quality regulatory challenges were addressed through ecological design and the use of stormwater biofilters which integrated nicely with the ultimate plans for wetland restoration, reducing the costly need to do additional excavation of sediments in existing biofilter areas.

A comprehensive park plan was designed – integrating passive recreation and access and active recreation with the conservation, ecosystem restoration and water quality design systems in the new park plans. The restoration was also focused on bringing back some of the rare coastal wetlands, dunal systems, and various types of migratory bird habitats that have otherwise been eliminated on the Chicago landscape.

All told, Taylor Creek Restoration Nurseries produced more than 350,000 native, local-ecotype plants for what has become one of the largest native plant restorations in Chicago history (Fig. 2).

Figure 2. Northerly Island image showing native, local-ecotype flowering plants.

Chicago's Jackson Park Ecosystem Restoration

Chicago's Jackson Park, designed by Frederick Law Olmsted as the centerpiece of the 1893 World's Columbian Exposition, is undergoing a holistic landscape renewal as part of a five-year, multimillion-dollar restoration led and funded by the U.S. Army Corps of Engineers and the Chicago Park District in partnership with non-profit Chicago Project 120 (Fig. 3).

Figure 3. Chicago's Jackson Park restoration with native wildflowers and park bridge in the background.

AES was selected to lead the project as general contractor because of its extensive ecological landscape expertise. AES re-created and restored natural features (e.g., lagoons, trails, and a historic feature, the Wooded Island) to promote pedestrian circulation throughout the park, provide aesthetic features, and enhance natural habitat. Their restoration work commenced with brushing and herbicide treatments to remove invasive plants followed by replanting the park with more than 500,000 native herbaceous plants and 11,000 trees and shrubs. Eroded lagoon shorelines were rehabilitated with grading followed by planting with 90,000 carefully selected wetland plant species. Reconfigured and refurbished dunes along Lakeshore Drive were subsequently planted with specialized Great Lakes dune vegetation. Maintenance—involving invasive plant control, mowing and burning—are ongoing.

Taylor Creek Nurseries staff worked hand-in-glove to meet U.S. Army Corps specifications for plant material. In recognition of AES plant propagation skills, USACE botanists purposefully expanded the restoration palette, including difficult native species and genotypes not typically propagated for restoration. The specific project palette, as well as production capacity, were major horticultural challenges.

SUMMARY

It is the very variety and complexity and diverse geographical locations of AES ecological land restoration projects that have driven the innovative growth in native plant horticulture that today characterizes Taylor Creek Restoration Nurseries.

In more than 30 years, Taylor Creek Restoration Nurseries has grown from a small operation growing a few thousand plants and with seed production beds started literally in the backyard to today's operation which handles more than 2 million plant plugs per year and manages approximately 250 acres of native seed production. But in addition to sheer increased capacity, TCRN has contributed significant innovations in native plant horticulture.

These include:

1) Protocols for successfully scheduling extensive, multi-year landscaping projects.
2) Techniques for creating remote nursery operations for large projects.
3) Adapting commercial shipping racks and containers to native propagules.
4) Development of creative partnerships both as clients and for production.
5) Experimental field trials of plantings to assess plant responses to new conditions such as extreme soil types, temperatures, pollutants, etc.
6) Strategies for propagating "difficult to grow" species (e.g., highly specialized requirements, low fecundity, low disease resistance) including tissue culture.
7) Intermixing of small and large projects in company work flow to maintain a steady client base in spite of unexpected delays and other challenges.
8) Techniques to cope with storage and long-term maintenance of product.
9) Balancing "ecologically ideal" species lists with "horticulturally possible" results.
10) Maintaining genetic diversity of product at all scales.
11) Tracking trends and responding to new markets such as garden centers and non-profits.

We anticipate additional innovative growth in native plant horticulture, continually driven by AES's equally innovative approaches to ecological restoration and land use.

Anatomy of a Gravel Garden

Jeff Epping

Olbrich Botanical Gardens, 3330 Atwood Ave, Madison, Wisconsin 53704 U.S.A.

jepping@cityofmadison.com

Keywords: Gravel gardens, sustainable plant communities, ecologically designed gardens, landscape design

INTRODUCTION

The Gravel Garden Concept

When you conjure up an image of a "gravel garden", it probably isn't very fun and colorful — a sea of hot, dry gravel, some harsh craggy rocks and a handful of plain green, sparsely growing plants come to mind? This image might be right on the money, if we were talking about mountainous alpine rock gardens, but the gravel gardens that we've created here at Olbrich are different...very different.

Our gravel gardens are chock full of tough, lush, and colorful plants that grow harmoniously together to create vibrant plant communities. The only time you see gravel is when we cut them back in spring before the plantings come to life for another season. One of the greatest things about our gardens beyond their beauty is their ease of care and inherent environmental compatibility with our climate. Once established, these sustainable plant communities thrive on natural rainfall alone and require no fertilizer or other chemicals to grow. Given a little time to grow and fill in, they will look similar to traditional perennial gardens, but require about 80% less maintenance.

The gravel garden concept is not new and most keen gardeners might associate it with one of the greatest gardeners of our time, Beth Chatto. Chatto converted a gravel parking lot to a beautiful drought tolerant garden over 25 years ago and wrote a book on the topic in 2000. Her gardens in Essex County, England thrive, despite a meager 20 in. of rain that falls each year. Beth's beautiful garden has influenced many a gardener over the years, me included, and I have hundreds of beautiful photos from my two past visits to show for it.

In recent years, horticulturist Cassian Schmidt, Director of Hermannshoff Garden, located outside of Heidelberg, Germany, has been at the forefront of the gravel garden movement. Hermannshoff is known for its innovative ecologically designed gardens, where functionality and practicality are as important as beauty. Schmidt began experimenting with gravel gardens in the late 1990s, with a goal to create beautiful, low maintenance gardens. I first became aware of his concepts via noted Wisconsin plantsman, Roy Diblik. Roy, a good friend and one of the best gardeners I have ever known, consulted with us on our first gravel garden at Olbrich. He willingly shared his knowledge gained from Cassian upon visiting Hermannshoff with Piet Oudolf and by creating his own gravel garden in 2007, at Northwind Perennial Farms, his retail nursery and garden center near Lake Geneva, Wisconsin.

Creating the Gravel Garden

So, how does it work? First off, we need to select a site with decent soil drainage and relatively sunny exposure. Once we have that, we begin by removing any existing herbaceous vegetation either mechanically or with the aid of a non-selective herbicide like glyphosate. If there are existing trees or shrubs, they can be left in place and incorporated into the gravel garden, just remove any mulch and organic matter and loose soil beneath the plants.

The next step is to create a solid border or hard edge around the perimeter of the garden if one doesn't already exist. Stone, boulders and pavers are often used to make an edge, but the foundation of a building, driveway, existing sidewalk or concrete curb, all make good edgers. Just be sure that the permanent border is tall enough to retain the 4-5" of gravel that will be placed atop the soil – we shoot for at least 6-7 inches.

Now for the gravel — not just any gravel, but a washed ¼ to ⅜ in. quartz or granite chip gravel works best. Every piece of gravel is relatively the same size, so it remains loose and never packs tight — visualize marbles in a jar. We use a local gray colored angular quartzite or rounded pea gravel, but any hard stone (never a softer stone that would break down over time) of similar size is fine. If the planting bed is large, it's a good idea to drive in stakes about every 5–6 ft and make a mark on the stake about 5 in. above the soil line. These stakes will help ensure a consistent gravel depth of 4–5 in. throughout the garden, which is important, since it acts as the barrier to weed seed germination, so we definitely don't want to create shallow gravel areas. The uniform gravel depth almost eliminates weeding the garden in the future.

Once the gravel is in place, you are ready for the fun part — planting! Selecting drought-tolerant, deep rooting species, innately tolerant of our hot, dry Midwest summers is essential. The ideal sized perennial for planting is grown in a 3½-in. to 4½-in. pot, which typically has a root ball depth of 4 in. Larger pots can also be used, but they will be more difficult and time-consuming to plant. Smaller pots or plugs can also be used, but they take longer to root into the soil, therefore making them more vulnerable to drought stress during the establishment period. Plants should be spaced anywhere from 10 to 15 in. on center, depending upon their mature size at maturity. The goal is to cover the gravel with vegetation and create a living, interacting community of plants, so a tighter than usual spacing is best.

When planting, be careful not to spill any potting mix into the gravel, since it may become a potential seed bed in the future. Soil or organic matter build-up in the gravel is our enemy and we always look to keep that from happening with the gravel garden system. No special tools are needed for planting, just a good pair of gloves that will allow you to dig through the gravel and put the root ball in place. When setting the plants, the crown of the plant should be flush or just slightly below the top of the gravel. Roy shared a good tip with us: remove the top inch or so of soil from the root balls before planting. In the process of removing that soil, you are also removing the majority of dormant weed seeds that may germinate after planting. You will get most of the seed, but not all — the weeds we have pulled the first 2 years, which hasn't been many, have been in the crowns of the perennials, not in the gravel between plants.

Once the garden is planted, you will have what looks like a sea of gravel with a few dots of green here and there. Even though you see more gravel than plants the first season or two, don't despair, that quickly changes as the plants grow to maturity. In 2–3 years, you will enjoy a beautiful tapestry of lush, colorful plants.

Caring for your gravel garden is a bit different than tending to your conventional perennial beds and borders. Maintenance will be much easier in the long-run, but getting the garden off to a good start is critical to its long-term success. The key to a good start is water and lots of it. This seems contrary to the whole gravel garden concept, but that requirement changes as the plants root-in, grow and reach maturity and maximum drought tolerance.

Even though the plant species selected for the garden are naturally drought tolerant, they need to root into the soil below the gravel to be that way. Keep in mind that the only moisture available to the plant is in the soil mass that was planted with it. The surrounding gravel has little to no water holding capacity. Think of the plant as still being in the plastic pot that it was initially grown in and water it as if it were. In the beginning, that may mean daily watering during sunny, hot, windy days. As the plants grow and root-in, the watering becomes less frequent. The best way to know when it's time to water is by monitoring the plants daily and water when you see signs of water stress — most often, off-colored and wilting foliage. A good way to make it easy on yourself and your plants is to set-up a semi-permanent irrigation system that can be left in place for the first season. Hoses and impact sprinklers are relatively inexpensive and worth the investment. We do like to water by hand from time to time, but that isn't always possible, so the inexpensive hoses and impact sprinklers are well worth the investment.

Beyond that, the maintenance is limited to an annual clean-up in spring before new growth begins to emerge from the gravel garden plants. This clean-up should be very thorough since any debris left behind will allow organic matter to build-up in the gravel and become a medium for weed seeds to germinate. We cut the plants back with pruners, pick-up the plant material to compost elsewhere and then go over the area with a blower to remove any material we weren't able to pick-up by hand. When you compare the maintenance of the gravel garden with traditional perennial plantings and lawns, well, there's almost no comparison – the gravel garden wins hands down.

Propagation and Production Changes at Johnson's Nursery

Ben French

W180 N6275 Marcy Road, Menomonee Falls, Wisconsin 53051 U.S.A.

bfrench@johnsonsnursery.com

Keywords: Propagation, business challenges, nursery growth, field production, container production, hardy plants, *Quercus*, *Acer*

INTRODUCTION

It's all about opportunity!

In the upcoming years, every company will be facing business challenges, following the trends of just about all the other older industries. Sales at our company are growing year over year since the recession, and our own production can hardly keep up. Some of our local business competition disappeared during the recession.

Industry Trends

High product demand and a small labor pool has forced us into a state of increased efficiency, we must prioritize activities to make the most money with the folks we have. We only take the best work and only produce the plants we're best at growing.

Other local nurseries we deal with have done similarly and we are developing more distinct product lines and methods that separate us and keep our individual values higher than if we were competing in the exact same category.

Another trend that might take a strong hold on our industry would be the "Amazon model" of business, where plant producers can list all of their inventory together and customers can immediately see all of the products and their variations in once spot (online) and make a choice of where to get their goods.

JOHNSON'S NURSERY FOCUS

Johnson's focus traditionally has been larger ball-and-burlap trees and shrubs, and we have a strong presence in that market. But where does a smaller container grower like me fit in?

Johnson's propagation department was founded with the idea that if we couldn't get it, we would grow it ourselves. In the 1980s, the focus was on trees and shrubs that our company had difficulty sourcing.

The 1990s was a time of increased focus on Johnson's specialties. This included many new plants developed by Mike Yanny, who is known for his work on Wisconsin-hardy plants in many genera including *Carpinus, Viburnum,* and *Malus*.

It was in the later years that a container growing portion was added and we began to grow many of the popular container plants used by landscapers. These commodities were certainly easy to produce, but we were only slightly more competitive than buying in a finished plant from one of the huge container plant factories. These include Bailey's and Midwest Groundcovers, which are both located within the Midwest.

By 2010, the propagation department had grown to include the production on perennials, shrubs and trees from quart sized pots to #10 containers. And our field production included whip trees and bare root shrubs produced primarily for our own use in larger product sizes.

Right around this time our company was feeling the pressure of the recession. New development was down, city budgets were trimmed, and many salaries were frozen. Positions were eliminated. Some of our local competition went out of business.

Mike Yanny, essentially the founder of the propagation department at Johnson's, formed his own business designed to remain in close contact with the nursery and handle all the plant selection aspects of the business.

CHANGES SINCE 2015

I was promoted to managing the department in 2015. Since taking over the department there have been significant production changes. More native plant production, increased container tree production, increased interest in grafting specialty plants and production of plants in the winter months.

At Johnson's approximately half of our workforce was laid off for the winter. As it became more difficult to keep good people that would tolerate a 3- or 5-month layoff, an increased need was seen for productive winter work. This need combined with the fact that Wisconsin currently has many tree removals due to emerald ash borer, and that many tree care companies doing the removals also buy from us. This resulted in a pretty big opportunity.

We decided to try and grow some plants in winter. The first crops we decided to try ended up being vegetables and fruits (Fig. 1). The theory behind this was that it was going to be a nice source of revenue and marketing exposure for us in winter. We installed a large wood burner that supplied hot water to four growing structures. Two of these structures were for production of evergreen and hardwood cuttings along with some evergreen grafts. The other two were going to be the fruit and veggie houses, with an emphasis on greenhouse grown, vine-ripened tomatoes. We grew some mean tomatoes! They were selling close to $5 per pound at some local grocers and we had the production schedule designed to supply a steady flow with staggered plantings and ripening times. Our little greenhouses were supplying 190 pounds per week for about 4 months.

Figure 1. Produce grown in our greenhouses during winter.

A financial analysis was done on our little experiment, and we were surprised to learn that our limiting factor was mostly electricity and less than expected demand. Even though we did a great job growing these things, we weren't going to be profitable at the scale we were at. At a larger scale, the models looked better, but the decision was made to halt tomato and vegetable production and go back to our roots — propagating woody ornamentals for our own production. We modeled out growing liners in the greenhouses as a source of savings instead of a source of revenue.

Our nursery had traditionally been an outdoor grower, so there wasn't a great depth of knowledge when it came to greenhouse growing. We decided to focus on plants we had trouble with in the past. Many of these species were small-seeded ones that we've always wanted to grow but were difficult to grow in our outdoor seed beds. These included birch, *Cercidiphyllum* tree, *Viburnum*, *Ginkgo*, *Magnolia*, *Staphylea*, and many perennials. When we added up the value of our ornamental plant starts, it landed very close to a $12,000 savings in yearly plant purchases (Fig. 2).

Other projects include seed grafting of *Aesculus* in the very earliest part of spring (March). Hardwood cuttings collected in January have resulted in some pretty nice plants as well. As we are growing many natives from seed, some stand out with superior production characteristics and we are able to take cuttings and propagate these smaller experimental crops at a slower time of year. This is an important intermediate step in the selection of nativars and taking labor pressure off the main production season.

Figure 2. Winter production of plants we found difficult to grow outside.

Native perennials have sold well for years, but we would typically buy in nearly finished plants and pot them up for early summer sales.

Using some of the space in the greenhouses in winter, we can start perennials that are traditionally difficult for us to find or finish in time for summer sales (Fig. 3).

Figure 3. Examples of native perennials (*Aralia, Arisaema, Amsonia*) started in the greenhouses in winter that are traditionally difficult for us to find or finish in time for summer sales.

Quercus (oaks) have long been an important product for Johnson's Nursery. We have been working to grow oaks with fibrous root systems for nearly 30 years. The idea behind this is to make a more transplantable oak. This starts at the emergence of the radicle on a chitted acorn. We treat them chemically and manually prune the roots at every stage in the nursery. This results with an oak tree with an unnaturally fibrous root mass — resulting in increased transplant success and faster establishment.

Growing our own oaks also allows for us to screen for alkaline tolerance, which is a big problem in Southeast Wisconsin, where our soil pH can be near 8 and sometimes higher. In the case of swamp white oak, we prefer propagating its hybrid form *Quercus* × *shuettei* over the straight species. This hybrid has many of the ornamental features of swamp white oak, but also often includes inherited alkaline soil tolerance of the bur oak parent (Fig. 4). The hybrid between white and bur oaks, *Q.* × *bebbiana*, is our next area of focus, and we are currently trialing many plants that feature the look of a white oak, but have durability and growing characteristics of a bur.

Figure 4. Variance in alkaline soil tolerance in *Quercus* × *shuettei* groups.

Japanese maples (*Acer palmatum*) in our region are marginally hardy. Many varieties aren't hardy through most winters and the hardiest of the variations seem to be hurt more than we'd like to see on a tree. The hybrids between Korean maple (*A. psuedosieboldianum*) and Japanese maple (*A. palmatum*) are of great interest to us. Ideally, we will be getting many of the ornamental features of a Japanese parent combined with the increased hardiness of the Korean maple (Fig. 5). When we grow out the hybrid seedlings, many desirable characters are transferred from the Japanese parent. We look for seedlings with interesting foliage colors and forms, as well as a rapid growth rate.

Figure 5. Hybrids seedilings between Korean maple (*A. psuedosieboldianum*) and Japanese maple (*A. palmatum*) showing variation.

THE FUTURE

As we look to keep moving our businesses forward, I would encourage all of you to consider the road less travelled by. Keep your people evolving and growing, just like the plants we look to sell. The business model of Johnson's Nursery is rare these days, by having in-house propagation all the way through landscape install. By maintaining all these different facets, our company can offer an unmatched degree of service and knowledge about our plants.

When someone works with us, they are purchasing solutions to the problems that they might have bought from someone else.

Some of our most well-known products, including the transplantable oak, came about because of attention to common problems experienced by various clients. Keep your eyes open and observe what you can do to solve your client's issues.

The production of native plants at our nursery is going to be our next big shift. The sale of natives has roughly doubled in the past ten years and the sales of native container material has increased about 4 times over. If you can promote local ecotype material and reduce national competition and make an even bigger difference for the world.

As the newer generations are coming into their buying powers, I foresee a larger emphasis on how a plant can be useful as a living thing in the world and a decreased emphasis on bloom size and color. Form follows function and as the architects of what plants are out there for people to purchase, we will be held to a higher standard on what our plants can do for adding ecological as well as ornamental value to our surroundings.

Opportunities in Commercial Hemp Production

Robert L. Geneve

Department of Horticulture, University of Kentucky, Lexington, Kentucky 40546 U.S.A.

rgeneve@uky.edu

Keywords: *Cannabis*, CBD, cannabidiol, fiber, grain.

INTRODUCTION

Industrial hemp (*Cannabis sativa*) has been used for food and fiber for centuries (Courtwright, 2001) and was an important crop to Kentucky (Figure 1) and other U.S. States prior to the 1950s. Subsequently, the U.S.A. imposed restrictions on *Cannabis* and related products until the 2014 Farm Bill began the process of making industrial hemp a legal agricultural commodity (Johnson, 2015). To separate recreational and medicinal *Cannabis* from industrial hemp, the 2018 Farm Bill clarified the legal definition of industrial hemp as containing less than 0.3% THC (the major psychoactive cannabinoid in *Cannabis*).

Figure 1. Historical images of industrial hemp production in Kentucky before 1945. A. Industrial hemp harvest in central Kentucky. B. Processing hemp for fiber.

Industrial hemp plants have been traditionally used for fiber or the seed (grain) utilized in food crops. The additional use of hemp for extraction of CBD (cannabidiol) oil has recently increased the demand for hemp plant production. Opportunities related to the hemp industry that involve traditional horticultural production and propagation practices suggest involvement in the hemp industry could be potentially profitable. A basic understanding of hemp morphology, production and propagation can be important in deciding whether to enter the industrial hemp market.

HEMP MORPHOLOGY

Although there are exceptions, most agricultural hemp plants are dioecious and produce male and female flowers on separate plants (Fig. 2). Most seed-propagated crops will produce approximately an equal number of male and female plants. Both male and female plants are required for seed production, but only female plants are useful for cannabinoid production. The majority of cannabinoids including THC, CBD, and CBG are produced in the bracts and sepals subtending flowers in the inflorescence (Fig. 3). The highest production of cannabinoids is in female inflorescences that are prevented from producing seeds. It is important therefore to roque-out male plants prior to flowering in seed-propagated crops or to isolate production fields from clonal female plants to avoid unwanted wind-borne pollen.

Figure 2. Dioecious hemp plants. A. Plant with a terminal male inflorescence. B. Terminal female inflorescence without seed production. C. Female inflorescence producing seeds.

Figure 3. Female floral production of CBD. A. Hemp inflorescence showing cannabinoid secretion crystalizing on the foliar surfaces. B. CBD is produced in trichomes on the leaf surface.

PRODUCTION SYSTEMS FOR CBD

Before initiating any production system, growers must check and usually register with their State Department of Agriculture. Hemp production protocols differ between States and some States require mandatory orientation before permits are issued. In 2019, USDA has issued regulatory guidelines and tolerances for State testing of hemp crops prior to harvest to ensure they meet the legal definition of hemp (0.3% THC). Growers will also have to stay current with changing or evolving rules regarding labeling and use of pesticides. As a new agronomic crop, there are few pesticides currently labeled for use in hemp. Field selection can be important for successful crop production and where perennial weeds are a problem, they should be controlled the year prior to planting hemp.

Agricultural systems currently employed for CBD crops include both field and greenhouse production. Field production can be initiated by direct seeding, seedling transplants or clonal cuttings. Greenhouse production is generally from clonal cuttings. Direct seeding is a cost-effective propagation method but there are issues with seed source, stand establishment and unless growers are using feminized seeds, labor will be required to roque male plants. Seedling transplants can be mechanically transplanted using conventional cone-style planters and this improves stand establishment. There is the same issue with sex determination in transplants as there is with a direct-seeded crop. Cuttings are more costly, but stock plants have been selected as high CBD-producing female plants and therefore cutting-propagated plants offer the highest yield potential per unit of planting space.

Field production varies depending on the propagation source and weed management strategy. Direct-seeded crops are produced on an agronomic scale with close in-row and between-row spacing. Crops produced from seedling or cutting transplants are most efficiently mechanically set with equipment that transplants and runs irrigation simultaneously. Row spacing reflects mechanical vs. hand weeding equipment and whether transplanting will be into plastic mulch (Fig. 4).

Figure 4. Hemp production systems. A. Irrigated clonal field with minimal row spacing. B. Irrigated and plastic mulch field on wide spacing. C. Greenhouse production in containers.

Propagation Systems

There is currently high demand for quality hemp propagules from germplasm selected for high CBD production. There are potential opportunities for entry into the propagation hemp market without significant alterations in infrastructure beyond those typically employed in other horticulture production.

Seed production in the U.S.A. is expanding mostly on the west coast. There can be issues with seed quality and availability of regionally adapted selections. There has also been issues with growers purchasing feminized seed that do not yield the desired high percentage of female seedlings. Feminized seed must be produced in isolation from female plants fertilized with a known pollen source. Typically, selected female plants are induced to produce male flowers employing an anti-ethylene compound like silver thiosulfate to become a suitable pollen donor. Properly pollinated, female plants have seeds that produce predominantly female seedlings. Any unwanted pollen contamination reduces the percentage of female plants in the seed-propagated progeny. Properly feminized seeds are currently in demand and command a high value for the seed producer.

There are greenhouse producers that are moving from traditional flower plug and transplant production to producing hemp seedlings. There has been a trend for large CBD producers to partner with greenhouse growers to produce seedling transplants. There is currently only a small demand for organic transplants, but that is anticipated to increase as CBD production diversifies based on market demand.

High input production systems like traditional or organic field production on irrigated / plastic mulch fields or year-round greenhouse production currently require reliably high CBD producing female clones from cuttings. There is currently a demand for cutting propagated material and this demand may increase as the price for CBD adjusts to the demands of the market. High quality cutting production should include a clean stock plant program. Several tissue culture firms are beginning to offer clean stock to large scale cutting producers. In addition to initiating from and maintaining clean stock plants, cutting growers must consider daylength control for seasonal cutting production to avoid photoperiodic flowering in stock plants. Night interruption lighting is sufficient to maintain vegetative stock plants. Stock plant production for cuttings is currently domestic, but growers should consider the potential for offshore entry into this market when considering their long-term business model.

CONCLUSION

Industrial hemp for CBD production is a classic example of an "emerging" high value crop. There are opportunities for early adopters to profit from the high demand for CBD. However, the supply and demand for CBD is already beginning to moderate and growers need to assess the risk before entering this market. There are opportunities for traditional horticulture enterprises that currently produce or propagate flower or woody crops to enter this market with limited change in existing infrastructure. Limited entry into this market may yield short-term profitability with acceptable risk while the long-term market demand becomes clearer. Growers should also consider lateral movement into future regional markets for medical or recreational THC-cannabis or other non-psychoactive cannabinoids like CBG (cannabigerol) that may enter the market and may not require significant changes in production techniques.

Literature Cited

Courtwright, D.T. 2001. Forces of Habit: Drugs and the Making of Modern World. Harvard University Press.

Johnson, R. 2015. Hemp as an Agricultural Commodity. Congressional Research Service Report 7-5700. http://www.fas.org/sgp/crs/misc/RL32725.pdf

The Economics of Biophilia: How Green Infrastructure Fosters Economic Development

Charlie Hall

Department of Horticultural Sciences, Texas A&M University, 2133 TAMU, College Station, Texas 77843-2133 U.S.A.

c-hall@tamu.edu

Keywords: Green industry, economics, green infrastructure.

INTRODUCTION

There is little doubt that the green industry supply chain has experienced unprecedented growth, innovation, and change over the last several decades. However, recent slower growth in demand and tighter profit margins point to a maturing market. Survival in the next decade will require a progressive mindset and a willingness to strengthen existing core competencies or develop entirely new ones, which may involve greater firm-level risk. If the green industry can position itself in such a way that its products/services are considered to be *necessities* in people's lives and not mere *luxuries*, that is the best mitigation strategy against recession and weather-related risks it can employ.

The objective of this presentation was to provide documentation of the economic development benefits of green infrastructure. Green infrastructure refers to ecological systems, both natural and engineered, that act as living infrastructure. Green infrastructure elements are planned and managed primarily for stormwater control, but also exhibit significant economic, environmental, and health and well-being benefits.

ECONOMIC BENEFITS

Beautification Draws Customers, Reduces Shopper Stress, And Enhances Store Appeal

In today's economic climate, businesses are seeking out new methods to effectively maintain their customer base. What few businesses realize is that the simple addition of landscaping to a store front can make a significant difference in how a customer perceives the store. The positive environment created by aesthetically pleasing landscaping not only welcomes customers inside, but also enhances their perception of the level of quality of products and services offered by that store. Consumers are willing to pay higher prices, travel further to, and shop longer in a store that they perceive to be a quality establishment. Adding flowers or shrubs to the exterior of a business can boost sales and widen the consumer base to include more customers.

In addition, ornamental plants are an effective way to revitalize a stale business, bring in additional customers, and improve the overall environment of the store. Furthermore, the presence of ornamental plants reduces shopping-related stress and customers feel more at ease in a store, which is a leading factor in increasing the total amount of time they spend shopping. Landscaping is an effective way to invite more customers to a store and to keep them coming back to experience the positive environment they encounter there.

Boost Occupancy Rates

Landscape amenities represent an effective tool to boost the occupancy rate of apartments and other commercial buildings. People enjoy aesthetically pleasing areas and are much more likely to choose to live and work in buildings with attractive landscapes, so there is a high correlation between nice landscaping and high occupancy rates of buildings. It therefore becomes more economically feasible to invest money in landscaping because the return in the number of tenants and amount of rent money they represent far exceeds the cost of installing ornamental plants.

Tourism Revenue

Parks and botanical gardens have been shown to be effective tourist attractions in both large and small cities, and consequently, a strategic source for generating tourism revenue. Tourist expenditures on food, transportation, lodging, etc. represent additional sources of revenue for local businesses that provide these services. Moreover, new revenue from outside the region adds to the overall financial prosperity of a city's economy. Admissions and ancillary dollars from botanical parks also augment municipal revenue through the community's attendance at special events and recreational attractions held on park premises. This revenue gets filtered back into the economy and adds to the financial prosperity of the community.

Reduced Health Care Costs

Through the direct use of the free or fee-based recreational services, residents of an area with urban green spaces benefit from improved physical fitness. Residents are much more likely to exercise if there is an aesthetically pleasing community park or other green space nearby for them to utilize. Healthy people spend less money each year on medical and other associated health care costs, and thus have more disposable income. By saving money on health care, residents tend to increase their spending on other services and goods.

Increased Property Values

Home improvements can add significant value to a property, but may not always yield a 100% return on money invested, with the exception of landscaping. Landscaping yields, on average, a 109% return on every dollar spent, much more so than other home improvements. Home buyers respond positively to homes with professionally landscaped and manicured lawns, and consequently perceive a higher property value. With landscaping, it is possible to increase its value to potential buyers and receive a significant positive monetary return on this external home improvement investment.

Tax Revenue Generation

Properties that are close to parks, botanical gardens, and other urban green spaces generally have a significantly increased property value compared to properties that are not. As such, they indirectly increase municipal revenues generated through its property tax base. The amount of additional property tax revenue generally means that urban green spaces are self-sustaining since cities can use the additional property tax income to pay for upkeep and improvements.

Reduced Street Repairs

Planting trees along paved residential roads reduces temperatures on the ground and helps increase the longevity of the asphalt. When asphalt gets continually heated by the sun causing high temperatures, it tends to break down faster, requiring more frequent maintenance that can cost a great deal of money, and be a source of inconvenience for residents. The simple installation of shade trees greatly reduces the sun's effect on the asphalt, lowering the temperature of the ground and lengthening the asphalt's life span. This ensures that streets stay well-paved and usable for longer periods of time, saving the city money on maintenance costs, and providing residents with shade over pedestrian walkways.

ENVIRONMENTAL (ECOSYSTEMS SERVICES) BENEFITS

Carbon Sequestration, Improved Air Quality

Trees and other ornamental plants are crucial to the sequestration of carbon from the earth's atmosphere and play an important role in reducing the urban and rural carbon footprint. Research has shown that large trees can absorb significant amounts of carbon from the atmosphere each year, as the tree absorbs carbon dioxide and release oxygen through respiration. As such, trees and other landscape plants serve as an important tool in improving air quality in cities. Trees and landscaped areas absorb particulate matter in the air that would otherwise pollute the atmosphere. Increasing urban green spaces will also help to mitigate a city's already existing poor air quality, which can oftentimes be so poor that it adversely affects the population. Trees and landscape plants also help to absorb odors and potentially harmful chemicals in the air.

Attract Wildlife and Promote Biodiversity

Cultivating urban green spaces is an effective way to help wildlife thrive in otherwise inhospitable areas. Providing animals, a natural refuge away from highly populated areas minimizes the likelihood of animal populations being adversely affected by urban sprawl and vice versa. Maintaining urban biodiversity is an important aspect of keeping a city environmentally friendly and this biodiversity is sustained through the balance between ecosystems, populations, and their habitats, and ensuring that local animal populations have a place to live is a key aspect of sustaining this biodiversity.

Energy Savings

Planting trees and other ornamental plants around a building can significantly reduce the sun's radiation effect on the temperature of the outer walls and lower the associated cost of energy for heating and cooling. In turn, this means that heating and cooling the building will require less energy, reducing the environmental impact that the building will have on the community. Thus, planting trees around a building or business is not only a positive step towards reducing energy consumption, but it also has a significant financial benefit as well.

Reduce Heat and Cold Damage

Trees offset the effects of heat waves and cold air, creating barrier between the walls of the building and the elements. Not only does this affect the building's temperature, it also reduces the effects of other negative effects of the weather, such as gradual wearing away of the walls. Surrounding a building with trees and landscaping is an effective way to protect it from the elements and, at the same time, lower heating and cooling costs.

Offset Heat Islands

Most cities are largely composed of cement and asphalt, which absorb heat from the sun's rays during the daylight hours. This can make city life uncomfortable and hot, especially for people who choose to walk along pedestrian walkways. The inclusion of urban green spaces in a city's landscape can offset the urban heat islands that often occur. Trees and other landscaping plants absorb heat and sunlight, reducing cooling costs and creating a much more enjoyable urban environment for residents of a community.

Reduce Noise Pollution

Trees and urban green spaces can help to reduce the amount of noise in neighborhoods, benefitting both human and animal populations. Noise and light pollution can cause animals to shift activity patterns, urban noise pollutants can cause physiological stress, and the loss of top predators can cause mesopredator release. Landscaped areas absorb noise pollution (sound waves considered to be abrasive and annoying to residents). By installing natural landscaping, noise pollution levels can be significantly reduced. This improves the quality of life for both the human population in a community and the local wildlife who may be affected by unwanted noise.

Reduce Soil Erosion

Planting trees and cultivating landscaped areas is an effective way to reduce the amount of soil lost due to erosion. Tree and other ornamental plant roots create an infrastructure below the surface that helps anchor the soil in place during torrential downpours and holds twenty times more soil than traditional tilled soil. Above the surface, plant foliage and surface cover help to reduce the wind's negative effects on topsoil, reducing the amount of soil lost as dust in the air. Urban green spaces also protect the fertility of the soil, ensuring that it remains sufficiently supplied with nutrients and minerals that enhance plant life.

Reduce Storm Water Runoff / Improved Water Quality

Green spaces absorb water in two ways: above the surface through the leaves and below ground through the root system. By absorbing water, trees and plants reduce the amount of runoff that the city has to deal with, pump out, or purify after significant rainfall events. This reduces the cost of storm water treatment plants and saves the city money. Landscaping is a natural solution to reducing storm water runoff. Landscaping also improves water quality by reducing the amount of dirty storm water that a city has to purify makes current purification methods more effective. Moreover, landscaping absorbs some of the pollutants in the water, meaning that there is less pollution in the water that the city has to purify. Plants and trees improve water quality by ensuring that current purification methods can effectively treat reasonable amounts of runoff.

Reduce Urban Glare

Urban glare is the excessive reflection of sunlight off reflective surfaces such as windows and buildings. Green spaces reduce urban glare in cities by absorbing light, strategically placed landscaping becomes an important tool for city planners to capture the light and deflect it so that the light is not so harsh and unpleasant for residents and drivers, improving the quality of life in the community.

Windbreak

Urban green spaces can reduce the harmful effects of wind in cities by slowing the wind and greatly diminishing its strength, helping to preserve delicate natural environments which could be harmed by high speed winds. Minimizing wind strength also reduces soil erosion due to wind gusts, keeping soil healthy and hospitable for new growth. Moreover, it reduces the need for extensive heating and cooling in buildings by moderating the effects that wind would have on temperature.

WELL-BEING BENEFITS

Health and Recreation

Parks and urban green spaces impact people's health by providing them with an inexpensive (often free) and convenient recreational service. There is a positive correlation between the presence of a park in a neighborhood and the level of physical activity of the residents; people are much more likely to exercise when there is a no-cost, aesthetically pleasing area or facility for them to use. As a result, residents of neighborhoods with beautiful parks are more likely to be healthy since their increase in exercise makes them less susceptible to physical ailments and more resilient against minor illnesses. As a result, these residents do not spend as much each year on health care and medical treatment because they require fewer of these services. Healthy people are happier people; residents who exercise often have excellent overall health and therefore have a more positive mental outlook.

Reduce Community Crime / Community Cohesion

When residents feel greater pride in the beauty of where they live, they are much less likely to detract from it (either by graffiti or endangering people within it). Communities that choose to clean up their parks and beautify crime-ridden neighborhoods have less crime and fewer criminals to deal with. Parks can positively affect the community be reducing criminal acts and bringing residents together. Cohesion in the community is critical to the success of the community as a whole, and this can be achieved through unifying people around a park or botanical garden. Parks decrease incentives for people to commit crimes in the community, and at the same time help to bring neighbors together. They can also increase local political activism. Urban green spaces inspire people to come together and fight for what they know is holding them together as a community.

Traffic Safety / Driver Satisfaction

Beautifying traffic medians not only improves the aesthetics of the roadways, it also affects driver attitudes. Studies show that drivers are more at ease on roadways with natural landscaping and are much more inclined to think positively about the community that they are driving through if the roadways are beautiful. Furthermore, adding trees to roadways creates a sort of natural obstruction which could reduce the likelihood of cars crossing medians into oncoming traffic lanes. This improves driver safety and makes the community a safer place for everyone to live.

Upgrade Effect

As more businesses and neighborhoods take on the task of beautifying their surroundings, other competing areas will be forced to follow suit. In other words, as a community works to better itself, other parts of the area will be forced to upgrade as well to keep drawing people in; this phenomenon is known as the upgrade effect. The upgrade effect positively affects everyone, because it keeps communities from ignoring the benefits of landscaping and developing green spaces, it forces competition and keeps the area looking beautiful. Neighborhoods will be encouraging each other to keep beautifying the landscaping, setting off a cycle of self-improvement that has positive ripple effects outwards to all sectors of the community.

SUMMARY

This paper has summarized the substantial peer-reviewed research has been conducted regarding the manner in which green infrastructure enhances the quality of life for consumers by providing significant economic benefits, environmental amenities in the form of eco-systems services, and a plethora of health and well-being benefits. This research should be strategically incorporated into both industry-wide and firm-specific marketing messages that highlight the quality of life value proposition in order to maintain the industry's sense of value and relevance for consumers of the future.

Buzzsod: A Cool-Season Grass Dominant Meadow Mix for Wildflower Sod Production

John Kaszan[1], Erik Ervin[1], Sue Barton[1], and Deborah Delaney[2]

[1]University of Delaware, Plant and Soil Science, Newark, Delaware 19716; [2]University of Delaware, Entomology and Wildlife Conservation, Newark, Delaware 19716 U.S.A.

jkaszan@udel.edu

Keywords: Meadow plantings, cool season grasses, warm season grasses, wildflowers

INTRODUCTION

Most meadow plantings tend to be warm-season grass dominant (Weaner and Christopher, 2016). There are more species of native cool-season grasses, however, found in mid-Atlantic meadows (Latham and Thorne, 2007). Replacing native warm-season grasses as the dominant grass type in meadow mixes with native cool-season grasses may open new methods of production for meadow establishment. When establishing meadows, weeds can be a massive hindrance (Weaner and Christopher, 2016). A well-developed wildflower sod root mass may ease the burden that weeds put on the establishment of native plant communities. Even if cool-season grasses cannot persist in a wildflower sod, they may allow for new means of propagating and establishing native plant communities.

There is market interest for a native wildflower sod. Focus groups were used to determine the market viability of wildflower sod amongst homeowners and landscape professionals (Barton et al, 1996). Weed pressure was one of the main concerns of both homeowners and landscape professionals (Barton et al, 1996). Wildflower sod development has been documented with articles published detailing various stages of wildflower sod production species selection (Johnson and Whitwell, 1997), soil types (O'Brien and Barker, 1997), and relevant patents that have been granted by the US Patent and Trademark Office (Milstein, 1987; Molnar, 1991). Neither patent seems to have yielded any lasting investment in the products.

Preliminary trials of wildflower sod have utilized species that lack taproots (Johnson and Whitwell, 1997). Given the depth of roots common to many grassland/meadow plants (USDA-National Resource Conservation Service, 2004), this considerably reduces the number of viable species for wildflower sod production. Still, several species have been identified as suitable for sod production (Johnson and Whitwell, 1997), however, these species were not being used as part of a mix.

One method that has proven useful in the harvest and production of sod is growing on plastic (Fig. 1). Utilizing plastic underneath of a fertilized compost based growing media was able to reduce the production time of several warm-season grass sod species (Cisar and Snyder, 1992). Sod grown over plastic also transplanted faster, developing greater root mass faster, and had root mass similar to that of field grown sod at the time of harvest (Cisar and Snyder, 1992).

Figure 1. Sod over plastic 3 months after sowing. Sod is between red lines.

Additionally, bermudagrass sod grown on plastic was found to have greater tensile strength than conventionally produced field grown bermudagrass of the same cultivar (Penn State Center for Sports Surface Research, 2014). The proven success of growing over plastic in sports field turf production show promise in for wildflower sods by increasing the concentration of roots, as well as providing a less invasive method of harvest. Utilizing plastic should also prevent the majority of weed seed from contaminating the wildflower sod during production.

MATERIALS AND METHODS

Two meadow mixes were designed using *Festuca rubra* L. ssp. *rubra* (creeping red fescue) and *Poa palustris* L. (fowl bluegrass) as a cool-season grass foundation to accompany several species of native forbs. Besides the two cool-season grasses, all other species included in both mixes were present at the same rates. Mixes were sown at a ½×, 1×, and 2× of the suggested rates in October using a completely randomized design on 4 m^2 plots.

Rates were calculated using the recommended rate of seed to establish a stand by the NRCS (2009). A quarry blended sandy loam was spread to 3 cm depth overtop of 4 mil plastic as a growth medium and to reduce weed pressure. Similar methods are used in the production of sports turf, primarily using bermudagrass (Penn State Center for Sports Surface Research, 2014). The sowing was covered with a row cover from October until mid-March. There were four replications of each treatment. The complete mixes can be seen in Table 1.

Table 1. The amount of Pure Live Seed (PLS) applied per plot for each of the rates of seed. Mixes including one of the cool-season grasses are shown in green.

Species	Common Name	kg PLS per ha	Rate	1x PLS per plot (mg)	1/2x PLS per plot (mg)	2x PLS per plot (mg)
Poa palustris	Fowl Bluegrass	3.4	60.00%	816.00	408.00	1632.00
Festuca rubra	Creeping Red Fescue	13.5	60.00%	3240.00	1620.00	6480.00
Coreopsis lanceolata	Lanceleaf Coreopsis	7.8	3.50%	109.20	54.60	218.40
Echinacea purpurea	Purple Coneflower	13.5	6.00%	324.00	162.00	648.00
Gaillardia pulchella	Indian Blanket	13.5	2.50%	135.00	67.50	270.00
Monarda fistulosa	Wild Bergamot	4.0	3.50%	56.00	28.00	112.00
Penstemon digitalis	Foxglove Beardtongue	11.2	6.00%	268.80	134.40	537.60
Solidago nemoralis	Gray Goldenrod	11.2	3.00%	134.40	67.20	268.80
Rudbeckia hirta	Black-eyed Susan	0.5	2.50%	5.00	2.50	10.00
Pycnanthemum virginianum	Virginia Mountainmint	11.2	3.50%	156.80	78.40	313.60
Agastache foeniculum	Anise Hyssop	11.2	2.50%	112.00	56.00	224.00
Symphyotrichum laeve	Smooth Aster	11.2	6.00%	268.80	134.40	537.60
Schizachyrium scoparium	Little Bluestem	11.2	1.00%	44.80	22.40	89.60

Sod sown using the ½× rate failed to produce viable transplants, so only the 1× and 2× rates were transplanted. Sod that was deemed suitable for harvest was cut into 46 cm × 224 cm strips and transplanted on a silt loam that was prepared with glyphosate (two applications at a 2% solution of 48.8% glyphosate) and dethatched. Supplemental water (8 cu. cm) was provided for the first 2 weeks after transplant and was only provided once during drought conditions in August.

Species richness (of species from the mixes) was measured at harvest and four months after transplant and was compared using a student's t-test at α=.05. Transplanted sod received one mowing to a height of 13 cm 2 weeks after transplant. Biomass of spontaneous vegetation was taken from 5 cm above the crown and dried for 6 days. Only vegetation within the treatments was sampled. Spontaneous vegetation (primarily weeds) that appeared between different treatments was not considered when calculating spontaneous vegetation biomass.

RESULTS

The rate of seed had no effect on biomass of spontaneous vegetation that occurred among the transplanted sod. The dominant grass species of the two mixes did affect the impact of spontaneous vegetation (p=0.025), seen in Figure 2. Spontaneous vegetation species richness was not found to differ significantly between treatments (p=0.3761). All treatments were just as likely to see the same number of species in the spontaneous vegetation.

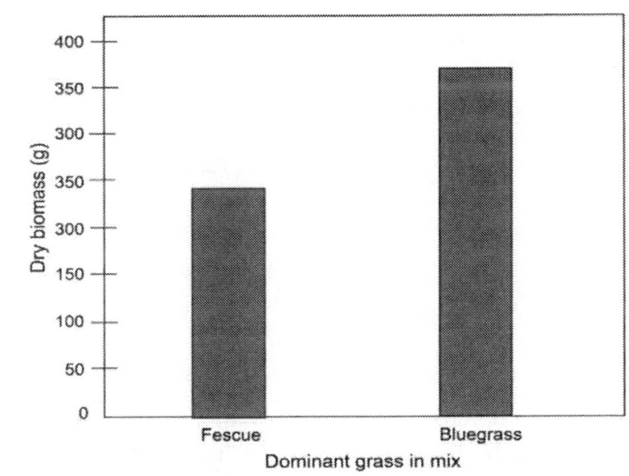

Figure 2. Difference in end of season biomass of spontaneous vegetation between the two mixes. α=.05

Differences in species richness at the time of harvest were not found to be significant (p=.1256). Species richness was found to be significantly different between mixes at the end of the study (p=.0306) with a mean richness of 2.18 for the fescue mix and 3.75 for the bluegrass mix. When species richness was examined under a single model, the mix (p=.004) and the mix*rate interaction (p=.0406) were found to be significant (Figure 3). The month was not found to be significant (p=.3816), showing that there was not a significant decrease in species richness between transplant and our final observations. Utilizing more seed did not help establish a more species rich sod in all cases.

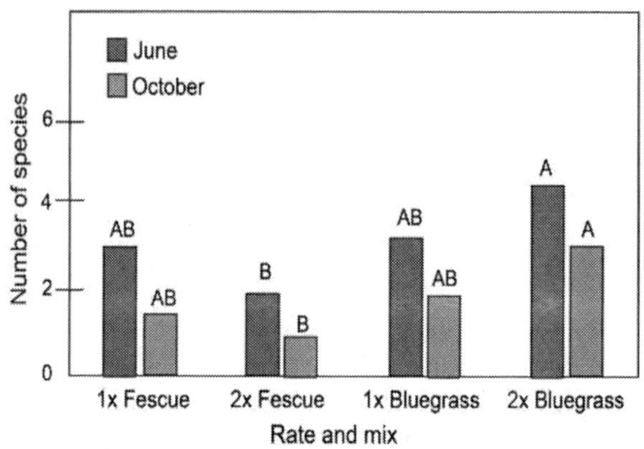

Figure 3. Species richness of transplanted sod at transplants (June) and at the end of season (October). Only species from the mixes were considered. α=.05

DISCUSSION

The fall sowing time strongly favored cool-season grass establishment while on the plastic. Though there were numerous forbs that germinated early on, we hypothesize that they failed to survive the winter due to the constant freezing and thawing of the shallow soil base overtop the plastic (Fig. 4). The experiment will be repeated with a spring sowing (mid-March) to see if forb establishment is more successful. A longer observation period is also worthwhile to test the durability of species richness.

Figure 4. Numerous forbs can be seen within the fescue mix. This photo was taken 3 months after sowing. Most of these forbs did not survive the winter.

The data suggests that the experiment would likely benefit from more repetitions all utilizing the 1× rate, as the 2× rate did not provide any reliable benefits compared to the 1× rate. Though the bluegrass dominant mix had contained higher biomass levels of spontaneous vegetation, this may be due to the transplant shock it suffered, or due to how the fescue began to lay flat under its own weight, further obscuring the soil surface from the light that weed seeds would require to germinate (Fig. 5).

Other methods of preparing the transplant site, such as preemergent weed control, could also be investigated to see how they may interact with newly transplanted wildflower sod. There are still a lot of unanswered questions and methods to be investigated before we can confidently establish wildflowers in sod for restoration or commercial use. However, we demonstrated enough success to indicate that sod remains a viable possibility for future wildflower meadow establishment.

Figure 5. Bluegrass (left of flag) and fescue (right of flag) sod two months after transplant. The bluegrass can be seen going dormant, while the fescue becomes matted. Several forb species can be seen in the bluegrass dominant sod. Weeds (primarily *Digitaria* sp.) can be seen gaining a foothold as well. Species pictured include *Solidago nemoralis* Aiton, *Monarda fistulosa* L., *Agastache foeniculum* (Pursh) Kuntze, *Symphyotrichum laeve* (L.) Á. Löve and D. Löve, and *Pycnanthemum virginianum* (L.) T. Dur. and B.D. Jacks. ex B.L. Rob. and Fernald.

Literature Cited

Barton, S., J. Mercer, and C.J. Molnar. 1996. Using focus groups to determine market potential for wildflower sod. HortTechnology 6(3):271–276.

Brown, R.N., Percivalle, C., Narkiewicz, S., and DeCuollo, S. 2010. Relative rooting depths of native grasses and amenity grasses with potential for use on roadsides in New England. HortScience *45*(3):393–400.

Cisar, J.L., and Snyder, G.H. 1992. Sod production on a solid-waste compost over plastic. HortScience *27*(3):219–222.

Johnson, A.M. and Whitwell, T. 1997. Selecting species to develop a field-grown wildflower sod. HortTechnology 7(4):411–414.

Latham, R., and Thorne, J.F. 2007. Keystone grasslands: reforestation and reclamation of native grasslands, meadows, and savannas in Pennsylvania State Parks and State Game Lands. Wild Resource Conservation Program, Office of Conservation Science, Pennsylvania Department of Conservation and Natural Resources, Harrisburg, Pennsylvania.

Milstein, G. 1987. US4941282A. Retrieved from https://patents.google.com/patent/US4941282.

Molnar, C. J. and J. R. Molnar. 1991. US5224290A. Retrieved from https://patents.google.com/patent/US5224290

O'Brien, T. A., and A. V. Barker. 1997. Evaluating composts to produce wildflower sods on plastic. J. Amer. Soc. Hort. Sci. *122*(3):445–451.

Penn State's Center for Turf Sports Research. 2014. Carolina Green bermudagrass testing. Center for Sports Surface Research, State College Pennsylvania.

USDA-National Resource Conservation Service. 2004. Native plant guide for streams and stormwater facilities in Eastern Illinois. USDA-Natural Resource Conservation Service, Chicago, Illinois.

USDA-National Resource Conservation Service. 2009. Understanding seeding rates, recommended planting rates, and pure live seed (PLS). United States Department of Agriculture, Natural Resource Conservation Service, Alexandria, Louisiana, Plant Materials Technical Note No. 11.

Weaner, L. and Christopher, T. 2016. Garden revolution: how our landscapes can be a source of environmental change. Timber Press, Portland, Oregon.

Assessing Ornamental Species and Cultivars for Invasiveness in Wisconsin DNR

Kelly Kearns, Tara Bergeson, and Jason Granberg

Wisconsin Department of Natural Resources, 101 S Webster Street, PO Box 7921, Madison Wisconsin U.S.A.

tara.bergeson@wisconsin.gov

Keywords: Invasive plants, invasive species council, risk assessment

INTRODUCTION

Invasive plants can cause significant harm to lands and waters by displacing and sometimes eliminating native species; changing ecological structure, composition, and/or function of native plant communities; negatively impact agriculture, recreation, tourism and other factors; and may cause harm to human health.

The Wisconsin invasive species rule (Wis. Adm. Code ch. NR 40) is aimed at preventing new invasive species from getting to Wisconsin, enabling quick action to control or eradicate those here but not yet established and educating the public about which species are or may become invasive and encouraging their control and limiting their spread.

There are two categories of regulated species. For both groups no one may transport, transfer (including buying, selling or offering to sell) or introduce regulated species in Wisconsin. The categories include:

PROHIBITED: Those species likely to become invasive and which are either not currently known in the state, or only in small stands, such that eradication may be feasible. DNR may order the control of prohibited species.

RESTRICTED: Those species which are invasive and are already established in the state such that state-wide eradication may not be feasible. Control is recommended but not required.

Exemptions Include:

- Specific cultivars may be exempt.
- There is a phase-out period for restricted plants to allow instate growers to reduce their inventory of regulated species.
- Permits may be available for research and education.

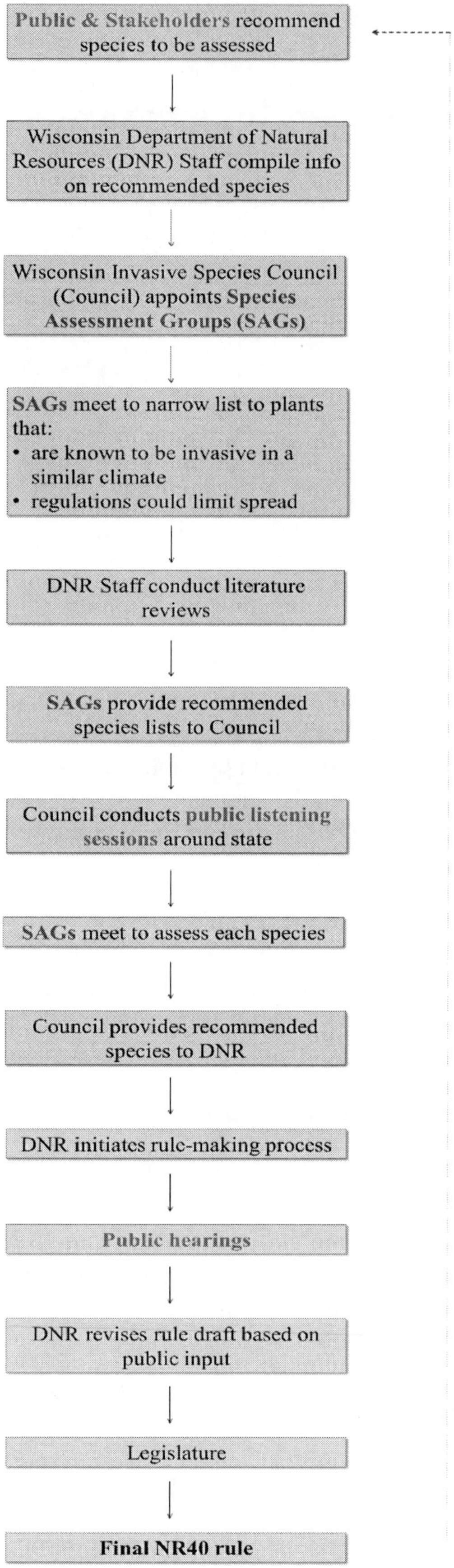

- Transportation may be done for disposal or identification purposes.

Wisconsin Invasive Species Council (WISC or the Council)

The Wisconsin Legislature created the 12-member Wisconsin Invasive Species Council to assist the WDNR in establishing a statewide program to control invasive species. Among their duties, it has an advisory role to focus on current and future rules, statutes, and administrative policies to prevent the spread of and control of invasive species, particularly through NR40.

Species Assessment Groups (SAGs)

SAGs are appointed by the Council. They are made up of stakeholders and experts who are familiar with the species that a particular SAG is assessing. Examples include:

- Nursery owners, propagators, botanic garden staff, land managers, restoration ecologists, horticulture professors and others.

For assessing plants, there are five separate **SAGs**:

- Woody ornamentals.
- Herbaceous ornamentals, including grasses.
- Plants with agronomic uses.
- Plants with no known commercial use.
- Aquatic plants.

Criteria to Be Assessed for Each Species (as per state statute 23.22)

- Potential to cause economic or environmental harm or harm to human health.
- Species distribution and abundance in Wisconsin and in other states with similar climates.
- Likelihood of establishment and spread in WI if introduced.
- Potential for eradication or control.
- Socio-economic value of the species.

Enforcement

- Wisconsin Department of Agriculture, Trade and Consumer Protection conducts regular inspections of all licensed nursery growers and retail nurseries. Inspectors look for regulated species and inform the nursery of any found. Regulated plants are returned to source or destroyed. Nursery staff are informed of plants that are restricted but within the phase-out period.
- If nurseries do not comply with NR 40, WDNR begins stepped enforcement process.
- Out-of-state suppliers and on-line nurseries selling regulated plants into Wisconsin are informed that must stop sales to Wisconsin buyers.

What Propagators Can Do to Assess and Prevent the Marketing of Invasive Species or Cultivars

- Grow new species and cultivars out to maturity in the nursery.
- Assess abundance and seed viability for all new species and cultivars.
- Avoid selling or marketing species with invasive tendencies.
- Use PlantRight Plant Risk Evaluation Tool that uses climate matching to help horticultural growers to screen plants to ensure they won't become invasive in specific regions. <www.greeninfo.org/work/project/plant-risk-evaluator-tool>
- Breed for sterility. Be aware of cross-pollination between cultivars.
- Share findings with other propagators and growers.

Dümmen Orange® Basewell™ Trial at Midwest Groundcovers LLC

Louis Manzella

Midwest Groundcovers LLC, PO Box 748, St. Charles, Illinois 60174, U.S.A.

lmanzella@midwestgroundcovers.com

Keywords: Dümmen Orange® Basewell™ products, adventitious root initiation, asexual propagation, unrooted cuttings

INTRODUCTION

Unrooted cuttings can be very difficult to root for many species of plants. They require time, space, and controlled environments for successful propagation. A new solution to propagation is known as Basewell™. Developed by Dümmen Orange®, Basewell™ products are bare-root cuttings with advanced root development that can be stuck directly in their final containers, eliminating the step of propagation. Cuttings arrive with a small root system and a cartridge at the base to give the cuttings good posture and a head start when transplanting. Some major benefits to Basewell™ include a more simplified process to propagation, space optimization by sticking in final containers, and reduction in labor costs. Basewell™ products are adaptable to automation and they offer a wide range of annuals and perennials to diversify production planning and maximize production turns (Dümmen Orange). Production of Basewell™ products is very simple. Bareroot cuttings are received, plants are stuck in their final containers, and then produced. The objective of the Basewell™ trial was to determine if taxa in 32-cell trays will be ready for transplant after 3 weeks. Furthermore, to determine production time for Basewell™ stuck in trade 1-gal containers.

MATERIALS AND METHODS

The trial consisted of nine taxa: *Leucanthemum maximum* 'Sweet Daisy Birdy', *Leucanthemum maximum* 'Snow Cap', *Salvia nemorosa* 'Spring King', *Coreopsis* × *hybrida* 'Little Bang Enchanted Eve', *Coreopsis verticillata* 'Moonbeam', *Coreopsis verticillata* 'Zagreb', *Phlox paniculata* 'Flame Pro Violet Charme', *Phox stolonifera* 'Sweet Seduction Blue', and *Sedum reflexum* 'Angelina', all of which were free samples. Each taxon was split between two container sizes (32-cell tray and trade 1-gal container) and two soil mediums

(Flat and Pack mix and Nursery soil mix). Flat and Pack mix consists of 50%–60% peat moss, 25%–35% pine bark, and perlite. Nursery mix consists of 70% pine bark, 20% peat moss, and 10% compost. Cuttings were provided shade and misted three times for the first day, and two times for the 2nd and 3rd day. Three samples from each container size where chosen to represent the total population of each taxon. For the first 3 weeks, pictures were taken at the end of each week to get a visual representation of root development. After 3 weeks, taxa in 32-cell trays were evaluated for transplant based on root development in soil media. Basewell™ plants were pruned back July 16th (Week 3) and August 15th (Week 6) to increase branching. August 23rd, plants in 32-cell trays were transplanted to trade 1-gal containers and grown next to Basewell™ plants stuck initially in their final containers. Production time for Basewell™ initially stuck in their final container was concluded September 13th (ten weeks). Production time for transplanted Basewell™ was concluded September 28th (12 weeks).

RESULTS AND DISCUSSION

Trade 1-Gallion Containers with Nursery Mix

Trade 1-gal containers with nursery mix were stuck with nine different Basewell™ taxa June 28th, 2019. Plants were watered, labeled, and spaced pot to pot. Misting applications were applied with one pass by an overhead boom. Mist was applied to cuttings for a period of 3 days along with an overhead shade curtain to provide plants with a better environment for rooting. After 1 week, most plants from each taxon were upright and establishing in their containers. *Leucanthemum* 'Sweet Daisy Birdy' showed some signs of leaf curl after 1 week, but plants recovered quickly, and no losses were recorded for the cultivar throughout the trial.

After 2 weeks of production, the only taxa with losses were *Salvia* 'Spring King' with one loss and *Sedum* 'Angelina' with five losses. Losses to sedum were likely due to water management rather than quality issues with the Basewell™ cuttings. Nevertheless, no other losses were recorded after the 2nd week for each taxon.

By the end of the 3rd week, most taxa had shown vigorous root development to the base of the container (Fig. 1).

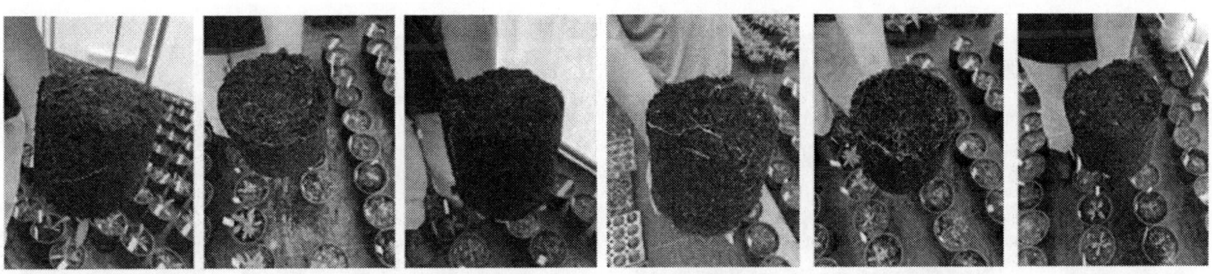

Leucanthemum 'Sweet Daisy Birdy' | *Coreopsis* 'Moonbeam' | *Phlox* 'Pro Flame Violet Charme' | *Salvia* 'Spring King' | *Coreopsis* 'Little Bang Enchanted Eve' | *Sedum* 'Angelina'

Figure 1. Root development to the base of the container after 19 days.

Phlox 'Pro Flame Violet Charme' and 'Sweet Seduction Blue' had slower root development with roots extending half way down the container. The only taxon to not show any root development was *Sedum* 'Angelina'. Pictures of samples from *Salvia* 'Spring King' after eleven days showed branching roots extending throughout the container. Root development was a major focus for the first 3 weeks to compare progress of Basewell™ stuck in their final containers with Basewell™ stuck in 32-cell trays. A major benefit to Basewell™ is the potential to eliminate the step of propagation. It was very clear after 3 weeks that all Basewell™ taxa, with the exception of *Sedum* 'Angelina', established very quickly in trade 1-gal containers with nursery mix.

Total production time for trade 1-gal containers with nursery mix was determined to be 10 weeks for each taxon (Fig. 2). The conclusion was made after the populations of *Leucanthemum* 'Sweet Daisy Birdy', *Salvia* 'Spring King', *Coreopsis* 'Little Bang Enchanted Eve', *Coreopsis* 'Moonbeam', *Coreopsis* 'Zagreb', and *Phlox* 'Pro Flame Violet Charme' were observed in full bloom, and taxa had grown throughout their entire containers. Taxa that were not in full bloom after 10 weeks were *Leucanthemum* 'Snow Cap', *Phlox* 'Sweet Seduction Blue', and *Sedum* 'Angelina'. These taxa had strong vegetative growth throughout 10 weeks, but did not send out flower buds. It is likely that time of planting, lighting, and short period of growth may have limited flower production for those taxa.

Figure 2. Growth of trade 1-gal containers with nursery mix after 10 weeks for nine taxa

Trade 1-Gallion Containers with Flat and Pack Mix

Trade 1-gal containers in Flat and Pack mix showed very similar results to trade 1-gal containers in nursery mix. For 3 days, taxa were misted and provided supplemental shading. After 1 week, taxa were upright and establishing in their containers.

After 2 weeks, losses were recorded for the following taxa: *Leucanthemum* 'Snow Cap' with one loss, *Salvia* 'Spring King' with two losses, *Coreopsis* 'Little Bang Enchanted Eve' with one loss, and *Sedum* 'Angelina' with four losses. More losses were noted in flat and pack mix compared to nursery mix, but no significant difference.

Pictures were taken Week 3 to highlight the development of root system. All taxa except for *Phlox* 'Sweet Seduction Blue' and *Sedum* 'Angelina' had branching roots stretching the entire pot. Root development in Flat and Pack mix after 19 days had no visual difference compared to nursery mix soil (Fig. 3).

Coreopsis 'Little Bang Enchanted Eve' *Phlox* 'Pro Flame Violet Charme' *Leucanthemum* 'Sweet Daisy Birdy'

Figure 3. Root development in flat and pack mix after 19 days.

Basewell™ grown in flat and pack mix were produced the same as Basewell™ grown in Nursery mix, though pictures and notes were not taken to determine total production time because we do not grow the selected taxa in flat and pack mix. Instead, we typically use nursery mix to produce perennials in 1-gal containers or larger, and flat and pack mix for containers smaller than 1-gal containers.

32-Cell Trays with Nursery Mix

Basewell™ stuck in 32-cell trays with nursery mix responded well to shade and misting applications for the first 3 days. Plants were upright in their trays and cartridges were firm in soil indicating good root development (Fig. 4).

Losses from the 2nd week were as follows: *Coreopsis* 'Moonbeam' with one loss, *Sedum* 'Angelina' with three losses, *Leucanthemum* 'Snow Cap' with two losses, and *Leucanthemum* 'Sweet Daisy Birdy' with one loss. Losses from each taxon were likely due to cultural practices such as overwatering or not enough mist rather than quality of Basewell™. No additional losses took place after the 2nd week. Root notes were also taken the 2nd week. Samples from *Salvia* 'Spring King' had the most vigorous roots after just 11 days of growing. Samples from *Leucanthemum* 'Snow Cap' and *Phlox* 'Pro

Flame Violet Charme' had rooted to the bottom of the cell with strong roots holding the soil. *Sedum* 'Angelina' had the slowest root development of all taxa in 32-cell tray. Too much water may have slowed root development leading to weaker roots with *Sedum* 'Angelina'. Overall, Basewell™ taxa showed tremendous vigor in nursery mix after eleven days of growing.

Coreopsis 'Little Bang Enchanted Eve' — *Phlox* 'Pro Flame Violet Charme' — *Phlox* 'Sweet Seduction Blue' — *Coreopsis* 'Moonbeam' — *Coreopsis* 'Zagreb'

Leucanthemum 'Sweet Daisy Birdy' — *Leucanthemum* 'Snow Cap' — *Salvia* 'Spring King' — *Sedum* 'Angelina'

Figure 4. Basewell™ cuttings stuck in 32-cell trays with nursery mix after 19 days.

Pictures were taken after the 3rd week to determine if Basewell™ taxa in 32-cell trays were ready or not for transplant into final containers. All Basewell™ taxa except *Sedum* 'Angelina' were fully established in cell with roots stretching to the bottom of the cell after 2 to 3 weeks.

Basewell™ continued to grow in 32-cell trays up to the beginning of Week 8 when they were transplanted to trade 1-gal containers. This is a limitation to determining production time of Basewell™ stuck in 32-cell first and then transplanting to final container because plants were not transplanted immediately after they were ready for transplant. On that note, Basewell™ products allow producers to stick directly into final containers eliminating the step of transplanting. The importance of determining time to transplant was to compare it with root development of plants stuck in their final containers. Basewell™ taxa stuck in trade 1-gal containers had a 2- to 3-week advantage over taxa stuck in 32-cell trays with respect to time establishing in final containers. This 2 to 3-week advantage provided Basewell™ in trade 1-gal containers a more developed root system than plants in 32-cell trays to feed the plant above the soil line. The trial concluded for transplanted Basewell™ September 28th. Transplanted Basewell™ had grown for four weeks reaching similar width and height of Basewell™ stuck initially in trade 1-gal containers.

32-Cell Trays with Flat and Pack Mix

Basewell™ in 32-cell trays with Flat and Pack mix responded well to shade and misting application after 3 days. Results from Flat and Pack in comparison to plants in nursery mix showed no visible difference. After 1 week, plants were upright and establishing.

After 2 weeks, the only taxon to have losses was *Sedum* 'Angelina' with four losses. Losses were likely contributed to cultural practices such as overwatering rather than quality of Basewell™. Both *Leucanthemum* and *Phlox* taxa had strong roots holding the soil. Soil was very loose with *Coreopsis* tasa, though roots were still vigorous. *Salvia* 'Spring King' had the fastest growing roots of all taxa, but roots did not hold the soil as well as in nursery mix. Samples from *Sedum* 'Angelina' showed roots extending a quarter of the depth of cell.

Basewell™ taxa were evaluated the 3rd week to determine if taxa were ready for transplant. Taxa in Flat and Pack mix showed similar results to taxa in nursery mix. *Phlox* 'Flame Pro Violet Charme' and 'Sweet Seduction Blue' showed stronger root development compared to phlox in nursery mix (Fig. 5). Samples held the soil better and stretched the entire length of cell.

The 32-cell trays in Flat and Pack mix were transplanted the beginning of week eight into trade one-gallon containers with nursery mix. The same limitation applies to Flat and Pack mix as nursery mix. The trial finished September 28th after plants had grown for 4 weeks and reached similar height and width to Basewell™ stuck initially in trade 1-gal containers.

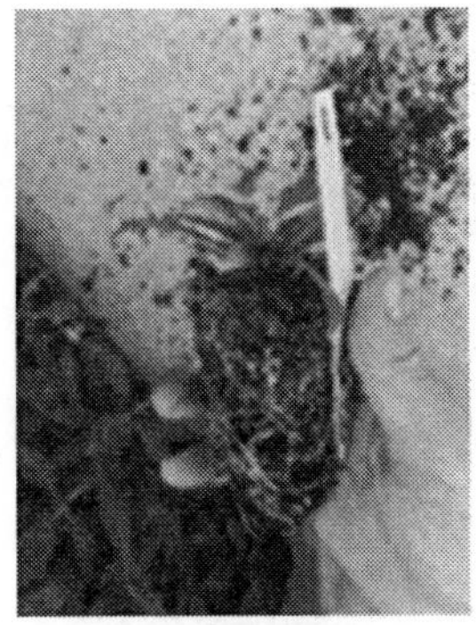

Phlox
'Pro Flame
Violet Charme'

Phlox
'Sweet
Seduction Blue'

Figure 5. *Phlox* 'Flame Pro Violet Charme' and 'Sweet Seduction Blue' showed stronger root development in Flat and Pack mix.

CONCLUSION

Basewell™ taxa showed tremendous potential to establish and mature quickly in trade 1-gal containers with nursery mix.

Salvia 'Spring King' had branching roots extending to the bottom of trade 1-gal containers after 11 days. *Leucanthemum* and *Coreopsis* taxa had roots extending to the bottom of the container after 19 days. *Phlox* taxa were not as vigorous after 19 days with roots half way down the side of the container. Overwatering may have slowed down root development for *Phlox* taxa. *Sedum* 'Angelina' had the slowest root development in trade 1-gal containers with nursery mix. Factors such as overwatering and container size may have limited growth and development. Samples of *Salvia* 'Spring King', *Coreopsis* 'Little Bang Enchanted Eve', and *Phlox* 'Pro Flame Violet Charme' in 32-cell trays with nursery mix had beautiful established roots ready for transplant after 11 days. All taxa except *Sedum* 'Angelina' were ready for transplant after 2 to 3 weeks.

To conclude, Basewell™ taxa were consistent in size, flowering, and all but *Sedum* 'Angelina' maintained losses at 6% or less. In relation to our typical production time for select taxa, Basewell™ taxa took on average 2–3 weeks more time. However, better management of certain factors such as time of planting, water and fertility management, plant growth regulation, and standards for saleable products are potential ways to decrease production time and increase turnover. The Dümmen Orange® Basewell™ trial demonstrated that Basewell™ is very simple, cuttings can be directly stuck in final container with almost no losses, and it has the potential to save nurseries space, labor, and production steps.

Literature Cited

Dümmen Orange. 2019. A groundbreaking rooting technology. Grow Basewell www.growbasewell.com/site/en

Foliar Indole-3-Butyric Acid Rooting Hormone Application and Cost Analysis

Jeffrey C. Martindell

Decker's Nursery, 6239 Rager Road, Groveport, Ohio 43215 U.S.A.

jmartindell@deckersnursery.com

Keywords: Rooting hormone, cost analysis, woody plants, hormone application method

INTRODUCTION

Indole-3-butyric acid (IBA) is an auxin that promotes root initiation. Specifically, in this trial Hortus IBA Water Soluble Salts were used as an overspray of hormone on 32,464.5 sq. ft. At Decker's Nursery, foliar rooting hormone is used on both hard and softwood cuttings with astonishing results. The information presented is from a complete overspray of hormone on hardwood winter cuttings.

MATERIALS AND METHODS

Cuttings were direct stuck into 18-count, 32-count, and 50-count trays. The trays are stuck, watered, sprayed with foliar IBA rooting hormone, and covered by a homemade Dutch style greenhouse tent (Fig. 1).

Cuttings are all given three initial foliar sprays applied at 24-hr. intervals. To ensure uniformity and to push certain crops along, there will be future supplemental reapplications given. These reapplications are applied every 2–3 weeks or as needed.

Figure 1. Homemade Dutch style greenhouse tent.

Over the following weeks and months constant monitoring and care is given to the cuttings. Tents are ventilated, watered, and checked for progress. In the time you will start to notice cell differentiation and callusing on the basal end of the cuttings (Fig. 2).

Figure 2. Callus developing at the base of a conifer cutting.

This is an indicator that the rooting hormone is active within the cutting and roots are on their way. Please note that some cuttings like taxa of some *Chamaecyparis* will create a large callus but never root. This indicator is helpful but not true for all plant taxa. In time you should start noticing root development (Fig. 3).

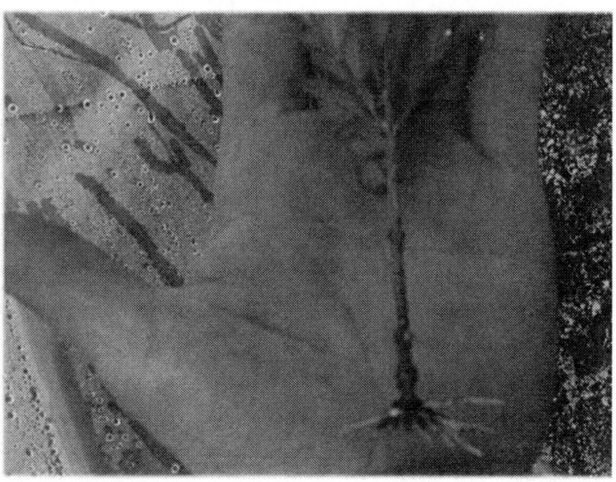

Figure 3. Root development on a boxwood cutting.

Root development from foliar hormone can go overboard. With excess hormone and humidity there can be development of adventitious roots along the entire stem. Here is an example of a softwood cutting from an unrelated crop showing excessive adventitious roots; note the roots covering the entire cutting (Fig. 4).

Figure 4. Rooted cutting showing excessive adventitious root development.

Spray Protocol

- Use distilled water for mixing IBA solution.
- Use a flag marker to mark each day's sticking progress to track the 3-day spray rotation.
- Measure IBA powder to mix in 1-gal of distilled water increments for desired parts per million.
- Unused hormone is kept in the sprayer overnight. Add additional hormone in gallon increments.
- All hormone applications occur in early morning; Stomata are open, and cuttings are generally not in moisture stress.

Backpack Sprayer Used

The electric backpack sprayer used is a Dramm BP-4 (Fig. 5, right). This backpack sprayer is versatile and makes applying hormone inside or outside of a tent fast, and user friendly (Fig. 5, center and left).

Figure 5. Dramm BP-4 sprayer (left), applying hormone inside or outside of a tent (center and left).

COST ANALYSIS

Area

- Hormone was reapplied at 2000 PPM, 40 grams in 1 gallon of distilled water.
- Treated north houses 1-12, each containing 2 tents. (90.5 ft × 11.5 ft = 1,040.75 sq. ft./tent)
- Treated south houses S1 - S2 west, S5 west ½.
    - 11.5 ft × 186 ft = 2139 sq. ft. per tent)
- Total area treated
    - 1,045.75 × 24 = 24,978 sq. ft. north houses
    - 2139 × 3.5 tents (2139 sq. ft.) = 7,486.5 sq. ft. south houses
    - 32,464.5 sq. ft. total area treated.

Solution Rate

- 40 grams per 1-gal distilled water = ~2000 ppm
- Total solution required is 18 gal
    - 18 gal × 40 grams = 720 grams applied (note: a 3-gal electronic backpack sprays about five short tents)
    - 32,464.5 sq. ft. / 18 gal = 1803 sq. ft. per gallon of solution. (about 1½ tents)
    - 32,464.5 / 720 grams = 45 sq. ft. per gram of IBA.

Hormone Cost Per Application

- $0.45 per gram of IBA
  - 720 × $0.45 =$324 per application at 2000 ppm
  - $0.45 per 45 sq. ft. = $0.01 per sq. ft.
  - Each tray in propagation is about 1.6 sq. ft.
    - Each hormone application costs about $0.016 per tray per application.
    - 18 plants per tray, $0.0008 per plant per application.
    - 32 plants per tray, $0.0005 per plant per application.
    - 50 plants per tray, $0.00032 per plant per application.

Hormone Cost Per Season (Per Plant)

- 3 initial applications at sticking
  - 18 trays, $0.0024
  - 32 trays, $0.0015
  - 50 trays, $0.00096
- 3 supplemental applications (2–3-week intervals after sticking)
  - 18 trays, $0.0024; total = $0.0048 per plant (½ of a penny)
  - 32 trays, $0.0015, total = $0.003 per plant (1/3 of a penny)
  - 50 trays, $0.00096; total = $0.00192 per plant (1/5 of a penny

Survival and Growth of Hickories and Pecan After Containerization and Field Planting

Brandon Miller and Nina Bassuk

School of Integrative Plant Science, Cornell University, Ithaca, New York 14853 U.S.A.

bmm244@cornell.edu

Keywords: *Carya,* hickory, propagation, containerization, pecan, production systems

INTRODUCTION

Hickories (*Carya*) include many stately, native trees, that offer superior ornamental and adaptable features with great promise for application in managed landscapes, especially urban environments. Immense interest exists in effectively producing these trees, however, due to their lag-phase shoot growth and strong development of a taproot with minimal fibrous-root branching, these trees exhibit resistance to standard growing techniques and reduced transplant success. New commercial products such as modified nursery containers are touted as better alternatives to traditional production techniques. If these new products are effective, they provide new opportunities for developing hickory crops for nursery production. We questioned whether traditional field production, standard plastic containers, or new fabric nursery pots could be used to effectively grow bare-root whips of hickories and northern pecan.

Our objectives were to:

1. Characterize the growth and viability of bare-root whips of hickories and northern pecan under different production regimens.
2. Assess species differences after containerization and field planting.

METHODS

Species studied: *Carya cordiformis* (bitternut hickory), *C. illinoinensis* (northern pecan) *C. laciniosa* (kingnut hickory), and *C. ovata* (shagbark hickory).

Treatments: Bare-root whips (12 in.–24 in.) of each species were purchased and either lined out in a field or grown out in standard plastic #10 containers or Root Trapper Series II grow bags (#10 equivalent). There were 12-single plant replicates/species with standard #10 and Root Trapper Series II, and 14-single plant replicates of *C. illinoinensis, C. laciniosa, C. ovata* (12-single plant replicates *C. cordiformis*) in field.

RESULTS AND CONCLUSIONS

This study characterized the differences between hickory species when purchased as bare-root whips and as they are grown out using three different production systems. General trends of increases in both caliper and height were observed in plants grown with standard #10 or Root Trapper Series II containers that were not consistent with plants grown in the field. At least one instance of within species differences was observed with each taxon in response to the production treatment(s). In each example, one or both of the container treatment(s) resulted in higher values of growth (caliper or height) compared to field-grown plants. Most plant mortality occurred with pecan and shagbark and was associated with the field-production treatment. Overall, growth was similar between standard #10 and Root Trapper Series II containers. Based on these data, we recommend nursery producers consider container production over field culture of bare-root liners when growing *Carya cordiformis* (bitternut hickory), *C. illinoinensis* (northern pecan) *C. laciniosa* (kingnut hickory), and *C. ovata* (shagbark hickory) (Fig. 1).

Acknowledgements: This work is supported by the USDA National Institute of Food and Agriculture, McIntire Stennis/Smith-Lever project 1020775.

Figure 1. Left to right: *Carya cordiformis* (bitternut hickory), *C. illinoinensis* (northern pecan) *C. laciniosa* (kingnut hickory), and *C. ovata* (shagbark hickory) shown in standard #10 plastic container (top row), and Root Trapper Series II (bottom row).

A Genome Size Survey of the Blue and White Fruited Dogwoods (*Cornus* L.)

Joseph Pryzdia[1] and Kim Shearer[2]

[1]Plant Breeding Intern, The Morton Arboretum, 4100 Illinois Route 53, Lisle, Illinois 60532; [2]Tree and Shrub Breeder, The Morton Arboretum. USA

kshearer@mortonarb.org

Keywords: *Cornus*, genome size, biotechnology, research, flow cytometry

INTRODUCTION

Consisting of approximately 58 different species of shrubs, small trees, and to a lesser extent herbaceous perennial, dogwoods (*Cornus*) are considered to be a greatly valued landscape plant with a range covering much of the temperate and subtropical regions of the Northern Hemisphere. Species belonging to *Cornus* have been frequently cultivated, bred, and selected with respect to their pronounced four-season attributes including attractive flowers, fruit, bark, foliage, and form (Cappiello and Shadow, 2013) (Fig. 1). According to the Census of Horticultural Specialties (USDA, 2014) sales of dogwoods within the United States accounted for greater than $27.8 million.

Research resolving a species-level phylogeny in dogwoods has shown that four distinct clades exist within *Cornus*. These include the big-bracted dogwoods (BB), the dwarf dogwoods (DW), the cornelian cherries (CC), and the largest amongst the clades, the blue- or white-fruited dogwoods (BW) (Xiang et al., 2006). Within the BW group, there are further taxonomic divisions into subgenera including *Kraniopsis I*, *II* and *III*; *Mesomara*; and *Yinquania*. Reported ploidy for species of BW covered in this survey are diploid with a base chromosome number of $x = 11$ in *Kraniopsis I* and *II* and $x = 10$ in *Mesomara* (Darlington and Wylie, 1956).

Genome size (nuclear DNA content) has been shown to support taxonomic relationships in *Cornus* (Shearer and Ranney, 2013). Shearer and Ranney found that relative genome sizes were taxonomically distinct in the BB, DW, and CC clades. It was also determined that all species belonging to these clades were diploid with the exception of a tetraploid *C. canadensis* and a triploid hybrid cultivar *C.* 'KN30-8' Venus® dogwood.

Figure 1. Inflorescence of *Cornus alba* (Frank Vincentz, Wikicommons) **(**top left)*;* blue fruit of *C. amomum* ssp. *obliqua*, ©The Morton Arboretum **(**top right)*;* red winter stems of *C. sanguinea*, (Jonathan Ballinger/*Glowing dogwoods*/CC BY-SA 2.0/cropped) **(**bottom left); white fruit of *C. racemosa* ©Deborah J.G. Brown **(**bottom right).

To complete the survey of *Cornus*, the objectives of this study were to complete a relative genome size survey for species of the BW clade, confirm ploidy is consistently diploid, and further explore the relationship between genome size and taxonomic classification.

Acknowledgements

Thanks to the Daniel P. Haerther Charitable Trust Foundation for their generous funding of the New Plant Development Program that supported the plant breeding internship and to Mike Yanny of JN Plant Select who provided plants to be used in this survey.

MATERIALS AND METHODS

Samples of expanding leaf tissue and vegetative buds were collected from 25 Morton Arboretum accessions and four selections provided by J.N. Plant Select. A total of 14 species are represented in addition to cultivars and hybrids of the BW clade of *Cornus*.

A 5 mm^2 sample from each of these individuals was co-chopped with 5 mm^2 of *Pisum sativum* 'Ctirad' (2C = 8.76 pg) in polystyrene petri dishes with a razor blade and 400 µl of nuclei extraction buffer (CyStain ultraviolet Precise P Nuclei Extraction Buffer; Partec, Münster, Germany). The nuclei suspension was filtered through 30-µm nylon filters and stained using 1.6 mL 4′, 6-diamidino-2-phenylindole (DAPI) staining buffer (Cystain ultraviolet Precise P Staining Buffer; Partec) (Fig. 2).

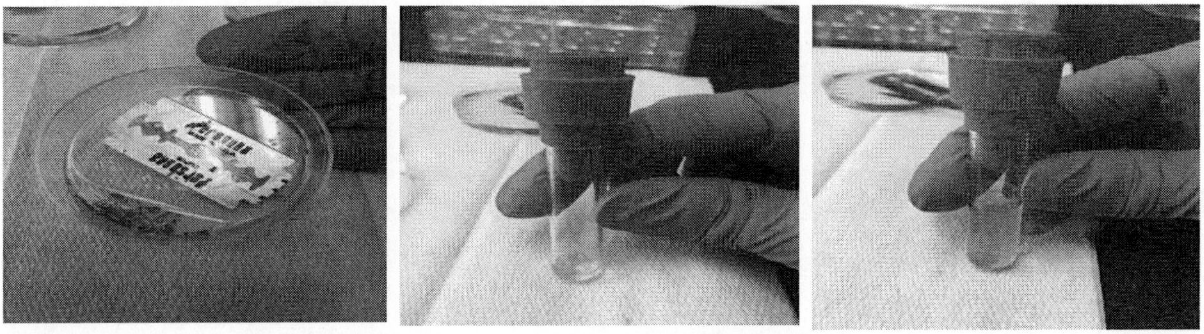

Figure 2. Sample preparation of *Cornus racemosa.* Chopped leaf tissue from both *C. racemosa* and the internal standard *Pisum sativum* 'Ctirad' immersed in 0.4 ml of nuclei extraction buffer (left). Contents of chopped leaf tissue from *C. racemosa* and the internal standard *Pisum sativum* 'Ctirad' filtered through a 50-µm nylon filter center). Addition of 1.6 ml of DAPI staining buffer to filtered chopped leaf tissue from *C. racemosa* and the internal standard *Pisum sativum* 'Ctirad' (right).

Relative 2C genome sizes were determined using flow cytometry equipped with a UV excitation lamp. A minimum of 3000 nuclei were analyzed for each sample and for each individual, three samples were prepared. Relative genome sizes were calculated using the equation

$$2C = DNA\ content\ of\ standard\ X\ \frac{Mean\ fluorescense\ value\ of\ sample}{Mean\ fluorescense\ value\ of\ the\ standard}$$

Statistical Analysis

One-Way ANOVA was used to determine statistical significance of monoploid genome size (1Cx) data at the species and subgenus levels with *Kraniopsis* separated into the two subgenera *Kraniopsis I* and *Kraniopsis II.* Tukey's honestly significant difference (HSD) was used for means separation. R package agricolae was used for statistical analysis (de Mendiburu, 2019).

RESULTS AND DISCUSSION

All individuals surveyed were diploid (*Kraniopsis,* $2n = 2x = 22$; *Mesomara*, $2n = 2x = 20$). Relative 2C genome sizes for the BW clade ranged from 1.70 pg (*C. controversa*) to 2.51 pg (*C. sanguinea*) (Figs. 3 and 4). Mean monoploid genome size of the three subgenera *Kraniopsis I* (1C*x* = 1.15 pg), *Kraniopsis II* (1C*x* = 1.07 pg) and *Mesomara* (1Cx = 0.92 pg) were found to be significantly different (Table 1). However, when considering mean 1Cx values of individuals in *Kraniopsis I* and *Kraniopsis II,* there is overlap between the two subgenera at the species level, suggesting that they are not necessarily significantly distinct in genome size as two separate subgenera.

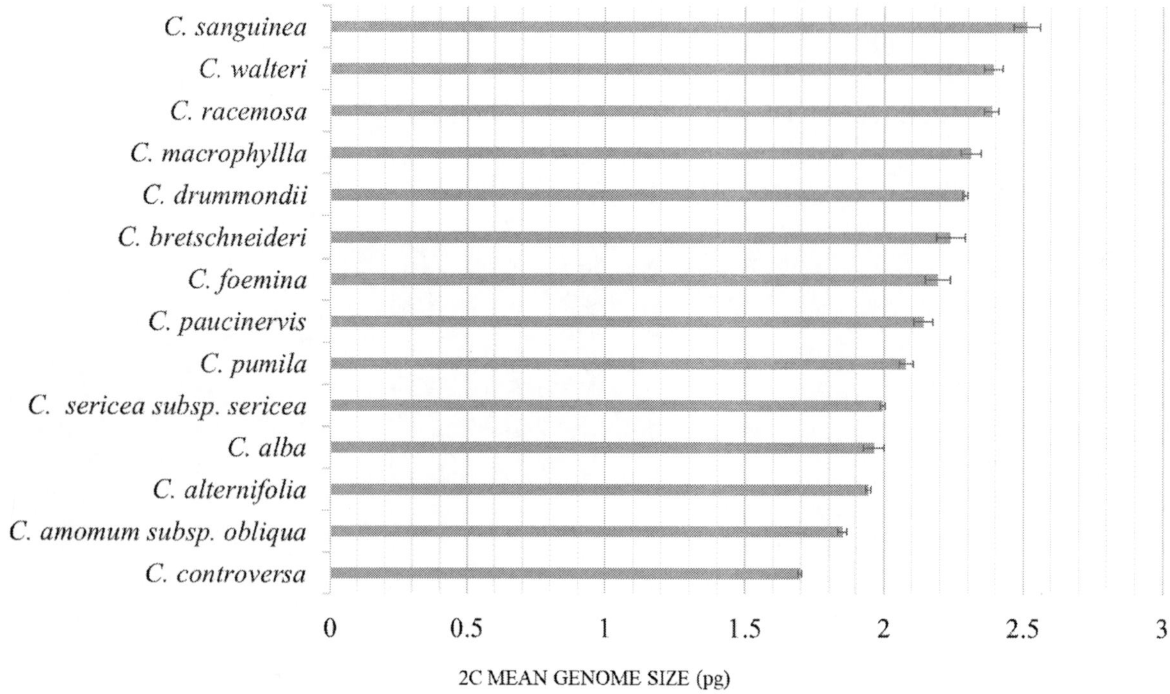

Figure 3. Relative 2C genome size of 14 *Cornus* species determined using flow cytometry analysis of nuclei stained 4′, 6-diamidino-2-phenylindole with *Pisum sativum* 'Ctirad' as an internal standard (2C = 8.76 pg).

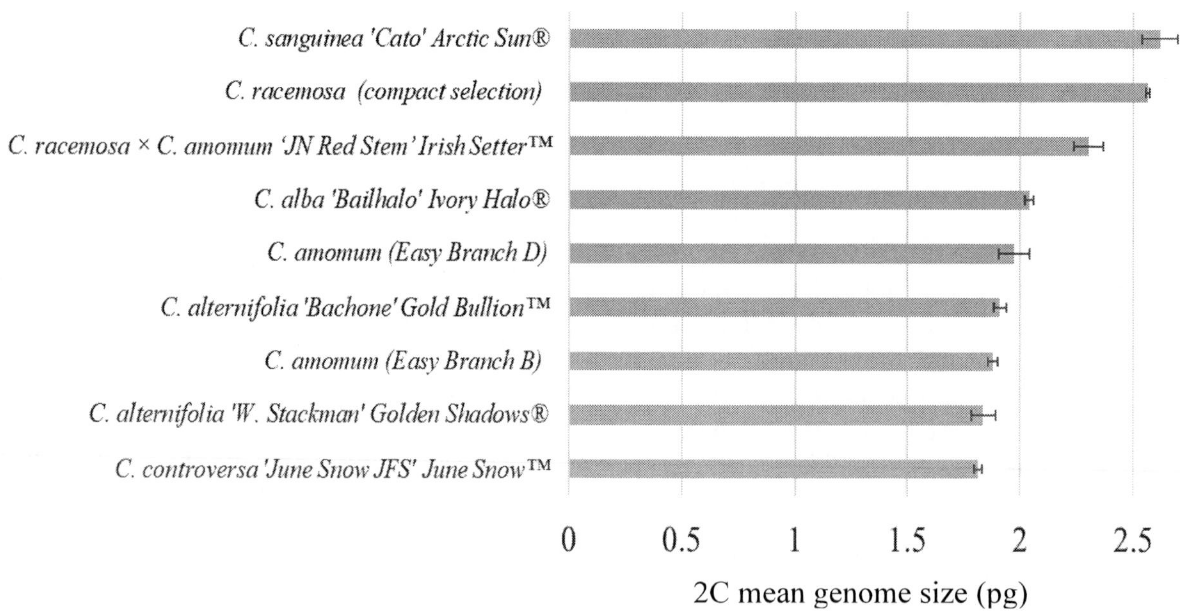

Figure 4. Relative 2C genome size of 9 *Cornus* cultivars determined using flow cytometry analysis of nuclei stained with 4′, 6-diamidino-2-phenylindole with *Pisum sativum* 'Ctirad' as an internal standard (2C = 8.76 pg).

Table 1. Relative genome sizes and ploidy levels for *Cornus* species, hybrids, and cultivars of the BW group determined using flow cytometry.

Subgenus	1C× (pg) / Taxon	Source/Accession[z]	1C×(pg)	Origin/Provenance[y]
Kraniopsis I	1.15±0.03 A[x]			
x = 11	*Cornus drummondii*	Mor117-24*1	1.15±0.00	Great Plains and Midwest US; W, Jasper, Missouri
	C. foemina	Mor279-84*3	1.07±0.04	Southeast and Eastern US; G
	C. foemina	Mor279-84*6	1.14±0.05	G
	C. foemina	Mor279-84*7	1.09±0.01	G
	C. racemosa	Mor181-92*2	1.19±0.01	Midwest and Eastern US; W, Will County, Illinois
	C. racemosa (compact)	JNPS01	1.28±0.00	G
Kraniopsis II	1.07± 0.01 B			
	Cornus alba	Mor105-79*9	0.98±0.02	Siberia/East Asia; W, Erfurt, German Democratic Republic
	C. alba 'Bailhalo', Ivory Halo® dogwood	Mor1319-2004*1	1.02±0.01	G
	C. amomum ssp. obliqua	Mor724-79*13	0.93±0.01	Midwest and Eastern US; W, Will County, Illinois
	C. amomum (Easy Branch B)	JNPS02	0.94±0.01	G
	C. amomum (Easy Branch D)	JNPS03	0.99±0.03	G
	C. bretschneideri	Mor303-90*5	1.12±0.03	Northeast China; W, Shanxi, China
	C. macrophylla	Mor503-2001*2	1.16±0.02	Eastern Asia; W, Mt. Maiji, Gansu, China
	C. paucinervis (*C. quinquenervis*)	Mor569-63*4	1.11±0.03	Central China; G
	C. paucinervis (*C. quinquenervis*)	Mor569-63*9	1.08±0.02	G
	C. paucinervis (*C. quinquenervis*)	Mor569-63*10	1.02±0.01	G
	C. pumila	Mor642-54*2	1.01±0.02	Unknown; G
	C. pumila	Mor642-54*3	1.05±0.03	G
	C. pumila	Mor719-79*4	1.06±0.01	G
	C. sanguinea	Mor196-81*3	1.26±0.02	Eurasia; W, Loire, France
	C. sanguinea 'Cato', Arctic Sun® dogwood	Mor270-2006*2	1.31±0.04	G
	C. sericea ssp. sericea	Mor464-85*6	1.00±0.00	North America; W, Lake County, Illinois
	C. walteri	Mor342-2002*1	1.20±0.02	East Asia; W, China
Putative	1.15±0.03 A			
hybrid	*Cornus* 'JN Red Stem' Irish Setter™ dogwood	JNPS04	1.15±0.03	G

Mesomara x = 10	0.92±0.02 C *Cornus alternifolia*	Mor295-2002*1	0.97±0.00	Eastern North America; W, Marinette, Wisconsin
	C. alternifolia 'Bachone' Gold Bullion™ dogwood	Mor695-2006*4	0.96±0.01	G
	C. alternifolia 'W. Stackman', Golden Shadows® dogwood	Mor595-2001*3	0.92±0.03	G
	C. controversa	Mor388-2007*2	0.85±0.00	East Asia; W, Liaoning, China
	C. controversa 'June Snow JFS', June Snow™ dogwood	Mor471-2009*1	0.91±0.01	G

There was one putative hybrid (*C.* 'JN Red Stem' Irish Setter™ dogwood) measured in this study. While the mean monoploid genome sizes of the putative parents (*C. racemosa* and *C. amomum*) are considered significantly different based on Tukey's HSD means separation (Table 2), the mean monoploid genome size of the putative hybrid was not found to be intermediate between the two by means separation. However, the putative hybrid was found to be intermediate with other taxa that fell within the Tukey's HSD Group E (Table 2). After personal correspondence with Mike Yanny, it has been determined that C. 'JN Red Stem' Irish Setter™ is most likely a hybrid between *C. racemosa* and *C. drummondii.*

The monoploid genome size of *Mesomara* was significantly smaller than that of *Kraniopsis I* and *II.* This reflects the smaller base chromosome number for this group.

CONCLUSIONS

Flow Cytometry is an efficient method for determining genome size in the BW group of *Cornus*. While cytological analysis of chromosome number was not completed, measured genome size seems to align with previous reports of diploid for taxa represented in this study. Monoploid genome size measured may support taxonomic relationships presented by Xiang et al. (2006) with further investigation and greater sampling.

The single putative hybrid measured in this study was found to be intermediate with two species with distinct relative genome sizes. As with the BB dogwoods, the potential for flow cytometry as a screening method for hybrids amongst the species of the BW could be of value for breeders.

Additional surveying of cultivated material may reveal unknown or unreported hybrids in the commercial industry. The smaller base chromosome number of *Mesomara* is evident in the relatively smaller monoploid genome size of 1Cx = 0.92 pg. Further investigation of cultivated material and species representative of taxonomic gaps including *Kraniopsis III* and *Yinquania* is needed to complete a genome size survey of *Cornus* L.

Table 2. Groupings of mean monoplid genome size values based on mean separating using Tukey's HSD at $\alpha = 0.05$.

Taxon	Mean monoploid genome size 1C× (pg)	Groups
Cornus sanguinea 'Cato' Arctic Sun®	1.31	a
C. racemosa (compact selection)	1.28	ab
C. sanguinea	1.26	abc
C.walteri	1.20	abcd
C. racemosa	1.19	abcd
C. macrophyllla	1.16	bcde
C. racemosa (putative hybrid)	1.15	bcde
C. drummondii	1.15	cde
C. bretschneideri	1.12	def
C. foemina	1.10	def
C. paucinervis	1.07	efg
C. pumila	1.04	fgh
C. alba 'Bailhalo' Ivory Halo® dogwood	1.02	fghi
C. sericea subsp. *sericea*	1.00	fghi
C. amomum (Easy Branch D)	1.00	ghi
C. alba	0.98	ghi
C. alternifolia	0.97	ghij
C. alternifolia 'Bachone' Gold Bullion™ dogwood	0.96	hij
C. amomum (Easy Branch B)	0.94	hij
C. amomum subsp. *obliqua*	0.93	ij
C. alternifolia 'W. Stackman' Golden Shadows® dogwood	0.92	ij
C. controversa 'June Snow-JFS' June Snow™ dogwood	0.91	ij
C. controversa	0.85	j

Literature Cited

Cappiello, P. and D. Shadow.2005. Dogwoods: The genus *Cornus*. Timber Press Portland Oregon.

Darlington, C.D. and Wylie, A.P. 1956. Chromosome atlas of flowering plants. MacMillan New York, New York.

de Mendiburu, F. 2019. Agricolae: Statistical procedures for agricultural research. R package version 1.3-1. https://CRAN.R-project.org/package=agricolae

Shearer, K., and Ranney, T.G. (2013). Ploidy levels and relative genome sizes of species, hybrids, and cultivars of dogwood (Cornus spp.). HortScience *48*(7):825–830.

United States Department of Agriculture. 2014. Retrieved from https://www.nass.usda.gov/Publications/AgCensus/2012/Online_Resources/Census_of_Horticulture_Specialties/

Xiang, Q.Y.J., Thomas, D.T., Zhang, W., Manchester, S.R., and Murrell, Z. (2006). Species level phylogeny of the genus *Cornus* (Cornaceae) based on molecular and morphological evidence — implications for taxonomy and Tertiary intercontinental migration. Taxon *55*(1):9–30.

Propagation of Hardy Begonia (*Begonia grandis* subsp. *evansiana*) from Seeds and Tubers

Zoe Schroeder, Sharon Kester, and Robert L. Geneve

Department of Horticulture, Lexington Kentucky 40546, U.S.A.

rgeneve@uky.edu

Keywords: *Begonia grandis*, seed germination, dormancy, hardy perennials, tubers

Abstract

Hardy begonia (*Begonia grandis* subsp. *evansiana*) has received a revival in popularity as a shade-tolerant, summer flowering perennial. It produces seeds and aerial tubers late in the summer under short daylengths. There is limited information on the propagation of hardy begonia related to seed or tuber propagation. Seed harvested from dried capsules collected from fruit showing some observable natural desiccation had a high percentage of filled seeds and these seeds germinated in less than 1 week. Aerial tubers form in leaf axils and tuber size varied based on node location on the plant. Tubers of various sizes were placed in plastic bags and placed at 10°C for 0 to 10 weeks. Tubers did not form plants without a chilling treatment. Plant formation began after 4 weeks of chilling, but the highest plantlet formation occurred after 8 and 10 weeks of chilling.

INTRODUCTION

Hardy begonia (*Begonia grandis* subsp. *evansiana*) is one of the tuber-producing begonia species and is the only reliably frost-hardy begonia (USDA Zone 6). Hardy begonia has received a revival in popularity as a shade-tolerant, late summer flowering perennial. It apparently survives winter temperatures as basal or dispersed aerial tubers. Aerial tubers form late in the summer under short daylengths and hardy begonia is one of the few begonias to form these aerial structures. Aerial tubers form in leaf axils and tuber size varies based on node location on the plant. Larger tubers generally form in the center of the axillary node and this tuber can be subtended by two smaller tubers. Hardy begonia may also persist and spread by seeds. Flowers form in late summer and fruits may not reach full maturity before frost. The objective of this study was to investigate the dormancy requirements for seed and tuber propagation of hardy begonia.

MATERIALS AND METHODS

Seed Afterripening and Germination

Seeds were harvested from swollen seed capsules following petal abscission and allowed to dry for several weeks in the capsules under ambient lab conditions. Seed capsules were also harvested later in the season from capsules that had undergone natural desiccation and initial seed dispersal. Dry seeds were germinated within 2 weeks of harvest or allowed to afterripen for 40 days in paper bags at ambient temperature and lab humidity.

Tuber Chilling

Mature tubers were collected in mid-October from hardy begonias grown outdoors. Tubers were separated by size (small or large) and placed in petri dishes filled with dry vermiculite sealed with parafilm. Dishes were placed in a cooler at 10 °C for 0, 2, 4, 6, and 10 weeks. Following chilling treatments, tubers were placed on the surface of a moist Pro-Mix® substrate and moved to short day conditions (8-h light). Approximately 100 small and 60 large tubers were used for each treatment. Shoot emergence from tubers was measured 2 and 5 weeks after tubers were removed from the cooler. Sprouted tubers were moved to greenhouse conditions to observe growth.

RESULTS AND DISCUSSION

Seed Afterripening and Germination

Seeds harvested from dried capsules collected from fruit prior to observable natural desiccation contained seeds that germinated slowly and at low percentages. Seeds harvested from dried capsules collected from fruit showing observable natural desiccation had a higher percentage of filled seeds and these initially germinated slowly (>2 weeks) apparently showing a degree of nondeep physiological dormancy. These seeds allowed to afterripen for 40 day germinated at high percentages in less than 1 week.

Tuber Chilling

Non-chilled tubers or tubers chilled for 2 weeks failed to initiate shoot development. Shoot formation began in tubers after 4 weeks chilling (<10% for both small and large tubers). The highest plantlet formation occurred after 6- or 8-weeks chilling with small tubers producing shoots at approximately 85%, while large tubers produced shoots between 65% and 75% (Fig. 1). The need for chilling treatment may support the idea that hardy begonia overwinters as dispersed aerial tubers, which will sprout in the following spring. Larger tubers produced larger plantlets compared to plants developed from small tubers (Fig. 2). Tubers displayed polar development and produced one shoot with one or more leaves at the tuber apex.

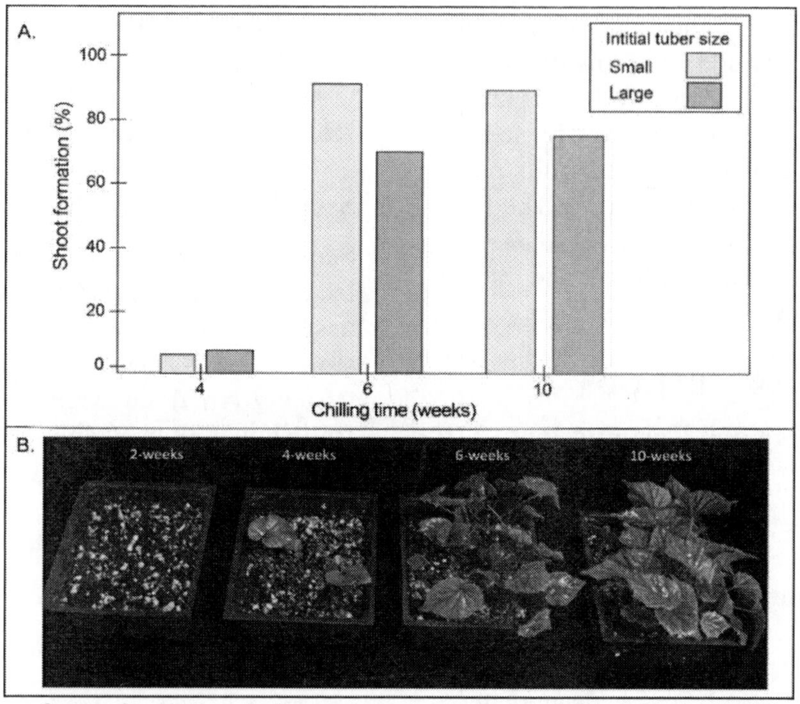

Figure 1. A. Shoot formation in hardy begonia tubers moist chilled for up to 10 weeks. B. Shoot formation after 5 weeks in tubers chilled between 2 and 10 weeks.

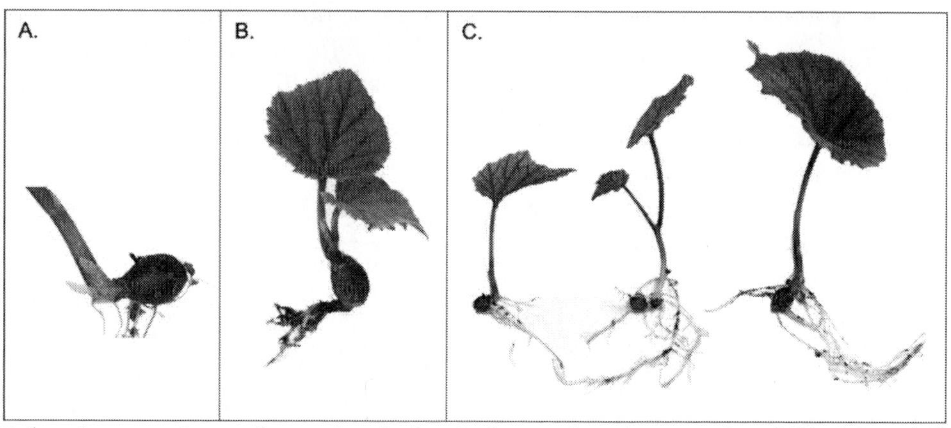

Figure 2. Polar shoot and root formation in hardy begonia tubers moist chilled for eight weeks. A. Shoots emerge from the apical end of the tuber. B. Apical shoot emergence is followed initially with roots from the basal end of the tuber. C. Eventually adventitious shoots will also form on the emerging stem.

Literature Cited

Agulló-Antón, M.A., Sánchez-Bravo, J., Acosta, M., and Druege, U. (2011). Auxins or sugars: what makes the difference in the adventitious rooting of stored carnation cuttings? J. Plant Growth Regul. *30*:100-113.

Normanly, J., Slovin, J.P., and Cohen, J.D. (2004). B1. Auxin biosynthesis and metabolism. In: Davies PJ (ed.) Plant hormones. Biosynthesis, signal transduction, action. Kluwer Academic Publishers, Dordrecht, pp 36–62.

A Legacy of Elm Improvement at Morton Arboretum

Kim Shearer

The Morton Arboretum, 4100 Illinois Route 53, Lisle, Illinois 60532 U.S.A.

kshearer@mortonarb.org

Keywords: Elm, *Ulmus*, breeding, hybridization, new plants

INTRODUCTION

The Morton Arboretum

In 1922, after some consultation with Director of Arnold Arboretum Charles Sprague Sargent, Joy Morton founded the Arboretum on a 400-acre private estate in the country side of Lisle, Illinois just west of Chicago. Morton was best known for his founding of the Morton Salt Company but wanted to leave behind a lasting legacy that would have a positive impact.

Morton was the son of J. Sterling Morton, originator of Arbor Day and Secretary of Agriculture under President Cleveland, and Caroline Joy French, an avid gardener. Their family motto "Plant Trees" proved to be a driving force for Morton as he approached retirement. By the time Morton passed in 1934, the Arboretum had already grown to 735 acres and included an herbarium, library, collection of living plants, nurseries, and a staff to manage it all.

Today, nearly 100 years later, the Morton Arboretum is a well-visited 1700-acre tree museum and lab with a mission focused on collecting trees and shrubs from around the world, displaying them in naturally beautiful landscapes, promoting the planting of trees, and making the world a greener, healthier and more beautiful place.

Acknowledgements

Thanks to Kris Bachtell for sharing photos, documents, experience, and expertise.

Thanks to Jim Ault for managing the Chicagoland Grows® program that is the introduction vehicle for Morton Arboretum elm selections and for providing photos.

ELMS AND DUTCH ELM DISEASE IN THE CHICAGO REGION

Elms (*Ulmus* L.) are a widely distributed genus of trees with approximately 35 species found across the temperate and subtropical regions of the northern hemisphere and a species reported in the southern hemisphere on the Indonesian island of Sumatra (Fu et al., 2004; Melville and Heybroek, 1971; Press et al., 2000; Sherman-Broyles, 1997; Todzia and Panero, 1998). There are 10 species of elm native to North America (Sherman-Broylese, 1997), but historically the most widely planted elm in the United States was *Ulmus americana* L.

In the early 1930s, Dutch elm disease (DED) was introduced to the United States by a shipment of logs imported from Europe (Marcotrigiano, 2017). The disease, *Ophiostoma ulmi* , is a fungus vectored by the European bark beetle which burrows into the host tree and creates galleries in the vascular tissues where the fungus is deposited and spreads. As American elms were widely planted as street and park trees all across the country as well as being native to floodplains, the disease spread rapidly and wiped out much of the urban and suburban tree canopy. Beginning as early as the mid-1930s, newspapers such as the Chicago Tribune began raising the alarm on the widespread loss of trees in the east as the disease proceeded to spread to the Midwest (Chicago Tribune, 1934). By 1970, the Chicago region was reported to have lost more than 50 thousand trees. At the time, there were 750 thousand to one million American elms planted in the region, and municipal foresters expected an additional loss of 50 thousand trees by 1972 (King, 1971).

GEORGE WARE, THE THORNHILL ELM, AND BUILDING GERMPLASM

As DED was ravaging the trees of American neighborhoods, parks, and forests, the Morton Arboretum was focused on elevating research at the institution and hired Dr. Marion Trufant Hall, a well-rounded and experienced scientist, as director in 1966. Dr. Hall, charged with the task of increasing the research capacity of the Arboretum slowly and steadily, reached out to his good friend and former classmate Dr. George Ware. In 1968, Dr. Hall hired Dr. Ware to be the Ecologist and Dendrologist (Doty, 1987).

Four years into Dr. Ware's tenure at the Arboretum and after some time observing the impact of DED on the landscape, he noticed a particular tree that was thriving with lush, healthy foliage and an arching canopy reminiscent of the beloved American elm. This tree would come to be referred to as the "Thornhill Elm" (Morton Acc. No. 2352-24*1), and would be the inspiration for initiating the Elm Improvement Program at the Arboretum. During the initial phase of acquiring plant material for establishing live collections at the Arboretum, a packet of seed labeled *Ulmus crassifolia* was sent to the Morton Arboretum from Arnold Arboretum (Jamaica Plains, Massachusetts) in 1924. Said seed was propagated, and trees were planted out on the grounds. Eventually, it would be identified as a hybrid of *U. japonica* × *U. wilsoniana* (synonym, *U. davidiana* var. *japonica*).

What Dr. Ware recognized in this tree was the potential for developing elms with DED resistance perhaps due to the coevolution of the Asian elm species with the pathogen. By 1980, Ware had begun the propagation and evaluation of this tree and was in search of more germplasm to build a breeding program. In the Journal of Arboriculture, Ware would publish two articles in 1980, both with a focus on describing the qualities necessary for trees to survive in urban environments and the potential of broadly distributed Asian elm species not yet introduced to the American nursery industry (Ware, 1980a; Ware, 1980b). As outlined by Ware, these traits were tolerance of extreme temperatures, drought and flood, high winds, blizzard, and "hostile" soils. These "hostile" soils that Ware referred to are high in pH, have poor aeration, and hardly any organic matter present. And while it is noted that these are common conditions of landscapes in the Midwestern and Great Plains landscapes, it was equally noted that these are the very conditions trees face in developed landscapes irrespective of the "native" region.

In addition to providing a list and description of Asian elm species to consider incorporating into elm breeding programs (Table 1), Dr. Ware would call for the enrichment of North American public garden collections.

Table 1. A list and description of elms suggested as potential urban trees by Ware Journal of Arboriculture and Landscape Plant News (1980a, 1980b, 1996b). Distribution and description adapted from Ware papers.

Species	**Geographic distribution**	**Ware description**	**Selections available in U.S.A. nursery trade?**
Ulmus davidiana Planch. (syn. includes *U. japonica, U. wilsoniana, U. propinqua*)	China, Manchuria, Japan, Korea, Mongolia, Siberia	Variation in habit; tolerant of hostile conditions	Yes; many introductions made in the past couple decades
U. glaucescens Franch.	Kansu Province (China), Northern China	Small tree; small leaves, fine texture; yellow to orange fall color; tolerant of urban conditions based on distribution	No
U. laciniata Göpp.	Humid areas of Manchuria, northern China, Korea, Siberia, and Japan	Small to medium tree; potential drought hardiness; lobed leaves; *Zelkova*-like branching	No
U. macrocarpa Hance	China, Mongolia, Korea, and Siberia	Strong wood; shrub to medium sized tree; adapted to humid/arid regions; tolerant of "hostile" conditions	No
U. parvifolia Jacq.	China, Korea, Japan	Tolerant of drought, pollution, poor soils; attractive lace bark; glossy leaves	Yes; many introductions made in the past couple decades

This was due to the limited provenance representation available at the time. As Ware highlighted, Asian elms were broadly distributed across the Asia and Eastern Europe, yet taxa were only represented by a handful of accessions representing a limited number of populations. This ultimately would result in a completion of five plant exploration expeditions in China and three in what was then considered the Soviet Union; and many collaborative efforts with fellow tree researchers around the world including foresters in China who would send wild collected seed to Ware that would be included in the breeding program. Today, in large part due to Ware's efforts, the Elm Collections of Morton Arboretum include 355 individual trees representing 25 species, 17 unnamed hybrids, 38 cultivars, and a grex making it one of the most comprehensive elm collections in the world.

The Elm Improvement Program

As Ware was building a rich breeding germplasm collection, he began evaluating seedlings in the Morton Research Field (now named Ware Field). With the assistance of his technician Mike Spravka and then propagator Kris Bachtell, Ware began selecting seedlings and having them cloned for the development of an elm breeding isolation block. Selection criteria that Ware developed included DED and elm yellows tolerance, pest resistance, cold hardiness, vigor, and red foliage. In a group of Asian elm seedlings, Ware noted the red pigmentation of emerging leaves. This new trait inspired a breeding objective for improved fall color in elms that generally displayed a range of yellow fall color. In addition to selecting chance seedlings from wild collected germplasm, Ware made controlled crosses to develop new breeding parents and identify potential selections for the nursery industry.

By 1995, Ware would finally send scion material to cooperating nurseries for propagation and field production evaluations. These nurseries included wholesale liner producer J. Frank Schmidt & Son working in collaboration with Keith Warren, and Microplant Nurseries, Inc., working in collaboration with Gayle Suttle. At the time, there were no Asian elms available through the nursery industry. In the time since, the Morton Arboretum has introduced five elm cultivars through the Chicagoland Grows® plant introduction program, all developed through chance seedling selection and controlled crosses conducted by Ware himself (Table 2), and today propagated via microcuttings, softwood cuttings, and grafting.

Elms Today

All across the country, Asian elms are planted in the developed landscapes of cities and suburbs. These include more than 13 cultivars of elms in what is known as the *U. davidiana* complex (*U. japonica, U. wilsoniana, U. propinqua*) (Dirr and Warren, 2019) in addition to at least 13 cultivars of the lacebark elm (*U. parvifolia*). The original "Thornhill Elm" once distributed to the Morton Arboretum by the Arnold Arboretum now is present in the landscape of the city of Boston, coming full circle in its journey through tree cultivation. In the years of 2017-2018 alone, there were approximately 97 thousand liners of Morton introductions sold in the wholesale market.

Table 2. A list and description of elm cultivars developed at Morton Arboretum by the Elm Improvement Program led by Dr. George Ware. Note that *Ulmus japonica, U. wilsoniana* are taxonomic taxa that make up the *U. davidiana* species complex, but they are listed here as the original species for the sake of simplicity. Information found in this table is adapted from Chicagoland Grows® Plant Release Bulletin #44.

Cultivar and trade name	Parentage/origin	Traits	USDA Hardiness Zones	Dimensions (feet)
Ulmus 'Morton' Accolade™ elm	Chance seedling *U. japonica* × *U. wilsoniana*	Vase-shaped habit and vigorous grower; foliage fine-textured, dark green, and glossy with yellow fall color; DED and elm yellows resistance; resistant to elm leaf beetle	5 – 8	**20 year** 30 ft H, 15 ft W **Mature** 50 – 60 ft H 30 – 40 ft W
Ulmus 'Morton Plainsman' Vanguard™ elm	Chance seedling U. *japonica* × *U. pumila*	Relatively upright branching and rounded habit in youth; requires corrective pruning to avoid included bark; dark green foliage with yellow fall color; DED and elm yellows resistant; susceptible to elm leaf beetle, Japanese beetle, and leaf miner.	5 – 7	**Mature** 45 – 50 ft H 40 – 50 ft W
Ulmus 'Morton Glossy' Triumph™ elm	Controlled cross *U.* Accolade™ × *U.* Vanguard™	Grower favorite due to ease of training; lustrous dark green foliage with yellow fall color; upright oval form that ages to vase shape; strong branching; excellent DED resistance; moderate pest resistance	4 – 9	**Mature** 50 – 60 ft H 40 – 50 ft W
Ulmus 'Morton Stalwart' Commendation™ elm	Controlled cross *U.* Accolade™ × (*U. pumila* × *U. carpinifolia*)	Symmetrical arching branches, upright oval habit; large, dark green leaves with yellow fall color; rapid growth and broad adaptability; excellent DED resistance; moderate susceptibility to elm leaf beetle, Japanese beetle, and gypsy moth.	(4)5 – 9	**Mature** 50 – 60 ft H 40 – 50 ft W
Ulmus 'Morton Red Tip' Danada Charm™ elm	Chance seedling *U. japonica*	Rounded habit in youth maturing to large and elegant vase-shape; fast grower; glossy green foliage with red-pigmented new growth; yellow fall color; excellent resistance to DED and elm yellows; moderate susceptibility to Japanese beetle and elm leaf beetle.	(4)5 – 9	**Mature** 60 – 70 ft H 50 – 60 ft W

The Next Generation of Ware Elms

While Ware worked diligently to make the American landscape green again through the introduction of Asian elm selections, he also was sure to secure a future for the next generation of elms. In addition to generously distributing elm seedlings to public gardens, researchers, and breeders around the United States, and perhaps beyond, Ware planted a breeding population in a safe space where selected trees would have a chance to mature, flower, and cross pollinate. The composition of this breeding population includes the *U. davidiana* complex, *U. macrocarpa*, *U. minor*, *U. parvifolia*, *U. pumila*, and *U.* 'Sapporo Autumn Gold'.

In 2018, 3000 seeds were collected from three female parents (*U. davidiana* complex) selected for their form and clean foliage. Seeds were collected as they were just beginning to ripen and samaras were turning yellow. Only seeds showing full development of embryos through the samara were sown. Seeds were sown immediately in Ray Leach Cone-tainers™ and germinated summer 2018. During the fall of 2018, seedlings began to shut down and display fall color exhibiting primarily yellow fall color with a minority of seedlings exhibiting variations of red fall color. In 2019, 300 seedlings were selected and grown on in containerized production. Trees will be spring planted in 2020 and further evaluation of fall color will continue in the field.

The Lacebark Elm

While Ware generally focused on the development of elms in the *U. davidiana* complex, he did publish a note in Landscape Plant News (1996a) regarding the lacebark elm (*U. parvifolia*.) This note was written following a USDA-sponsored US-China Scientific and Technical Exchange Program that included Dr. Raymond Guries and Dr. Weidong Wu (UW-Madison), Dr. Mark Widrlechner (USDA-Ames), and Joan Smalley. The objectives of this journey were to photograph, seed-collect, and determine the natural range of *U. parvifolia*. While in a following note of the same publication (1996b) Ware notes that *U. parvifolia* would likely not be hardy in northern states, at the Morton Arboretum these trees were grown and evaluated. Today, there are 16 *U. parvifolia* and three unnamed hybrids of *U. parvifolia* in the Morton collections that have survived multiple polar vortex events.

The lacebark elm is broadly adapted tolerating heat and some cold, flooding and drought, hostile soils, and humid and arid conditions. It can be seen planted all over the southern United States from the east coast to the west coast. While it has not been planted as widely in the Midwest, it is already listed as a weed of concern in Wisconsin (Hoffman and Kearns, 1997). With this in mind, the new plant development program at Morton Arboretum has a lacebark elm breeding program focused on developing triploid selections with cold hardiness in the Midwest. We will focus on using cold hardy individuals present in the Arboretum collections and hybridizing with a tetraploid individual developed and shared by breeder Dr. Tom Ranney of the Mountain Crop Improvement Lab at North Carolina State University.

Literature Cited

Dirr, M and Warren, K. 2019. The Tree Book. Timber Press, Portland, Oregon.

Doty, C. 1987. Marion Trufant Hall, third director of The Morton Arboretum: A twenty-year retrospective: 1966 – 1986. The Morton Arboretum Quarterly *22* (4):49–54.

Fu L, Xin Y, and Whittemore AT. 2004. *Ulmus* In: W. Zheng-yi and PH Raven (eds.). Flora of China vol. 5. Missouri Botanical Garden Press St. Louis, Missouri. http://www.efloras.org/florataxon.aspx?flora_id=2&taxon_id=134116

King, S.S. "Dutch elm disease spreads westward" *New York Times*. Sep. 7, 1971. Accessed online 2019 <nytimes.com>

Melville, R. and Heybroek H.M. 1971. The elms of Himalaya. Kew Bulletin 26(1):5–28.

Press, J.R., Shrestha, K.K., and Sutton D.A. 2000. Annotated checklist of flowering plants of Nepal. The Natural History Museum, London http://www.efloras.org/florataxon.aspx?flora_id=110&taxon_id=134116

Sherman-Broyles, S. 1997. *Ulmus* In: FNA Editorial Committee(ed.) Flora of North America vol. 3. Oxford Univ. Press New York, N.Y. http://www.efloras.org/florataxon.aspx?flora_id=1&taxon_id=134116

"The elm in danger" *Chicago Daily Tribune*. Nov. 9, 1934. Accessed online 2019 <newspapers.com>

Todzia, C.A. and Panero, J.L. 1998. A new species of *Ulmus* (Ulmaceae) from southern Mexico and a synopsis of the species in Mexico. Brittonia *50*:343–347.

Ware, G.H. 1980a. Little-known Asian elms: urban tree possibilities. J. Arbor. *6*:197–99.

Ware, G.H. 1980b. In search of new kinds of elms. J. Arbor. *6*:233–37.

Ware, G.H. 1996a. Notes on elms observed on a trip to China. Landscape Plant News *7* (10):4–6.

Ware, G.H. 1996b. New elms for urban landscapes. Landscape Plant News *7* (10):6–8.

Turning Science into Business: A 40-Year Perspective on Micropropagation

Gayle R. Ladiser Suttle

Microplant Nurseries, Inc., Gervais, Oregon 97026 U.S.A.

gayle@microplantnurseries.com

Keyworks: Commercial micropropagation, biotechnology, business development

INTRODUCTION

The talk I presented to the Eastern Region this year was a pictorial story of our journey at Microplant over the past 40 years in the field of commercial micropropagation. My talk covered some of the most important developments, disasters, and discoveries we have experienced along the way. Microplant is owned by Treco® and Peter K. McGill.

In the late 1970s, there were several labs in America producing volumes of house plants, ferns, orchids, rhododendrons and azaleas using micropropagation (tissue culture) techniques. There was some micropropagation of peach almond trees going on in Italy, and University research papers being presented on a host of different kind of trees, but nobody in the world was attempting to mass produce large volumes of apples, pears, cherries, and shade trees such as maples, birch, plum, and crabapple via tissue culture (Fig. 1).

A partnership between Pete McGill, Treco®, Stark Brothers® and Adams Rootstock was formed and Microplant Nurseries, Inc. was launched in 1980 with that goal in mind.

Figure 1. Rooted microcutting of maple.

I arrived on scene that year a bright-eyed 25-year old college graduate with Science (Botany) and Agriculture (Soils) degrees, hired as "Chief Tissue Culture Technician" ready to create a million plantlets the first year. Within a few month's it was very clear that the current technology was not ready to make that happen. It eventually did though.

Fast forward from that seemingly impossible goal to our present-day situation where, over the years, we have produced several hundred million young trees, shrubs, grasses, perennials, bulbs, and small fruits for the horticultural nursery trade. We have partnered with growers, farmers and researchers to bring some of the most exciting new plants into landscapes and fields around the world.

LESSONS LEARNED FROM TRIAL AND ERROR

It has not been easy. As every person who has ever attempted to micropropagate any amount of anything knows, (and will be happy to tell you), it is very difficult to do this reliably. As we have grown our Company from two people to over 70 people, we have learned from trial and error, making many mistakes, and discovering some important lessons. Here are a few examples:

Cold Storage Is A Powerful Tool

We have found that we can store certain plants for many months at 2–4 °C. This smooths out production peaks and saves labor during maintenance of crops. Also, *Malus* and *Pyrus* actually benefit from a cold treatment of 2–4 °C for at least 1,000 h prior to acclimatization, especially when the greenhouse step is done under cold short-day conditions (Fig. 2).

Figure 2. Pear rootstock OH×F-333 planted at the same time during cool/short days. Left image is without and right is with vernalization period.

Inorganic Salts

Inorganic salts play the most important role in media optimization for difficult to grow plants (Fig. 3).

Figure 3. *Acer* on basic formula (left) and *Acer* after computer aided salts optimization (right).

Cutting Plants is an Art

The discovery of where to cut and how to trim plants for maximized multiplication or rooting is critical to successful optimization (Fig. 4). There can be nuanced differences even between very closely related genetics.

Figure 4. The cutting of an apple rootstock over several generations.

SOME BASIC BUSINESS HINTS

Here are a few key nuggets we have learned along the way that guide us every day in growing our business:

1) Know what your goal is.
2) Check your progress toward the goal often.
3) Adjust if necessary and let the customer know.
4) Make it easier to do things right, harder to do things wrong.
5) Do it excellently and have fun doing it.
6) In business, it needs to work for the customer. If it doesn't work for them, you have no business.
7) Do it once well. Baling wire works great but put in on knowing it likely needs to last forever.
8) Invest yourself in IPPS. It will provide you with ideas, information and most of all friends who will inspire you, console you and bring you great joy in your journey together.

In conclusion, over the past four decades I have seen micropropagation become a major mainstream power tool used throughout the nursery industry worldwide to solve huge problems. It will be exciting to see what happens next.

New Plant Forum 2019 – Eastern Region IPPS

Charles Tubesing, Moderator
Holden Arboretum, Ohio USA

ctubesing@holdenfg.org

PRESENTER

Steve Castorani
North Creek Nurseries, Inc., Landenberg, Pennsylvania 19350 U.S.A.
steve@northcreeknurseries.com

Iris versicolor 'Purple Flame' blue flag iris

We're fortunate to have the splendor and garden grace of Mt. Cuba Center in our backyard. We're even more fortunate to be sharing their new, sultry selection of *I. versicolor* with the trade — welcome *Iris* 'Purple Flame'. Discovered by a staff gardener along a pond edge on-property —

this drumroll-worth selection marches to its own beat as it vibrantly announces the return of spring.

An extraordinary performer in Mt. Cuba's gardens (and our trial evaluations as well!), we've been awed by the lasting impression of 'Purple Flame'. Plant along the water's edge, where roots receive consistent moisture to achieve best early season color before foliage slowly transition to green.

Plainly stated, it's a show stopper. Beyond other *Iris* selections on the market, 'Purple Flame' has a richer, more intense cast to stems and flowers, yes. However, the true beauty lies in the irresistible foliage. Emerging in March, the purple flame-like foliage provides an unparalleled performance of vivid and intense eggplant purple foliage followed by an encore of rich and abundant flowers. Height is 2–3 ft. with a USDA Hardiness Zone 2–7.

Lysimachia lanceolata var. *purpurea* lance-leaved loosestrife

Lysimachia lanceolata var. *purpurea*, or purple lance-leaved loosestrife, is a native plant endemic to North America east of the Mississippi. The wine-red foliage of *Lysimachia* slowly creeps in the garden and grows 1 -2 ft tall with an equal spread. Throughout the summer, the diminutive flowers twinkle and bloom a bright yellow, contrasting with the maroon leaves. Lance-leaved loosestrife is a quiet plant, settling along garden borders and along hillsides and at the base of trees or shrubs. What we love about this plant is its sunny tough-as-nails attitude, heat tolerance, and how it steadily spreads it rhizomes, covering the ground and suppressing weeds while supporting pollinators and wildlife. It's easy to grow, delightful all year round, and is the living embodiment of 'green mulch'. We couldn't ask for more!

The short stature, continuous sunny bloom, and deep wine coloring of this native plant make *L. lanceolata* var. *purpurea* a new favorite groundcover, winning over our R&D Department by how it keeps weeds at bay and is a carefree plant. Height is 12-24 in. and spread is 12–24 in.; USDA Hardiness Zone 4–8.

PRESENTER

Bill Hendricks
Klyn Nurseries, Perry, Ohio 44081 U.S.A.
bhendricks@klynnurseries.com

Carpinus betulus 'Lucas'

'Lucas' has a dense form with a narrow, columnar habit staying narrower than *Carpinus betulus* 'Fastigiata'. The rugose foliage remains dark green through the growing season, turning brown in autumn. The dried foliage remains on the tree until pushed off in late March to early April as buds swell. Other upright cultivars of *C. betulus* hold foliage into winter but drop much earlier than 'Lucas'. The retained foliage can be an advantage where a winter screen is preferred. Selected in Belgium. 'Lucas' has a projected height of 20–30 ft and spread of 10–12 ft; hardy to Zone 4.

Diospyros virginiana 'WFHgodel', **Golden Delight™ persimmon**

The tree was selected from a group of seedlings and displayed unique traits in size of fruit (larger), flavor, and early ripening. In northeast Ohio the flavorful fruit ripens in early October with no astringency. Native persimmon fruit typically ripens after frost and in northeast Ohio this happens around the end of October to early November. Trees are vigorous when grown from grafts or budding (height: 35–60 ft, spread: 20–35 ft) and can be trimmed into a shapely tree with a broadly pyramidal form. Hardy to Zone 4.

***Carpinus caroliniana* 'MY1', Autumn Fire™ American hornbeam**

Autumn Fire™ American hornbeam brings a new form to our native American hornbeam. Not only is it a true fastigiate form with crisp green summer foliage but also sports amazing shades of burgundy, orange, and red foliage in the fall with color that can persist into November. It has proven to be a vigorous grower and brings a new possibility for design where a narrow upright small to medium sized tree is needed. Selected as a seedling from native trees in Lake County, Ohio. Height is 25 ft and spread is 6–8 ft; hardy to Zone 3.

Brent Horvath
Intrinsic Perennial Gardens, Inc.,
10702 Seaman Rd, Hebron, Illinois 60034 U.S.A.
BrentH@intrinsicperennialgardens.com

Bouteloua gracilis 'Honey Blonde' PPAF

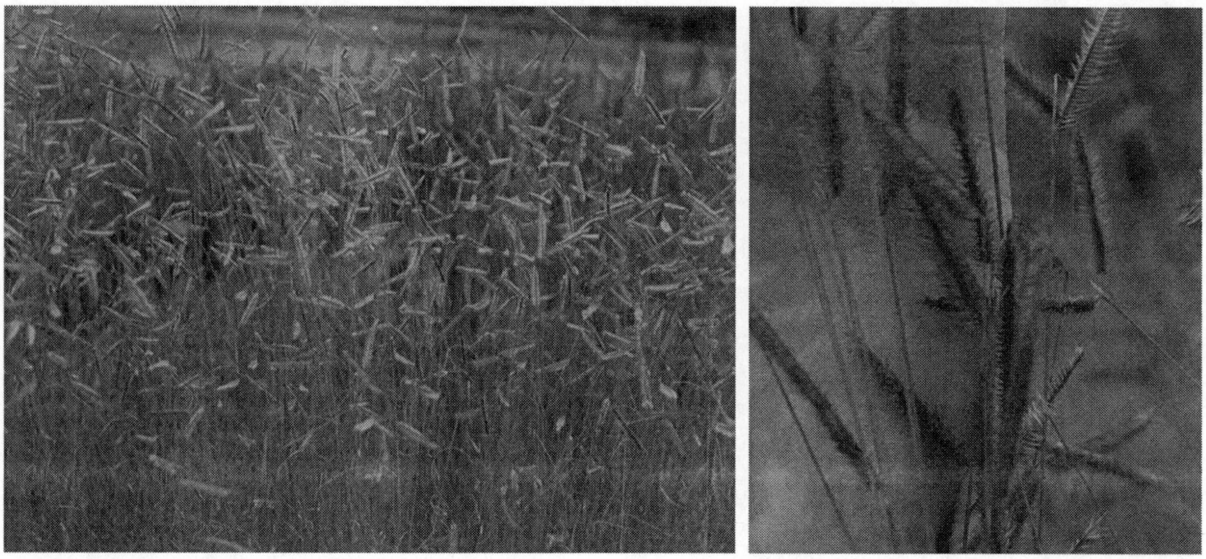

An Intrinsic Introduction. Hardy and adaptable plant with good vigor has a similar size and stature compared to 'Blonde Ambition'. Upright wire like stems stand 30 in. plus and begin blooming with eyelash like flowers, holding yellow pollen in June. The tan colored seed heads lighten with age holding on into winter. Full sun, average to well-drained soil.

Panicum virgatum 'Prairie Dog' PPAF

An Intrinsic Introduction. Selected for a more reserved habit growing 4–5 ft tall with clean blue foliage on upright stems. Heavy flowering plants bloom on the top third of the plant. Full sun, most any soil.

Pennisetum alopecuroides 'Love and Rockets' PPAF

An Intrinsic Introduction! After 19 years of breeding *Pennisetum* we finally have a small red flowered fountain grass in league with 'Hameln'. Dark green foliage blends nicely with the wine-red flowers softening in color by fall. Mature plants reaching around 24 to 30 in. tall. Red tipped foliage at times adds another dimension. Full sun. Most any soil but wet.

Michael Yanny
JN Plant Selections, LLC, Milwaukee,
53225 Wisconsin U.S.A.
jnplantselections@gmail.com

Aesculus × *bushii* 'Aaron1' PP29,092,
Mystic Ruby™ buckeye

Mystic Ruby™ buckeye was found in a seedling block of *Aesculus glabra* at Johnson's Nursery in Jackson, Wisconsin by Aaron Jambura. It is thought to be a hybrid between *A. glabra* and *A. pavia*. The plant bears showy, reddish pink flowers in mid to late spring. It is a relatively small-scale tree, growing to about 18 ft tall and 8 ft wide in 20 years. Mystic Ruby™ buckeye has better than average scorch and blotch resistance. It is fruit bearing and is probably the hardiest red flowering *Aesculus* in existence. It survived the polar vortex in Wisconsin in 2013 and -27 °F in 2019. A young, single plant located about 20 miles east of Fargo, North Dakota overwintered fine in the landscape after experiencing temperatures of about -40 °F in January of 2019. Licensing is being handled by Upshoot.

Buxus 'Prostrate 3' PP29,574, Flat-tery™ Boxwood

Flat-tery™ boxwood is a low growing boxwood for use as a groundcover or accent in shady locations. The plant can grow to 24 in. tall and 4 ft wide in about 25 years. It is maintained by a light annual top shearing and a minimal side shear to keep its round shape. Plants are nearly flat-topped. It maintains a good green winter color and is hardy to Zone 4. No discernable tolerance to boxwood blight has been observed as yet. Selected by Michael Yanny. Marketing and licensing are handled by Upshoot.

Rhus aromatica 'Fine Texture Compact Select A' PP28,669, Lacette™ Fragrant Sumac

Lacette™ was selected by Michael Yanny for being a well branched, compact plant with small leaves that give it a fine textured look in the landscape. Plants will get 4 ft tall by 7 ft wide in about 8 years with minimal shearing. An annual, heavy spring shearing can keep plants lower for use in mass plantings. Lacette™ makes a nice hedge in tough urban sites. Fall foliage colors can be red to maroon in most years. The plant is resistant to eriophyid mite gall damage which is common on many *R. aromatica*. This plant roots readily from softwood cuttings and is not as problematic as 'Gro-low'. Care should be taken when cutting plants because exposure to the sap can cause dermatitis in some individuals. Marketing and licensing is being handled by Plants Nouveau.

Syringa pubescens subsp. *patula* 'JN Upright Select' PP28,959, Violet Uprising™ Manchurian Lilac

An upright, barrel shaped lilac that is useful as a hedging plant. Individuals can grow to 6 ft tall by 4–5 ft wide when sheared annually. It has violet colored inflorescences in mid to late spring that fade to white. The flowers have a sweet lilac fragrance. Plants should be hedged immediately after flowering to remove the spent flowering stalks. Late pruning will result in cutting off next seasons flower buds. Violet Uprising™ Manchurian Lilac can have attractive maroon and yellow fall foliage colors. It's likely hardy from Zones 3–7. It roots readily from softwood cuttings in late spring. The plant was selected by Michael Yanny. Marketing and licensing are handled by Plants Nouveau.

Tilia americana 'Kromm' PPAF, Sweet Street™ Linden

Sweet Street™ is a tight pyramidal form of American linden with fragrant flowers. This cultivar should make an excellent street tree. It shows good resistance to Japanese Beetle feeding in southern Wisconsin. The plant was select by Darrell Kromm of Reeseville Ridge Nursery from a Dodge County, Wisconsin provenance. Licensing is being handled by Plants Nouveau.

Breeding, Selection and Evaluation and Propagation in the NDSU Woody Plant Improvement Program

Todd P. West

North Dakota State University, Plant Sciences, Loftsgard Hall 266E, Fargo, North Dakota 58102, U.S.A.

Todd.P.West@ndsu.edu

Keywords: Breeding, new plants, woody plants, winter hardy

INTRODUCTION

Woody plant evaluations at North Dakota State University began in 1954. In 1971, Dr. Dale E. Herman initiated the Woody Plant Improvement Program (WPIP). Woody plant introductions begin in 1986 with the collaborative release of Meadowlark forsythia (*Forsythia* 'Meadowlark') with South Dakota State University, Arnold Arboretum, and North Carolina-Regional Plant Introduction Station. To date, this program has released 58 woody plant selection into the ornamental nursery trade.

Diversity is important and there is a great need for adapted, winter hardy, pest resistant woody plants suitable for use in the northern U.S.A. and prairie Canada. Many of the current commercially available nursery cultivars are not suitable for USDA cold climatic Zones 3 and 4, lower annual moisture availability, and higher soil pH levels.

There is also a need to increase plant diversity in response to disease and insect pest issues and loss of adapted genera and species (*Betula* spp., *Fraxinus* spp., *Picea* spp. and *Ulmus americana*).

The WPIP has two primary goals:

1) Evaluate unreleased or released cultivars from the nursery trade to determine usability in the United States Northern Great Plains
2) Increase diversity through selecting and/or breeding new woody plants suitable for the U.S.A. Northern Great Plains.

The NDSU WPIP program woody plant selections are ideal for urban climate conditions. Urban soils are typically compacted, dry, and have a high pH (>7.5). North Dakota is considered to be the eighth driest state with respect to annual precipitation in the United States. North Dakota (ND) soils are typically alkaline with a pH >8.0.

The WPIP has nine research evaluation sites in North Dakota (Fig. 1.) There are three primary research evaluation sites:

1) NDSU Horticulture Research Farm (HRF) and Dale E. Herman Research Arboretum (DEHRA) (Absaraka, North Dakota, USA; Lat:46.9859, Long: -97.3549)
2) Research plots (Fargo, North Dakota),
3) NDSU Langdon Research Extension Center (Langdon, North Dakota; Lat: 48.7631, Long: -98.3713).

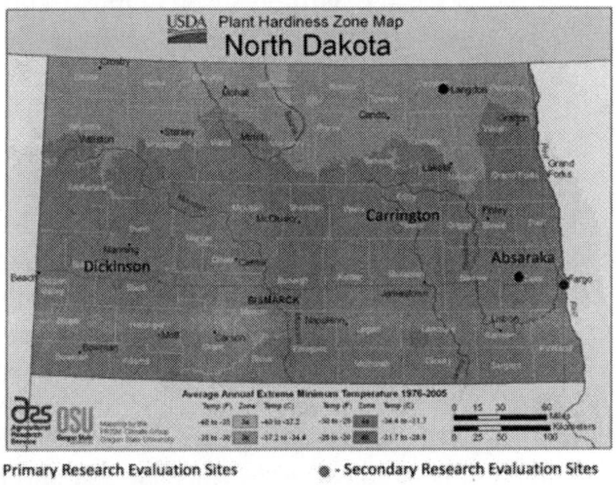

Figure 1. Evaluation site locations in ND for the NDSU Woody Plant Improvement Program.

The other secondary sites include Williston, Grand Forks, Dickinson, and Bismarck, ND. The first two primary sites are located in a USDA plant hardiness Zone 4a while the NDSU Langdon REC is classified as a hardiness Zone 3b. This past winter (2018-2019) was the coldest on record for the last 25 years. Most of ND reached winter average temperature lows that were less than -34 °C. The NDSU HRF and DEHRA has a low of -38 °C putting this location into a USDA hardiness Zone of 3b. the Langdon REC was classified with dropping into a USDA hardiness Zone of 2b.

EVALUATIONS AT NDSU WPIP

The NDSU WPIP has evaluated 200+ genera and 3,000+ species and cultivars of trees and shrubs. Over 9500+ accessions obtained, evaluated since planting began in 1974. The largest and most diverse woody ornamental plant collection in North Dakota and the Northern Great Plains is located at the NDSU HRF and DEHRA with a total of 80 acres (~32 ha). The NDSU WPIP program focuses on obtaining germplasm materials based on the Köppen-Geiger climate classification which places North Dakota in a *Dfb* with a warm-summer humid continental climate; coldest month averaging below 0 °C, at least 4 months averaging above 10 °C, and no significant precipitation difference between seasons. This *Dfb* classification allows for matching other locations with the same *Dfb* classification around the world to focus on suitable sources of germplasm.

The NDSU WPIP is involved with several woody plant evaluations including cultivar comparison with several industry cooperators and private breeders. It is also the northern site for hybrid maple evaluations (*Acer palmatum* × *A. pseudosieboldianum*).

For plant evaluation, selections and breeding, germplasm is collected from three different methods including:

1) Foreign and domestic seed sources (growing out seedling populations and selection individuals with superior attributes),
2) Plant breeding (tradition breeding including F2 populations to observe segregation of traits including hybridizing with both intra and interspecific hybridization),
3) In vitro tissue culture utilizing somaclonal variations, embryo rescue and mutagenesis.

Three plant improvement methods utilized are: Selections by landscape observation, mass selection (seed source and seed lot variation), and breeding (both traditional and mutagenic).

GENERA OF PRIMARY BREEDING FOCUS

The NDSU WPIP is focusing on breeding four primary genera: *Acer*, *Magnolia*, *Sambucus*, and *Ulmus*. The primary goal of all of the breeding work is on increased hardiness and secondarily on aesthetic improvements. With the large germplasm collection located at the NDSU HRF and DEHRA, there are many accessions that have shown outstanding hardiness and make excellent parents for improvement through breeding efforts. These include Spring Welcome® magnolia (*Magnolia* × *loebneri* 'Ruth'), Fall Grandeur™ red maple (*Acer rubrum* 'Minnkota'), *Sambucus nigra* 'TS14019' (prostrate form), and Northern Empress® Japanese elm (*Ulmus davidiana* var. *japonica* 'Burgundy Glow').

Magnolia breeding objectives focus on flower tepal color, introducing any color from *M. acuminata* hybrids coupled the hardy Spring Welcome® selection (white flower color). Maple (*Acer* spp.) breeding objectives are utilizing known hardy and environmental tolerant selections to develop a better adapted Freeman maple (*A.* × *freemanii*). The current selections, such as Autumn Blaze, do not have consistent performance with respect to pH tolerance and hardiness. Utilizing a red maple selection that is known to be pH tolerant and have outstanding hardiness would be better suited for a Freeman maple hybrid selection.

Elm breeding objectives focus on crossing Northern Empress® Japanese elm (outstanding burgundy fall color and other attributes) with Hallelujah lacebark elm (*Ulmus parvifolia* 'Hallelujah') which has outstanding ornamental bark.

Ornamental breeding research at NDSU includes developing freeze test procedures for earlier hardiness screening, traditional breeding efforts (making interspecific crosses with cold hardy species and hybrids), developing molecular markers for breeding selection and propagation protocols (micropropagation, grafting and vegetative cuttings). Freeze tests and molecular markers will assist in reducing time, efforts and costs with selection of desirable progeny. Measuring electrical conductivity of damaged cells from organs such as vegetative buds, flower buds and stem tissue, allows to screen for winter hardiness. Recently, research has been conducted on establishing a micropropagation protocol for Northern Empress® Japanese elm to reduce basal callus and increase axillary shoot proliferation utilizing nutrient salt formulation screening and use of anti-auxin compound p-chlorophenoxyisobutyric acid (PCIB).

RECENT INTRODUCTIONS

Some outstanding recent selections that have come out of the NDSU WPIP include:

Summer Flare™ Japanese tree lilac (*Syringa reticulata* 'SumDak'), Northern Empress® Japanese elm, (*Ulmus davidiana* var. *japonica* 'Burgundy Glow'), Cinnamon Curls® dwarf Korean birch (*Betula costata* 'CinDak'), Lavaburst® Ohio buckeye (*Aesculus glabra* 'LavaDak'), Emerald Flare® birch (*Betula tianshanica* 'EmerDak'), and Fireflare Orange® Mollis azalea (*Rhododendron* × *kosteranum* 'FireDak'). Several future selections include: *Cercidiphyllum japonicum* (Japanese katsuratree) and *Acer truncatum* × *A. platanoides* hybrids (Figs. 2 - 6).

Figure 2. Summer Flare™ Japanese tree lilac (*Syringa reticulata* 'SumDak')

Figure 3. Northern Empress® Japanese elm, (*Ulmus davidiana* var. *japonica* 'Burgundy Glow')

Figure 4. Cinnamon Curls® dwarf Korean birch (*Betula costata* 'CinDak')

Figure 5. Lavaburst® Ohio buckeye (*Aesculus glabra* 'LavaDak')

Figure 6. Fireflare Orange® Mollis azalea (*Rhododendron* × *kosteranum* 'FireDak')

The Perfect Plant

Michael Yanny

JN Plant Selections, LLC, Milwaukee, Wisconsin, 53225 U.S.A.

jnplantselections@gmail.com

Keywords: Poem, nursery plant, propagation

INTRODUCTION

Mike Yanny recited a poem at the end of the New Plant Forum as a whimsical look at the "Perfect Plant" for the propagator, nursery production and landscape use.

I have it.
In my imagination.
It's beautiful! — ALL the time.
My little shrub is 3 feet tall.
It gets there fast
And that is all.
From that point on,
It's never 4
And absolutely never more.
It grows to 3 feet and STOPS!

Perfectly Perfect™, that's its name — TM, trademark.
It blooms from May through August with huge red, fragrant flowers.
Come September, loads of delicious, pea-sized, porcelain blue berries mature.
Deer, turkeys, squirrels, bears, chipmunks, cedar wax-wings, robins, cardinals, and others, gather for the feast.
It's a natural wonder!
Of course, my shrub is native,
to everywhere in the world!
So, it's not invasive.

Oh—Oh, the leaves on Perfectly Perfect™ are purple,
All season—
Until they turn bright, fire engine red in fall,
before dropping in mid to late October.

Perfectly Perfect™ has no disease problems.
It resists vole, rabbit, and deer browsing,
And is hardy from zone 1 to 15!

From a production standpoint,
my shrub is the best that has ever been grown.
It roots easily from cuttings.
Growers typically get 110% rooting from softwood cuttings
taken any time from May to September.
A traditional mist set-up works great.
May cuttings should be rooted in a week.
From there they go to a #1 container to fill out by the end of year one.
The next spring, they are moved up to a #3 container.
By October you have 24 in. × 24 in. finished plants.
It's a grower's dream-plant!!!

Perfectly Perfect™ is even more perfect for the end user.
It can be used in all situations.
It grows in any soil,
—all exposures—
And never needs watering!
And best of all, Perfectly Perfect™ greets homeowners with a warm hello
when they step out on the patio in the morning and sip that first cup of coffee.
Yep—it can talk!!!!!
Perfectly Perfect™ is the perfect plant and then some!
Everyone should have one!!!!
At least that's what I imagine ....

PROCEEDING'S PAPERS

SOUTHERN REGION OF

NORTH AMERICA

Dr. Fred T. Davies, Jr., Regional Editor

Forty-fourth Annual Meeting - 2019

Baton Rouge, Louisiana USA

Technical Sessions of International Plant Propagators' Society-Southern Region of North America Annual Meeting

Elliott Hallum

Mountain Creek Nursery, 941 Scenic Hills Drive, McMinnville, TN 37110, USA

ehallum@gmail.com

Keywords: Annual Meeting, Southern Region of North America (SRNA)

SUMMARY

The 44th Annual Meeting of the International Plant Propagators' Society-Southern Region of North America (SRNA) convened at 7:30 am on 14 October 2019 at the Crowne Plaza Baton Rouge, Baton Rouge, Louisiana, with President Elliott Hallum presiding.

PRESIDENT ELLIOTT HALLUM

President Hallum welcomed everyone to Baton Rouge, Louisiana for the 44th Annual Meeting of the SRNA. He thanked Local Site Committee Chair, Allen Owings and his committee and volunteers for the long hours in arranging the excellent tours, hotel, other planning activities and all their attention to detail. There are 230 registered for this conference, which is great!

He welcomed students, first time attendees and new members, asking them to stand and be recognized. Hallum thanked the Executive Committee, and Bobby Green's Sponsorship Committee, which raised $50,650 in cash and $10,000 in-kind sponsorships; this was outstanding! Hallum encouraged the membership to visit and show their support of our sponsors during the meeting. He encouraged all members to make new members and first-time attendees feel welcome — share with them and seek from them. He called for good questions and enthusiastic participation at the Tuesday night question box.

Hallum announced that the SRNA is in its third year of the Southern Region Educational Endowment, which will be discussed in greater detail later in the meeting. With a base donation of $20,000 from an anonymous donor. The Education Endowment balance is now $70,057 – and growing. It will greatly enhance our region's ability to support students and early career professionals – and ensure continued quality of the outstanding educational programs our region is known for. He thanked Kevin Gantt for leading the endowment. Charlie Parkerson challenged the membership to match his $4400 ($100 for each year he as been at a SR-IPPS meeting) for the Endowment. Hallum encouraged attendees to contribute to the challenge – and move the Endowment forward.

Hallum announced that this is the seventh year the SRNA has participated with the European Region in the Early-Career Propagator Exchange program between the two regions. He recognized Vicki Enderby from the European Region, who was hosted by Christine Coker and Patricia Knight of the SRNA. Shea Keene of the SRNA was our designee to the European Region. Both of these early-career professionals had an incredible exchange experience in our respective regions.

This is the eighth year for the Vivian Munday Young Horticultural Professional Scholarship Work Program (Vivian Munday Scholarship). We currently have four young professionals: Yongjun Yue (University of Georgia), Julia Rycyna (University of Florida), Gabriel Campbell (University of Florida) and Jordan Baylor (University of Georgia) - who are making a strong contribution to this year's program.

Hallum recognized the new members and asked them to come forward to be recognized. Hallum thanked Program Chair and 1st Vice-President, Brie Arthur, for the excellent program and slate of speakers she assembled (Fig. 1).

Figure 1. President Elliott Hallum (right) with Brie Arthur (left), Program Chair of the 2019 Baton Rouge, Louisiana 44th annual conference.

PROGRAM CHAIR BRIE ARTHUR

Program Chair Brie Arthur welcomed all members, guests and students. She thanked the membership for the opportunity to serve them, and then reviewed the scheduled program. The Question Box, scheduled for Tuesday evening, was to be moderated by Drs. Judson LeCompte and Sandy Wilson. She then introduced the first moderator, Siddhi Jadhav from Greenleaf Nursery.

Favorite Perennials for Pollinators

Heather Alley and Lauren Muller

State Botanical Garden of Georgia, University of Georgia, Athens, Georgia 30605, USA

alley@uga.edu; lvmuller@uga.edu

Keywords: Biodiversity, growers, larval host plants, native plants, perennials, pollinators, propagation

Summary

The State Botanical Garden of Georgia (SBGG) has a long-standing commitment to promoting the use of Georgia native plants in landscapes. Native plants are the foundation of nature's food web upon which all other wildlife depends. The SBGG grows over 175 species of grasses, forbs, trees, and shrubs for annual plant sales and native garden installations across Georgia. The honey bee's decline is threatening to impact the production of many food crops. This has been widely publicized; however, less well publicized are the decline of native bees, which is adversely affecting ecosystem stability. Gardeners have an increasing appreciation of nature and the importance of native plants in the food web. We list some SBGG recommended native perennial plants which are excellent pollen sources for domesticated and native bees, as well as other pollinators.

INTRODUCTION

The State Botanical Garden of Georgia (SBGG) recognizes the critical role native plants play in preserving nature as we know and love it. We have a long-standing commitment to promote the use of Georgia native plants in all of Georgia's landscapes (Fig. 1). Mounting scientific evidence confirms that native plants are the foundation of nature's food web upon which all other wildlife depends. At the (SBGG), native plants are propagated for restoration projects at the garden and across the state (Fig. 2). The SBGG grows over 175 species of grasses, forbs, trees, and shrubs for annual plant sales and native garden installations across Georgia (Fig. 3).

Connect to Protect

- **Support** wildlife
- **Encourage** schools, businesses and parks to include native plants in their landscapes.
- **Provide** teaching materials to foster an appreciation for the connections between plants, animals, and people.

Figure 1. The State Botanical Garden of Georgia has programs to support wildlife, provides teaching materials and encourage the use of native plants in the landscape.

Figure 2. At the State Botanical Garden of Georgia, native plants are propagated for restoration projects at the garden and across the state. Some of our partners in habitat recovery in forests, floodplains, and grasslands include the Fish and Wildlife Service, Georgia Dept. Natural Resources, and U.S. Forest Service.

Figure 3. The State Botanical Garden of Georgia grows over 175 species of grasses, forbs, trees, and shrubs for annual plant sales and native garden installations across the state.

Native plants are becoming ever more popular as their role in sustaining nature is revealed. The honey bee's decline is threatening to impact the production of many food crops. This has been widely publicized; however, less well publicized are the decline of native bees, which is adversely affecting ecosystem stability. The Monarch Butterflies' was listed as an endangered by the U.S. Fish and Wildlife in 2015. This resulted in national shortages of the native Monarch host plant, Milkweed, which was in great demand by consumers to plant in their gardens as Monarch butterfly food source. The Milkweed/Monarch Butterfly co-dependency illustrates the connection between plants and animals. This example is changing the mentality of many consumers as gardeners. As such, gardening with native plants is not a fad. Gardeners have an increasing appreciation of nature and the importance of native plants in the food web.

However, native plants are hard to find. Commercial growers are often uncertain what plants are technically native, as "native" may be interpreted differently by people. Another obstacle to native plant availability is the lack of information on how to produce them or what species to grow.

PROPAGATION METHODS

Each of the species listed in this paper may be propagated using the same germination, cutting and potting mixes, and fertilizer routines. The propagation, material needs and production conditions are explained below.

Growing media and fertilization

With the exception of wetland plants, native perennials need good drainage or root rot can develop. The Southern U.S. warm summer nights [> 21 °C (>70 °F)] and high humidity increase the risk of root rot. To achieve good drainage, we rely on composted pine bark. The ideal bark particle size is between 1.3 to 2.5 cm (0.5 and 1-in.). The mix will retain too much water if it is too fine, and bigger chunks do not adhere well to roots. The basic native plant perennial mix was adapted from Bill Cullina's mix at the New England Wildflower Society. Where Cullina uses peat moss, we use a modified standard potting mix that has a peat moss base with additional pine bark and perlite for extra drainage. We use a potting mix that has about 65% composted or aged pine bark, which is typical.

Native Perennial Media

- 3 parts composted or aged pine bark
- One-part potting soil
- 0.75 lb. dolomitic lime per cubic foot
- 1 cup of slow release fertilizer (10N :10P: 10K) per cubic foot

Fertilizer is critical as potting mix nutrients are exhausted after a month or so. We have tried various brands and ratios of fertilizer and not noticed much difference among brands. Hen-manure-based organic fertilizer with 5:3:3 (5N-1.3P-2.5K) ratio incorporated into the Native Perennial Mix also works well. After potting, fertilize up to once a week or as needed with a balanced 300 ppm liquid fertilizer.

A standard potting mix with about 40% processed pine bark may be used for starting seedlings. After about 3 months, seedling will start needing additional fertilizer as the potting mix loses its starter charge of fertilizer. Growing mediums sold as "germination mix" tend to stay too wet for many native plants.

For cuttings, the following mix works great for the species listed in this manual. Because our greenhouse tends to stay too warm in the ideal months for rooting most cuttings (May-June), we have started keeping cuttings under fluorescent lights (with a timer for 18 hours of light per day) in a room set at 70 degrees Fahrenheit. Keep the cuttings in a sealed plastic bag or under clear plastic covers. Substrate was 3-parts perlite/1-part peat.

Environmental Conditions

There is a fair amount of leeway in the timing of sowing seeds for the species included in this manual. Seedlings do best with soil temperatures around 21 °C (70 °F). You do not need a greenhouse to raise Georgia native plants, but horticultural heating pads are helpful for getting seeds going in early spring. We time seed stratification periods so that seeds come out of the refrigerator in spring or summer. All seeds can be stratified in damp wet sand in a refrigerator at 2-5 °C (35-41 °F). A 40% shade cloth or greenhouse whitewash is recommended - even for full sun plants. Shade keeps plants cooler and helps prevent desiccation in the heat of the summer. For shade plants, use a 50% or 60% shade cloth, or the shade of tall trees. Let your native plants go dormant in the winter, so they regrow in the spring.

FAVORITE GEORGIA NATIVES FOR POLLINATORS

Asclepias tuberosa **– Butterflyweed.** This wildflower is a must-have plant for the biodiversity-conscience gardener (Fig. 4). This drought-tolerant and deer-resistant perennial works well in sunny borders, meadows and as single specimens. Few plants have such a vibrant orange color. This species is the larval host plant for the Grey Hairstreak, Monarch, and Queen Butterflies, Milkweed Tussock Moths, and several others. It a nectar source for bees, butterflies, and hummingbirds.

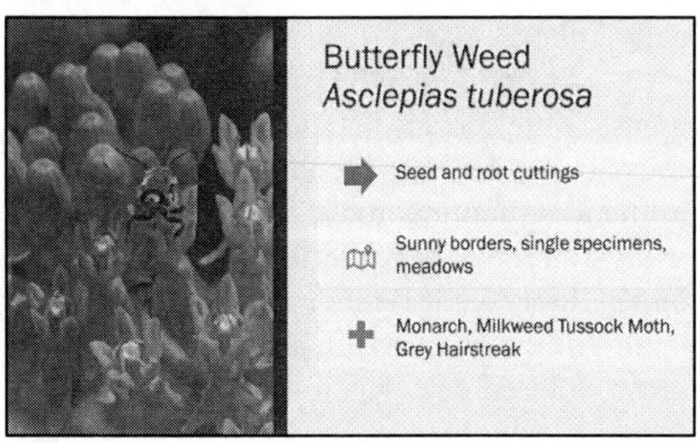

Clinopodium georgiana **– Georgia Calamint.** Georgia Calamint is an underutilized small shrub with a desirable mounding growth habit that is easy to grow (Fig. 5). Plants bloom reliably and profusely in the fall, serving as a nectar source for native bees and other pollinators.

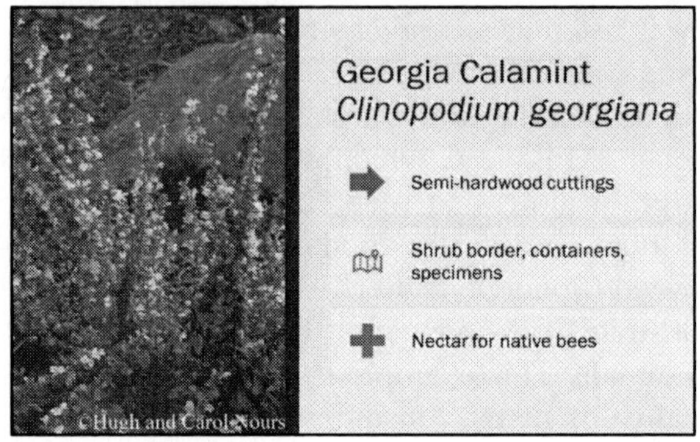

Chrysopsis mariana **– Maryland Goldenaster.** A particularly useful plant for xeriscaping, containers, and sunny borders (Fig. 6). In summer, this diminutive aster shoots up a 0.6 m (2-ft.) flowering stem topped with cheerful flower heads. It has interesting leaves covered with gauze-like webby hairs. Native bees and other pollinators frequent the flower. It is the host plant for Camphorweed Cucullia moth.

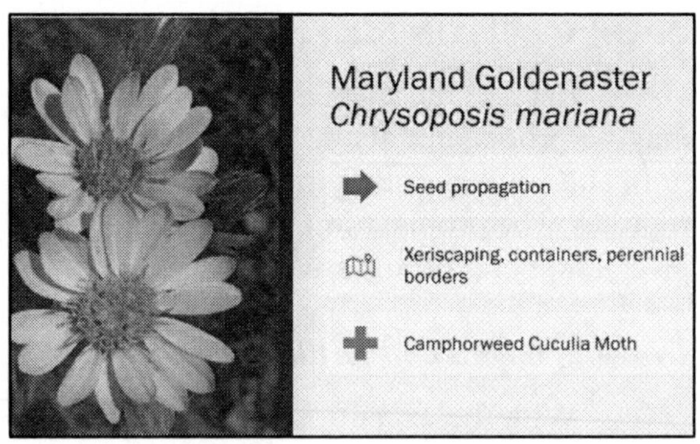

***Conradina canescens* – False Rosemary, Wild Rosemary.** Wild Rosemary is a small evergreen woody shrub with needle-like gray-green leaves that smell like lavender when crushed (Fig. 7). This plant looks great as a specimen, shrub border, or in a container. It blooms abundantly in April, supporting native bees and other pollinators.

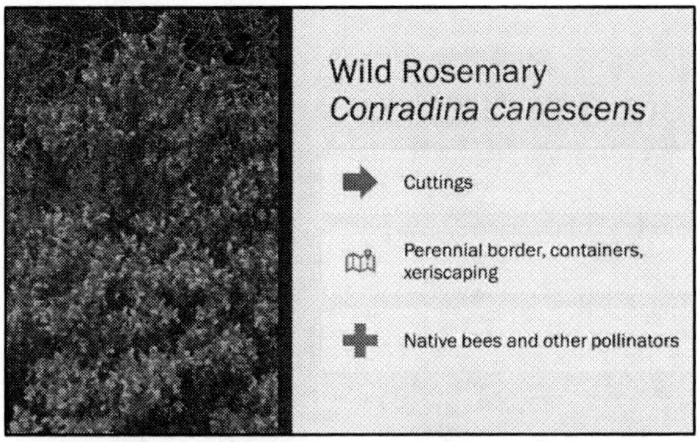

***Eryngium yuccifolium* – Rattlesnake Master.** It is a distinctive member of the carrot family with rosettes of long, gray sword-shaped leaves and ball-shaped flower heads resemble that of thistles - with spiny bracts and tiny white flowers (Fig. 8). This unusual plant is easy to grow and has a striking appearance that lends itself to specimen plantings and containers. Rattlesnake Master is the larval host plant for the Black Swallowtail and its flowers attract many native bees.

***Helianthus atrorubens* – Appalachian Sunflower, Purple-disk Sunflower.** This sunflower is found in grasslands and roadsides. It is tolerant of a variety of soil types and drought-resistant (Fig. 9). Stems can reach 1.5-1.8 m (5-6 ft) tall, making it a great candidate for the background of perennial beds or meadows in full sun. Daisy-shaped flowers have dark reddish-purple centers for which it is also called purple-disk sunflower. It is a good source of nectar for bees and butterflies. It is the larval host plant for the Silvery Checkerspot and 72 other native butterflies and moths! In the fall - birds will feed on its seeds.

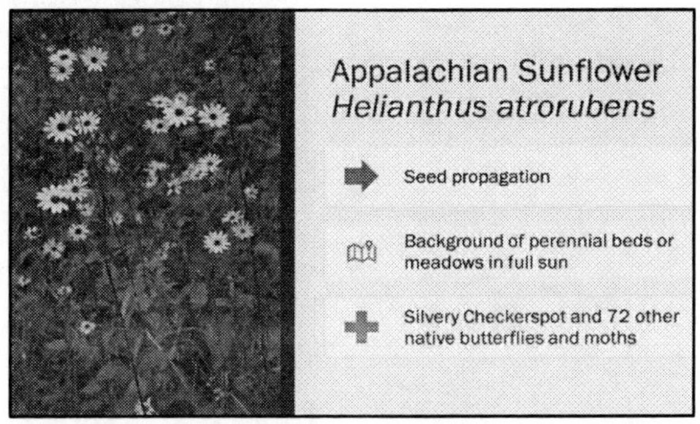

***Ionactis linariifolius* – Stiff-leaved aster.** This is another great option for sunny borders, containers, or rock gardens (Fig. 10). This aster may be few to many stemmed, forming a spray or mound, with each reddish stem covered in small, stiff, needle-shaped leaves which turn an attractive golden brown in the fall. Flower heads are small [1.5 cm (0.5 in. wide)], with lemon yellow centers and baby blue rays. Its flowers are a nectar source for bees, butterflies, and moths.

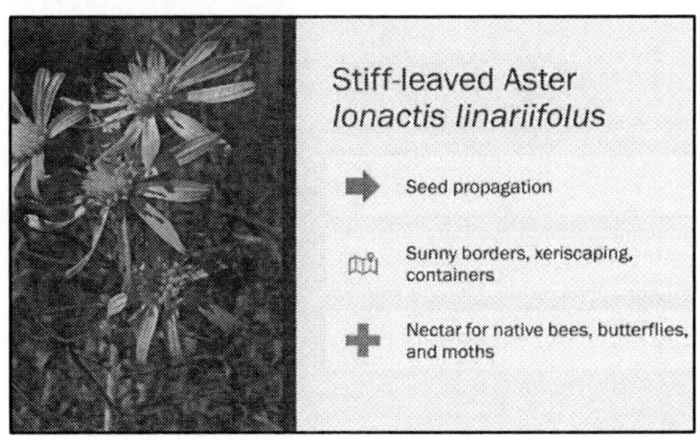

***Liatris microcephala* – Small Headed Blazing Star.** It is a great option for sunny borders or as single specimens. Blazing star grows erect, tall spikes [0.6-1.2 m (2-4 ft)] and is densely packed with tiny, bright purple flowers (Fig. 11). The vibrant flowers are tiny, yet borne by the thousands - providing abundant pollen and nectar for bees and butterflies from July-September. It tolerates a range of soils from well-drained to clay - but becomes too large in rich soil, thus requiring staking. Its flowers provide nectar to many butterflies and moths.

***Scutellaria incana* – Downy or Hoary Skullcap**. It inhabits shady, upland woods. Downy Skullcap does well under a variety of soil types, from full-sun to part-shade. In late summer, clusters of tubular [2.5 cm (1-in.)]-long, purplish-blue flowers put on a great show (Fig. 12). Like most mint relatives, Skullcap is deer resistant. It supports bees and other pollinators.

***Solidago petiolaris* – Downy Goldenrod.** This is one of the prettiest, smaller goldenrods (Fig. 13.). Its flowers are in dense, spike-like clusters at the top of stems, creating a yellow plume from August to October. It is great for a sunny border, meadow, or prairie. Its flowers are a great source of late-season nectar and pollen for butterflies, moths, and bees.

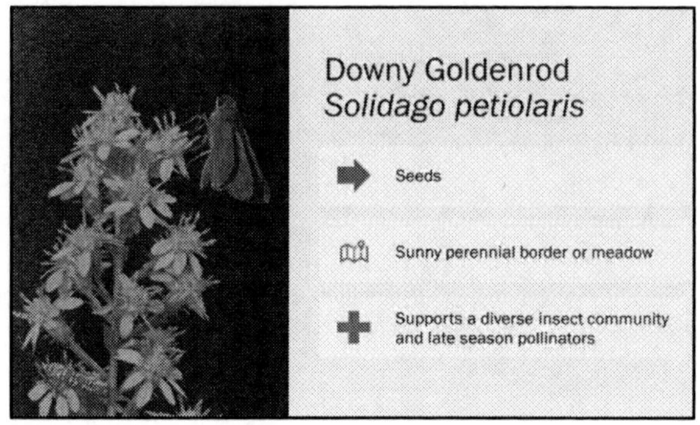

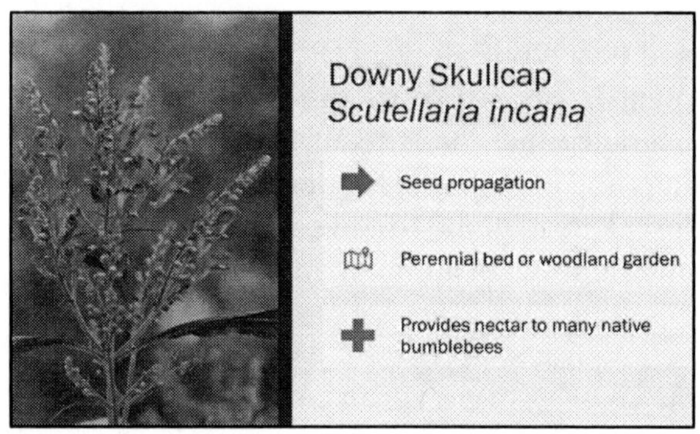

The Changing Garden Paradigm: Perceptions of One Extension Agent

Gary R. Bachman

Mississippi State University, Coastal Research and Extension Center, 1815 Popps Ferry Road, Biloxi, Mississippi 39532, USA

Gary.bachman@msstate.edu

Keywords: consumer gardening information, container growing, millennials, urban and consumer horticulture, vegetable gardening

Summary

There is increased interest in gardening, vegetables, and sustainably grown product with today's consumers. Consumers want smaller garden footprints, such as container growing, in their more confined patios and balconies. The gardening consumer wants to grow some of their own food but are concerned with food security and production. The modern gardener wants recommendations, which includes social media, workshops, garden shows and festivals. To maintain market share, the green industry needs to address the shifting perceptions and expectations of the modern gardener.

INTRODUCTION

During the past 20 years that I have been involved in urban and consumer horticulture -I have noticed a shift towards increasing landscape and garden interest. Data collected by the National Initiative for Consumer Horticulture (https://consumerhort.org/) suggests that interest in vegetable gardening has increased 17% (2014-2019). Much of this is being driven by the 63% increase in the 18 to 34 demographics – the millennials!

But the modern gardener has constraints that gardeners from previous eras did not encounter. Among these constraints are having jobs outside of the home and perceived garden labor requirements. There is also a shift in garden size. The modern urban gardens have smaller footprints, restricted by ever decreasing house lot sizes.

Much of this increased interest is from new, younger gardeners who tend to be more highly educated and digitally connected. They want what they perceive to be simple pleasures past gardeners experienced and enjoyed. But they don't want those big gardens. Container growing is highly attractive - especially considering the smaller garden footprints, maybe having only a patio or balcony.

The new gardening consumer wants to grow some of their own food and is worried about food security and production. They want organically grown, but are not interested in an organic food religious experience. They want locally grown food and want to know where their food comes from. With much of the food supply being transported long distances, any news of disruptions (weather, disease, etc.) is a concern - especially in large cities where food has to be shipped in every day.

The sources in which the modern gardener gets their information changing. It used to be the Home & Garden Television (HGTV) was the go-to source. But what happen to the "Garden" in HGTV? Looking at the show lineup for the past 10 years, there has been a steady influx of home remodeling, home flipping and real estate shows at the expense of gardening. Any mention of gardening is only a tiny fraction across all the current shows.

So, where do the modern gardeners get their information? This digital gardening generation is also looking towards social media. Social media is a wonderful vehicle to get information out quickly to a wide swath of the gardening public. But it is also organically propagates misinformation just as quickly.

These gardeners like to go directly to the source. There has been an increase workshops, garden shows, and festivals that are asking the garden and landscape influencers to be speakers. And the attendance of these events of increasing every year, giving attendees the opportunity to interact with experts.

Consumers do not want to plant on a trial and error basis. The modern gardener wants recommendations. This feeds into the expansion of branded plants. From national programs like Proven Winners and First Editions to state programs like the Mississippi Medallion Winners and Louisiana Super Plants.

This means that if the horticulture industry wants to keep market share and customers -then addressing the changing and shifting perceptions and expectations of the modern gardener should be of primary concern.

Disclaimer

This presentation is not based on any actual data, but from my interactions addressing gardening questions and face-to-face interactions in my role as an Extension agent.

Naturalistic Planting at the Plantery

Jared Barnes

Stephen F. Austin State University, Department of Agriculture, Nacogdoches, Texas, 75964, USA.

barnesj@sfasu.edu

Keywords: Ecological and matrix planting, naturalistic design, landscape design, pedagogy, perennials, plant survival strategies, teaching

Summary

Naturalistic planting continues to garner interest in the horticulture industry due to the desire to have sustainable landscapes that require minimal inputs. Naturalistic design, also known as designed plant communities, ecological planting, and matrix planting, is an approach to urban horticulture that is increasing in popularity due to a variety of factors. This includes minimizing the high cost of maintaining traditional landscapes, interest in green infrastructure to mitigate environmental issues, and utilizing plants that create habitat for pollinators and other wildlife. In my Herbaceous Plants class at Stephen F. Austin State University (SFA), we teach students principles of ecological design and then apply those concepts through the inception and installation of a 93 m^2 (1000 ft^2) planting.

INTRODUCTION

Naturalistic design, also known as designed plant communities, ecological planting, and matrix planting, is an approach to urban horticulture that is increasing in popularity (Fig. 1). The driving factors are reduced funding for high-maintenance landscapes, opportunities in developing green infrastructure to mitigate environmental issues, incorporating plant communities that ecologically benefit pollinators and other wildlife (Oudolf and Kingsbury, 2013). These new approaches require well-educated horticulturists that have some experience doing naturalistic designs and installations - as well as propagating appropriate species for their demand (Fig. 2).

Figure 1. A naturalistic planting in a highly urban setting.

Figure 2. Stephen F. Austin (SFA) student grow houses used for teaching instruction and growing plants for student sales.

Hence, our horticulture program at Stephen F. Austin State University (SFA) is teaching students about sustainable approaches to landscaping in my Herbaceous Plants class (Fig. 3). This paper will describe our efforts to teach students about naturalistic planting and give them hands-on experience in the Plantery, our agriculture department's teaching gardens, micro-farm, and grow houses.

Figure 3. Stephen F. Austin (SFA) teaching gardens, maintained by students.

TEACHING ECOLOGY

Naturalistic design relies heavily on the understanding and application of ecological concepts beyond the normal considerations of light, soil, water, etc. when planting. Instead of viewing plants as single ornaments in the landscape, naturalistic design focuses on seeing the plants as part of a community that has function and purpose in the landscape (Rainer and West, 2015).

We start my Herbaceous Plants course discussing broad ecological concepts. One ecological concept that students learn in Herbaceous Plants is about survival strategies. Survival strategies, also called the universal adaptive strategy theory, which entails three survival strategies that plants that plants utilize (Grime, 1977).

Competitor plants (low disturbance, low stress) are vigorous growers that outcompete other plants around them (Fig 4). They exhibit rhizomes that allow for expansion. *Stress-tolerators* (low disturbance, high stress) usually take 3–7 years to flower, and typically form a storage organ (bulb, tuber, or corm) that allows them to endure harsh conditions (freezing, heat, drought, etc.) (Fig 5).

Figure 4. Competitors (low disturbance, low stress) are vigorous growers that out-compete other plants around them.

Figure 5. Stress-tolerators (low disturbance, high stress) usually take 3–7 years to flower, and they often form a storage organ like a bulb, tuber, or corm that allows them to endure harsh conditions (freezing, heat, drought, etc.).

Ruderals (high disturbance, low stress) are short-lived plants that evolved to deal with frequent interruptions at the ground level (fire, flooding, animal trampling, etc.) (Fig 6).

Figure 6. Ruderals (high disturbance, low stress) are short-lived plants that evolved to deal with frequent interruptions at the ground level (fire, flooding, animal trampling, etc.). They set copious amounts of seed.

Hence, they frequently produce copious amounts of seed. Survival strategies can be visualized in Grime's C-S-R triangle, and similar to the soil texture triangle, no one plant species perfectly fits one survival strategy (Grime, 1977; Pierce et al., 2013) (Fig. 7). Often, they have a dominant strategy, but they can show additional adaptations. For example, *Asclepias tuberosa* has a tuberous root and can live for many years, but it will produce wind-dispersed seed when cross pollinated.

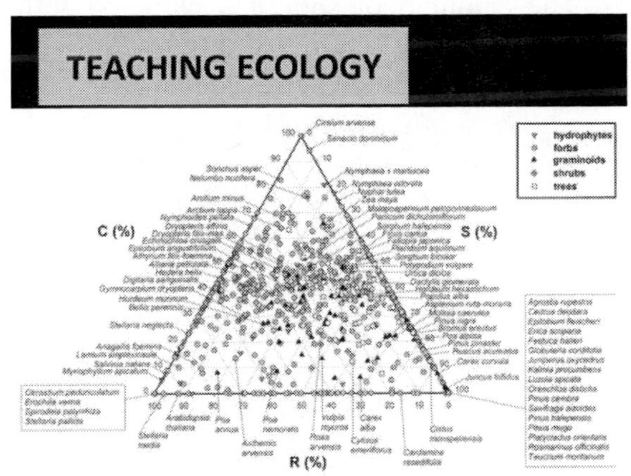

Figure 7. Plant survival strategies can be visualized in Grime's C-S-R triangle. Similar to the soil texture triangle, no one plant species perfectly fits one survival strategy.

To teach about survival strategies in class, we cover concepts in class, and then for an activity, students in small groups are given lists of eight herbaceous plant species. They then determine characteristics of competitors, stress-tolerators, and ruderals by reading about the plant's growth characteristics online with their cell phones. Afterwards, students are allowed to come and pin their species name on a triangle on the board. When all students have placed each plant where they believe it fits in Grime's triangle, we discuss their placement on the board.

TEACHING DESIGN

Design is a cornerstone of public horticulture, and no matter how wild they appear - naturalistic plantings must be aesthetically pleasing and fulfill a purpose whether it be for perception, ecology, or function (Rainer and West, 2015). For a few weeks in class, we cover principles of design concepts about color, texture, form, etc. After basic design concepts are learned, I then cover five design principles of naturalistic design based on the work by Rainer and West (2015):

1. Plants should occur in related populations instead of being planted as isolated individuals.
2. Stress is an asset instead of a liability.
3. The ground plain should be carpeted with vegetation by layering plants.
4. Plantings must be perceived as attractive and thus must be legible.
5. A planting is managed and allowed to evolve instead of maintaining a *status quo*.

One of the aspects we cover in great depth was principle three about layering plants. Rainer and West (2015) state there are four layers that exist within a naturalistic planting. The structural layer is the tallest species in the planting. These plant species often hold interest for three-plus months during the year - and thus form the structure of the planting. They can either be woody or herbaceous in nature. The seasonal filler layer is shorter in stature and provides good color and seasonal interest. Often, these bloom in five-to-seven waves of color during the year from early spring flowering to frost. They are coordinated to bloom together and play off design themes (color, texture, form, etc.). The purpose of the dynamic filler layer is to have self-sowing ruderal species that occur to quickly fill gaps that may appear during disturbance or senescence events in the planting. These first three layers serve to provide some aesthetic characteristics and thus are labeled as design layers. The fourth layer, the matrix or groundcover layer, is primarily functional in nature and serves to cover the ground with vegetation. This layer is the biggest difference between naturalistic plantings and traditional intensive horticulture plantings where plants cover the ground, rather than mulch.

Once we finish with layers of naturalistic design, we discuss three approaches to installing naturalistic designs (Oudolf and Kingsbury, 2013). Randomized mixes are sown by seed and allowed to develop based on the environmental conditions present. This approach is more cost-effective because it uses seed and limited plugs – rather than large scale transplanting. However, it is dependent on quickly generating biomass to cover the site - to reduce weed growth (Oudolf and Kingsbury, 2013). A second approach, modular designs, is where a small module planting [perhaps 3.7 x 6.1 m (12 × 20 ft)] are developed with a planting scheme and then tiled by rotating the design, inverting the design, and changing out species along environmental gradients (Diblik, 2014). This approach allows the horticulturist the opportunity to plant a large-scale planting using a module and gives a random order to the design.

Designed intermingling is the most complex approach for naturalistic design where ever plant is purposefully placed in a mix. This approach relies heavily on the designer having a strong plant knowledge to understand whether or not species will go well together - since many plants occur and are planted to appear as if random (Oudolf and Kingsbury, 2013). Because of the difficulty of this approach, large blocks or drifts are planned to allow for easier design, installation, and management.

For our planting, the students chose to install a designed intermingled planting using plants in blocked arrangements for the site. The site in the SFA Plantery where this year's planting was installed. Its dimensions were approximately 93 m^2 (1000 ft^2) in area and trapezoidal in shape. The area was 21 by 2.4 m (70 ft long by 8-ft wide) on the narrow end and flaring out to 6.1 m (20 ft) on the wider end. On the narrow end, the site receives full sun, which grades to a partial-sun woodland habitat on the wider end. The site is moderately trafficked by students and visitors to campus.

To compose the flora for the site, students were given availability lists from Hoffman Nursery, North Creek Nursery, and Southwest Perennials. These wholesale nurseries grow southeastern native plants. Students were instructed to list four primary plants, six seasonal fillers, four dynamic fillers, and four matrix species - and submit their selections online.

From their assemblage, I compile the student lists into one master list, and in lab the students hashed out what species would be used from this master list. Using design concepts for color, texture, form, and seasonality, students decided to primarily use cool-colored plants (blue, purple, and pinks) with some neutral-colored white flowering species for spring and summer. For fall they chose to juxtapose blue/purple with it's complementary color yellow. Students decided to grade the height of the plants from short on the east end where it was most narrow to taller species as the path got wider and approached the woodland habitat. An example was using the shorter *Symphyotrichum oblongifolium* 'Raydon's Favorite' on the narrow end of the bed and the taller *Symphyotrichum* 'Bill's Big Blue' on the wider side.

Students were then given outlines of the planting and allowed to create structural, seasonal filler, dynamic filler, and matrix layers. After some discussion, we synthesized their designs together into one final plan (Table 1). Plants were then either ordered from their respective companies, or stock was pulled from plants propagated by students in the Herbaceous Plants class during earlier labs in the semester.

TEACHING PLANTING AND INSTALL

Planting preparations began on 9 April 2019, by clearing the site of existing vegetation and leveling the site for planting during a two-hour lab. Students placed stakes in the ground that marked every 1.5 m (5-ft) so that the design could be transferred from paper to the ground
(Fig. 8). On April 16 students planted the structural layer and then the seasonal layer, and on April 23 students planted the matrix layer. This layer planting order is usually followed to make sure there is space for the larger plants before the lower species and groundcover layer is installed. In total, three two-hour lab sessions and one one-hour class period were used for planting. Students were then allowed the opportunity in a later lab to reflect on the process through a discussion session.

In reflection, six months after the install, the majority of the species we planted survived, and our goal to close the ground plain using vegetation was achieved (Fig. 8). In this initial planting, many primary plants have yet to reach their full size and potential, but we anticipate their increased presence in the coming year. Overall, this planting is a great gateway for educating students and the public about naturalistic design.

Figure 8. Naturalistic landscape design and planting using coded planting maps (above left), matching the design with the site (bottom left), seeding and using seedling plugs (above right), and the seasonal change in the naturalistic site (bottom right).

Figure 9. The Stephen F. Austin (SFA) Plant.ed. with Jared Barnes https://meristemhorticulture.com

CONCLUSION

Naturalistic planting continues to garner interest with consumers and the horticulture industry. There is increasing demand to have sustainable landscapes that require minimal inputs. In our Herbaceous Plants class at SFA, we teach students principles of ecological design and then apply those concepts through the inception and installation of an approximately 93 m^2 (1000 ft^2) planting. In sharing these teaching methods., I hope to inspire other educators on approaches of how naturalistic planting can be taught in the classroom while also giving students hands-on experience.

Literature Cited

Diblik, R. (2014). The Know Maintenance Perennial Garden. Timber Press, Portland, OR.

Grime, J.P. (1977). Evidence for the existence of three primary strategies in plants and its relevance to ecological and evolutionary theory. Amer. Natural. *111*: 1169–1194.

Oudolf, P. and Kingsbury, N. (2013). Planting: A new perspective. Timber Press, Portland, OR.

Pierce, S., Brusa,G., Vagge,I., and Cerabolini, B.. (2013). Allocating CSR plant functional types: The use of leaf economics and size traits to classify woody and herbaceous vascular plants. Funct. Ecology *27*:1002–1010.

Rainer, T. and West, C. (2015). Planting in a post-wild world. Timber Press, Portland, OR.

The Times, They Are a Changin'

James Berry

J. Berry Nursery, 201 Private Road 5180, Grand Saline, Texas 75140, USA

jim@jberrynursery.com

Keywords: Adaption to change, automation, climate change, consumer preferences, digital communication, green industry, marketing, metathesiophobia, new crops, plant breeding, personnel development, propagation, technology, sustainable growing practices, tissue culture

Summary

The thriving horticulture world has seen its share of changes. How we as an industry have evolved over the years is simply the beginning of bold, bright changes the future holds. Advancements in technology, especially the science of plant-breeding and increasingly automated operations are even more striking. Our industry can adapt and evolve to create the landscape - both literally and figuratively - to ensure longevity and success. Where will we position ourselves as an industry? Technology affects the labor force, climate change affects consumer habits and so on; change is here and accelerating. New crops and their specific regulations have created opportunities for technologies and growing operations never seen before. We must change our market outlook on customer bases - from local to state, region, country and beyond. Climate change has the potential to alter the horticulture industry in a way no other outside threat has before. The industry must remind consumers that plants are a partial solution to environmental change and global warming. The curse of cloud access to information is the loss of credibility of science and truth - dismissed by bloggers and trolls disguised as news sources. While we are competitors, our green industry relies heavily on one another to succeed. Our industry can remain united to speak truth to those unwilling to protect our environment - lead by example in instituting positive change in regulation, land and water use, and sustainable growing practices. The future is bright, and change is constant. We must remain diligent as partners going forward in uncertain times.

INTRODUCTION

The ancient philosopher Heraclitus said, "We both step and do not step in the same rivers. We are and are not." Explaining change in the context of dipping one's toe into the moving waters of a river is an indictment of the river meandering through the unique topography of the land. We also change from our first eager steps in to the water as a child to becoming more measured, thoughtful and experienced as we age. The thriving horticulture world, both academic and industrial, has seen its share of changes over the years. Just like that child dipping into the always new river, horticulture has evolved through changes in technology, the labor force, globalization, consumer awareness and participation - and now more than ever, political uncertainty and climate change.

Dr. Lowell Catlett, a futurist and change embracer, gave a presentation over twenty years ago which still stands out to me today. He shared and warned that markets were about to become increasingly different and that change would continue and accelerate. Just recently, he wrote that more new data is created every day than existed in the entire world less than fifty years ago.

Change is both simple and complex, quick and sometimes excruciatingly slow, inexpensive and a checkbook buster, exciting and terrifying. How we as an industry have evolved over the years is simply the beginning of the bold, bright changes the future holds. Studying the history of gardening, commercial and consumer horticulture and floriculture in the United States is fascinating. Seeing where the industry has been only shows the potential for innovation is ever increasing. Advancements in technology, especially the science of plant-breeding and increasingly automated operations are even more striking - when comparing today with past generations. Just one hundred years ago, the industry would have been buzzing about a brand-new mechanical animal to pull a plow: the farm tractor. In that short hundred years, technology has hit its stride and is now in overdrive, turning over the new ground of the future.

Metathesiophobia is the fear of change. No one likes to be accused of being afraid, but we as an industry have some tough choices to make in the coming years and uncertainty can and will overwhelm some of our colleagues. People are naturally hesitant to change, reluctant to try something other than: "It's the way we've always done it." Death and taxes are not the only constants; "change" needs to be added. (Fig. 1). Change is frightening sometimes. Our industry can adapt and evolve to create the landscape, both literally and figuratively - to ensure longevity and success.

"....you better start swimmin' or you'll sink like a stone, for the times they are a changin'......."

– Bob Dylan

The 3 constants: death, taxes & Change!

- Jim Berry

Figure 1. The challenge of adapting to change for the green industry.

It is time to look toward the future, to changes on the horizon. Where will we position ourselves as an industry? Where will we have our voices heard to affect change positively? From our vantage point at J. Berry Nursery, here are a few of those areas of opportunity. While each category will have a great deal of overlap - technology affects the labor force, climate change affects consumer habits, and so on - change is constant and accelerating (Fig. 2).

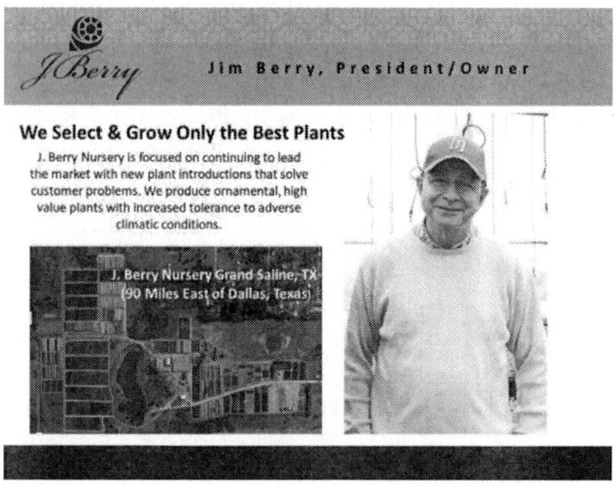

Figure 2. Jim Berry of J. Berry Nursery, Grand Saline, Texas

The previous mention of the farm plow is an excellent example. A simple, single-bladed, walk-behind tool powered by a domesticated animal led to the remarkable horsepower of fuel-driven tractors where farmers rode and guided heavy equipment - championing a future of artificial intelligence, robots and precision agriculture. Before long, one could plow a field from miles away! Automation will continue to become a hotter and hotter topic in academic and industrial circles as technology improves, becomes more reasonably priced and attainable by a larger cross-section of producers. When asked by the Greenhouse Grower Survey if automation was on the investment horizon for growers, 67% said yes in 2019, up from 59% in 2018. In the past few years, automation from the beginning of a plant's journey to a consumer's garden has changed how many nurseries propagate. It is broadening the thought process of approaching one task at a time - to adapting nursery-wide automation at scale (Fig. 3).

Figure 3. The importance of discovery, science, technology, new plants and markets - adapting to change. Automation is changing nursery production from propagation to adaption of nursery-wide automation at scale.

New crops and their specific regulations have created opportunities for technologies and growing operations never seen before. *Cannabis* is not just being grown solely indoors, but under a new sort of security most nurseries never thought necessary or possible. Additionally, the cannabis industry has strained certain segments of the supplier market, creating shortages in building materials, operational needs, and even labor. Artificial intelligence now allows nurseryman to monitor plant progress while drones and radio frequency identification (RFID) tags are no longer the toys we remember from trade shows past. They are now tools of the green industry.

Propagation of selected plants has shifted toward tissue culture where working conditions in laboratories are more desirable than traditional nursery practices. Robotics and biological sanitation mitigate environmental variables that could cause plant loss and help producers comply with regulatory and trade laws worldwide. All this enables "clean" plants to be sold and distributed globally. While still a large investment for most producers, tissue culture propagation may become a necessity for appropriate crops, as opportunities grow and regulations change (Fig. 4).

Figure 4. Adapting to controlled enviroment agriculture (CEA), such as tissue culture propagation for appropriate crops.

Skilled and unskilled labor costs continue to rise. With over a half a million estimated nursery jobs in the United States, the current labor market is making it more difficult for many nurseries to become and remain fully staffed. Migrant workers face new challenges now that the federal government has failed to maintain guest worker and temporary foreign worker programs. Government policies are more hostile toward individuals and groups the nursery industry has traditionally relied on for labor sustainability. Temporary worker programs have experienced restructuring, delays and increased scrutiny under past presidential administrations - while the present politics of illegal immigration and undocumented migrant workers have become markedly polarized. As the industry looks to offset the potential loss of unskilled labor, technological advancements in automation, broadly and task-based, offers some relief - but also increases the need for skilled labor to maintain more complicated new equipment.

LEAN and LEAN flow metrics have improved efficiency and reduced waste in nursery production but require a similar pledge to technology. Uninspired commitment to LEAN principals can result in a disjointed process which could negatively impact production and morale with nursery employees. The future of the nursery workforce is a mix of highly skilled personnel with technological backgrounds and efficient, competent management which can lead labor processes and execution - despite the changing political landscape (Fig. 5).

Figure 5. The importance of nursery personnel – selected from diverse backgrounds and experiences; empowering people through inclusion in decision making and encourageing their continuing education.

The changing attitudes toward how Americans shop, both wholesale and retail, have impacted the industry in a variety of ways. The days of administrative staff traveling to trade shows to do little more than take orders have mostly ended, thanks to e-commerce. This is creating new opportunities for seeing and touching future offerings, building and cultivating relationships and educational opportunities at trade shows and events (Fig. 6). Trade show attendees and exhibitors can now be more focused on specific objectives and spend more time cultivating business relationships, trust and reliability - rather than simply writing orders.

Figure 6. Trade shows are the opportunity promote products, engage, guide and educate clientele.

Training personnel to understand customer issues and opportunities has changed as technology opens the doors for virtual shopping and planning - before the actual show begins. Digital media and the internet allow exhibitors to create a presence and reputation - while not so long ago, decades of leg work might have been required to achieve the same goal. Future trade shows will likely become more virtual, allowing information and education to flow more freely across the industry. The professional grower and trade organizations, like IPPS, are now more in touch thanks to email, social and digital media, e-Learning and the lines of communication more easily accessible for everyone. Organizations that have adapted to the changing technological capabilities can now disseminate information - on larger scales and more quickly to professionals.

Rapid urbanization has changed the way consumers shop and how producers bring items to market. While the first nurseries appeared a couple hundred years ago, the current ways consumers connect with producers is quite different. Big box retailers have only been offering wide selections of plants for the past 30 or so years, and now the internet and e-commerce have altered previous brick and mortar and catalog sales. Urban dwellers have necessitated a new retail market, where plants are delivered potted in a decorative container -with virtually no post-delivery work on the part of consumers. Customers want instant gratification of a finished product.

Many people looking for plants for their home are now focused on décor over gardening, altering the traditional offerings of some nurseries and retailers. Brick and mortar shoppers are likely using the internet as inspiration long before they step foot on a nursery or retail garden center. Online nurseries are growing quickly with large capital pouring in from investors seeing the trends and consumer habits of today's home gardeners. Shipping methods have improved, allowing once easily damaged live goods to be transported across country through a variety of delivery methods (Fig. 7).

Figure 7. The dynamic, changing marketplace of the green industry.

Public awareness of climate change and the push for sustainable materials are enabling development of new growing substrate. New containers made from recycled, recyclable and biodegradable resources are taking center stage to the once black-plastic-pot dominated industry. With a focus on better containers, consumers now have an opportunity to purchase branded plants, supplying home gardeners diversified offerings combined with premium packaging. The first woody ornamentals to arrive to market with a brand name appeared in the 1980's. Now, J. Berry Nursery and many other producers provide consumers with a buying experience beginning with a branded product and continuing with supporting items, care advice, plant use ideas and world-class customer service. Because today's customer is especially internet-savvy - suppliers now must be able and willing to provide consumer support, follow-up information and care - beyond what previously would be limited to a plant tag.

Plant tags with RFID technology can track an item through the nursery process, through the retail process and beyond - gleaning valuable information for producers, garden centers and consumers. All-in-all, the internet and technology continue to build a culture where the world is smaller. Producers can more efficiently target emerging and new markets. Everyone has more access. The future is already telling its story today. Plant producers and nurseries are recognizing that simply producing a good plant is not enough. We must learn how to put plants in the context of what is important in peoples' lives: family, health and happiness (Fig. 8).

Advancements in communication and travel has made our global community immeasurably more reachable. Nurseries around the world can share data, successes and failures, and reach new markets more easily than ever. Investment into emerging economies is on the rise, as well as opening doors in other major world economies for producers in the United States. We should look to the world for industry allies, understanding their horticultural businesses and history - so to make more informed and wiser decisions domestically.

Figure 8. Connecting with the needs of consumers: family, health and happiness.

One brief example: The Chinese government has a unique ability, even with its faults, to anticipate and plan for problems. Witnessing China's large-scale urban housing construction developments in person is remarkable, not to mention the opportunity for a plant or two on every patio. The middle class in China is burgeoning, creating enormous opportunities for market share in the nursery and horticulture industries.

European producers have responded to consumers' desires for smaller retail marketing and footprint. Plants and plant products are highly packaged, decorated often as standalone gifts. Many indoor plants are accepted as a live decoration with a set shelf-life. This helps secure a customer for life or repeat customer - at minimum. We must change our market outlook on customer bases - from local to state, region, country and beyond.

Perhaps the one area of significant change is climate and global warming. Scientists around the world have been sounding the alarm for years about reducing carbon footprints, concerns about sea-level rise and its connection to an increasingly warming planet. Many nurseries and producers around the country and world have already begun taking steps to be a more sustainable and responsible climate partner. Water-wise systems, reducing fossil fuel usage, adaption of sustainable packaging and containers are not always easy changes. However, they are necessary as the evidence of human-impacted environmental change reaches a crescendo.

Climate change has the potential to alter the horticulture industry in a way no other outside threat has before. As plant producers, we rely on a certain amount of weather and climate stability to produce our products. Hotter days and longer summers are causing growers to reevaluate product mixes. Standards once never questioned, like geraniums and petunias in European window boxes, cannot hold-up to the soaring temperatures many locations are experiencing. The push toward tropical varieties in areas once thought to be off-limits are allowing for exciting new markets. The industry must remind consumers that plants are a partial solution to environmental change and global warming. Advocating plantings of many sorts convince consumers they can help cool urban heat islands, provide shade and housing for animals, remove toxins from the air and supply life-sustaining oxygen for people.

Sadly, we live in a divided society. Divisions exist on so many fronts from political and religious viewpoints to socio-economic status and personal backgrounds. The curse of access to information is the loss of credibility of science and truth - dismissed by bloggers and trolls disguised as news sources. During my time as IPPS President in the 1990's, we were fortunate to carry out a broad cultural and horticultural tour that took participants across our great region from Texan cities of San Antonio and Houston, across Louisiana with stops in Lafayette, New Orleans and then on to Mobile, Alabama. Each area shared its extraordinary personality, customs and celebrated differences. While happy to share unique traits of each area with my guests - deep down we all shared great national pride and a oneness evident to international travelers. How would those same eyes view us today? Would they see us as disparate tribes?

While we are all competitors, our green industry relies heavily on one another to succeed. Our industry can remain united to speak truth to those unwilling to protect our environment. We can lead by example in instituting positive change in regulation, land and water use, and sustainable growing practices. The future is bright, and change is accelerating. We must remain diligent as partners going forward in uncertain times.

On an optimistic note, we must decide how change is going to steer our industry and our business decisions going forward. The choice is quite simple: one day or day one. There is a great quote from the novelist L.P. Hartley: "The past is a foreign country; they do things differently there." We will do things differently in the future - "for the times, they are a changin'".

Taming the Wild *Stewartia*

Timothy M. Boland and Todd J. Rounsaville

Polly Hill Arboretum, 809 State Road, West Tisbury, Massachusetts 02575, USA

tim@pollyhillarboretum.org

Keywords: Asexual propagation, native trees, plant collections, seeds, Stewartia

Summary

The Polly Hill Arboretum (PHA) began working with native stewartia in 1967. Our founder, Polly Hill, was devoted to growing trees from seed. In 2006, the Polly Hill Arboretum was recognized as the Nationally Accredited Collection holder for stewartia. This status has guided our collection development, particularly on focused seed expeditions, which began in 2007. The PHA has been successful growing both North American species from seed, however, overwintering survival and transplanting of juvenile plants has proved more challenging. New insights into winter storage of seedlings is beginning to shed light on this problem. Experimentation with overwintering rooted cuttings has revealed that plants have preferred temperature and chilling requirements. These new overwintering protocols have thus far yielded positive results. Recent work with tissue culture has also shown promising results with both species. Future work includes grafting superior clones of our native stewartia onto Asiatic species in an effort to overcome the problematic issues of overwintering, transplantability, and better resistance to soil borne pathogens. Our Plant Collections Network (PCN) development plan outlines our next phase work with stewartia over the upcoming several years. The results of this work will be shared as we continue to bring these exceptional small flowering trees into commercial production.

INTRODUCTION

The commitment to building Polly Hill Arboretum's (PHA) stewartia collection is based on our founder Polly Hill's history with the genus and our own desire to encourage the cultivation of these superb small-flowering trees in home gardens. Polly Hill began her experimentation with mountain camellia (*Stewartia ovata*) in 1967 with wild seed sent from Colonial Williamsburg, Williamsburg, VA. In 1982, she began propagating and cultivating the silky stewartia (*S. malacodendron*). These two species represent the only extant stewartia in North America, as the center of diversity for this genus is eastern Asia.

PARTNERSHIP LEADS TO NATIONAL COLLECTION

In 2006, Polly Hill Arboretum's Stewartia collection was recognized as a Nationally Accredited Collection by the North American Plant Collections Consortium, now known as the Plant Collections Network (PCN). Administered by the American Public Garden Association, the PCN is a recognized standard of excellence in plant collection management. Accreditation as a national collection demonstrates a garden's enduring commitment to global efforts to save plants. Today, over 70 stewartia trees representing 22 taxa can be found in PHA's collections, many of them Polly Hill introductions.

PHA director emeritus Stephen Spongberg produced the monograph on the genus *Stewartia* in 1974. This publication served as the authoritative resource for much of our work. In 2015, then curatorial intern Victoria Stewart used Steve's monograph to develop a web-based pictorial key creating a comprehensive resource for the genus. This work, combined with our efforts to research, collect, and grow the collection, resulted in the International Society for Horticultural Science appointing PHA as the international cultivar registration authority for the genus. This designation positions us to work with plant breeders, commercial growers, and other arboreta to further expand the availability and use of this beautiful group of trees (Boland, 2017).

SPECIES PROFILES

<u>*Stewartia malacodendron*</u>

Silky stewartia is native to the coastal southeastern states, occurring from Texas to Virginia. Although this species is Globally Secure (G4), its limited distributed in some states has yielded local designations of Vulnerable (S3) in Mississippi, Florida, North Carolina, and Virginia, Imperiled (S2) in Louisiana, Alabama, and Georgia, and Critically Imperiled (S1) in Texas and Arkansas (Nature Serve, 2019; Fig. 1).

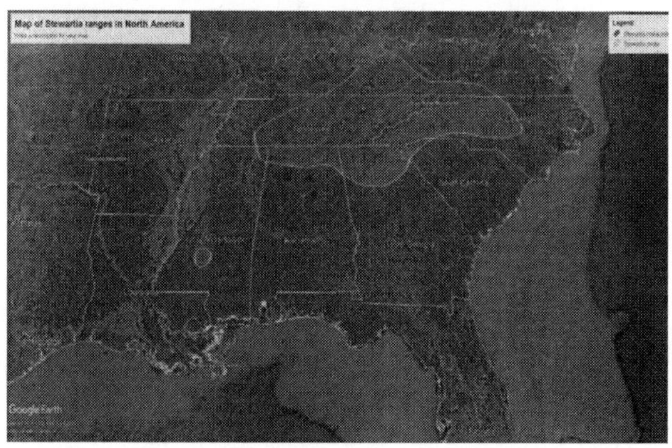

Figure 1. Approximated range map of *Stewartia malacodendron* and *Stewartia ovata* in the United States. Points represent populations that were sampled by Polly Hill Arboretum (PHA), with specimens currently active in the living collection.

Spongberg (1974) notes that although *S. malacodendron* is found primarily within the coastal plain in the southeastern U.S., it can also be found in some piedmont localities as well as in the mountains. It prefers low rich woodlands and is typically growing along stream banks and slopes of ravines. In our experience, silky stewartia grows poorly in deep shade, and flower/seed production suffers as a result. It has been hypothesized that this species would have been more widespread historically, but its natural propensity to grow in rich floodplain terraces conflicted with the favorability of these same sites to European settlers and farmers. Thus, many populations are presumed to have been destroyed when land was cleared for agriculture. This is similar to the now (functionally) extinct family member, *Franklina alatamaha*.

Flowering of silky stewartia occurs primarily between May and June, and this species is noted for its purple-colored anthers. Spongberg (1974) notes that *S. malacodendron* is remarkable uniform across its range, much unlike *S. ovata* and other Old-World species. Pollination is performed by insects, particularly bees, but individuals of most stewartia species appear self-fertile and even isolated specimens are capable of producing viable seeds.

To read about the singular cultivar that Polly Hill selected of the silky stewartia see: http://www.pollyhillarboretum.org/plants/stewartias-2/stewartia-malacodendron/

Figure 2. *Stewartia malacodendron*. (A). Flower; (B) Capsule; (C) Seed.

Stewartia ovata

Mountain stewartia, much like silky stewartia, is concentrated to the southeastern states. As its name suggests it is more concentrated in mountainous sites. Also listed as Globally Secure as a species, state conservation status for *S. ovata* includes designations as Vulnerable (Kentucky, Georgia, North Carolina), Imperiled (Alabama, South Carolina, Virginia) and Critically Imperiled (Mississippi). The species is often listed in Florida, where it is classified as 'not ranked/under review', however the only known specimen from Florida is from a cultivated (nursery) site. *Stewartia ovata* is typically found in partially shaded, moist ravines and gorges from 200-1100 meters. The challenging terrain of these sites make them less amenable to land-use conversion, but habitat loss and fragmentation remain the greatest threat to this species.

The flowering window for *S. ovata* is shifted approximately one month later than *S. malacodendron*. There appears to be little, if any, sympatric populations of the two species that are extant, and no hybrids have been reported. Some variation exists in the flowers of *S. ovata* however, and anther color may vary between white, yellow, and purple (Fig. 3).

Polly Hill selected three cultivars based on anther and filament variations; see: http://www.pollyhillarboretum.org/plants/stewartias-2/stewartia-ovata-mountain-stewartia/

Although the name *S. ovata* var. *grandiflora* has been published on the basis of flowers having 5-8 petals and purple filaments, it is generally not recognized as a distinct taxonomic unit, but rather as a typical phenotypic variance.

In the landscape, *S. malacodendron* grows as a large shrub with a rounded crown, whereas *S. ovata* is a small to medium sized tree. Both species have rather ordinary bark in comparison to the multicolored, papery, and exfoliating bark of many Asian species such as *S. monadelpha* and *S. pseudocamellia*. Some degree of hybridization appears to be possible within stewartia, for example *S. pseudocamellia* has been crossed with both *S. monadelpha* and *S. ovata*. Nevertheless, *S. malacodendron* remains a phylogenetic outlier, distantly related to all the other deciduous taxa (Lin et al., 2019). The subtropical evergreen cohort of stewartia (formerly known as *Hartia*) are in fact nested between *S. malacodendron* and all other species.

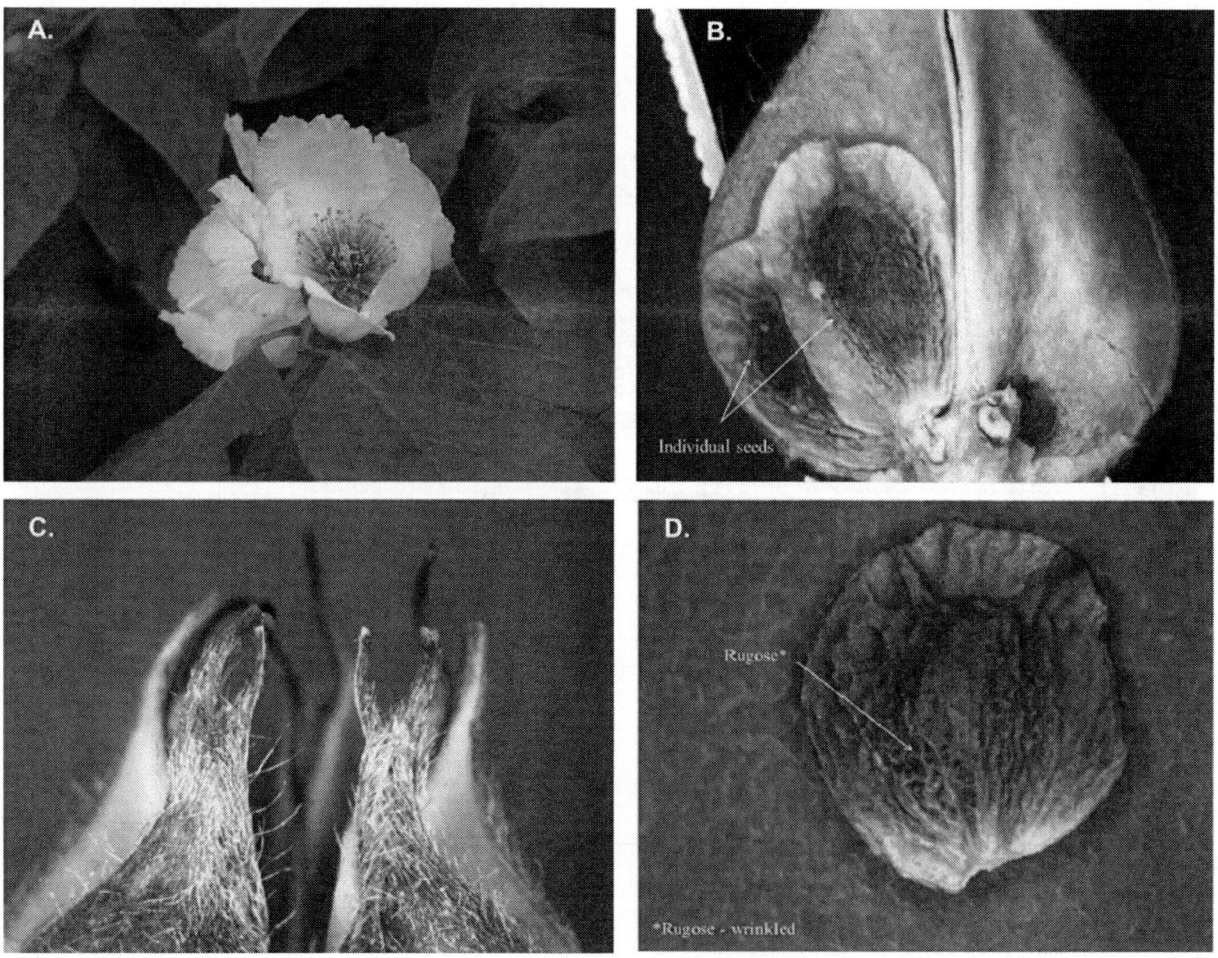

Figure 3. *Stewartia ovata*. (A). Flower; (B) Capsule with seeds; (C) Capsule with forked locule apex (persistent styles); (D) Seed.

SEED EXPEDITIONS

On our past trips to the southeastern United States over the last 12 years, only a small percentage of seed has germinated, and only a few have made it into the collections. Seed propagation is difficult: the woody seeds rarely germinate within a year of collection. They require 3-5 months of warm stratification followed by 3 months of cold stratification to overcome internal dormancy mechanisms (Dirr and Heuser, 2006). Polly Hill understood this lengthy process and sowed seeds directly in the ground, patiently letting nature do its work.

Our beginning work with seed propagation of Stewartia began in earnest in 2007 with the opening of a new greenhouse. The same year, PHA Executive Director, Tim Boland was introduced to plantsman Jack Johnston, of Clayton, GA. Mr. Johnston has nearly 40 years of stewartia observations throughout the Southeastern United States. He has an uncanny ability to spot trees along the river ledges where their naturally occur. Using a hooked walking stick he has mastered the art of pulling down sun-soaked branches to harvest seeds. Jack grows several trees in his home garden as well as a northern property he owns in Macon County, North Carolina.

The Polly Hill Arboretum is part of an informal collaborative Stewartia Working Group (SWG) which consists of the following members: Mt. Cuba Center, Delaware; Birmingham Botanical Gardens, Alabama; Yew Dell Gardens, Kentucky; and Smithgall Woodland Garden (Atlanta Botanical Garden), Georgia. The primary objective is to collect both species from throughout their natural range. Jack Johnston has guided many of these expeditions.

INTERNATIONAL CULTIVAR REGISTRATION AUTHORITY

In 2016, the International Society for Horticultural Science appointed the PHA as the international cultivar registration authority (ICRA) for the genus. This designation positions us to work with plant breeders, commercial growers, and other arboreta to further expand the availability and use of this incredibly beautiful group of trees (Boland, 2017). Currently, the PHA has four registered cultivated varieties of the two native Stewartia species with plans to registered more recent discoveries.

SEED PROPAGATION

Woody capsules of both species open and release seeds under dry conditions, which naturally occurs between late September and early November. Individual seeds become sclerified before natural dehiscence for both species (Fig. 2 and 3).

For harvesting purposes, collecting is best accomplished by gathering capsules just prior to natural dehiscence, and storing them in a paper bag until seeds are released. It has been noted that seed of *Stewartia malacodendron* and *S. ovata* require a warm/cold stratification period for dormancy breaking and germination (Dirr and Heuser, 2006). Seed sowing has traditionally relied upon two methodologies: directly sowing seeds into the ground or sowing directly in free draining Anderson flats. In general, germination is irregular and the majority of seedlings emerge after the second winter, roughly 20 months after sowing the first fall. Given this extended stratification period, it is advantageous to use fine, wire-screened cages to exclude rodent seed-predators.

VEGETATIVE CUTTINGS

The other principal method of growing our native *Stewartia* is stem cuttings. We have had success rooting cuttings taken between mid-June and July using 8000 ppm Hormodin 3® (IBA) rooting powder. Cuttings are placed in Anderson Flats using a 1:1 ratio of perlite and peat. Although we have had success rooting cuttings, overwintering survival is a challenge noted by many growers. Temperature and duration of chilling are two important factors of overwintering, and previous studies have helped elucidate the optimal protocols for stewartia. For example, Nair et al. (2008) found that overwintering Japanese stewartia (*S. pseudocamellia*) at 5°C resulted in 65% survival, while stem cuttings held at -30° and -12° C experienced 100% mortality. Furthermore, 22% of cuttings survived when held at 21° C, effectively lacking cold storage altogether (Nair et al. 2008). These results further clarified by experimental analysis of *S. ovata*, for which chilling at 6° C (for 0, 2, 4, 6, 8, or 10 weeks) was positively correlated with overwintering survival (Curtis et al. 1996). At PHA we have adopted these ostensibly optimal conditions (~12 weeks at 5° C) for species and cultivars of *S. malacodendron* and *S. ovata* with 70-100% cutting survival. There is also evidence that fertilization of stem cuttings with 200 ppm N results in greater shoot growth following overwintering, but N addition had no effect on survival (Curtis et al., 1996). We have observed best results when allowing cuttings to remain in their original rooting container and medium for the overwintering period. The cuttings are then transplanted to a new container well after bud-break the following year.

TISSUE CULTURE

PHA has been working cooperatively on stewartia propagation with Heather Gladfelter, a researcher at the Warnell School of Forestry at University of Georgia in Athens. We sent Heather seed of *Stewartia ovata* 'Red Rose' and *S. malacodendron* 'Delmarva' in 2013. She extracted the embryos and grew them in a Petri dish using plant growth regulators to encourage the formation of roots and shoots. This process has shown very promising results and we plan further collaborative experiments with UGA in the future (Gladfelter et al., 2019).

FUTURE WORK WITH STEWARTIA

This past July the PHA began a new phase of *Stewartia* propagation with a look towards further optimizing overwintering and transplanting stem cuttings of key cultivars. In a collaborative partnership with Heritage Seedlings and Liners Inc, Salem, Oregon, PHA sent its most vigorous and best flowering selections to be chip budded this past August onto Japanese Stewartia, *Stewartia pseudocamellia*. Japanese stewartia has seemingly broad soil adaptability and is less prone to soil borne pathogens than our native species (Mark Krautmann, personal communication). Additional wood will be sent to be dormant grafted this winter season. Moving forward with our efforts involving the Plant Collection Network; a plan has been developed that further defines our mission by advancing botanical research, plant conservation, curatorial excellence, and collaborative partnerships working with this fantastic group of trees (Rounsaville, 2019). It is our hope that this next stage of testing, and our continued efforts to sample wild stewartia from across their ranges will help bring new cultivated varieties into the marketplace.

Acknowledgements

The authors wish to thank the many people who helped the Polly Hill Arboretum (PHA) on our quest to introduce native stewartia into today's gardens. A huge debt of gratitude goes to our early collection partners in the southeastern U.S.: the exceptional plantsman Jack Johnston of Clayton, Georgia, who guided multiple trips in pursuit of stewartia seed; and collaborators Rick Lewandowski (Shangri La Botanical Garden and Nature Center) and Fred Spicer (Chicago Botanic Garden).

We are also grateful for the assistance of Thomas Clark, PHA's former Curator (2006-2017); Thomas Murphy, graduate student (Austin Peay University, Clarksville, Tennessee); Victoria Stewart, (San Francisco Botanic Garden); Mark and Jolly Krautmann, Heritage Seedlings and Liners Inc, Salem, Oregon; Philippe de Spoelberch, founder Arboretum Wespelaar, Belgium; Heather Gladfelter, graduate student (University of Georgia, Athens, Georgia); and finally, Dr. Stephen A. Spongberg, PHA Director Emeritus and the author of the monograph of *Stewartia* published in 1974.

Literature Cited

Boland, T.M. (2017). PHA's Stewartia Collection, Patience, Persistence, Pride. Meristems *19*:8-9.

Curtis, D.L., Ranney, T.G., Blazich, F.A. and Whitman II , E.P. (1996). Rooting and subsequent overwinter survival of stem cuttings of *Stewartia ovata*. J. Environ. Hort. *14*:163-166.

Dirr. M.A. and Heuser, C.W. (2006). The Reference Manual of Woody Plant Propagation. From Seed to Tissue Culture, 2nd. Ed. Varsity Press. Cary, North Carolina.

Gladfelter, H.L., Johnston, J, Wilde, D. and Merkle, S. (2019). Propagation of *Stewartia* for the Nursery Industry: Somatic Embryogenesis of Rare Species and Elite Cultivars. Comb. Proc. Intl. Plant Prop. Soc. *69*: (In press).

Lin H., Hao, Y., Li, J., Fu, C., Soltis, P.S., Soltis, D.E., and Zhao, Y. (2019). Phylogenetic conflict resulting from ancient introgression following species diversification in *Stewartia* s.l. (*Theaceae*). Molec. Phylo. Evol. *135*: 1-11.

Nair A., Zhang, D. and Smagula, J. (2008). Rooting and overwintering stem cuttings of *Stewartia pseudocamellia* Maxim. relevant to hormone, media, and temperature. HortScience *43*:2124-2128.

Spongberg, S.A. (1974). A review of deciduous-leaved species of *Stewartia* (*Theaceae*). J. Arnold Arboretum *55*:182-214.

Rounsaville, T.J. (2019). Collection development plan; nationally accredited Stewartia collection™. https://www.publicgardens.org/resources/polly-hill-arboretum-collection-development-plan-nationally-accredited-stewartia. Accessed October 1, 2019.

Crapemyrtles - Past, Present and Possibilities

Stan Brown

Plants of Redlick, Highway 818, Clarksville, Arkansas 72830, USA

blossomberrynursery@gmail.com

Keywords: Breeding, *Lagerstroemia*, plant breeders, woody plants

Summary

My breeding efforts with crapemyrtle which has spanned 60 years. Important crapemyrtle breeders include Otto Spring, Donald Egolf, David Chopin, Mike Dirr, Carl Whitcomb, Cecil Pounders, David Chopin, Bob Hambuchen, Dr. John Creech, and Ewa Nelson. Some of my efforts at Plants of Redlick have been developing compact forms with dark leaves. I have three distinct compact forms that may prove beneficial in providing new selections. For extending flowering, I am also working on improved length of flower bloom. Repeat blooming is related to sterility.

INTRODUCTION

A little background before we get started on the specifics of my breeding efforts with crapemyrtle (*Lagerstroemia*) which has spanned 60 years. I grew up in the Northwest corner of Arkansas and was exposed to horticulture at an early age. My dad was a fruit grower and had a nursery (Earl Brown Nursery) focused on fruits. Before enrolling at the University of Arkansas, I worked for Dr. Jack Hull starting in 1960. On my first day I was in a screen house making crosses on blackberries. Under his direction I learned everything about pollen, seed processing and plant evaluation. As a result of this, I gained an appreciation of plant breeding.

Crapemyrtle since its introduction has long been associated with the Southern U.S. Its ability to survive in heat and drought and provide a long bloom season of bloom has endeared it to landscapes where it is winter hardy. In the U.S., the genetic base was narrow, and most selections were for flower color. Some of the first efforts to modify the crapemyrtle 'family' was finding and developing more compact forms. Mr. Otto Spring of Okmulgee, OK was among the first to exploit this trait. He developed compact forms that were later marketed by Monrovia Nursery as their Petite series. Mr. Spring described 'dwarf' as under 0.9 m (3 ft), 'midget' as under 0.4 m (15-in.) and 'wee wee' as under 10-13 cm (4 to 5 in.) from dormant pruning.

The University of Arkansas (UA) using material from Mr. Spring introduced a series of four crapemyrtles of which the most popular was 'Victor', a smaller growing red-flowered form. When the UA introductions were released, Dr. Al Einert said they just grew slower and were not true dwarfs.

Dr. Donald Egolf at the U.S. National Arboretum used a dwarf red from Otto Spring in his program to develop 'Chickasaw', a true dwarf. J.B. Fitzpatrick's nursery in Texas contributed several crapemyrtles as 'Low Flame', 'Royalty' and 'Pink Ruffles'. David Chopin introduced a series of low growing crapemyrtles. His 'World's Fair' selection made a hit at the 1984 World's Fair in New Orleans and several varieties are available today. Dr. Michael Dirr introduced the Dazzle^R and Magic series with lots of appeal and many flower colors; Cherry Dazzle was one of the best.

The next major advancement was the introduction of new species, notably the introduction of *Lagerstroemia faurei* and *L. subcostata* complex. Dr. Egolf was working on disease resistance and used the new species to start an explosion of new varieties. Most of these were tree types with improved disease resistance, attractive bark, and various flower colors.

My first major breeding effort used the first released hybrids and *L. subcostata* sent as a plant introduction (PI) to the University of Arkansas. From these crosses a few selections were made. When 'Tonto' was released it was crossed with the best reds available including 'Victor', 'Okmulgee' and two red selections, a dark and a bright red obtained from Mr. Spring. 'Freedom' (96-50) was the best selection from this population and is used as a parent in nearly all of my selections made since.

The next major advancement in the crapemyrtle family was Dr. Carl Whitcomb's new cherry red flower color. Again, a flurry of attempts was made to spread this color resulting in many new selections.

The latest new trait that has been impacting breeding programs is the 'black' dark leaf form found and expanded on by Dr. Cecil Pounders. My efforts now are developing compact forms with dark leaves. I have three distinct compact forms that may prove beneficial in providing new selections.

Another area that can be improved upon is the length of bloom. Dead heading has long been used to extend flowering, but that method is not practical with a woody plant like crapemyrtle. Seedlessness would provide the same effect. A crapemyrtle I received from Otto Spring, which I refer to as O.S. Blue. It is nearly seedless, and it offers heavy panicle re-blooming. I have never experienced a selection with more bloom over a longer season. 'Freedom' is a panicle rebloomer also which contributes to its long season of bloom.

Polyploidy may provide some sterility. One selection I have appears to be sterile and blooms over an extended season with poor seed set. Early Bird^{TM} White has a trait of late flowering which is a result of axillary buds on leafy branches. This trait has been passed to F1 and F2 populations with increased bloom in late season.

As Dr. Dave Creech recommended several years ago, there are other opportunities with new selections having a good red flowered form and 'Natchez'-like bark. There is need for these tall tree type with dark leaves, and enhanced fall color. With few exceptions' disease resistance is still poor in all pure *indica* varieties. Insect resistance can be found for Japanese beetles and work shows aphids have preferences. While screening efforts are being made, we need to find resistance to the crapemyrtle bark scale (CMBS).

Dr. Michael Dirr was recently quoted in a national magazine saying: “A major frustration in plant breeding is insufficient time to determine the true worth of a particular selection.” Oh, how true this is with woody plant breeding!

Acknowledgements: I would like recognize the effort of those have preceded me and set a solid foundation for our collective efforts: Otto Spring, Dr. Donald Egolf, David Byers, Dr. Carl Whitcomb, Dr. Michael Dirr, Dr. Cecil Pounders, David Chopin, Bob Hambuchen, Dr. John Creech, Ewa Nelson, and Dr. Jim Robbins for helping with this presentation and manuscript.

Growing Greener Production Opportunities for Nurseries

Steve Castorani

North Creek Nurseries, Inc., 388 North Creek Road, Landenberg, Pennsylvania 19350, USA

steve@northcreeknurseries.com

Keywords: Ecosystem services and green infrastructure, green industry, native perennial plants, plant mitigation, sustainable landscapes, urban sprawl

Summary

Our current development model of urban sprawl and uncontrolled development is recognized as unsustainable by planners, architects and environmentalists. There are opportunities for nurseries in the Ecosystem Services and Green Infrastructure. Man-made, designed landscapes must evolve by reducing energy inputs while increasing biodiversity and plant density. These methods will also reduce the threats posed by invasive species. Additional opportunities present themselves to landscape contractors that are willing to remove invasive species and revegetate land using native plants. Perennials selected have the ability to help slow runoff, accelerate infiltration and enhance the evaporation of stormwater runoff. The inclusion of plants in these engineered systems also creates an opportunity for increased biodiversity by increasing the variety of plants that support pollinator species.

INTRODUCTION

North Creek Nurseries is a wholesale propagation nursery located in Southeastern Pennsylvania. Our product lines are perennials, grasses, ferns, vines, and a few shrubs. Our primary focus is the production of native plants and their cultivars. We focus on varieties that are native to the eastern region of the US with a strong emphasis on plants native to the Mid-Atlantic.

Contrary to many commonly held opinions of the economic health of the green

Industry, I am here today to tell you that I believe that there are great growth opportunities for nurseries today, and well into the future. Now let's take hemp for an example. There is a lot of "buzz" surrounding this emerging industry. We can dedicate a whole lecture about this market, but I am not going to talk about that today. There is certainly a lot of chatter about the revival of the houseplant industry, especially among millennials, but I am not going to talk about that either. Likewise, there has been a strong market and renewed interest in edible plants recently. Some people refer to this as the new gateway to gardening. Well, I am not going to speak about this trend either.

I am here to speak about opportunities for nurseries in ecosystem services and green infrastructure. This is not a talk about the "Green New Deal!" Our current development model of suburban sprawl and uncontrolled development is recognized as unsustainable by planners, architects and environmentalists. Man-made, designed landscapes must evolve by reducing energy inputs while increasing biodiversity and plant density (Table 1). These methods will also reduce the threats posed by invasive species. Additional opportunities present themselves to landscape contractors who are willing to remove invasive species and revegetate land using native plants. It will take a concerted effort by government and the private sector along with a robust sales and marketing initiative. State governments are beginning to demand this level of ecological planning. Public awareness is evolving, and expectations are on the rise!

These opportunities will be focused primarily on the following three areas of development and redevelopment sites:

1. the reduction of stormwater runoff
2. the infiltration of that water using engineered systems
3. mitigation techniques that include plants

Table 1. Opportunities for the Green Industry

- Suburban sprawl and uncontrolled development are not sustainable. Landscape practices have to evolve to reduce inputs, add biodiversity and increase plant density.
- Opportunities abound for landscape contractors to remove invasive species and re-vegetate land using native plants. It will take marketing and sales efforts.
- This would be a worthwhile and be a benefit to nurseries as well as landscape contractors.
- Reduction of stormwater runoff and infiltration mitigation is a key driver
- Increasing biodiversity - regeneration of plants that support pollinator species - Larval food source, nectar, and pollen.
- Use of green mulches to reduce overall maintenance cost and increase plant sales
- Plants have numerous health and wellness benefits. They aid us in achieving a healthy lifestyle.
- Trees will be planted to reduce the effects of CO_2 plus provide shade to cool our environment.
- Habitat restoration is happening throughout the country.
- All these will drive plant sales and add perceived value to our products.

Installing plants, especially native perennial plants is the preferred method of mitigating storm water runoff. Perennials are being specified as having the ability to help slow runoff, accelerate infiltration and enhance the evaporation of stormwater runoff. The inclusion of plants in these engineered systems also creates an opportunity for increased biodiversity by increasing the variety of plants that support pollinator species. These plants offer larval food sources, nectar, and pollen to birds and insects (Fig. 1).

Figure 1. Wildlife corridors connect ecosystems.

Who is driving this demand?

The following is a list of influencers that are specifying these systems. It's imperative that nurseries, as well landscape contractors align themselves with these groups and individuals in order to take advantage of these opportunities (Table 2).

Table 2. Who is Driving Demand?

- Landscape Architects, Engineers
- Public Water and Sewer Authorities
- State Department of Plant Industries through Department of Agriculture
- Highway Departments
- City Planners
- NGO's
- Public Gardens, Arboreta & Nature Conservancies

Landscape achitects, engineers, public water and sewer authorities, state departments of plant industries through departments of agriculture, highway departments, city planners, non-governmental organizations, public gardens, arboreta & nature conservancies to name a few. At North Creek we market directly to these groups and make personal connections with influencers by inviting them to visit our nursery. We also support, sponsor and attend many of their trade meetings. We offer "Lunch & Learns" at their home office as well.

As an example of this potential, the city of Philadelphia is slated to spend upward of $10 billion over the next 10-15 years (Fig. 2). This proposal, which several experts called the nation's most ambitious, reimagines the city as an oasis of rain gardens and green roofs with thousands of additional trees and porous pavement. The vision now is to "peel back" the city's concrete and asphalt and replace them with plants – via rain gardens, green roofs, heavily planted curb extensions, vegetated swales in parking lots, and mini-wetlands (Fig. 3).

Similarly, the city of Lancaster, Pennsylvania is in the midst of a surge in spending to add green infrastructure to its city streets. Chicago, Illinois incentivizes developers to add green infrastructure and green roofs on their buildings. By designing with these systems, they are assured an expedited process to move their project forward.

Figure 2. Philadelphia is slated to spend upward of $10 billion over the next 10 -15 years. This proposal, which several experts called the nation's most ambitious, reimagines the city as an oasis of rain gardens and green roofs with thousands of additional trees and porous pavement.

Many customers that are specifying plants for green infrastructure projects do not have a plant or ecology background, so it is important that you provide resources for them to gain the knowledge needed to be successful in establishing these landscapes.

Figure 3. The idea is to "peel back" the city's concrete and asphalt and replace it with plants – via rain gardens, green roofs, heavily planted curb extensions, vegetated "swales" in parking lots, and mini-wetlands.

Make the connection: create learning laboratories and offer garden tours.

Over the years we have invested into extensive green infrastructure at our facilities. We lead customers on tours through our "Learning Laboratory". The following engineered systems hold countless opportunities for nurseries and landscape contractors to sell more plants and obtain lucrative contracts. At North Creek we can show our clients examples of green infrastructure such as: Riparian Buffers, Constructed Wetlands, Detention Basins, Rain Gardens, Bioswales, Green Roofs, Wetlands and Stormwater Management systems (Figs. 4, 5, 6).

Figure 4. Establishing a green roof in Landenberg, Pennsylvania.

Figure 5. Layered landscapes (left) and green mulch (right).

Figure 6. Constructed wetlands.

The use of Landscape Plugs™ in green infrastructure

In recent years and as more green infrastructure projects have been completed, the use of Landscape Plugs™ is preferred over larger container plants (Fig. 7). Landscape architects, designers and contractors are specifying plugs on larger projects due to their ease of use, rapid establishment and economy. Landscape Plugs™ have these advantages: quick root establishment, suitable for challenging sites, easy to transport and install, available native species, cost effectiveness. In addition, plugs establish more reliably and faster than seed mixes.

Layered Landscapes, or the use of plugs as green mulch (ground covers)

Here is yet an additional opportunity for nurseries to grow and sell more plants. Installing plants more closely in a matrix design allows them to knit together, reducing weed seed germination and eliminating open ground. Green mulch is significantly more sustainable as it eliminates the need for wood mulches. Planting green mulches is a great technique often used for covering ground in detention basins (Fig. 5). The use of perennial groundcovers with evergreen basil foliage reduces weed penetration and aids in infiltrating storm water.

- Cost Effective
- Quick Root Establishment
- Suitable for Challenging Sites
- Easy to Transport and Install
- Eastern US Native
- Species and Cultivars
- Establish Faster than
- Seed Mixes

Figure 7. Landscape plugs 5.1 x 12.7 cm (2 x 5-in.).

- Native perennial plants preferred
- Local ecotypes – increases biodiversity
- Pollen and nectar sources
- Variety of bloom times
- Variety of shapes and colors
- No doubles or pollen-less varieties
- Succession planning

Figure 8. General criteria for selecting preferred native perennial plants.

Figure 9. Plants are not a luxury. They are a vital part of what keeps our world in balance.

Biodiversity sells more native plants

Increasing your nurseries offerings of native plant species will add additional sales opportunities and expand your customer base. Understanding the ecology and connection between plants, insects and animals, especially birds will position your firm as the experts in this growing market (Fig. 8). Plan your production so there are plants available all year for this market. When supplying plant species, remember to include plants that provide nectar and pollen throughout the growing season, early spring through fall.

An additional concern is the importance of growing your plants with pollinator friendly practices (Fig. 9). Plant nectar/pollen bearing species may not be treated with chemicals that will affect feeding pollinators. Do your research and commit to producing plants that are truly free of harmful chemicals that could translocate into pollen and nectar.

For those of you who are landscape architects and designers, it is important to establish habitats in rough borders, hedgerows and shelter beds. Develop corridors that expand and connect important pollinator habitat patches. Focusing on native plants and cultivars will increase the season long forage capacity for honey bees and native bees. Provide nesting sites for native bees. Provide nectar resources and habitat for butterflies which will then provide butterfly larvae for birds. Eliminate pesticides and reduce/stagger mowing practices.

Plants offer numerous health and wellness benefits for us as well. They aid in achieving a healthy life style. Trees will continue to be planted to reduce the effects of CO_2 (carbon sequestration) while providing shade to cool our environment. Habitat restoration is happening throughout the country. These combined efforts will drive plant sales and add perceived value for our products.

In conclusion, I want to remind you that plants are not a luxury, they are a vital part of what keeps our world in balance. There is a world of opportunity out there for the entrepreneur who continues to seek out new opportunities.

IPM Approaches for the Management of Chilli Thrips and Crapemyrtle Bark Scale

Yan Chen[1], Rodrigo Diaz and Dennis Ring[2], Michael E Merchant[3], Erfan Vafaie[3], David Held[4], Mengmeng Gu[5], Gary W. Knox[6], Yu Zhang[7], Jim Robbins[8], and Derald Harp[9]

[1]LSU AgCenter Hammond Research Station, [2]LSU AgCenter Department of Entomology, [3]Texas A&M AgriLife Extension, [4]Department of Entomology and Plant Pathology, Auburn University, [5]Department of Horticultural Sciences, Texas A&M University (TAMU), [6]University of Florida IFAS, [7]Department of Agricultural Economics, TAMU, [8]University of Arkansas-CES, [9]Department of Agricultural Sciences, TAMU-Commerce.

yachen@agcenter.lsu.edu; mgu@tamu.edu

Keywords: Crapemyrtle, crapemyrtle bark scale, *Scirtorhtips*, *Lagerstoemia*

Summary

Both chilli thrips (*Scirtothrips dorsalis*) and crapemyrtle bark scale (*Acanthococcus lagerstroemiae* (Kuwana) are among the top insect pests challenging the ornamental horticulture industry both in the nursery production and landscape management of ornamental species. The regional integrated pest management (IPM) team works have been conducted over the past fifteen years for chilli thrips and five years for crapemyrtle bark scale to gain knowledge on their biology and test the effectiveness of various cultural, biological and chemical management options. In this presentation, basic biology and current management recommendations are discussed to provide options for industry pest management professionals when considering a programmatic approach to manage these two pests.

INTRODUCTION

Both chilli thrips (*Scirtothrips dorsalis*) and crapemyrtle bark scale (*Acanthococcus lagerstroemiae* (Kuwana) are among the top insect pests challenging the ornamental horticulture industry both in the nursery production and landscape management of ornamental species.

CHILLI THRIPS

Over the past decade, chilli thrips has become more established in the southeastern U.S. economic losses due to this pest have been reported from many sectors of the horticultural industry such as landscape ornamental plants, production nurseries (of ornamentals, vegetables, and fruit trees), and field grown vegetables and fruits (Fig. 1).

Figure 1. Chilli thrips (left) in comparison of size and shape with western flower thrips (right).

Among a large number of ornamental species that are suitable hosts, Knock Out roses, cleyera, camellia, bottle brush, duranta, distylium, Indian hawthorn, ligustrum, pomegranate, and viburnum are some of the woody species that are more susceptible. Infestations on herbaceous plants such as begonia, coleus, snapdragon, zinnia, coreopsis and verbena have also been reported.

Chilli thrips is very small, only 0.016 to 0.024 inch in length. Compared with the more common western flower thrips, chilli thrips are about one-quarter of the size of western flower thrips, with a more bulged abdomen (Fig. 1). All thrips inhabit secluded areas on plants, such as unopened flower and terminal buds, thus reduces exposure to insecticide applications. They also superficially resemble some other thrips species that, if mistaken, may result in poor control because the insecticides selected may not be efficient against chilli thrips. Damage usually shows up on young leaves first but both young and old leaves can be bronzed, curled and distorted, which may look like herbicide burn or leaf rust. Severe infestation can defoliate or stunt plant growth (Fig.2).

CRAPEMYRTLE BARK SCALE (CMBS)

Known scientifically as *Acanthococcus lagerstroemiae* (Kuwana), the crapemyrtle bark scale is a relatively new insect found principally on crapemyrtles (*Lagerstroemia* spp.) across the Southeast and is a member of the bark or felt scale family (Hemiptera: Eriococcidae). Native to Asia, CMBS was first noticed in a north Dallas, Texas, suburb in 2004. By 2019, the insect had been reported from twelve states. This exotic scale causes heavy honeydew deposits followed by a disfiguring layer of dark black sooty mold which severely diminishes the landscape value of this important ornamental plant. Heavy infestations of CMBS reduces the size of panicles, delays flowering, and kills small twigs on crapemyrtle. In its native range, this bark scale was reported infesting plants from 16 genera in 13 families, many of which are economically important crops, most notably persimmon, pomegranate, fig, and blackberry. In the U.S., the scale has been recorded feeding on American beautyberry (*Callicarpa americana*) and St. John's wort (*Hypericum perforatum*) in the landscapes. Host susceptibility experiments conducted at TAMU with different varieties of crapemyrtles did not suggest any resistance or tolerance among *L. indica* and its hybrids (Wang et al., 2018).

Figure 2. Foliage injury symptoms on various woody ornamental species and herbaceous (pepper) crops of economic importance.

This scale is easy to identify because it is the only bark scale know to occur on crapemyrtle. The adult females appear as white or gray felt-like encrustations on small twigs to large trunks, with early infestation often appearing near the leaf nodes of a branch, and bleeding pink when squeezed (Fig. 3). Up close, CMBS is white to gray in color and approximately 2 mm (0.08 inch, a bit longer than the thickness of a dime) in length (Wang et al., 2016). Careful examination may reveal dozens of pink eggs or crawlers under some of the larger white scale covers. Most gardeners will be alerted to CMBS by black sooty mold which appears on the bark.

The presence of sooty mold may confuse the diagnosis since that is also commonly associated with a significant aphid or whitefly problem. The distribution of CMBS is frequently updated on https://www.eddmaps.org/cmbs with future distribution range closely related to the distribution range of its primary host, crapemyrtles (Wang et al., 2016). Landscape professionals and homeowners are encouraged to report new infestation to local extension offices so its spreading can be monitored.

The USDA APHIS has determined that a CMBS quarantine was not justified because it is already established in the U.S. Therefore, no targeted surveys for the crapemyrtle bark scale have been conducted or are planned.

However, in some states (Arkansas, Louisiana, Oklahoma and Tennessee), there is currently a stop-sale on crapemyrtle whenever this scale insect is found during routine inspections of nurseries and retail sites as a quality pest, not as a regulated quarantine pest.

Figure 3. Crapemyrtle bark scale is easy to identify on infested trees with a close inspection and will bleed pink when pinned with an object.

IPM OPTIONS FOR MANAGING CHILLI THRIPS

The key to managing chilli thrips is to detect and treat the pest before a high population builds up. Chilli thrips prefers to feed on tender plant tissue, thus more outbreaks have been reported in mid-May and/or September to early October when many landscape plants having their new growth. Therefore, it is important for landscape managers or home gardeners to periodically check plants during these months. Tapping foliar terminals over a sheet of white paper will dislodge thrips that can be examined with a hand lens.

In production nurseries, because plants are routinely pruned to promote branching and new growth, chilli thrips can be problematic throughout the growing season. It is critical for nursery growers to be familiar with early damaging symptoms such as leaf curl and distortion of susceptible plant species or setting up sticky cards/tapes to monitor chilli thrips population in high risk crops on a weekly basis.

For cultural practices, sanitation is important for both production and landscape management of this pest. Cleaning up debris from infested plants and removing weeds are important because chilli thrips overwinter as adults in leaf litter or weeds. This is especially critical for nurseries that have had chilli thrips infestations in the past. When treating a local outbreak, either in nursery or landscapes, severely infested branches should be cut and bagged for disposal. We have also found that high nitrogen and phosphorus contents in plant leaves contribute to higher numbers of chilli thrips on Knock Out roses. Applying fertilizer lightly, such as a split application at the recommended rate, may avoid promoting chilli thrips reproduction.

The use of pyrethroids, organophosphates or other broad-spectrum insecticides is not recommended for controlling chilli thrips in landscape plants because of their potential effects on beneficial species, including minute pirate bugs (*Orius spp.*), lacewings and predatory mites or spiders that help prevent outbreaks of chilli thrips as well as other pests.

The AgCenter and Univ. of Florida research team found that a rotation between spinosad and the entomopathogenic fungi *Metarhizium brunneum* (Met52) and *Beauveria bassiana* (BotaniGard 22WP), or insect growth regulator (azadirachtin, Molt-X), and horticultural oils (such as the ultra-fine oil or SuffOil-X) reduced chilli thrips populations by 88% to 95%. These products are considered "soft" on beneficial arthropods and are available to commercial landscape professionals. For home gardeners, insecticides containing spinosad, such as Conserve, can be rotated with ultra-fine oil or soapy water to treat infested plants during thrips active months (Aristizábal et al., 2016).

For nursery growers, insecticides containing abamectin (Avid), acephate (Orthene), chlorfenapyr (Pylon, greenhouse only), flonicamid (Aria), imidacloprid (i.e., Marathon), spinetoram (XXpire), spiromesifen (JUDO) and spinosad (Conserve) can help control chilli thrips in production nurseries. Rotation among different classes and modes of actions is recommended to reduce the risks of developing insecticide resistance and the outbreaks of secondary pests.

IPM OPTIONS FOR MANAGING CRAPEMYRTLE BARK SCALE

Based on several years of field and greenhouse testing, we have found that crapemyrtle bark scale can be controlled with the right insecticides and application methods. The scale is especially well-controlled using soil applied insecticides called neonicotinoids. However, the research team on CMBS has most recently found out that concentrations of two neonicotinoids, imidacloprid and dinotefuran in pollen collected from crapemyrtle trees treated in previous fall, winter and early spring are alarmingly high. While data is still being summarized for publication, we are strongly recommending you consider other management options before applying neonicotinoids because of the possibility that it can be detrimental to pollinators that collect and bring pollen back to their hives.

Our current recommendations include: Avoiding buying infested plants – inspect nursery stocks carefully for signs of CMBS before the purchase. Avoid plants with sticky leaves; black, sooty trunks; or white scale insects (that bleed pink when crushed) on stems and trunk.

The most susceptible life stage of CMBS to low-impact chemicals is the crawler stage, when the first instars are mobile and not covered by the protective wax. Crawler population monitoring by removable double-sided sticky tapes over years have indicated that late April and early May is the time period that crawlers are coming out as their first peak presence in southeastern Louisiana. If this population is detected, horticultural oil or insect growth regulators (IGR) such as pyriproxyfen and buprofezin as foliar sprays can provide 50% to 100% control crawlers.

If trees are heavily infested, pressure-wash the trunk and reachable limbs with water can help remove the sooty mold and many female scales and egg masses, then using a soft brush and mild solution of dishwashing soap can provide some control to the rest. Washing will make other contact type chemicals more efficient.

Some lady beetle species, especially the twice-stabbed lady beetle, are important predators of crapemyrtle bark scale but cannot eradicate the scales or control heavy infestations. Certain insecticides such as carbaryl (i.e., Sevin) and many of the pyrethroid insecticides are harmful to lady beetles and may worsen the scale problems. Research on identifying parasitic wasps for biological control options is on-going and shedding light on biological solutions in the future.

Systemic insecticides in the neonicotinoid class have provided good control of CMBS in most cases. These products are more effective when applied to the root zone as a soil injection or drench, than as a spray applied to the foliage. Good control can be achieved with applications of clothianidin, dinotefuran (Safari), imidacloprid (Merit or Bayer Advanced Garden Tree and Shrub Insect Control), or thiomethoxam (Meridian). When using this class of insecticides, be sure to read all label directions carefully. Most neonicotinoid insecticides prohibit use when plants are in bloom to minimize risks to bees.

Acknowledgement: The Crapemyrtle Bark Scale research is supported by USDA SCRI Award No. 2017-51181-26831.

Literature Cited

Aristizábal, L.F., Arthurs, S.P., Chen, Y., Cherry, R.H., and Cave, R. (2016). Efficacy of biorational insecticides applied against chilli thrips, *Scirtothrips dorsalis*, infesting roses under nursery conditions. J. Appl. Entomol. *27*. DOI:10.1111/jen.12340

Wang, Z., Chen, Y., Gu, M., Vafaie, E., Merchant, M., and Diaz, R. (2016). Crapemyrtle bark scale: a new threat for crapemyrtle, a popular landscape plant in the U.S. Insects *7*:78.

Wang, Z., Chen, Y., Diaz, R. (2018). Temperature-dependent development and host range of crapemyrtle bark scale. 2019. Florida Entomol. *102*:181-186.

IPPS European Exchange 2018

Lindsay Day

Saunders Brothers Nursery, 2717 Tye Brook Highway, Piney River, Virginia 22964, USA

LDay@saundersbrothers.com

Keywords: Early-Career Professional Exchange Program, European Region-IPPS (ER-IPPS), Southern Region of North America-IPPS (SR-IPPS), gardens, retail and wholesale nurseries

Summary

"Too seek and share" is not only the motto of the International Plant Propagation Society (IPPS), but a foundational principal that the horticulture industry has cultivated for many years. The SR-IPPS (Southern Region of North America) embodies this message with its "Early-Career SR-IPPS Professional Exchange Program" with the IPPS European Region (ER-IPPS). In 2018 I was selected to represent the SR-IPPS as their delegate. This paper is my report.

INTRODUCTION

"Too seek and share" is not only the motto of the International Plant Propagation Society (IPPS), but a foundational principal that the horticulture industry has cultivated for many years. The SR-IPPS (Southern Region of North America) embodies this message with its "Early-Career SR-IPPS Professional Exchange Program" with the IPPS European Region (ER-IPPS). In 2018, I was selected to represent the SR as their delegate. It was an honor to be chosen and my experience was one I will cherish for a lifetime. The relationships built and knowledge that was shared will benefit me throughout my career in horticulture.

The program is a two-week exchange where you are hosted by several industry members and given the opportunity to see different elements of horticulture in Europe. It concludes at the annual European IPPS conference. In 2018, the conference was located in Breda, Netherlands. The exchange program began in the United Kingdom and my hosts treated me with tours and visits to many nurseries and gardens on our way to Breda.

The trip was scheduled to begin on Monday October 1, 2018, but as an avid traveler, I decided to take extra time to visit a friend, and fellow plantsman Colm O'Driscoll in Dublin, Ireland. Colm was a delegate from the ER-IPPS that visited Virginia and the SR-IPPS in 2014 and spent some time at Saunders Brothers. I flew through the night on Friday and he picked me up from the airport early Saturday morning. After a day of exploring Glendalough and the Wicklow Hills NP, we drive over to see the personal garden of renowned gardener, Jimi Blake. His garden is a stunning collection of rare and novel plants, combined in manty different styles, including a walk in the woods that leads to a view of the Wicklow Hills.

After my weekend in Dublin, I caught a flight to London and a train out to Worcester to meet my first host, Karl O'Neil. After a lovely dinner, I got to explore the charming town of Malvern. The following morning, we went to Bransford Webbs Nursery where Karl and his wife work. We were joined by Sophie Lewis, who had been the ER-IPPS delegate to the SR-IPPS conference the year before. Brandsford Webbs was an impressive nursery and greenhouse operation that specialize in *Hebe* production and breeding. Most of their plants were sold to garden centers or other retailers. It was interesting to see how different the market was and how plants were branded in the United Kingdom. Unlike the U.S., there is a lot of in-house branding rather than major national brands.

After a lovely tour, I left with Sophie as we headed to her home in Shropshire, stopping at garden retailers on our way. One of the garden centers we stopped at was Webb Garden Center which is a major buyer of Brandsford Webbs Nursery. Garden centers in the U.K. are much larger and more product-diverse than garden centers in the U.S. Many of these retailers had clothing, food and other products for sale.

The next morning, we visited Stokesy Court Estate where a husband and wife had turned the old walled garden into a cut flower nursery. Sophie has a special interest in cut flower production, so we both enjoyed learning about it

We left Stokesy Court and went to Frank P. Matthews Nursery where Sophie works full time. Frank P. Matthews is a tree nursery that grows fruit and ornamental trees for retailers. There were acres and acres of trees waiting to be grafted, or freshly grafted. Nick Dunn, a plantsman that has been part of Frank P. Matthews for many years took Sophie and I up to his experimental block of apple varieties where I tasted some of the most delicious and most unique apple varieties I've ever seen!

From Frank P. Matthews Nursery we went to Sophie's "flower patch" which is a small nursery that she is producing cut flowers in her spare time. It was stunning, and you could see how dedicated and passionate she is about horticulture.

The next morning, I said goodbye to Sophie and caught a train to Chichester where I met my next host, Tim Lawrence-Owens. Tim and his Wife, Annette, welcomed me into their home., The next morning, I went to work with Tim at Walberton Nursery. Walberton is part of the Farplants group which oversees sales and marketing for four nurseries that are side by side. These four nurseries work together, each specializing in different types of crops. Many of them were highly mechanized and it was impressive to see what seemed like miles of conveyors and potting lines. All four nurseries ship from one main dispatch area. It was empty at the time, but you could imagine how busy it must get there in the spring.

The next day was a rainy Saturday, so Tim, Annette, and I went over to West Dean

Gardens and wandered through a stunning illustration of English style gardening.

Sunday morning, I said goodbye to Chichester and Tim and I went to the Royal Horticulture Society's Wisley Gardens outside of London where we met my next host, Richard McKenna. Although we only had two hours to explore this magnificent garden, I was still able to see a lot. One of the most notable parts of the garden was the "Thinking Outside the Box" knot garden.

Boxwood have been devastated in Europe over the past few years, first from Boxwood Blight and now from the Boxwood Tree Moth. Boxwood is one of my favorite shrubs, so I was fascinating to see an entire knot garden constructed with boxwood alternatives.

After Wisley, Richard and I went back to his home in Kent, where his wife Zaza and his three children Valentina, Luca, and Marco greeted us. We enjoyed a lovely evening at their house where we had a fire and roasted marshmallows. The next morning, I went to work with Richard to Provender Nursery, a re-wholesale nursery outside of London where many landscapers and retailers can come and purchase plants. It was a much different set up than I was used to, but it was fascinating to see.

The next morning, we left for the IPPS conference in Breda. We left by car and took the tunnel under the channel over to France. On our way to Breda, we visited three nurseries. Richard does a lot of buying of novel plants from different nurseries in Europe. Most of my horticulture experience has been with wholesale nurseries so it was interesting to see how different re-wholesale is.

My favorite nursery we visited was Van der Poel Nursery in Boskoop. This charming, little, Dutch horticulture town is set on channels where there are nurseries and greenhouses all lined up next to one another. William Van der Poel showed us around his beautiful nursery that was filled with unique plants. He had been a breeder and worked with Proven Winners and he showed us his trial garden. We wandered around until the sunset and we left for Breda.

Wednesday marked the first day of the IPPS conference. The morning was filled with technical talks from growers and breeders all over Europe, followed by nursery tours in the afternoon. We visited two nurseries, the first specializing in tea camellias and some other perennials. One of the most fascinating part about this nursery was the fact that they had spent a lot of resources to establish a marketing campaign around their products, making them very desirable to the consumers and sold at a higher price point. One of these products was simply moss grown in containers, yet it was marketed to be used in fairy and terrarium gardens and they sold thousands. The second nursery was a tree nursery where they showed us how to dig and wrap trees that they shipped all over Europe.

The second day of the conference we enjoyed more talks in the morning. Many of the talks focused on sustainability and the use of plastics in horticulture and how we can use better material and reduce waste. The afternoon was filled with more tours including Florensis, a gigantic greenhouse operation that utilized very high-tech automation. We also visited a nursery that specialized in *Leucothoe* and were selling the for $99 to the retailer. After our nursery tours we enjoyed a banquet as a group.

The next morning finished with a few more talks and then the conference was over. It was insightful to see different operations and understand the culture around the horticulture industry in different parts of the world. I found horticulture and gardening to be valued higher in Europe than in the U.S. It was eye opening and I hope these are lessons I can take with me throughout my career.

After the IPPS conference I spent more time in the Netherlands and Belgium visiting boxwood growers, and then took a train down to Tours, France to attend a Boxwood Pest and Disease Conference.

All in all, I spent a total of 21 days traveling around Europe, experiencing the horticulture industry in many facets. This was an experience of a lifetime and I am incredibly grateful to IPPS for supporting me and giving me this opportunity and am grateful for all my hosts. This was a truly unforgettable experience and one I will cherish forever.

Observations from the LSU AgCenter Hammond Research Station Trial Gardens

Jeb S. Fields and Jason Stagg

Louisiana State University AgCenter, Hammond Research Station, 21549 Old Covington Hwy. Hammond, Louisiana 70403, USA

JSFields@agcenter.lsu.edu

Keywords: Landscape horticulture, landscape sustainability, plant evaluation, trial gardens

Summary

The Louisiana State University Agricultural Center's (LSU AgCenter) Hammond Research Station is the Center of Excellence for the Louisiana Green industry. As such, we strive to provide the most relevant and beneficial research and programming aimed to help the industry grow. Our mission is to enhance production efficiency and landscape sustainability through research, extension, and educational outreach - as well as to evaluate and promote specialty crops for the nursery, landscape, and garden center industries in Louisiana. The station houses research efforts into environmental nursery production, sustainable landscape practices, landscape entomology, and landscape horticulture. Most notably, the Hammond Research Station is home to the Hammond Trial Gardens, which were featured on the International Plant Propagators Society Southern Region tour and brochure. The Hammond Trial Gardens occupy approximately 40 acres and house well over 1000 different plant varieties, trials and demonstrations, annually. Plant trials range from seasonal bedding plants to ornamental shrubs and trees to edibles.

THE GARDENS

The most prominent garden at the Hammond Research Station is the Allen D. Owings Sun Garden (Fig. 1), which houses the majority of the bedding plant trials throughout the year. This garden is often the highlight of the station, and the first garden observed when entering the property. Aside from seasonal color trials, many root hardy tropical and perennial landscape plants are trialed in the Sun Garden. Moreover, this is the location of many of our sustainable landscape demonstration gardens, including edible landscaping, pollinator gardens, and rain gardens.

Figure 1. The LSU AgCenter Hammond Research Station, Allen D. Owings Sun Garden bedding plant trials.

The second most visible garden within the Hammond Trial Gardens is the Margie Yates Jenkins Azalea Garden (Fig 2.). Established in 2006, the Margie Y. Jenkins Azalea Garden houses some of our understory trials and many of our flowering shrub collections. This garden was initially designed around many of the historically important shrubs and small trees that thrived in the Louisiana nursery and landscape industry throughout the years. In addition to the heritage plantings, a great deal of our azalea trials and collections are located within the Margie Jenkins Azalea Garden.

Figure 2. The LSU AgCenter Hammond Research Station, Margie Yates Jenkins Azalea Garden.

The Piney Woods garden (Fig. 3) is the largest garden at the Hammond Research Station. The Piney Woods Garden is the home for many of our long term displays and demonstrations, as well as serving as a repository for many of the woody ornamental collections that have relevance to the industry. Many of the long-term woody ornamental trials are located here.

Figure 3. The LSU AgCenter Hammond Research Station, Piney Woods Garden.

UNDERUSED PLANT MATERIALS

Much of the plant material we trial at the Hammond Research Station is or has been associated with the nursery and landscape trade throughout the Southeastern United States. However, there are many plants that perform fantastically in our trials that do not have the recognition or receive the exposure that they deserve. These plants thrive in our Louisiana landscapes and would have a positive impact if assimilated into the trade. None of the plant varieties listed here were developed at the LSU AgCenter Hammond Trial Gardens; instead these are plants we have collected from amazing breeders and plant experts throughout the Southeast U.S. We believe they deserve more consideration for use in landscapes and should be in wider production.

Aloysia virgata – Known by the common name sweet almond verbena. Root hardy tropical shrub with sandpaper-like silver leaves. This is a large shrub that will grow upwards of 2.4 m (8 ft.) tall. The flowering panicles give off fantastic fragrance and are irresistible to pollinators. This should be grown in full sun (Fig. 4).

Figure 4. *Aloysia virgata.*

***Camellia* 'Snow Blizzard'** – This is a cross between *Camellia pitardii* and *C. fraterna* and is one of our favorite camellia hybrids at the Hammond Research Station. These shrubs are evergreen and can grow quite large with numerous small spicy-scented white flowers covering their branches (Fig. 5).

Figure 5. *Camellia* 'Snow Blizzard'.

***Camellia* 'Tiny Princess'** – Like 'Snow Blizzard' this is a hybrid of *C. fraterna*. Small blooms cover the branches of this camellia; however, these have a delicate pink color instead of white. This camellia has an open spreading habit with a slight weeping form to its branches resulting from the weight of its numerous blooms (Fig. 6).

Figure 6. *Camellia* 'Tiny Princess'.

Camellia edithae – This fantastic camellia species features pink double blooms. It will grow approximately 1.8 m (6 ft.) tall. This is one of the latest blooming camellias in the garden, with bloom time in late spring (Fig. 7).

Figure 7. *Camellia edithae.*

Camellia fraterna – A few older specimens of this camellia are currently on display in our Azalea Garden. An abundance of small white blooms make this as a stunning camellia. The bark develops an orange hue and the branches can develop a weeping habit (Fig. 8).

Figure 8. *Camellia fraterna.*

***Cestrum* x 'Orange Peel'** – This is one of the best cestrums at the Hammond Research Station. A hybrid of *Cestrum diurnum* and *C. nocturnum*, this large tropical shrub will take the heat and can quickly grow to 1.2 – 2.4 m (4-8 ft.) tall by 0.9-1.5 m (3-5 ft.) wide, making it an ideal background planting. Dark green, deer-resistant narrow leaves provide good contrast for the numerous golden yellow/light orange tubular blooms. Best planted in full sun with good drainage. Begins blooming in late spring until first frost. Cestrum is easily propagated by cuttings. This selection can be a deciduous shrub in warmer parts of Zones 8 and 9 but may die back to the roots during cold winters of Zone 8 (Fig. 9).

Figure 9. *Cestrum* x 'Orange Peel'.

***Gardenia jasminoides* 'Martha Turnbull'** – This heavily flowering gardenia was collected from Rosedown Plantation in St. Francisville, Louisiana. The single blooms are shaped like a pinwheel and are approximately 7.6 cm (3 in.) in diameter. The blooms last from early spring through the summer and result in orange hips during the winter. This is a great, quick growing, profuse flowering gardenia that provides plenty of the classic gardenia fragrance (Fig. 10).

Figure 10. *Gardenia jasminoides* 'Martha Turnbull'.

***Gardenia jasminoides* 'Variegata'** – This fantastic gardenia exhibits large slender variegated leaves that stand out more than the fragrant double bloomed flowers. Grown in part shade, this gardenia makes a show-stopping, bright addition to the landscape (Fig. 11).

Figure 11. *Gardenia jasminoides* 'Variegata'.

***Lantana* x 'Grandpa's Pumpkin Patch'** - Vibrant flower clusters in shades of orange and dark yellow bloom from spring to frost. This is a large lantana that makes an excellent background planting, growing 1.2-1.5 m (4-5 ft.) tall by 1.2-1.5 m (4-5 ft.) wide. Excellent butterfly attractant. Propagated by cuttings. Reliable perennial in Zones 8 and 9 (Fig. 12).

Figure 12. *Lantana* x 'Grandpa's Pumpkin Patch'.

***Magnolia* 'Anilou'** – One of the best of the yellow magnolia hybrids at the Hammond Research Station, this tree has long-lasting bright yellow blooms that start in late spring. This deciduous magnolia also exhibits a columnar form and fits perfectly in the landscape (Fig. 13).

Figure 13. *Magnolia* 'Anilou'.

***Magnolia* 'Butterflies'** – This hybrid magnolia is a deciduous flowering magnolia that has buttery yellow flowers which hold their color throughout the bloom period. The tree is pyramidal shaped and performs well throughout the Gulf South. The tree reaches 20 feet tall and blooms in late spring (Fig. 14).

Figure 14. *Magnolia* 'Butterflies'.

Magnolia 'Jon Jon' – This hybrid deciduous magnolia flowers somewhat later after many other deciduous magnolias and usually after danger of freezes. Incorporating this variety into landscapes keeps deciduous magnolia blooms longer into the spring. The thick blooms are a cream white with a hint of purple on the underside of the petioles. It has grown to a height of 6.1 m (20 ft.) at the Hammond Research Station (Fig. 15).

Figure 15. Magnolia 'Jon Jon'.

***Pentas lanceolata* 'Nova'** – This 1999 Georgia Gold Medal Winner is reportedly one of the hardiest and most vigorous pentas varieties. Grows to 0.9 m (3 ft.) tall by 0.6 m (2 ft.) wide in full sun. Large clusters of 7.6-10.2 cm (3-4 in.) rose-pink, star-shaped flowers appear atop dark green leaves from late spring through fall. Like most pentas, this plant is an excellent butterfly attractant. It is easily propagated by cuttings and can be pruned periodically to control growth. 'Nova' can be perennial in warmer regions of Zone 9 (Fig. 16).

Figure 16. *Pentas lanceolata* 'Nova'.

***Prunus mume* 'Fragrant Snow' and 'Peggy Clarke'** - Flowering apricots are a completely underutilized small to medium size ornamental trees in the landscape! These trees provide amazing late winter color and fantastic fragrance that permeates the gardens. 'Fragrant Snow' is a white blooming variety and 'Peggy Clarke' has pink blooms (Fig. 17A 'Fragrant Snow' and Fig. 17B 'Peggy Clarke').

Figure 17. *Prunus mume* (A) 'Fragrant Snow'; (B) 'Peggy Clarke'.

***Rhododendron* 'Red Luster'** - This is a striking, evergreen azalea that has wonderful variegated foliage and dense showy pinkish red blooms. The variegation is often in the form of light cream-colored margins (Fig. 18).

Figure 18. *Rhododendron* 'Red Luster'.

***Rhododendron* x 'Koromo Shikibu'** – With its long, narrow petals, this lavender colored indica azalea is a fantastic specimen that has been around for quite some time. 'Koromo Shikibu' is sometimes called a "spider" azalea, and has a sweet subtle fragrance. It has been described as a hybrid of *R. macrosepalum,* but true parentage is unclear (Fig. 19).

Figure 19. *Rhododendron* x 'Koromo Shikibu'.

Rostrinucula dependens – Often referred to as "Weeping False Buddleia," this deciduous shrub is a member of the mint family. This shrub does not die back to the ground in our climate and maintains its form and structure year after year. The weeping habit of unusual purple flowers on white racemes give it very unique look. It does best when grown in part sun and will grow to approximately 1.5-1.8 m (5-6 ft.) tall. This is a very unique and underutilized plant that should be featured more in southern landscapes (Fig. 20).

Figure 20. *Rostrinucula dependens.*

***Salvia farinacea* 'Rebel Child'**– This is a chance hybrid between 'Henry Duelberg' and 'Cedar Hill'. It can reach 0.6-0.9 m (2-3 ft.) tall and is a vigorous bloomer from spring through first frost. The bloom is bluer than 'Henry Duelberg', and it is an excellent pollinator attractor. 'Rebel Child' is one of the toughest salvias you can grow and is a perennial in our climate. Like most salvias, 'Rebel Child' prefers full sun. It is easily propagated by cuttings (Fig. 21).

Figure 21. *Salvia farinacea* 'Rebel Child'.

Somatic Embryogenesis of Rare *Stewartia* Species and Elite Cultivars

Heather J. Gladfelter[1,3 a], Jack Johnston[2], Dayton Wilde[1], and Scott Merkle[3]

[1]Department of Horticulture, University of Georgia, 1111 Plant Sciences Building, Athens, Georgia, 30602, USA; [2]P.O. Box 121, Lakemont, Georgia, 30552, USA; [3]Warnell School of Forestry & Natural Resources, University of Georgia, 180 E. Green St., Athens, Georgia, 30602, USA

[a]Fourth Place- Charlie Parkerson Graduate Student Research Paper Competition

Keywords: Liquid culture, somatic embryos, woody plant basal medium (WPM)

Abstract

The methods of somatic embryogenesis proved successful for five Asian and two North American Stewartia species. Immature embryos were used as explants and cultured on gelled WPM basal medium with and without PGRs (Lloyd & McCown, 1980). Within 3-4 weeks embryogenic callus was produced from the immature embryo explants and these contained tiny developing somatic embryos. The embryogenic callus was propagated via liquid culture and plated on maturation media (WPM basal nutrient) devoid of PGRs for production and development of somatic embryos. Following the maturation process the embryos were placed on germination medium which was the same as maturation medium devoid of glutamine but supplemented with 0.25% activated charcoal. After 3-6 weeks, the somatic embryos were germinated and continued to develop into somatic seedlings and were transferred to soil and acclimated to the greenhouse and natural environment.

INTRODUCTION

Stewartia is a genus of flowering shrubs and trees in the family Theaceae. Stewartia species are native to eastern Asia, with the exception of two species, *S. ovata* and *S. malacodendron*, which are indigenous to southeastern North America. All species have large and showy flowers that are 3-11 cm in diameter, usually with 5 white petals (Figures 1a, 1b).

Figure 1a. *Stewartia malacodendron* 'Delmarva' courtesy of The Polly Hill Arboretum, Tisbury, MA; **Figure 1b**. *Stewartia ovata* 'Red Rose' courtesy of The Polly Hill Arboretum, Tisbury, MA.

The bark of the trees flake and are very distinctive in that the colors range from smooth orange to yellow-brown. Despite having outstanding ornamental value and features, Stewartias are not readily available for landscaping in the horticultural trade due to difficulty with mass propagation (Struve & Lagrimini, 1999; Nair & Zhang, 2010). *S. pseudocamellia* (Japanese Stewartia) has been propagated *in vitro* from single node explants (McGuigan et al. 1997), but axillary shoots were not produced at a level applicable for commercial production. In addition to the commercial propagation of Stewartia, there is a need for developing propagation techniques for the conservation of North American species that are rare (*S. ovata*) or endangered (*S. malacodendron*). Preservation of these species through seed is limited by difficulty with collection, seed yield and viability, and germination requirements. The objective of this research was to examine somatic embryogenesis as a means to mass propagate Asian and North American Stewartia species and to provide material for cryopreservation of valuable Stewartia germplasm.

MATERIALS AND METHODS

Explant sources

Immature fruit was collected in August 2015 from cultivars of *Stewartia malacodendron* 'Delmarva', *S. ovata* 'Red Rose', and *S. ovata* 'Royal Purple' were obtained from The Polly Hill Arboretum (Tisbury, MA). Immature fruit of *S. pseudocamellia* var. *koreana*, a cold tolerant cultivar, was provided by the University of Maine in August 2016. Seed capsules of *S. sinensis*, *S. koreana*, *S. monadelpha*, and *S. pseudocamellia* were obtained from plants of a private collector in August 2015 and 2016. Wild material of *S. malacodendron* was collected from two sites in Georgia and one in Alabama and wild material of *S. ovata* was collected from one site each in North

Carolina and Georgia every August from 2012-2016.

Culture initiation

The disinfestation and culture initiation procedures for immature fruit of the Stewartia species employed the methods outlined for somatic embryogenesis of American chestnut (Merkle et al., 1991). Under sterile conditions and using a stereomicroscope, the immature fruit was dissected to retrieve the immature seeds containing the immature zygotic embryos, which were excised from the seed and placed on a semisolid nutrient medium with or without plant growth regulators (PGRs). The basal nutrient medium was Woody Plant Medium (WPM) (Lloyd & McCown, 1980) supplemented with 3% sucrose, 500 mg/L glutamine, and 3 g/L Gelrite, set at a pH of 5.65. The PGR treatments tested included picloram at a concentration ranging from 0.05 mg/L to 10 mg/L; indolebutyric acid (IBA) at concentrations of 0.1 and 1.0 mg/L; 2, 4-dichlorophenoxyacetic acid (2,4-D) at three concentrations, 1, 2, and 4 mg/L; 0.1 mg/L IBA in combination with 0.5 or 1.0 mg/L 6-benzylaminopurine (BAP); 0.1 mg/L IBA in combination with 0.5 or 1.0 mg/L meta-topolin; and 2 mg/L 2,4-D in combination with 0.2 mg/L meta-topolin. The immature zygotic embryos were cultured in the dark at 25 $\pm$ 2 °C.

Cell suspensions and embryo production

Somatic embryogenic tissues produced from explant cultures in the PGR study were placed in a 125 ml Erlenmeyer flask containing 30 mL of liquid WPM supplemented with or without the PGR(s) that produced the somatic embryogenic tissue. The liquid suspension cultures were grown at 25 $\pm$ 2 °C on a platform shaker in the dark at 100 rpm for 45 days. The cultures were fed with fresh medium every two weeks. Once one gram of tissue was produced, the embryogenic tissue was collected and size-fractionated using 100 µm pore size metal sieves. The pro-embryogenic masses (PEMs) recovered from the sieves were thoroughly rinsed with 150 ml of embryo maturation medium, WPM devoid of PGRs. The PEMs collected on the sieves were transferred to a new sterile 125ml flask with 30 ml of embryo maturation medium. The cultures were placed back onto the shaker in the dark for five days. Using sterile glass Büchner funnels and vacuum, the PEMs were collected by pipetting 10 ml of the suspension culture evenly onto nylon mesh (30 µm pore size) rafts. The nylon rafts were transferred to semisolid embryo maturation medium and placed in the dark at 25 $\pm$ 2 °C for one month.

Embryo maturation and pre-germination cold treatment

Once somatic embryos started to form on the nylon rafts, individual embryos were transferred to fresh embryo maturation medium to allow for further maturation. The embryos were incubated in the dark for one month. In some cases, the somatic embryos were exposed to a cold vernalization period whereby the plates were wrapped in foil and placed in a walk-in cooler at 4 °C for 30, 45, or 90 days.

Germination and Conversion

Following maturation and cold treatment the somatic embryos were placed on germination medium (WPM without glutamine and PGRs, with 0.25 mg/L activated charcoal) in an incubator with a 16:8 hour photoperiod, fluorescent lighting (30 µmol m^{-2} s^{-1}), and temperature set at 25 $\pm$ 2 °C until the embryos produced roots from the radicle and primary shoots from the apical meristem.

Transfer to soil and acclimation

Each somatic seedling was carefully transferred to a 10.2 cm (4-in.) square pot containing 1 part Fafard 3B: 1 part vermiculite soil mix, misted with a dilute solution of Miracle Gro (1/2 teaspoon/L tap water), and covered by an inverted GA7 vessel (Magenta Corp.) to create a humid micro-environment. The pots were placed in a dome-covered tray on moist perlite and incubated in a growth chamber with a 16:8 hour photoperiod, fluorescent lighting (30 μmol m−2 s −1), and temperature of 25 ± 2 °C. The somatic seedlings were gradually exposed to lower humidity by removing the GA7s after 2-3 weeks and gradually opening the vents on the lid of the dome tray. Once the dome was removed, plantlets were transferred to the greenhouse and eventually potted up into 1-gal pots containing 3 parts mini nugget bark chips: 1-part peat: 2 parts perlite, and top dressed with Osmocote slow release fertilizer. The final step was transfer of the plants to the shade house and monitoring of survival.

Cryopreservation

The cryopreservation method used with the Stewartia embryogenic cultures was described by Vendrame et al. (2001). Embryogenic tissue of each Stewartia species was grown in culture flasks on a shaker at 100 rpm in the dark containing liquid WPM maintenance medium with PGRs for at least one week. The day before freezing, the tissue was transferred to flasks containing the same medium plus 0.4M sorbitol. The culture flasks were incubated on the shaker overnight and the next day, they were placed in the cold room at 4 °C. After 30 minutes in the cold, the flasks were placed on ice until the cells were ready to transfer to cryovials.

Enough WPM maintenance medium containing 0.4 M sorbitol with DMSO was added to each culture flask so that the final concentration of DMSO in the medium was 5%. The cells were mixed thoroughly in the DMSO/sorbitol solution and 1.8 ml of cells were pipetted into the 2 ml cryovials and placed in a -1 °C "Mr. Frosty" freezing container (Nalgene, Rochester, NY). The freezing container was placed in an ultralow freezer set at -80 °C overnight. The next day the cryovials were put in a cryobox and placed in the liquid nitrogen freezer at -196 °C. Following removal from liquid nitrogen, the cryovials were transferred from the cryobox to a pre-frozen "Mr. Frosty" container at -80 °C for 1.5h. Cryovials were then placed in a floating tray in a 40 °C water bath for 2 minutes to facilitate rapid thawing. Once the cells were thawed, the embryogenic cell clusters were collected on sterile nylon mesh raft overlaying several layers of filter paper to soak up DMSO solution. The nylon rafts were then transferred to fresh WPM maintenance medium supplemented with PGRs. The transfer was repeated the next day for continued removal of remaining DMSO. Then, the cultures were placed in the dark at 25 ± 2 °C and observed over the following weeks for re-growth of the embryogenic cultures.Once the cells were successfully recovered, they were tested for ability to make somatic embryos following the cryopreservation treatment.

Statistical analysis

All statistical analyses were performed using R statistical package, R-3.5.1.tar.gz (2018-07-02). The frequencies of somatic embryogenesis induction by the different PGR treatments were compared using a 2-sample test for equality of proportions with continuity correction. The test variables for the germination and conversion response were tested for significance using the ANOVA function following transformation of data using the arcsine function. Differences among dependent variables were determined using Tukey's HSD multiple comparison test at a significance of α=0.05.

RESULTS

Induction of somatic embryogenesis

The North American species of Stewartia were more responsive to picloram than other PGRs in producing somatic embryos, although only at low concentrations. Table 1 shows that picloram concentrations above 2 mg/L proved to be ineffective for inducing somatic embryos.

Table 1. Percent induction of somatic embryogenesis in three Stewartia species, *S. malacodendron* 'Delmarva', *S. ovata* 'Red Rose', and *S. rostrata*, following exposure to a gradient of picloram concentrations from 0 to 10 mg/L in the nutrient medium. For each genotype, there were 10 zygotic embryo explants per picloram concentration.

	Picloram (mg/L)							
Species	0	0.05	0.1	1.0	2.0	4.0	8.0	10.0
S. malacodendron 'Delmarva'	100%	0	33%	0	0	0	0	0
S. ovata 'Red Rose'	0	10%	20%	0	0	0	0	0
S. rostrata	0	0	65%	0	25%	0	0	0

In fact, upon exposure to high concentrations of picloram, the tissue turned brown and died within a very short time. Interestingly, somatic embryogenesis was induced from the North American native cultivar *S. malacodendron* 'Delmarva' when exposed to no PGRs. Thirty-three percent of the immature embryos of this cultivar produced somatic embryos upon exposure to 0.1 mg/L picloram. Immature zygotic embryos of another North American native cultivar, *S. ovata* 'Red Rose', were most responsive at the lower concentrations of picloram, 0.05 and 0.1 mg/L, at frequencies of 10% and 20%, respectively (Table 1). Immature embryos from the Asian species, *S. rostrata*, produced somatic embryos upon exposure to 0.1 and 2.0 mg/L picloram at frequencies of 65% and 25%, respectively (Table 1).

Two additional PGRs, 2, 4-dichlorophenoxyacetic acid (2, 4-D) and indolebutyric acid (K salt-IBA or KIBA) were tested for induction of somatic embryos with several Asian species, including *S. monadelpha*, *S. pseudocamellia*, *S. rostrata*, and *S. sinensis*, along with the two North American native species, an Alabama cultivar of *S. malacodendron* and Polly Hill Arboretum cultivar *S. ovata* 'Royal Purple'. In one experiment, two concentrations of 2, 4-D (2.0 and 4.0 mg/L), one concentration of K-IBA (0.1 mg/L), the control with no PGRs, and two concentrations of picloram (0.1 and 1.0 mg/L), were tested for induction of embryogenesis. Interestingly, The Polly Hill Arboretum cultivar *S. ovata* 'Royal Purple' also produced somatic embryos without PGRs and upon exposure to 0.1 mg/L picloram and to 4.0 mg/L 2,4-D at frequencies of 33.3 percent, 14.3 percent, and 16.7 percent, respectively (Table 2). Induction of somatic embryogenesis was only observed on medium containing 2.0 mg/L 2, 4-D for the Asian species *S. sinensis* and the Blount County Alabama cultivar of *S. malacodendron,* at frequencies of 30 percent and 20 percent, respectively (Table 2).

Table 2. Percent induction of somatic embryogenesis in five Stewartia species, *S. malacodendron*, *S. monadelpha*, *S. ovata* 'Royal Purple', *S. pseudocamellia* var. *koreana* (UMaine), and *S. sinensis* following exposure to three different PGRs at various concentrations versus no PGRs in the nutrient medium. For each genotype, there were 10 zygotic embryo explants per PGR treatment.

	Plant Growth Regulators (mg/L)					
Species	No PGRs	0.1 Picloram	1.0 Picloram	2.0 2,4-D	4.0 2, 4-D	0.1 K-IBA
S. malacodendron	0	0	0	20%	0	0
S. monadelpha	0	20%	20%	50%	80%	0
S. ovata 'Royal Purple'	33.3%	14.3%	0	0	16.7%	0
S. pseudocamellia var. *koreana*	0	0	0	72%	75%	0
S. sinensis	0	0	0	30%	0	0

Three *Stewartia* species produced somatic embryos upon exposure to KIBA, *S. rostrata* (Table 3), *S. ovata* 'Otto, NC' and *S. koreana* (data not shown).

Table 3. Percent induction of somatic embryogenesis of *S. rostrata* following exposure to varying concentrations and combinations of auxins with and without cytokinins. The negative control is exposure to no PGRs.

Plant Growth Regulators (mg/L)	SE induction (%)
No PGRs	0
2.0 2,4-D	80
4.0 2,4-D	0
0.1 KIBA	80
0.1 KIBA + 0.5 meta-Topolin (mT)	20
0.1 K-IBA + 1.0 meta-Topolin (mT)	20
0.1 K-IBA + 0.5 BAP	80
0.1 K-IBA + 1.0 BAP	40
1.0 K-IBA	60

Table 3 shows a PGR experiment with *S. rostrata* where K-IBA alone or in combination with the cytokinins meta-topolin and benzylaminopurine (BAP) induced embryogenesis from immature zygotic embryo explants.

In this experiment, 2, 4-D was also tested at two different levels, 2.0 and 4.0 mg/L, as well as a no-PGR treatment. Chi-Square analysis with continuity correction showed no significant differences at $\alpha = 0.05$ for treatment comparisons for induction frequency of somatic embryogenesis. Eighty percent of the explants exposed to 2.0 mg/L 2, 4-D, 0.1mg K-IBA, and 0.1 K-IBA with 0.5 mg/L of BAP produced somatic embryos. No somatic embryos were produced on 4.0 mg/L 2, 4-D, nor when the medium was devoid of PGRs. Embryogenesis was only induced from 20 percent of the immature zygotic embryos when K-IBA at 0.1 mg/L was in used in combination with the cytokinin meta-topolin at 0.5 or 1.0 mg/L. However, K-IBA at 1.0 mg/L induced embryogenesis at a frequency of 60 percent, while K-IBA at 0.1 mg/L in combination with 1.0 mg/L BAP induced embryogenesis from 40 percent of the explants.

Figure 2 is a summary showing the various Stewartia species that produced somatic embryos and/or embryogenic callus from these PGR induction experiments. Figure 3 summarizes the various steps from the production of somatic embryos to germination and conversion to somatic seedlings in soil.

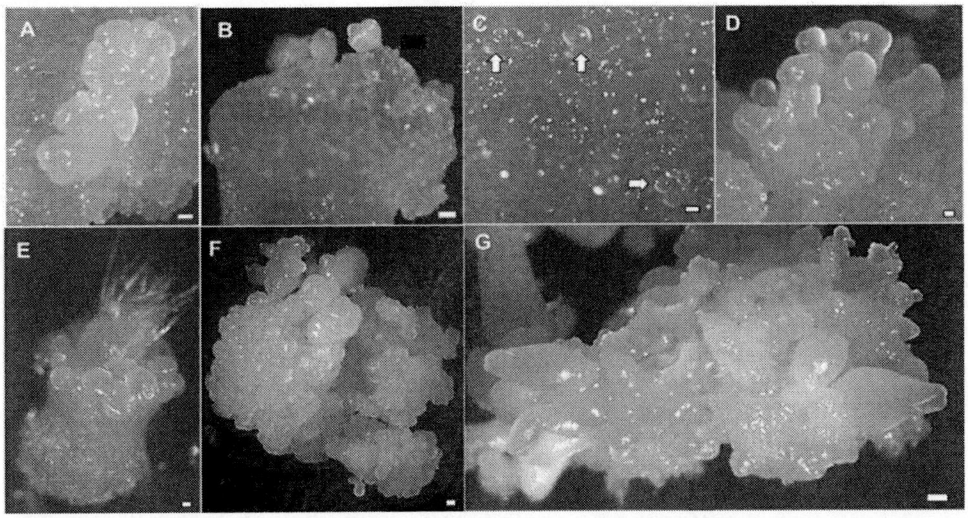

Figure 2. Somatic embryos and embryogenic callus of Stewartia species (A) *S. monadelpha* on WPM medium with 4.0 mg/L 2,4-D, (B) *S. malacodendron* 'Delmarva' with 0.1 mg/L picloram, (C) *S. rostrata* with 0.1mg/L picloram , (D) *S. pseudocamellia* var. *koreana* 'UMaine' with 2.0 mg/L 2,4-D, (E) *S. sinensis* with 2.0 mg/L 2,4-D, (F) *S. koreana* with 2.0 mg/L 2,4-D, and (G) *S. ovata* 'Red Rose' with 0.1 mg/L picloram.

Figure 3. Somatic embryogenesis of Stewartia species from explant to plants. (A) explant source is the immature zygotic embryo, (B) embryogenic callus proliferation, (C) liquid cell suspension, (D) embryo development, (E) embryo maturation, (F) conversion and germination, (G) *ex-vitro* acclimation, (F) 2 year-old tree derived from somatic embryogenesis.

Germination and Conversion

A comparison of cold versus no cold treatment was tested for several somatic embryo genotypes of *S. rostrata*. There is a comparison of the no cold (25 °C) versus cold (4 °C) treatment with the six *S. rostrata* genotypes, SR-1, SR-2, SR-4, SR-10, SR-31, and SR-32, with regard to the effect on percentages of germination and conversion of somatic embryos (Table 4). Analysis of variance results showed that the treatment effect for percent germination was significant ($p<0.0001$). There was a significant difference among genotypes following Tukey's HSD multi-comparison test ($p = 0.0309$) which was only due to a difference between two genotypes, SR1 and SR10. Analysis of variance results also showed that the cold (4 °C) treatment gave significantly higher percent somatic embryo conversion than the no cold (25 °C) treatment ($p< 0.001$).

Table 4. Comparison of no cold (25 °C) versus cold (4 °C) treatment with six *S. rostrata* genotypes, SR-1, SR-2, SR-4, SR-10, SR-31, SR-32, and the effect on percentage of germination and conversion of somatic embryos. For each genotype, there were 30 somatic embryo explants per treatment.

	Germination (%)		Conversion (%)	
Genotypes	25 °C	4 °C	25 °C	4 °C
SR-1	31.7 *c*	65.0 *a*	19.2 *c*	39.2 *a*
SR-2	20.0 *d*	46.7 *ab*	4.2 *d*	21.7 *b*
SR-4	24.2 *d*	55.8 *ab*	3.3 *d*	18.3 *b*
SR-10	21.7 *d*	41.7 *b*	4.2 *d*	10.8 *b*
SR-31	23.3 *d*	51.7 *ab*	3.3 *d*	14.2 *b*
SR-32	25.0 *d*	54.2 *ab*	5.0 *d*	17.5 *b*

Tukey's HSD multi-comparison test indicated by letters *a*, *ab*, *c*, and *d* and those letters that differ are significant at $\alpha=0.05$.

Cryopreservation

Somatic embryogenic cultures of 17 genotypes across six species of Stewartia were put into cryopreservation: *S. rostrata*, *S. koreana*, *S. monadelpha*, *S. pseudocamellia*, *S. pseudocamellia* var. *koreana* (UMaine-cultivar), and *S. ovata* 'Red Rose'. Cultures of four out of the six tested species were recovered from cryopreservation and five genotypes were recovered at 100%. Recovery was measured by growth of cells, maintenance by these cultures, and the ability to produce somatic embryos that were able to mature, germinate, and convert into plants.

DISCUSSION

Somatic embryogenesis was successfully induced from immature zygotic embryo explants of seven Stewartia species—five Asian and two North American. Additionally, three unique North American cultivars, *S. ovata* 'Red Rose', *S. malacodendron* 'Delmarva', and *S. ovata* 'Royal Purple' from Polly Hill Arboretum were successfully propagated by somatic embryogenesis using no or a variety of PGRs. Picloram at low concentrations, such as 0.05 and 0.1 mg/L, was more successful at producing somatic embryos with the North American Stewartia species. Most of the Asian Stewartia species responded best to 2, 4-D at concentrations of 2.0 or 4.0 mg/L in producing somatic embryos. This result is in contrast to results reported for the closely related genus *Camellia* in the Theaceae family. Somatic embryos of *Camellia japonica* were induced from immature zygotic embryos but used a different basal medium Murashige and Skoog (MS) (1962) and different PGR combination, BAP plus IBA (Vieitez and Barciela, 1990). Somatic embryo conversion and somatic seedling acclimatization to the greenhouse were achieved with all seven tested Stewartia species. Results with several of the species showed that a cold treatment enhanced germination and conversion of somatic embryos into somatic seedlings. Somatic embryogenic tissue from five Stewartia species was successfully recovered from cryopreservation - suggesting feasibility of long-term storage of valuable germplasm. This includes unique cultivars of Polly Hill Arboretum selected North American Stewartia species: *S. malacodendron* 'Delmarva', *S. ovata* 'Red Rose', *and S. ovata* 'Royal Purple'. This is the first report of somatic embryogenesis in the genus Stewartia.

Literature Cited

Lloyd, G. and McCown, B. (1980). Commercially feasible micropropagation of mountain laurel, *Kalmia latifolia*, by use of shoot tip culture. Proc. Intern. Plant Prop. Soc. *30*:421-437.

McGuigan, P.J., Blazich, F.A., and Ranney, T.G. (1997). Micropropagation of *Stewartia pseudocamellia*. J. Environ. Hort. *15*:65-68.

Merkle, S.A., Weicko, B., and Watson-Pauley, B.A. (1991). Somatic embryogenesis in American chestnut. Can. J. Forest Res. *21*:1698-1701.

Murashige, T. and Skoog, F. (1962). A revised medium for rapid growth and bio assays with tobacco tissue cultures. Physiol. Plant. *15*:473-497.

Nair, A. and Shang, D. (2010). Propagation of *Stewartia*: Past research endeavors and current status. HortTechnology *20*:277-282.

Struve, D.K. and Lagrimini, M. (1999). Survival and growth of *Stewartia pseudocamellia* rooted cuttings and seedlings. J. Environ. Hort. *17*:53-56.

Vieitez, A.M. and Barciela, J. (1990). Somatic embryogenesis and plant regeneration from embryonic tissues of *Camellia japonica* L. Plant Cell Tis. & Org. Cult. *21*:267-274.

Vendrame, W.A., Holliday, C.P., Montello, P.M., Smith, D.R., and Merkle, S.A. (2001). Cryopreservation of yellow-poplar (*Liriodendron tulipifera*) and sweetgum (*Liquidambar spp.*) embryogenic cultures. New Forests *21*:283-292.

Fertilizer Placement Effects on Weed Growth and Competition with Container-grown Ornamentals

Yuvraj Khamare[a], S. Christopher Marble, and Annette Chandler

University of Florida/IFAS, Mid-Florida Research and Education Center, Apopka, Florida 32703, USA.

marblesc@ufl.edu

[a]Third Place- Charlie Parkerson Graduate Student Research Paper Competition

Keywords: *Buxus microphylla*, eclipta (*Eclipta prostrata*), *Ligustrum lucidum,* weed control

Abstract

The objective of this study was to determine the effect of fertilizer placement on the growth of eclipta (*Eclipta prostrata*) and evaluate its competition effect in container-grown ornamental plants. Results indicated that subdressing at a depth of 7.5 cm resulted in a 50% decrease in eclipta growth in comparison with a topdress fertilizer treatment, but subdressing at 2.5 or 5 cm had no effect on eclipta growth. Growth of *Ligustrum lucidum* and *Buxus microphylla* were similar in pots that were subdressed at 2.5 or 5 cm, but growth decreased when pots were subdressed at a depth of 7.5 cm. Overall, results indicate that subdressing could be an effective weed management strategy but in order to prevent delays in production time, subdressing depth needs to be based on initial liner size.

INTRODUCTION

Weed control is one of the most costly and challenging aspects in nursery production. While weeds can cause growth reductions in field-grown ornamentals, they have an even greater impact on container-grown plants because nutrients and water are more limited compared with field crops. Previous research has shown that a single weed can reduce ornamental growth by 40 to 60% over one season (Berchielli-Robertson et al., 1990). Even if competition were not a concern, container plants must be weed-free in order to be marketable, thus weed thresholds are very low compared with other agricultural systems.

Due to the cost of hand weeding and the challenges associated with preemergence herbicides, more integrated weed management strategies are needed for nursery production. One potential weed management strategy is strategic placement of controlled-release fertilizer (CRF). Currently, most growers fertilize container crops by incorporating (thoroughly mixing fertilizer with potting substrate prior to potting) or topdressing (placing the entire allotment of fertilizer on the media surface after potting). Two alternative or strategic approaches to fertilization include dibbling and subdressing fertilizer (Stewart et al., 2018). Dibbling is accomplished by placing fertilizer just below the rootball of the plant while potting. This method has been shown to be highly effective for weed control. In a study evaluating the effect of fertilizer placement on prostate spurge (*Euphorbia prostrata* L.), topdressing caused an 888% growth increase compared with spurge growth in dibbled containers (Fain, 2004). While dibbling is effective, it is less commonly used due to a high concentration of fertilizer around the plant roots, which may cause phytotoxicity in some crops (Bir and Zondag, 1986).

Subdressing is another fertilization approach involving filling the pot approximately 50 or 75% with inert (non-fertilized) media, adding the fertilizer in a single layer, and then filling the remainder of the pot with inert media (Stewart et al., 2018). This results in the fertilizer being placed in a single layer, approximately two to three inches below the substrate surface. Crop roots can access the nutrients, but germinating weed seedlings cannot. This also offers an advantage over dibbling because subdressing is typically not associated with phytotoxicity issues (Broschat and Moore, 2003).

Preliminary data has shown that subdressing fertilizer at a depth of 2 inches can reduce the growth of spotted spurge (*Euphorbia maculata*) and large crabgrass (*Digitaria sanguinalis*) by over 90% in soilless pine bark substrates (Marble, unpublished data). If subdressing consistently reduced weed growth, it could be used as part of an overall weed management plan in container crops. However, all previous research has focused on subdressing in terms of only weed growth or crop growth, but not both (Broschat and Moore, 2003). In order to utilize subdressing in containers, more research is needed to fully understand the competitive dynamics of weeds and crops under this fertilization method. The objective of this experiment was to first, determine germination and growth of eclipta in pots that been topdressed or subdressed at three different depths. Secondly, our objective was to determine the competitive effects of eclipta on two common ornamental species fertilized via topdressing or subdressing at three different depths.

MATERIALS AND METHODS

All experiments were conducted at the Mid-Florida Research and Education Center in Apopka, FL during the 2019 summer. In all studies, the potting substrate was a pinebark:sand (8:1 v:v) that was amended with 3 kg m^{-3} dolomitic lime prior to use. In all cases, pots were fertilized using 35g of controlled-release fertilizer [Osmocote® Plus micronutrients 17-5-11 (8-9 mo.), ICL Specialty Fertilizers, Dublin, OH] based on manufacturer recommendations. Regardless of fertilizer placement, all pots were fertilized at the same rate. However, the control was not fertilized. All trials were conducted on a nursery pad in full sun, outdoor conditions and received 1.3 cm of overhead irrigation per day.

Eclipta germination and growth

The objective of this experiment was to determine if eclipta germination or growth is affected by fertilizer placement. Containers were filled with the aforementioned potting substrate and fertilized via topdressing, subdressing, or contained no fertilizer. Pots that were topdressed were filled with pine bark substrate, and 35 g of fertilizer was added to the container surface. For pots that were subdressed, the bottom portion of the container was filled either 2.5, 5, or 7.5 cm from the top with pine bark substrate, fertilizer was added, and then the remaining portion of the pot was filled with substrate. This resulted in subdressed containers with fertilizer being placed at depths of 2.5, 5, or 7.5 cm from the substrate surface. Another set of containers were included that only contained the pine bark substrate and no fertilizer was added to serve as a non-fertilized control. After filling, all containers were placed on a full sun container pad and 25 eclipta seeds were surface sown. After four weeks, weed counts were taken to determine differences in germination between the different fertility placements. A second study was installed at the same time following the same procedures except that after four weeks, pots were thinned (hand weeded) so that only one eclipta plant was in each container. The eclipta plant in each pot was allowed to grow for 12 weeks to determine the effects of fertilizer placement on eclipta growth. Data collected included eclipta growth index (average of plant height and two perpendicular width measurements), final shoot, and root dry weight determination at trial conclusion, which was 12 weeks after potting (WAP). Substrate pH and EC was also assessed on a bi-weekly basis throughout the trial. Both experiments were completely randomized designs with four single pot replications per treatment. All data were subjected to analysis of variance in JMP software (SAS Institute, Cary, NC) and means separation was performed using Tukey's Honest Significance Differences test at $P = 0.05$. For sake of brevity, only eclipta germination and growth index data is presented and discussed.

Eclipta competition with ornamentals

The objective of this experiment was to assess the competitive effects of eclipta on two ornamental species fertilized via topdressing or subdressing at depths of 2.5, 5, or 7.5 cm. Pots were filled and fertilized as described above except the non-fertilized control was not included. Uniform liners of ligustrum (*Ligustrum lucidum*) and Japanese boxwood (*Buxus microphylla*) grown in 5-cm plug trays were transplanted into separate sets of containers after filling. A separate set of containers were fertilized as described above but were left fallow and contained no ornamental plant. At 3 days after potting, eclipta seedlings that had been germinated in a separate set of pots were transplanted into half of the pots containing ligustrum and half the pots containing boxwood. Eclipta seedlings contained two true leaves, were approximately 1-cm in diameter, and had emerged two weeks prior to transplanting. In ornamental pots containing eclipta, the eclipta was placed 2.5 cm from the ornamental rootball. In fallow containers, eclipta was transplanted into the center of the container. This resulted in boxwood and ligustrum being fertilized via topdressing or subdressing at three different depths (2.5, 5, or 7.5 cm) and either containing one eclipta plant to assess competition or being hand-weeded on a bi-weekly basis and having no weed competition. Eclipta growth was also monitored both when grown in fallow containers and when it was grown in containers with ornamentals to determine if ornamentals had any negative influence on eclipta growth. Data collected included growth index measurements on ornamentals at 8 and 12 WAP, root and shoot dry weight determination for ornamentals and eclipta at

12 WAP, eclipta seed production at 12 WAP, and foliar nutrient concentrations at 12 WAP. To determine the significance of trends in ornamental or eclipta growth based on fertilizer depth, orthogonal contrast analysis was also performed and were considered significant at $P = 0.05$. Multiple comparison procedures were performed as described above to make individual comparisons of crop growth based on fertilizer depth. For sake of brevity, only shoot and root dry weight data are discussed for ornamentals and eclipta.

RESULTS AND DISCUSSION

Eclipta germination and growth.

Eclipta germination was similar in all treatments regardless of fertilizer placement, ranging from 56 to 66% (Table 1). However, subdressing at a depth of 7.5 cm resulted in a greater than 50% growth decrease in comparison with topdressed pots. Eclipta did not grow past the cotyledon stage in non-fertilized pots. These data indicate that subdressing could be an effective management method for eclipta, but depths of 7.5 cm would be needed.

Table 1. Eclipta germination and growth in response to fertilizer placement.

Placement[a]	Germination (counts/pot^{-1})[b]	Growth index (cm)[c]
Topdress	15.3 a[d]	27.1 a
Subdress 2.5 cm	16.6 a	25.8 a
Subdress 5.0 cm	14.0 a	20.1 ab
Subdress 7.5 cm	16.1 a	13.0 b
No fertilizer	15.6 a	0.0 c*

[a]Placement refers to the location of 35 g of 17-5-11 (8 to 9 month) controlled release fertilizer in the container.

[b]Eclipta counts at 4 weeks after surface sowing 25 seeds per container.

[c]Average of plant height and two perpendicular width measurements.

[d]Means followed by the same letter within a column are not significantly different according to Tukey's HSD ($P = 0.05$).

*Indicates a mean growth index of less than 1 cm.

The greater fertilizer placement at 7.5 cm is required because eclipta can grow deep roots soon after germination - even in the absence of fertilizer (Fig. 1). In these studies, results suggest eclipta roots reached the fertilized layer in contrast to other trials with spurge or crabgrass, which were unable to do so (Stewart et al., 2018).

Ornamental growth

Averaged over all fertilizer treatments, eclipta reduced shoot growth of boxwood by 5% and ligustrum by 16% over a 12-week period (Table 2). Boxwood growth was minimal in all treatments, likely because this is a slower growing species, thus competitive effects were not as evident in this short-term study.

Figure 1. Eclipta rooting in a pot subdressed at a 7.5 cm depth. Note the thin white roots growing into contact with fertilizer layer. Darker tan roots are *Ligustrum lucidum*.

In contrast, ligustrum shoot data showed that growth tended to decrease in a linear manner as fertilizer depth increased. When comparing individual treatments, ligustrum growth was similar in pots that were topdressed and subdressed at 2.5 or 5 cm, both with and without eclipta competition. Both crop species had similar root growth regardless of eclipta competition. While boxwood roots grew similarly regardless of fertilizer placement, ligustrum root growth tended to decrease as fertilizer depth increased.

Overall, this trial indicates that subdressing could be an effective management tool for eclipta if depths of at least 7.5 cm were used. Previous research has also shown that depths of 2.5 cm are effective for spotted spurge, crabgrass, bittercress (*Cardamine flexuosa*) and liverwort (Stewart et al., 2017). However, for eclipta, a depth of 7.5 was needed to reduce growth. This depth could slow the production time of ornamentals in smaller 3.8 L containers as was observed for ligustrum and boxwood in this trial.

For both species, liners had root balls of 2.5 cm wide by 2.5 cm deep. Liners had no contact with fertilizer, at least initially, when fertilizer was subdressed at 7.5 cm. When rootballs were in contact with fertilizer (topdressing or subdressing at depths of 2.5 or 5 cm), no differences in shoot growth was observed in comparison with the industry standard of topdressing. This indicates that subdressing depth should be based on initial liner size. While eclipta or other species may not be well controlled at shallower subdressing depths, many other weed species would be as outlined in previous research reviews (Stewart et al., 2017). Future research is planned in order to develop recommendations for subdressing depth based on liner size. This would allow growers to subdress fertilizer at proper depths that do not decrease crop growth. Evaluating other predominate weed species would also provide information to growers on which weed species could be controlled with subdressing, and the critical depth that is needed for a particular weed species. When combined, this information could be used by growers to implement subdressing to help reduce hand weeding costs and possibly eliminate one or more herbicide applications throughout the year to provide additional cost savings.

Table 2. Competitive effects of *Eclipta prostrata* on container-grown *Ligustrum lucidum* and *Buxus microphylla* as influenced by fertilizer placement.

Depth (cm)[c]	Boxwood: Shoot wt.[a] weeded	Shoot wt. w/eclipta	Shoot wt. decrease (%)[d]	Root wt.[b] weeded	Root wt. w/eclipta	Root wt. decrease (%)
0.0	21.8 ab[e]	20.3 ab	7	17.6 a	17.7 a	-1
2.5	23.0 a	22.5 a	2	17.8 a	18.1 a	-2
5.0	22.5 ab	20.1 ab	11	17.7 a	17.3 a	2
7.5	19.7 b	19.8 b	-1	17.2 a	17.3 a	-1
Mean	21.7 A[f]	20.7 B	5	17.6 A	17.6 A	0
Linear	NS[g]	NS		NS	NS	
Quadratic	**	NS		NS	NS	
	Ligustrum					
0.0	43.4 a	33.6 a	23	22.6 a	22.8 a	-1
2.5	38.2 a	32.6 a	15	21.9 ab	21.2 ab	3
5.0	34.1 ab	28.0 ab	18	19.7 bc	19.9 b	-1
7.5	26.1 b	24.5 b	6	19.3 c	20.1 b	-4
Mean	35.4 A	29.7 B	16	20.9 A	21.0 A	-1
Linear	***	***		***	**	
Quadratic	NS	NS		NS	NS	

[a]Shoot dry weight grown either weed free ("weeded") or in competition with one eclipta plant ("w/eclipta) over 12 weeks.

[b]Shoot dry weight grown either weed free ("weeded") or in competition with one eclipta plant ("w/eclipta) over 12 weeks.

[c]Depth of 35 g of Osmocote 17-5-11 (8 to 9 month) controlled release fertilizer. Depth of 0 cm is a topdress application.

[d]Percent decrease in growth resulting from eclipta competition over 12 weeks. Negative values indicate a percent increase in growth.

[e]Means within a colum and variable (shoot or root dry wt.) followed by the same lowercase letter are not significantly different based on Tukey's HSD at $P = 0.05$.

[f]Means within a row and variable (shoot or root dry wt.) followed by the same uppercase letter are not significantly different based on t-tests $P = 0.05$.

[g]*, **, and *** represent significant linear or quadratic responses at $P = 0.05$, 0.01, and 0.001, respectively based on orthogonal contrasts. NS = not significant.

Literature Cited

Berchielli-Robertson, D.L., Gilliam, C.H., and Fare, D.C. (1990). Competitive effects of weeds on the growth of container grown plants. HortScience *25*:77–79.

Bir, R.E. and Zondag, R.H. (1986). The great dibble debate: Test results raise more questions. Amer. Nurseryman *164*:59–60, 62–64.

Broschat, T. and Moore, K.K. (2003). Influence of fertilizer placement on plant quality, root distribution, and weed growth in container-grown tropical ornamental plants. HortTechnology *13*:305-308.

Fain, G.B., Paridon, K.L., and Hudson, P.M. (2004). The effect of cyclic irrigation and herbicide on plant and weed growth in production of Magnolia grandiflora 'Alta'. Proc. South. Nurs. Assoc. Res. Conf. *49*:37–39.

Stewart, C., Marble, C., Jackson, B., Pearson, B. and Wilson, C. (2018). Effects of three fertilization methods on weed growth and herbicide performance in soilless nursery sbstrates. J. Environ. Hort. *36*:133–139.

Stewart, C., Marble, C., Pearson, B., and Wilson C. (2017). Impact of container nursery production practices on weed growth and herbicide performance. HortScience *52*:1593–2000.

Following in the Footsteps of Legends! Lessons Learned Through Four Generations

Andrew King

Department of Horticultural Sciences, Texas A&M University, College Station, Texas 77843, USA

aking@tamu.edu

Keywords: Lynn Lowrey, multi-generation family business, Benny Simpson, Barton Warnock, native plants, plant collectors, wildscapes

Summary

Through four generations of horticulture, the King family has been influenced and inspired by a number of the legends of the horticulture industry. These included influential nursery professionals, plantsmen and academics that all had one thing in common: a deep love and respect for horticulture. The accomplishments of Lynn Lowery, Benny Simpson, Dr. Barton Warnock, and others shaped the practice of horticulture in the Lone Star State. These legends also taught many valuable lessons to many willing students along the way. Their influence and accomplishments are outlined in the current work.

INTRODUCTION

In 1915, J.B. King Sr. began growing strawberries in the sugar sands of east Texas (Fig. 1). With no other strawberry growers in the area, he developed a market for them by shipping the produce to Shreveport, Louisiana. Soon the idea became popular with many neighboring farmers, and J.B. realized that he needed to differentiate his farm. He began grafting and growing fruit trees and roses - and the King Nursery was born (Fig. 2). His slogan was: "Where the name of the firm indicates the quality of stock." In 1947, J.B. King Jr. was called into service at the nursery. He incorporated more ornamental plants into the nursery's inventory, until the early 1950s when it became a one-stop shop for retail customers. Even though the nursery has always depended on retail customers, J.B. Jr. had a wholesale sensibility. He was a tremendous propagator and preferred the solitude of plant production to the often-hectic nature of retail sales (Fig. 3).

Figure 1. The Kings' strawberry field in Tenaha, TX in the early 1920s.

Figure 2. J.B. King, Sr. and Katie King standing in front of the King Nursery sign in the late 1930s.

Figure 3. J.B. King, Jr. grafting *Cornus florida* in the late 1970s.

In 1979, J.B. Jr.'s son, Aubrey King, began full-time work on the nursery. Like his father, he was an excellent plantsman; however, he was also highly skilled in the art of sales (Fig. 4). Treating customers as students, he would spend hours with those that showed interest in horticulture. He placed his signature on the nursery by offering cutting-edge plant material, especially in the area of perennials and small trees. These men, three generations of horticulturists, passed to the next generation much knowledge and many lessons that were taught to them by experience and many great friends and mentors. This paper is an attempt to highlight these lessons and pay homage to those that helped to teach them.

SURROUND YOURSELF WITH INCREDIBLE PEOPLE

In the 1950s and 1960s, there were three horticulturists/botanists that scoured the Texas countryside in search of new and rare plants. These men rarely worked together, but their combined efforts changed the trajectory of horticulture in the Lone Star State.

Figure 4. Margaret King (J.B. King Jr.'s wife) and Aubrey King on the nursery in the late 1990s.

Lynn Lowrey, Benny Simpson and Dr. Barton Warnock were each uniquely qualified to make their own contributions - yet they shared a passion for improving the plant palette in a state rich with ecological variety (Fig. 5). Though each of them worked extensively throughout the state, it has been said that boundaries for their "regions of expertise" were understood. If the state were divided into thirds, Lynn would have been responsible for the southeast, Benny the northeast and west Texas would have been covered by Dr. Warnock.

Figure 5. From left to right: Lynn Lowrey, Benny Simpson and Dr. Barton Warnock on a plant collecting trip to Mexico in the late 1980s. (Photo courtesy Dr. Dave Creech).

Lynn Lowrey spent his career working for nurseries, generally around the Houston, Texas area - including owning and operating his own nursery. He quickly set himself apart from other nursery professionals based upon his interest in native plant material (Fig. 6).

Figure 6. Lynn Lowrey collecting *Scutellaria suffrutescens* in Mexico in the late 1980s. (Photo courtesy Dr. Dave Creech)

Selling native plants in the 1950s was a tough job, but Lynn proved to be up to the challenge. Along with these native plants, he also championed a more naturalistic landscape design, even going so far as to install them in some of Houston's most affluent neighborhoods. Along with many of the native plants that he introduced, this "wildscaping" concept is seen widely in Houston still today. Lynn touched the lives of many customers, but also many fellow horticulturists.

Along with his contemporaries, Benny and Dr. Warnock, Lynn influenced many younger horticulturists including Dr. William C. Welch, Dr. Dave Creech, Dr. Jerry Parsons, Jill Nokes and Greg Grant (Figs. 7 and 8). Plants that Lynn was at least partially responsible for introducing to the Texas nursery trade include *Quercus canbyi*, *Q. polymorpha*, *Q. risophylla*, *Lagerstroemia indica* x *L. fauriei* 'Basham's Party Pink' (possibly the first ever hybrid between these species) and *Scutellaria suffrutescens*, among many others.

Figure 7. Dr. William C. Welch at his home outside of Round Top, Texas in the mid-1990s. (Photo courtesy Greg Grant)

Benny Simpson began his career in 1954 at the Texas Research Foundation in Dallas, Texas. In 1972, as Texas A&M University assumed control of the Research Foundation center, he developed an interest in discovering native Texas plants and introducing them to the nursery trade. Although the previously mentioned "boundaries" were in place, Benny made many fruitful trips to the Trans-Pecos region and brought back plants to the northeast that he thought would thrive. Though he had many selections, some of his most popular plant introductions include *Chilopsis linearis* 'Dark Storm' and *Leuchophyllum frutescens* 'Green Cloud'.

Figure 8. Dr. Dave Creech awarding the "SFASU Senior Horticulture Student of the Year" to Dr. Andrew King. (Photo courtesy Dr. Dave Creech)

Dr. Barton Warnock studied botany at Sul Ross State Teachers College and later was on faculty there. He was the foremost collector of plants in the Trans-Pecos region, especially the region that became Big Bend National Park. Dr. Warnock was among the first to catalog and collect plants in this expanse after the federal government acquired the property. Once the park was opened, he continued to work there, taking students on collecting trips and instilling his love and passion for the plants of the region into others. He trained many well-known botanists and left his mark in many other ways. The Barton H. Warnock Environmental Education Center sits on the Big Bend Ranch State Natural Area and houses an arboretum filled with plants collected by Dr. Warnock himself.

Each of these men provided inspiration for nurseries all over the state of Texas, including the Kings. But more than their inspiration, the plants that they introduced were vital to shaping the industry.

The Kings were impacted personally through relationships with many other leaders in the horticulture world. A chance meeting with Greg Grant, a well-known horticulturist who lived only 20 miles from the King's nursery in Tenaha, set Aubrey on the path to discovering new plant material and many new friends. Greg introduced the Kings to Dr. William C. Welch, Dr. Dave Creech and Dr. Jerry Parsons, just to name a few. These horticulturists were responsible for expanding the plant palette, notoriety and imagination of many nurseries - including the Kings. Dr. Welch's style and class, Dr. Creech's adventurous plant collecting and Dr. Parsons' intense practicality, all influenced the Kings and their operation. This could be said by countless other firms as well.

PLANTS DONOT READ BOOKS

The aforementioned plant collectors certainly had their share of technical knowledge about the plants they were collecting. After all, each of them either wrote or inspired books about the plants of a given area in Texas. They realized however, that finding a plant that was "out of place" (e.g. an acid-loving plant in an alkaline soil; a cold-sensitive plant in a cold area) represented an opportunity rather than a conundrum. Perhaps they could not explain why that plant was thriving in said area - but they were certainly excited to see it and consider the possibilities of introducing the species into a new area.

Throughout childhood, I was taught that *Leucophyllum frutescens* and *Dermatophyllum secundiflorum* would not thrive, or even survive for long, in the southeast; yet as we toured nurseries throughout southern Louisiana, I saw both of them thriving in a humid, mesic environment. Virtually any horticulturist, that has been one for long, has their own story about finding a plant that is growing in the "wrong place." Often, we find that this is an artifact of an unintended artificial environment (e.g. structure providing protection for a cold-sensitive plant), but when it is not, it can be an exciting breakthrough in plant distribution. Ultimately it reminds us that plants cannot read, and what an author has written about them is merely from their experience.

THINK OUTSIDE THE BOX

Through the many years, the Kings have seen trends come and go, and the common theme is that nurseries are often left with large inventories that have fallen out of favor with consumers. At times, they simply took their loss and moved on to the next crop, dumping the previous crop to make room. Alternatively, there were times when they thought outside of the box and repurposed the previously trendy crop for another use. In the fall of 2017, I was reminded of this concept when grounds maintenance on campus at Texas A&M University decided to take a new approach to pruning the *Tecoma stans*. They had grown to approximately 15 feet tall by late October and were just beginning to bloom. Instead of cutting them back to the ground prematurely, they limbed them up, creating a yellow-blooming, multi-stemmed, small tree. The first time I observed them from afar they appeared to be the "holy grail": a yellow Crapemyrtle. Alas, they were not; but what they were, was beautiful and interesting, all because someone thought outside of the box.

EDUCATE

Through Aubrey King's tenure as owner/operator of King's Nursery, many people came to Tenaha for the hard-to-find plant material, many people came for the affordable prices, but most people came for his knowledge. Industry-wide, nursery professionals hold some of the most widely sought-after information in our society today. If you question the last statement, just tell someone that you are a horticulturist and then be ready for the barrage of questions that are sure to follow. Educating people about the nuances of horticulture is one of the true joys of the field. There is certainly some validity in limiting the interaction time between the professional and the consumer. This is a highly efficient model that can lead to great productivity and economic gain. However, there is also value in the interaction with consumers, and enriching their lives with the knowledge that you have acquired. It is not the way to do business, but it is a way to do business, and it is a great way to make lifelong friends and customers. Share your knowledge. Educate.

Ultimately, the nursery industry is a great one. I'll never forget my father's words: "The nursery business is a wholesome way to make a living." They are still true today and hopefully they always will be.

I Have a Drone, Now What Can I Do with It?

Joe Mari Maja[1*] and James Robbins[2]

[1]Clemson Univ., Edisto REC, 64 Research Road, Blackville, South Carolina 29817, USA
[2]University of Arkansas Cooperative Extension Service, 2301 S. University Avenue., Little Rock, Arkansas 72204, USA

jmaja@clemson.edu

Keywords: aerial images, drone, green industry, nursery, sUAS - unmanned aircraft system, UAV- unmanned aircraft vehicle

Summary

There will be increasing opportunities in using drones, small unmanned aircraft systems (sUAS), in green industry production systems. The potential usage of drone activity include: crop monitoring, (application of chemicals and nutrients, asset tracking and management), plant inventory, and marketing and sales. Information is included on types of aircraft, flight preparation, and image processing – stitching.

INTRODUCTION

Small Unmanned Aircraft Systems (sUAS) or 'drones' are an emerging technology that can be used to automate or augment certain operations in outdoor plant production systems. They may be used for a variety of activities including 1) crop monitoring (i.e. nutrients, water, pest, general health, indoor plant propagation), 2) application of chemicals and nutrients, 3) asset tracking and management (e.g., estimate volume of bark piles), 4) plant inventory, and 5) marketing and sales.

Until recently, the greatest challenge to wide-scale adoption of this technology has been regulatory issues. From a regulatory standpoint, horticulture applications of sUAS should be viewed as a low risk (minimally intrusive) flight environment. sUAS offer several advantages over other methods (i.e. satellite, manned aircraft) including: higher image resolution; autonomous flight path which can be repeated; 'as needed' use; lower flight altitude so image collection is less likely to be influenced by clouds; safe solution for dull/dangerous/dirty applications; reduction in time needed to prepare and initiate flights; and lower cost. The increased image resolution is cited most often as the primary reason for using sUAS.

AIRCRAFT

A challenge to sUAS users is the large number of aircraft options, each having their strengths and weaknesses, which need to be considered to achieve the best option for the user (Table 1). This discussion will focus on 'small unmanned aircraft systems' (sUAS), which are defined by the U.S. Federal Aviation Administration (FAA) as having a take-off weight of less than 55 lb. (~25 kg). It might be worth noting that frequently the terms UAS and UAV are used interchangeably; however, UAS (unmanned aircraft SYSTEM) more often refers to the 'aircraft plus sensor(s)' whereas a UAV (unmanned aircraft VEHICLE) is simply the aircraft (platform). There are two broad groups of UAV including rotary and fixed-wing. However, there are a small number of 'hybrid' aircraft that lift off vertically like a rotary and then reorient propellers to fly as a fixed-wing. This discussion will focus on rotary aircraft since that is the primary platform used in specialty crop plant production.

Beginning sUAS users are encouraged to purchase a 'practice' sUAS as a cost-effective (<$50) means to learn the basics of flying an aircraft using a remote control (RC). There are many criteria that must be considered when purchasing a sUAS including: 1) intended use (e.g., recreation, image collection), price, payload capacity, ability to interchange sensors, flight time, and object avoidance capability. Although advances in aircraft (e.g. lightweight, strong frames; improved battery life) has been made - the most significant change has been in the flight navigation software which enables for repeated and complex flight plans.

Remember that not all sUAS can carry additional payload and that modifications to the payload requires weight and balance considerations. The object avoidance sensors on sUAS must be considered when adding user-installed payload. Retractable landing gear reduces the potential for shadows and obstruction when capturing imagery. sUAS users need to evaluate whether purchasing an aircraft system is the best option for their needs or whether it is more cost-effective to hire an outside firm to collect and process imagery. Remember that regardless of the size (weight) of the sUAS, if it is used for commercial purposes, the operator must be Remote Pilot Certified by the FAA.

FLIGHT PREPARATION

Before flying a sUAS, it is essential that the pilot evaluate the airspace they will be flying in. Under 'Part 107' guidelines sUAS can fly up to 400 ft above ground level (AGL) in uncontrolled airspace (Classes G, E3, E4). To fly in controlled airspace (Classes A, B, C, D, and E-surface) sUAS pilots must get permission from the FAA. The first step in planning your flight is to obtain the latitude and longitude for the location where you will fly. While coordinates may not be necessary, they are a good piece of information to have. There are many ways to get coordinates, but one method is to use Google Earth or Google Maps. The next step is to access the FAA UAS Data Map or alternatively called the 'facility map' or 'grid map'. By entering the address of your flight location, you will immediately learn whether air traffic control (ATC) authorization is required, what altitudes are likely to be allowed, and how (e.g. LAANC or DroneZone) you can apply for flight authorization if required. Flight areas that are covered by the Low Altitude Authorization and Notification Capability (LAANC) will likely receive authorization faster than non-LAANC areas. Consulting the FAA Sectional Chart for your specific flight area before flying is recommended to give the pilot a full understanding of the airspace around the flight area. Current regulations require that a pilot must keep their aircraft within visual line of sight (VLOS).

Table 1. Information of rotary aircraft, fixed wing, sensors, imaging processing software and flight planning.

Rotary Aircraft

For rotary, in general, organized from lower to higher cost (as of June 2019):

Group 1: Practice drones, some examples: Blade Nano QX RTF ($60)
Holy Stone HS170 ($36)
Cheerson CX-10 Mini ($18)
Cheerwing Syma X20 ($27)

Group 2a:
Yuneec Breeze (fixed arms; 4K) $168 (eBay)
DroneX Pro: **$99**
Ryze Robotics Tello: $99 (DJI is using in the U.S. at their 8 training centers)

Group 2b:
DJI Spark: (fixed arms; $500 bundle Amazon)
DJI Mavic Pro: (collapsible blade arms) Basic: $788 (Amazon)
DJI Mavic Pro 2: $1,099 w/out RC & charger (B&H)
DJI Mavic Air: $854 (Amazon)
Parrot ANAFI: $599 basic (Amazon) $899 bundle (Amazon)
Yuneec Mantis Q (collapsible blade arms) $499 (Crutchfield)

Group 3:
DJI Phantom 3: (really was a workhorse mid-level drone; was replaced by Phantom 4; no longer sold by DJI in US) Basic: $549 (Amazon)
DJI Phantom 4 (models: Advanced; Pro)
Basic: $1,049 (no RC or charger) (B&H) ; DJI P4 Multispectral ($6,500 DJI)
Autel Robotics X-Star: $450 (eBay; used)
Autel Robotics EVO: $999 (Autel Robotics)
Yuneec Typhoon H: (retractable legs) $1,199 bundle (B&H)

Group 4:
DJI Inspire 2 (retractable landing gear; difficult to attach non-DJI camera/sensor) Basic: $2,599 (DJI) Bundle: $5,199 (SSEPhotoVideo). *Example of 2nd party plug and play multispectral sensor (Sentera 5 band, $3,500):* https://sentera.com/product/inspire-double- 4k-upgrade-crop-health-sensor/

Group Sa:
Mikrokopter (Germany): http://www.mikrokopter.de/en/home (while you can purchase pre-built these are mostly self-built kits)
DJI Matrice 100 (quad): Basic bundle (frame, 1 battery, battery charger, 4 blades, RC) $3,299 (NewEgg); same price at B&H. Battery: $139; Zenmuse X5S: $1,835; hard case: $429;

Bundle (frame, 4 batteries, hard case; RGB, multispectral): $8,300 (DroneNerds)

OJI Matrice 200 series (quad copter)

DJI Matrice 210 V2 Combo, $9,600 (Ad-vexure) DJI Matrice RTK 210 v2, $14,500 (Vertigo Drones)

Group Sb:

OJI Matrice 600 Pro (Hexa copter) (this is clearly a professional unit; retractable landing gear):

Basic (frame, 6 batteries, RC): $5,699 (B&H)

Fixed Wing

- **AgEagle** (Neodesha, KS) (models: RX-60; RX-47)
- **senseFly eBee X** (SenseFly was acquired by Parrot) ($13,000)

Sensors

Llewellyn Data Processing, LLC (maxmax.com)

Multispectral:

- MAPIR, Survey3 RGNIR (replaces Survey 2; modified GoPro): $400 (MAPIR)
- MicaSense RedEdge ($4,900)

Thermal:

FUR Duo: $999 (Wellbots)

Image Processing Software

- Microsoft Image Composite Editor (ICE).
- Agisoft Photoscan/Metashape.

Flight Planning

- FAA 'Visualize It' UAS Data Maps (or grid maps) (consult before flying).
- UAS Data Exchange (LAANC) (includes list of approved 3rd party partners).
- UAS flight planning (maps and NOTAM application).
- Pilot Certification and Aircraft Registration for Non-hobby Users of Small Unmanned Aircraft Systems (sUAS).

No endorsement is implied or discrimination intended for firms or references included or excluded from this list.

IMAGE PROCESSING – STITCHING

Besides using your sUAS to capture a single photograph or short video, it is likely many users will need to 'stitch' images together to capture a larger area of interest. The first step is to understand the overall workflow in capturing imagery using a sUAS. The process begins by understanding the airspace and receiving proper authorization where required. The next step is to plan specific parameters (e.g. altitude, orientation of sensor, flightpath, when and where to capture images) of the flight using waypoint navigation software. Once proper planning is complete you can fly the aircraft to capture images. In most cases, the images will be downloaded to a computer once the aircraft has landed. Be prepared for very large datasets. It is recommended that you view your Images shortly after flying to make sure the sensor was working and to edit images that are not needed or of poor quality.

For this discussion we will focus solely on stitching images together. There are many software options (e.g. Microsoft Image Composite Editor (ICE); Agisoft Photoscan; Pix4D) but for this discussion will focus on Image Composite Editor from Microsoft. ICE is easy to use and is free. It should be noted that ICE will not create a GeoTIFF file if you require your stitched image to be georeferenced. The time to stitch images will vary depending the number of images and software used and can vary from minutes to hours.

CONCLUSION

There is no doubt sUAS will be a useful technology in outdoor plant production, but users need to carefully evaluate their specific situation to determine what the best approach is. Some users may conclude that purchasing the sUAS and software is the best option while others may find that hiring an outside company to acquire and process the images is their best option.

Acknowledgement

The authors, who have been using sUAS for 10 years, wish to acknowledge the collaborative help from many people and organizations including the J. Frank Schmidt Family Charitable Foundation, Horticultural Research Institute (HRI), J. Frank Schmidt & Son Co., McCorkle Nurseries, Greenleaf Nursery Company, Bailey Nurseries, Cherry Lake, Parks Brothers Farm, Yule Tree Farm, Woodburn Nursery & Azalea, Willoway Nurseries, Hale and Hines Nursery and R.A. Dudley Nurseries to name a few.

Both authors are certified Remote Pilots. Robbins: #3952601; Maja: #3952164

Meadow Gardens: Grass Landscapes from Seed

Marc Pastorek

Pastorek Habitats LLC, 72322 Ingram Street, Covington, Louisiana 70435, USA

meadowmakers@gmail.com

Keywords: Grassland landscapes, grass garden landscapes, landscapes from seed, native gardens, natural garden design and management, naturalized gardens, prairie gardening in the Southern U.S., pushy perennials, seed propagation

Summary

Creating native meadow habitat gardens of perennial plant communities grown from seed has been an effective approach for developing sustainable landscape designs in city parks and open spaces. A grassland garden can easily contain fifty or more species. Optimum conditions occasionally yield as many as 15 species per square meter. Manual weeding of the garden is not recommended as this activity causes soil disturbance, which is to be kept to a minimum. Controlled burns reduce weedy species and enhance the growth of fire-tolerant (pyrogenic) species. Using seed from species that originate from similar ecological zones to your general locality will maximize adaptability and resilience of the grassland garden landscape. It is highly desirable to have grass species that establish and compete well in the landscape. Successful seed mixes for the U.S. Gulf Coastal Region include: *Andropogon* and *Schizachirium*, along with short-statured *Panicum*, *Dicanthelium*, and *Paspalum*. Grasses are the fabric that flowering plants are woven into. Reconstructing naturalized grassland landscapes is an effective horticultural approach to managing land for biodiversity, and equally important: reconnecting people with nature.

INTRODUCTION

The art of creating native meadow habitat gardens composed of perennial plant communities grown from seed - has been an effective approach for developing sustainable landscape designs in city parks and open spaces in the central Gulf coastal region of the U.S. These grassland gardens contain numerous species of plants that have coevolved, with a flowering phenology that lasts through the entire growing season - supporting butterflies, bees, beetles and a variety of other wildlife. The grassland gardens are permanent plantings that are nurtured through natural means, borrowing principles of natural land management - and often including the use of controlled burn or mowing annually as the main tool of the land manager (Noss, 2018).

Natural grasslands once made up a large portion of the continental U.S. (Hill and Barone, 2018). The predominant landscape in the Southeastern U.S was the natural grassland which is a fire dependent ecosystem. The North American Coastal Plain (NACP) was recognized as meeting the criteria of for the 36th global Biodiversity hotspot: more than1500 endemic vascular plants and greater than 70 percent habitat loss - a habitat on the brink of extinction.

Historical use in the U.S of the natural meadow landscaping or at least attempts to replicate a natural grassland in the urban environment goes back at least to the 1890's when Jans Jensen was creating prairie garden landscapes. Jensen "harnessed" the natural grassland of the Midwest to design and construct "prairie rivers" in the parks of Chicago, gardens of native plant associations - that became features of his designs. Jensen realized and shared an appreciation for the artistic and scientific blend of the natural grass landscape. In the 1930's Aldo Leopold and John Curtis, biologists at the University of Wisconsin, Madison, attempted a reconstruction of native prairie using seed and plants. Their focus was one of scientific study, of restoration and preservation (Leopold, 1949). These experiments became the foundation for the University of Wisconsin Arboretum, which features the natural plant communities of the region, using natural means to sustain them.

More recently, public park and arboretum designs have focused on using natural systems and natural mechanisms to reintroduce nature to the urban context. The work of Richard Hansen and Friedrich Stahl in their seminal book, *Perennials and the Garden Habitats,* describes the use of specific plants that persist and proliferate and make gardening easier by specifically designing for beauty and ease of maintenance (Hansen and Stahl, 1993). The Dutch garden designer Piet Oudolf has recently brought his use of highly designed grass dominant gardens to the U.S., using herbaceous perennials and grasses chosen for their ability to cover ground and suppress weeds as much as they are chosen for structure, texture, and color.

The natural grassland is one aspect that can be used to enhance biodiversity and civic enjoyment. Natural grassland gardens increase the public's knowledge and appreciation for biodiversity and the importance of preservation and conservation of the remnants of this precious habitat.

Regarding horticultural value and developing ornamental landscapes planting of prairie grasses and companion flowering plants are profoundly beautiful and remarkably dynamic (Stein, 1995; Wasowski, 2002; Werner and Christopher, 2016). They fit together as comfortably in the garden, as they do in the wild. Prairie garden landscapes propagated from seed change dramatically for the first five years; thereafter more subtly. Managing 0.4 to 2 ha (1 to 5 acre) grass garden is quite doable when it comes to grassland landscapes. The larger the project, the more ground is covered from seed and the more vignettes occur - little meadow gardens within the larger garden.

Landscapes such as the Storm King Sculpture Gardens in New York and the Highline Park, Manhattan, N.Y. demonstrate the cutting-edge appeal of landscapes that were designed to contain masses of native grasses that form dramatic sculptural vistas of beauty and bounty.

Much of the work described here by the author was inspired by the work of Charles M. Allen and Malcolm F. Vidrine's work on prairie ecology and their attempt to re-establishing a prairie landscape in the mid 1980's in Eunice, Louisiana (Vidrine, 2010). Their work was inspired by scientific work done with earlier prairie restorations with Tall grass prairie grasslands in the Midwestern U.S. Through Allen and Vidrine's work, basic seed and planting knowledge was gained and shared. From those shared bits of knowledge, the author has honed and sharpened some of the skills - but much remains to be learned and garnered. The design potential of using the numerous species in combination is limitless.

GETTING STARTED

Urban grass landscapes typically require seeding in order to become established since the integrity of the vegetation is typically not conservative in value. Rural gardens that are herbaceous in nature, in many cases, can be transformed using no seed - but simply by reintroducing fire. Identifying the plants on site may inform you by way of indicator species, or species that are of significance botanically - typical prairie herbs and/or grasses. An experiment was initiated in the community of Carriere, Pearl River County, Mississippi in 1999 where only controlled burns were introduced on two-year rotations, on a one-acre site dominant in Bahia grass. Very few conservative species existed. The result over 20 years has been the development of a rather beautiful, floriferous landscape. Active fire suppression is one of the leading factors in the loss of biodiversity in Southeastern U.S. (Noss, 2018)

One of the main considerations for using this design style is that it must be a sunny open area. The fewer trees in this landscape the better. Soils can vary greatly with poor clay or sandy soils, which are better candidates than those high in nutrients and organic matter. However, even rich soils can support a prairie landscape if proper preparation is taken before planting. Areas generally in urban conditions are often more difficult to establish than in areas where human impact has been minimal. Weedy exotic invasive species and a history of soil disruption are often common characteristics of the urban condition. Urban plantings need more planning and preparatory elimination of competitive species.

Preparatory tasks before planting can include tilling, applying specific herbicides and/or solarizing. The seed used to plant such a garden should be adaptable, meaning it germinates, thrives and persists for many or numerous years. Maintenance needs include the use of mowing, burning, and/or eradicating woody plant growth (Fig. 1). Typical forestry management techniques are modeled after typical herbaceous grassland management principles.

Figure 1. After a controlled burn, the ground is scorched but the perennial plants are rejuvenated.

Seed for planting projects can be purchased, but the best scenario is gathering seed from a high-quality natural area that is near your planting location. Doing this ensures seed adaptability and often yields species that are endemic to the area. Seed collections sites are good study sites as well - to observe and take in the growth dynamics over time of a natural local grassland species.

Planting can be rigidly structured or fairly loose regarding a preferred method, which depends upon the ultimate goals and growth requirements. The speed in which a prairie grassland establishes is often a major factor, as public perception is often critical and can lead to differing opinions. For this reason, the interpretive aspect of the garden is an important factor, showing the significance of reconstruction of natural grasslands to the urban environment.

Seeding can be as easy as plowing the soil and broadcasting seed and observing what happens. One should conduct different experiments with seeding within the planting area. The different results teach one the dynamics and the best practices for specific locations. In 50 experimental seed plantings established by the author on 3.2 ha (8 acres) since 1998, the resulting gardens have been particularly beneficial for establishing best management practices and observing establishment and revegetation patterns.

In gardens of high profile, where a landscape needs rapid transition, such as a city park or an impatient client is funding the work - strategies such as thorough soil preparation - repetitive plowing, herbicide applications, solarizing, heavy seeding rate, etc., can speed the process of garden maturation.

The management of the landscape is driven by natural succession, the replacement of one plant by another. Early establishing species are limited by individual viability and germination rates and their place in the successional clock of time - how well they compete for sunlight, moisture, and root zone. Grasses and flowering plants are important components, as they colonize and create a competitive environment - putting pressure on less aggressive species, while favoring growth of others. The term 'pushy perennials" was coined to describe the aggressive nature of some grassland species. Grassland perennial herbs and grasses live a highly competitive life.

A SEASON IN THE GARDEN

During springtime (late March and early April) along the U.S. Central Gulf Coast, prairie plants emerge from winter dormancy. The height of the vegetation is less than a 30 cm (1-ft). The grassland garden seems to contain few grasses, as grasses take on a secondary roll visually until late summer; they do add color (green), texture and form (Figs. 2 and 3).

Flowering occurs in drifts and in random patterns - throughout the growing season. By June, the prairie is growing taller as the grassland is in full swing, triggered by higher temperatures. By August, the height is around 0.6 m (2-ft), and some species are even taller. Inflorescences (flowering stalks) of Bluestem grasses grow vertically and are waist-high by late September.

Figure 2. A grass garden landscape with forbs, in May, Folsom, Louisiana.

Figure 3. The same grass garden landscape in mid-June, Folsom, Louisiana.

Much of the vegetation is knee-high with flowering stalks rising above (Figs. 4, 5 and 6). By November, the mean height of the grass garden is 0.9-1.2 m (3-4 ft) and grasses are fruiting. First-frost arrives about December 1 and all flowering and growth activity ceases. Bluestem grasses turn a golden yellow or bright red, depending on the species (Fig. 7). The grasses are the dominant feature throughout the winter period and into January and February - when mowing or a burn can be done. This takes the vegetation down to the ground, eliminating the previous year's growth. The cycle begins anew.

Figure 4. *Eryngium yuccafolia* and honey bee in Folsom, Louisiana.

Figure 5. Buckeye butterfly larvae on False Foxglove (*Agalinus*) in October, Folsom, Louisiana.

Figure 6. The Folsom, Louisiana grass garden landscape in October.

Figure 7. A grass garden landscape dominated by Little Bluestem, in December, Mandeville, Louisiana.

CONCLUSION

A grassland garden can easily contain fifty or more species. Optimum conditions can occasionally yield as many as 15 species per square meter. Manual weeding of the garden is not recommended as this activity causes soil disturbance, which is to be kept to a minimum. The first year of growth does not yield much in the way of ornamental value. Mowing can be used sparingly to keep the planting uniform and to keep sunlight on the smallest seedling plants during the first year. The best-case scenario is to not disturb the planting at all. Just observe. If burning is included in the management program, the biomass of grasses and flowering plants should be allowed to accumulate – so there is sufficient fuel for burns after the first year and subsequent years. Burning is a natural part of the grassland environment. Fire reduces weedy species and enhances the growth of fire-tolerant (pyrogenic) species.

Using seed that originates in your general locality insures selecting suitable ecotypes for greater adaptability and resilience. Incorporate grass species that create and maintain a competitive atmosphere in the landscape, such as *Andropogon*, *Schizachirium*, short-statured *Panicum*, *Dicanthelium, Paspalum*. Grasses are the fabric that the flowering plants are woven into. Taller grasses such as Switch grass and Gamma grass are used sparingly as accents or dramatic elements unless the design calls for more height.

The grassland landscape is dynamic: daily and seasonally changing. Reconstructing naturalized grassland landscapes is an effective horticultural approach to managing land for biodiversity. Besides the horticultural enjoyment, grassland gardens help preserve diverse, uncommon plant species and habitats. The wildlife activity that comes with plant community reveals a whole new reason to garden. Observing the garden over a period of many years is an education unto itself. Grassland gardens bring the observer back to the historical landscape of the past - focusing on life and death as the changing of seasons. Grasslands reconnect people with the nature.

For Flowering Phenology Information

Craine J.M. , Wolkovich, E.M., Gene-Towne, E., Kembel, S.W. (2012). Flowering Phenology as a Functional Trait in a Tall-grass Prairie. Flowering phenology as a functional trait in a tallgrass prairie. New Phytol. *193*(3):673-82.

Announcing the World's 36TH Biodiversity Hotspot: The North American Coastal Plain https://www.cepf.net/node/4422

Noss , R.F., Platt, W.J., Sorrie, B.A., et al. (2014). How global biodiversity hotspots may go unrecognized: lessons from the North American Coastal Plain Diversity and Distributions *21*: 236–244.

Table 1. A Prospective List of Species for Cajun Prairie Reconstruction[1,2].
[1]Species recommended for the coastal plain of Louisiana (some species apply to east Texas, Mississippi, and southern and central Alabama and eastwardly along the Atlantic coast to North Carolina and beyond.
[2]See Vascular Flora of the Cajun Prairie of Southwestern Louisiana (Allen et al. 2001). Use this paper as a guide to species that will be comparable to local and regional flora lists foundnear a given area.

	Monocots	
Andropogon gerardii	*Dicanthelium aciculare*	*Panicum dichotomiflorum*
Andropogon glommeratus	*Dicanthelium commutatum*	*Paspalum laeve*
Andropogon gyrans	*Dicanthelium dichotomum*	*Panicum laxum*
Andropogon morhii	*Dicanthelium scoparium*	*Panicum virgatum*
Andropogon ternarius	*Dicanthelium scabrusculum*	*Paspalum floridanum*
Andropogon scoparium	*Dichromena colorata*	*Paspalum laeve*
Andropogon virginicus	*Digitaria filiformis var. villosa*	*Paspalum praecox*
Anthaenantia rufa	*Eliocharis montevidensis*	*Paspalum plicatulum*
Aristida purpurascens	*Eliocharis quadrangularis*	*Rhynchospora corniculata*
Aristida dichotoma	*Eragrostis elliotii*	*Rhynchospora inexpansa*
Aristida longespica	*Eragrostis refracta*	*Rhyncospora glaberata*
Bothriochloa longipanicu-	*Eragrostis spectabilis*	*Rhyncospora globularis*
lata	*Erianthus gigantea*	*Scirpus cyperinus*
Carex glaucescens	*Erianthus strictus*	*Schizachyrium scoparium*
Carex vulpinoidea	*Eriocolon decangulare*	*Schizachyrium tenerum*
Cladium jamaicense	*Fuirena squarrosa*	*Scleria pauciflora*
Coelorachis cylindrica	*Juncus dichotomus*	*Scleria reticularis*
Coelorachis rugosa	*Juncus tenuis*	*Sorgastrum nutans*
Ctenium aromaticum	*Juncus marginatus*	*Sporobolus junceus*
Cyperus acuminatus	*Leersia orysoides*	*Steinchisma hians*
Cyperus erythrorhizos	*Muhlenbergia capillaris*	*Tridens ambiguus*
Cyperus haspan	*Muhlenbergia capillaris var*	*Tridens flavus*
Cyperus psuedovegetus	*expansa*	*Tridens strictus*
Cyperus oxylepis	*Panicum anceps*	*Tripsicum dactyloides*
Cyperus virens		
	Dicots	
Agalinus fasciculata	*Eurybia hemispherica*	*Pluchea foetida*
Agalinus viridis	*Euthamia leptocephala*	*Polytaenia nuttallii*
Aletris aurea	*Euthamia tenuifolia*	*Pycnanthemum albescens*
Baptisia spherocarpa	*Gailardia aestivalis*	*Pycnanthemum muticum*
Baptisia nuttalliana	*Gailardia aestivalis var*	*Pycnanthemum tenuifolium*
Bigelowia nudata	*flarovirens*	*Rhexia mariana*
Boltonia difusa	*Gnaphalium obtusifolium*	*Rhexia lutea*
Boltonia asteroides	*Guara lindhiemeri*	*Rhexia virginica*
Biden aristosa	*Guara longiflora*	*Ruellia humilis*

Bidens mitis	*Helianthus angustifolius*	*Rudbeckia hirta*
Buchnera americana	*Helianthus mollis*	*Rudbeckia grandiflora*
Cicuta maculata	*Heterotheca subaxillaris*	*Rudbeckia texana*
Chamaecrista fasciculata	*Hibiscus mosheutos*	*Sabatia campestris*
Coreopsis tinctoria	*Hibiscus grandiflorus Hype-*	*Sabatia gentianoides*
Coreopsis lanceolata	*ricum nudiflorum*	*Sabatia macrophylla*
Coreopsis linifolia	*Hydrolea ovata*	*Salvia azurea*
Coreopsis tripteris	*Hydrolea unifora*	*Scutellaria integrifolia*
Coreopsis pubescens	*Hyptis alata*	*Shrankia quadrivalis*
Chrysopsis mariana	*Kosteletzkya virginica*	*Silphium asteriscus*
Croton monanthogynus	*Lespedeza capitata*	*Silphium gracile*
Croton capitatus	*Liatris acidota*	*Silphium laciniata*
Dalea candida	*Liatris elegans*	*Solidago nitida*
Desmodium paniculatum	*Liatris spicata*	*Solidago odora*
Echinacea pallida	*Liatris pycnostachya*	*Solidago rugosa*
Erigeron strigusus	*Liatris squarrosa*	*Solidago sempervirens*
Eryngium yuccafolium	*Lobelia appendiculata*	*Strophostyles umbellata*
Eryngium integrifolium	*Lobelia floridana*	*Symphyotrichum dumosum*
Erythrina herbacea	*Lobelia puberula*	*Symphyotrichum concolor*
Eupatorium album	*Manfreda virginica*	*Symphyotrichum lateriflorus*
Eupatorium coelestinum	*Monarda fistulosa*	*Symphyotrichum patens*
Eupatorium hyssopifolium	*Monarda lindhiemeri*	*Symphyotrichum praealtus*
Eupatorium ivifolium	*Monarda punctata*	*Tephrosia onobrychoides*
Eupatorium perfoliatum	*Oxypolis filiformis*	*Teucrium canadense*
Eupatorium rotundifolium	*Passiflora incarnata*	*Vernonia gigantea*
Eupatorium xpinnatifidum	*Penstemon digitalis*	*Vernonia missourica*
Euphorbia corollata	*Pluchea comphorata*	*Vernonia texana*

Literature Cited

Hansen, R. and Stahl, F. (1993). Perennials and Their Habitats. Timber Press. ISBN 10: 0881922226

Hill, J.G and Barone, J.A.(2018). Southeastern Grasslands: Biodiversity, Ecology, and Management. University of Alabama Press. ISBN 978-0-8173-1988-5

Leopold, A. (1949). Sand County Almanac: And Sketches Here and There, Oxford University Press. ISBN-10:019505928X

Noss, R. F. (2018). Fire Ecology of Florida and the Southeastern Coastal Plain. University Press of Florida. Project MUSE muse.jhu.edu/book/59025

Radford, A.E., Ahles, H.E. and Ritchie-Bel, C. (1968). Manual of the Vascular Flora of the Carolinas. University of North Carolina Press. ISBN 978-0-8078-1087-3

Stein, S.B. (1995). Noah's Garden: Restoring the Ecology of our Own Backyards. Houghton Mifflin Harcourt. ISBN 0395709407

Vidrine, M.F. (2010). The Cajun Prairie: A Natural History. Cajun Prairie Habitat Preservation Society. ISBN 9780615368

Wasowski, S. (2002). Gardening with Prairie Plants. University of Minnesota Press. ISBN 978-0-8166-3087-5

Weaner, L. and Christopher, T. (2016). Garden Revolution: How our Landscapes can be a Source of Environmental Change. Workman Publishing. ISBN 9781604696165

Auxin Concentration and Cutting Submersion Duration Impact Survivability and Root Response of Florida Azalea

Jenny B. Ryals[a], Patricia R. Knight, Daryl R. Chastain, Lloyd E. Ryals III, Christine E. H. Coker, Gary R. Bachman, Jim M. DelPrince, Patricia R. Drackett, and Anthony T. Bowden

Coastal Research and Extension Center, South Mississippi Branch Experiment Station
P.O. Box 193, Poplarville, Mississippi 39470, USA

j.ryals@msstate.edu

[a]First Place- Charlie Parkerson Graduate Student Research Paper Competition

Keywords: Auxin application methods, basal quick-dip, Hortus IBA Water Soluble Salts™, IBA, immersion, propagation, *Rhododendron austrinum*.

Abstract

Florida azalea (*Rhododendron austrinum)* is a deciduous azalea native to northern Florida, coastal Alabama, southern Georgia, and southeastern Mississippi. To provide growers with relevant cutting propagation recommendations, the objective of this research was to determine optimal commercial auxin concentration and submersion timing on softwood stem cuttings. The auxin used was Hortus IBA Water Soluble Salts™ (Hortus IBA) at 0, 1000, 2500, 5000, 7500, or 10000 ppm IBA. Submersion durations were 0, 1, 6, 12, or 24 hours with 0 receiving a 5-sec basal quick-dip. Duration of submersion effected root percentage ($P<.0001$), number of roots ($P=0.01$), and average length of the three longest roots ($P=0.04$). There was an interaction between auxin concentration and submersion duration for root quality ($P=0.006$), cutting quality ($P<.0001$) and growth indices ($P<.0001$). Results indicate that softwood Florida azalea cuttings had a better rooting response when treated with a 5-sec basal quick-dip and auxin concentration was 2500.

INTRODUCTION

Deciduous azaleas are described by Dirr (2017) as being "among the most common woody flowering shrubs in the United States, with a myriad of shapes, sizes and flower colors". Although native to many areas of the world, 15 species are native to the Eastern United States (Dirr and Heuser, 2018; Hyatt, 2006). Florida azalea, (*Rhododendron austrinum* (Small) Rehder*)* was discovered by A. W. Chapman before 1865 and reported as a distinct species by John K. Small in 1913 (Galle 1967). Flowers are fragrant and range from pale yellow to orange in color with clusters of 8 to 15 blooms appearing in early spring and generally preceding or coinciding with emergence of dark green leaves that turn a yellow to bronze-orange color in the fall (Dirr, 2017; Hyatt, 2006,). Native range is across northern Florida, coastal Alabama, southern Georgia and southeastern Mississippi (USDA Hardiness zones 6b-10a) (Knight et al., 2005).

Propagation of native deciduous azaleas can be done by seed, cutting, and layering. Due to variability in seed grown azaleas, cutting propagation is preferred (Hyatt, 2006; Sommerville, 1998). However, deciduous azaleas are considered to be a difficult-to-root plant species. According to Dirr and Heuser (2006), slightly firm, 15.2 cm (6 in.) cuttings should be taken from the beginning to end of April. They recommend using a fungicide with 4000 ppm IBA; however, recommended auxin concentrations can vary with different cultivars. Hyatt (2006) recommends taking 5 to 8 cm (2 to 3 inch), softwood cuttings in late May to early June while the plants are actively growing. Bir (1992) took softwood cuttings after the new growth had ceased and treated with 1000-2500 ppm IBA.

Florida azalea has been reported to be easy to propagate according to Galle (1987) and Skinner (1961). Treatment of Florida azalea softwood cuttings with 10000 ppm K-IBA resulted in successful rooting (Knight et al. 2005). Rooting with lower rates of K-IBA occurred, however higher rates increased root number, length, and quality. Knight et al. (2001) also observed that while root ratings, lengths and numbers were similar for cuttings treated with 8000 ppm and 10000 ppm K-IBA, cuttings receiving 10000 ppm rooted 100%.

Developed in the 1940's, long soak immersions have been useful for some hard to root species (Doran 1957, Kroin 2016). Kroin (2016) states that long soak immersions are used to "improve the rooting of hard to root cuttings". Skinner (1937) applied a basal soak treatment from 8 to 48 hours on 45 different plants in the *Ericaceae* L. family, including *Rhododendron* L. Skinner (1937) observed that "some plants rooted satisfactorily without auxin treatment, but most exhibited an increase in average rooting over nontreated cuttings".

The objective of this study was to determine if different submersion durations across a range of IBA concentrations improves rooting response on very soft cuttings of Florida azalea.

MATERIALS AND METHODS

A completely randomized experimental design was utilized with five cuttings per treatment. Florida azalea cuttings were taken on 19 April 2019 from a native population at Crosby Arboretum in Picayune, MS (USDA zone 8b). Cuttings were taken around 6:30 am after a recent rain to ensure they were turgid to aid in reduction of transpiration stress on the cuttings. Using the method that was described by Jenkins (2007), cuttings were taken from tissue soft enough to be removed via pinching. This resulted in variable cutting sizes, however the average length of the cuttings was around 5 cm (2 in.) long. Immediately after pinching, cuttings were placed and stored in a cooler of water until being stuck in the respective treatments (Jenkins 2007). At sticking, cuttings were turgid and showed no signs of wilting or stress.

Based on previous studies, Hortus (Hortus IBA Water Soluble Salts™) was chosen as the auxin. IBA rates were 0, 1000, 2500, 5000, 7500, or 10000 ppm. Submersion durations were 0, 1, 6, 12, or 24 hours with 0 receiving a 5-sec basal quick-dip. Cuttings were wounded then submerged for each time interval, removed, and stuck into 100% perlite substrate in a 6.4 cm (2.5 in.) container. They were then placed under intermitted mist for 4 seconds every 6 minutes during daylight hours. Sixty days after sticking, it was noted that most all cuttings had callused, but formed no roots. At this time, mist intervals were reduced to 2 seconds every 10 minutes and a liquid application of 20-10-20 (Peters® Professional, J.R. Peters, Allentown, PA, USA) general purpose fertilizer at the rate of 50 ppm nitrogen was applied to try and encourage root growth.

Data collected after 120 days included rooting percentage, growth index (new shoots), cutting quality (1-5, with 1=dead and 5=transplant-ready cutting), total root number, average root length (of three longest roots), and root quality (1-5, with 1=no roots and 5=healthy, vigorous root system). Data were analyzed by JMP 14.1.0 Student Edition (SAS Institute, Inc., Cary, NC, USA). All parameters were analyzed by two-way mixed effects ANOVA using standard least squares.

RESULTS

There was an interaction between auxin concentration and submersion duration for root quality ($P=0.0056$), cutting quality ($P<.0001$) and growth indices ($P<.0001$). (Table 1). The 0-hour submersion (5-second quick-dip) resulted in higher cutting and root quality when compared to longer submersion durations. Hortus at a rates of 1000, 2500 or 7500 ppm IBA showed a higher cutting quality when applied at 0-hour submersion duration compared to IBA rates of 2500, 5000, 7500, or 10000 ppm applied at the 6, 12, or 24-hour submersion durations. Root quality was increased when 2500 ppm IBA was applied at the 0-hour submersion duration in comparison to IBA rates of 0, 1000, 2500, 5000, 7500, or 10000 ppm applied at 1, 6, 12, or 24-hour submersion durations. Growth indices also increased when 2500 ppm IBA was applied at the 1-hour immersion timing compared to IBA rates of 1000, 2500, 5000, 7500, or 10000 ppm applied at the 6, 12, or 24-hour immersion timings.

Root percentage ($P<.0001$), number of roots ($P=0.0101$), and average length of the three longest roots ($P=0.0415$) responded negatively to immersion duration treatments except for average length of the three longest roots for cuttings submerged for 6 hours. (Table 2). For all three parameters, the 0-hour immersion timing resulted in better cuttings compared to the other four timing treatments. Auxin rate did not have an effect on these three data parameters.

Table 1. Influence of auxin concentration and immersion duration and on root quality, cutting quality, and growth of Florida azalea.

Treatment	Root quality rating[y]	Cutting quality rating[x]	Growth index[w]
0 hr Immersion Control	1.5def[v]	3.2ab	4.6abc
0 hr Immersion Hortus 1000 ppm	1.6cde	3.6a	5.4ab
0 hr Immersion Hortus 2500 ppm	2.3a	3.6a	4.2abc
0 hr Immersion Hortus 5000 ppm	2.2ab	2.8bc	4.5abc
0 hr Immersion Hortus 7500 ppm	1.7bcde	3.6a	5.4ab
0 hr Immersion Hortus 10 000 ppm	1.6cde	3ab	5abc
1 hr Immersion Control	1.5def	3.2ab	5.3ab
1 hr Immersion Hortus 1000 ppm	1.5def	3.2ab	5.2abc
1 hr Immersion Hortus 2500 ppm	1.5def	3ab	5.7a
1 hr Immersion Hortus 5000 ppm	1.5def	3ab	5.1abc
1 hr Immersion Hortus 7500 ppm	1.4def	2.6bcd	3.8cd
1 hr Immersion Hortus 10 000 ppm	1f	1.4fg	0.8fg
6 hr Immersion Control	1.5def	2.6bcd	4.8abc
6 hr Immersion Hortus 1000 ppm	2.1abc	2.6bcd	4.3abc
6 hr Immersion Hortus 2500 ppm	1.3def	2.8bc	4.2bcd
6 hr Immersion Hortus 5000 ppm	1f	1g	0.0g
6 hr Immersion Hortus 7500 ppm	1f	1g	0.0g
6 hr Immersion Hortus 10 000 ppm	1f	1g	0.0g
12 hr Immersion Control	1.8abcd	3.2ab	5.5ab
12 hr Immersion Hortus 1000 ppm	1.6cde	2.8bc	4.2bc
12 hr Immersion Hortus 2500 ppm	1.2ef	1.8ef	1.9ef
12 hr Immersion Hortus 5000 ppm	1f	1g	0.0g
12 hr Immersion Hortus 7500 ppm	1f	1g	0.0g
12 hr Immersion Hortus 10 000 ppm	1f	1g	0.0g
24 hr Immersion Control	1.3def	2.2cde	2.7de
24 hr Immersion Hortus 1000 ppm	1.3def	2def	2.2ef
24 hr Immersion Hortus 2500 ppm	1f	1g	0.0g
24 hr Immersion Hortus 5000 ppm	1f	1g	0.0g
24 hr Immersion Hortus 7500 ppm	1f	1g	0.0g
24 hr Immersion Hortus 10 000 ppm	1f	1g	0.0g

[y]Root quality (1-5, with 1=no roots and 5=healthy, vigorous root system).

[x]Cutting quality (1-5, with 1=dead and 5=transplant ready cutting).

[w]Growth index=(width1+width2+height)/3.

[v]Means followed by the same letter are similar and not significantly different ($\alpha = 0.05$).

Table 2. Influence of immersion duration on root percentage, number of roots, and average length of the three longest roots of Florida azalea.

Comparison	Rooting (%)	Roots (no.)	(Length of 3 longest roots)/3 (cm)
0 hr Immersion	30a[y]	1.3a	0.6a
1 hr Immersion	0b	0b	0b
6 hr Immersion	10b	0.3b	0.2ab
12 hr Immersion	10b	0.1b	0.2b
24 hr Immersion	0b	0b	0b

[y]Means followed by the same letter are similar and not significantly different ($\alpha = 0.05$).

DISCUSSION

Rooting percentages ranged to 0 to 30% depending on treatment with overall rooting percentages of 9%. In other studies, Florida azalea rooting ranged from 60% to 90% (Knight et al 2001, Knight et al. 2005, Thompson 2018). Difference between rooting results could partially be attributed to other studies using older, less soft cuttings compared to this study. IBA rates for this study were determined based on studies using harder cutting types, but it appears that lower rates may be more beneficial with softwood cuttings. Hortus recommends concentrations not exceeding 400 ppm IBA when using a basal long soak (Kroin 2016). Auxins, if applied in excess, can inhibit plant growth and ultimately cause plant death (Eliasson et al. 1989). Treatments using over 2500 ppm IBA or treatments submerged in higher concentrations for over one hour performed poorly in this study.

Based on the results found in this study, it would appear that young new plant tissue cuttings performed the best overall when subjected to Hortus at a rate of 2500 ppm IBA at a 0-hour immersion (five-second quick-dip).

Literature Cited

Bir, R.E. (1992). Growing and Propagating Showy Native Woody Plants. Univ. North Carolina Press, Chapel Hill, NC.

Dirr, M.A. (2017). Dirr's Encyclopedia of Trees and Shrubs. Timber Press. Portland, Oregon.

Dirr, M.A. and Heuser, C.W. Jr. (2006). The Reference Manual of Woody Plant Propagation. Timber Press. Portland, Oregon.

Doran, W. L. (1957). Propagation of Woody Plants by Cuttings. Massachusetts Agricultural Experiment Station Bulletin, 491.

Eliasson, L., Bertell, G. and Bolander, E. (1989). Inhibitory action of auxin on root elongation not mediated by ethylene. Plant Phys. *91*:310-314.

Galle, F.C. (1967). Native and some introduced azaleas for southern gardens: kinds and culture. Amer. Hort. Mag. *46*:13-23.

Galle, F.C. (1987). Azaleas. Timber Press. Portland, Oregon.

Hyatt, D.W. (2006). Propagation of deciduous azaleas. Comb. Proc. Intl. Plant Prop. Soc. *56*:542-547.

Jenkins, M.Y. (2007). Rooting *Stewartia* and native azaleas using softwood cuttings. Comb. Proc. Intl. Plant Prop. Soc. *57*:646-647.

JMP® Version 14.1.0. SAS Institute Inc., Cary, NC, 1989-2019.

Knight, P.R., File, S.L., and Brzuszek, R.F. (2001.) Impact of hormone concentration for propagation of native Azaleas. Proc. South. Nur. Assoc. Res. Conf. *46*:365-367.

Knight, P.R., Coker, C.E.H., Anderson, J.M., Murchison, D.S, and Watson, C.E. (2005). Mist interval and K-Iba concentration influence rooting of orange and mountain azalea. Native Plants J. *6*:111-117.

Kroin, J. (2016). Hortus Plant Propagation from Cuttings: A Guide to Using Plant Rooting Hormones by Foliar and Basal Methods. Hortus USA. New York, New York.

Skinner, H.T. (1937). Rooting response of azaleas and other ericaceous plants to auxin treatments. Proc. Amer. Soc. Hort. Sci. *35*:830–838.

Skinner, H.T. (1961). Classification of Native American Azaleas. Amer. Rhododendron Soc., Portland, Oregon. Proc. Intl. Rhododendron Conf. 81-86.

Sommerville, E.A. (1998). Propagating native azaleas. J. Amer. Rhododendron Soc., *52*:126-127.

Thompson, P. (2018). Rooting response of deciduous azaleas, *Rhododendron* section *Pentanthera,* stem cuttings to mist regimes and media mixes. (Master's Thesis).

Earth-Kind™ Grapes: Low Input Grapes for the Backyard

Justin Scheiner, Jeff Floyd, Kenny Rollins, Greg Church, Stephen Hudkins, Jeff Raska, Steve George and Michael Cook

Texas A&M AgriLife Extension Service, Dept. of Horticultural Sciences, College Station, Texas 77843, USA

jscheiner@tamu.edu

Keywords: Earth-Kind, grapes, low-input culture, muscadines

Abstract

Earth-Kind™ Grapes research trials were initiated in Texas in 2015 and have since expanded to include seven sites across the state. These research sites represent a wide range of climatic and soils conditions with an aim of identifying well-adapted grapes that are suited for low-input culture. Beyond initial site preparation, no fertilizers are applied and no pesticides are used for the duration of the study. Data collection on vine vigor, nutritional status, yield, fruit composition, and overall health and appearance begins in year three when vines reach bearing age. Data collection is currently underway at four sites and will begin at three more sites in 2020. Observations of nutrient deficiency and incidence of pests and disease have made and will contribute to the overall determination of superior cultivars.

INTRODUCTION

Grapes represent one of few fruit crops that can be successfully grown across the entire state of Texas. However, the availability of well-adapted cultivars is limited, and cultivar selection is the single most important criterion for success. With a wide range of grapes to choose from, homeowners often plant cultivars they recognize, but most require intensive management to be successful. The most well-known grapes used for winemaking (e.g., 'Cabernet Sauvignon', 'Chardonnay', 'Riesling', 'Merlot'), raisins (e.g., 'Thompson Seedless'), table grapes (e.g., 'Flame Seedless', 'Thompson Seedless'), and juice (e.g., 'Concord', 'Niagara') are highly susceptible to a range of fungal diseases as well as the bacterial disease - Pierce's Disease (PD). PD is caused by the bacterium *Xylella fastidiosa* and it is endemic to the U.S. Gulf Coast. PD represents the single greatest limiting factor to grape production in Texas. All European (*Vitis vinifera*) and most hybrid grapes (*Vitis sp.*) are susceptible to PD. However, muscadine grapes (*Vitis rotundifolia*), which are native to Texas, are PD resistant or tolerant.

Due to a different chromosome number (2n = 40), muscadines are largely incompatible with other species of grapes (2n = 38), but they have been greatly improved through breeding. Over one hundred improved muscadine cultivars have been developed including self-fertile cultivars and most recently, seedless cultivars. Muscadines represent a good, disease resistant option for homeowners, but their tolerance of alkaline soils conditions has been largely undocumented. The goal of this research was to identify high quality, disease resistant grapes that can be successfully grown across Texas using Earth-Kind™ trialing methods (Harp et al., 2009). Earth-Kind™ is program started by the Texas A&M AgriLife Extension Service in 1990s to identify adaptable landscape plants through regional trialing, and without the use of pesticides.

METHODS AND MATERIALS

Earth-Kind™ Grapes Trial Sites

Earth-Kind™ Grapes trials have been established in six counties across the state of Texas in areas representing annual precipitation of over 127 cm (>50 in.) to less than 38 cm (<15 in.) (Fig. 1). Soil types across the sites range from acidic and sandy with pH values of 4.5 to alkaline and clay-based with pH values of 8.3. Plot establishment and maintenance is performed by Master Gardener organizations in each county. This gives the Master Gardener volunteers an opportunity to participate in research and learn about grape hands-on growing, while the researcher benefits from the potential to expand the scope and scale of the trials.

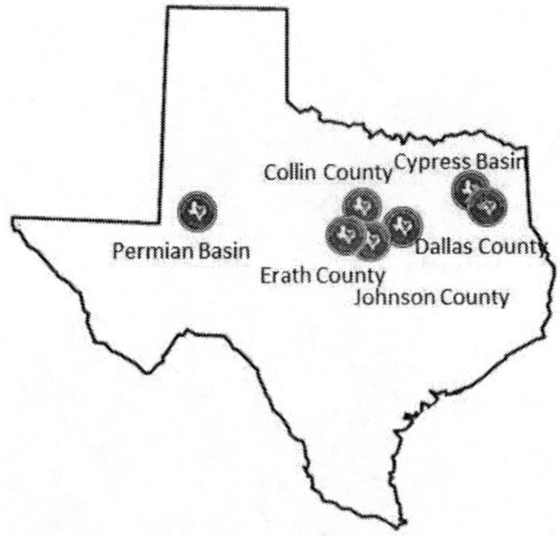

Figure 1. Earth-Kind™ grapes trial locations in 2019.

Plant Material

Nine grape cultivars, representing diverse parentage and plant characteristics, were selected for trialing (Fig. 2). The criteria used to select the grape cultivars included:

1. Pierce's Disease resistant/tolerant
2. Fungal disease resistant/tolerant
3. Self-fertile
4. Not patented
5. Attractive
6. Multiuse

'Champanel': Interspecific hybrid (*V. champinii* x 'Worden') developed by T.V. Munson in 1893. Champanel has medium to large clusters with large black slip-skin berries. Champanel makes excellent jelly and can be used to make a fruity flavored wine (Scheiner, 2019).

'Herbemont': *V. bourquiniana* hybrid developed by Nicholas Herbemont, 1771-1839. Herbemont has medium to large clusters with reddish-brown berries. Herbemont makes a white wine and may be used for jelly typically with purple or black colored grapes added for color.

Figure 2. The nine cultivars selected for trialing included muscadines, interspecific hybrid bunch grapes, and one muscadine/bunch grape hybrid.

'Lake Emerald': Interspecific hybrid ('Pixiola' x 'Golden Muscat') released by the University of Florida in 1954. Lake Emerald has large clusters of small green berries that turn yellow with advanced maturity. Lake Emerald was developed as a white wine grape cultivar and takes on a *V. labrusca* flavor when fully mature (Stover 1954).

'Lomanto': Interspecific hybrid ('Salado' x 'Pense') developed by T.V. Munson in 1902. Lomanto has compact clusters of medium-sized black berries. Lomanto makes an exceptionally deep colored red wine and jelly (Scheiner 2019).

'Miss Blanc': Interspecific hybrid 'Galibert 261-12' x ('Extra' x 'Marguerite' seedling) released by Mississippi State University in 1982. Miss Blanc has medium to large clusters of large white- to green-colored berries. Miss Blanc produces an aromatic white wine and may also be used as a juice grape (Overcash et al., 1982).

'Nesbitt': Self-fertile muscadine (*V. rotundifolia*) cultivar released by North Carolina State University in 1985. Nesbitt has large black berries that ripen over a period of three to four weeks and is primarily used for fresh eating and jelly (Goldy et al., 1985).

Southern Home: Complex hybrid ('Summit' x P-9-15) with muscadine and *V. vinifera* parentage. Southern Home was released by the University of Florida in 1994. It produces medium-sized clusters with small (for a muscadine) black berries that have a crisp texture and a muscadine flavor. Southern Home may be used for fresh eating or winemaking, and has very attractive foliage (Mortensen et al., 1994).

'Tara': Self-fertile muscadine (*V. rotundifolia*) cultivar released by the University of Georgia in 1993. Tara has medium to large bronze berries primarily used for fresh eating and jelly (Lane 1993).

'Triumph': Self-fertile muscadine (*V. rotundifolia*) cultivar released by the University of Georgia in 1980. Triumph has medium to large bronze berries with a pinkish hue that are primarily used for fresh eating and jelly (Lane 1989).

Experimental Design

Initial site preparation followed the methods described by Harp et al. (2009). In brief, existing vegetation in the planting row (1.5 m wide strip) was killed and removed prior to planting. A layer of finished compost 7.5 cm (3 in.) deep was tilled into the native soil (Fig. 3).

Figure 3. Prepared trial site immediately before planting and subsequent mulching.

Following planting, organic mulch was applied and subsequently maintained at a depth of 7.5 cm (3 in.) over the course of the study. Irrigation was applied via surface drip with a single emitter placed near the base of each vine. No additional soil amendments or fertilizers were added for the remainder of the study and no pesticides were applied in order to evaluate disease and pest susceptibility.

Vines were spaced at 2.4 m (8 ft.) between vines and 3 m (10 ft.) between rows. The experimental design was a randomized complete block with four, two-vine replications per cultivar. Vine establishment and training were completed during the first two years of the trial and data collection began in year three. The training system used consisted of a simple high-wire system with a single cordon wire position between 1.5 and 1.8 m (5 ft to 6 ft) high. All vines were cordon-spur pruned to approximately 20 buds per meter (3.3 ft) of cordon.

Data Collection

Data collection began in the third year after planting when vines reached bearing age/size. Data was collected on an individual vine basis to characterize vine vigor, nutritional status, yield, fruit composition, and overall health and appearance as follows:

1. **Vine vigor:** dormant pruning weight
2. **Vine health:** visual symptoms of chlorosis, incidence of pests and disease
3. **Phenology:** bud break, bloom, veraison, harvest
4. **Yield components:** cluster number, cluster weight, berry weight
5. **Basic fruit chemistry:** soluble solids and juice pH
6. **Subjective quality assessment:** flavor and preference

RESULTS AND DISCUSSION

Although data collection formally began in year three, all trial sites produced significant quantities of fruit in year two as a result of high vine vigor in most cultivars (Fig. 4).

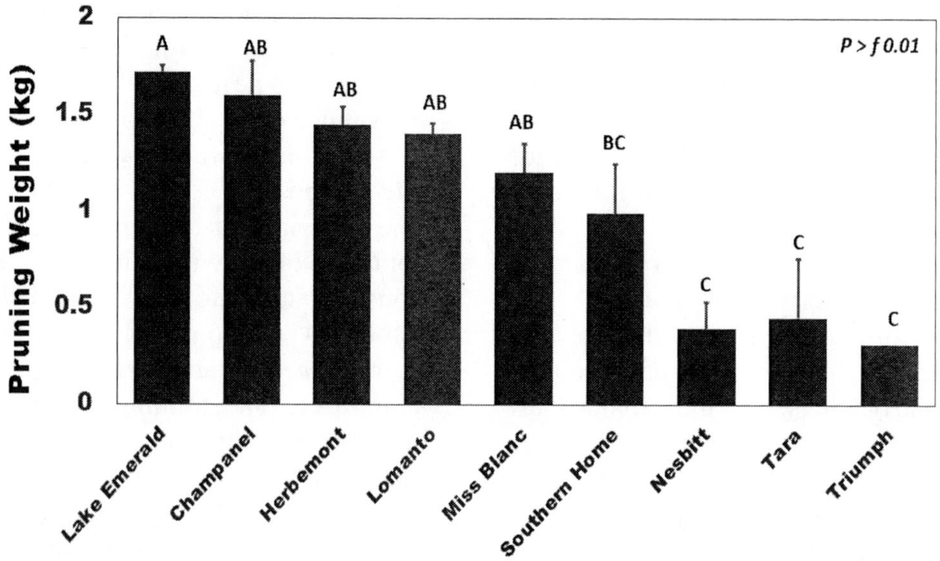

Figure 4. Mean pruning weights of two trial sites in second year after planting. Values are mean ± SE. Means indicated by different letters are significantly different at $p \leq 0.01$, Tukey's HSD.

While site establishment and data collection are still ongoing, the bunch grape cultivars have produced higher pruning weights than the muscadines as a result of greater cane caliper. All bunch grape cultivars under study filled the trellis in their second leaf, but the performance of the muscadines has been more site specific. On sites with moderately alkaline soils (soil pH > 8.0), shoot tip chlorosis has been consistently observed (Fig. 5). Future data collection will quantify shoot chlorosis through visual ratings and attempt to correlate it with vine vigor and fruit yield.

Figure 5. Iron chlorosis observed in 'Southern Home' and 'Nesbitt' on alkaline soil.

Harvest data has been recorded at three sites (Fig. 6). Total vine yield has been challenging to accurately ascertain in the muscadine cultivars due their asynchronous ripening pattern.

The total number of clusters per vine and mean cluster weight has been used to estimate yields and on average, 'Lake Emerald' has been the most productive cultivar with an estimated vine yield ranging from 4.5 to 8.5 kg/vine across sites (data not shown).

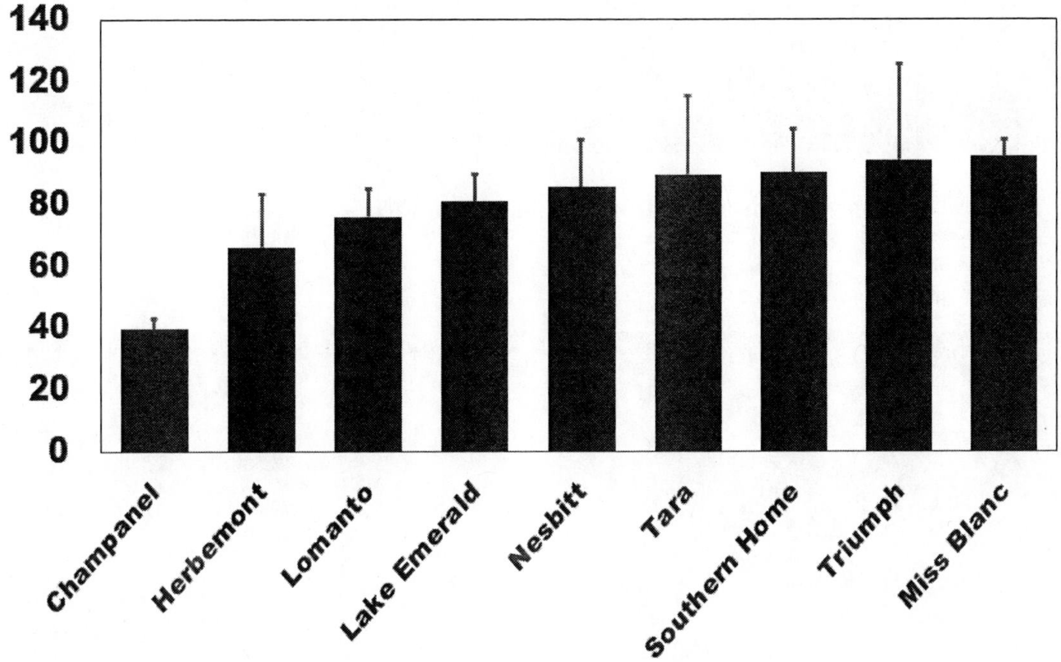

Figure 6. Mean number of clusters per vine in third year after planting for three sites. Values are mean ± SE. Data were not analyzed over sites due to significant cultivar x site interaction.

Observations of the incidence of pest and disease have included variable susceptibility to grape leaffolder moth (*Desmia funeralis*) (Fig. 7), black Rot (*Guignardia bidwelli*) (Fig. 8) and anthracnose (*Elsinoe ampelina*) (Fig. 9). 'Champanel' and 'Lake Emerald' grapes have been infested (≥ 50% of leaves) by leaffolder moth in mid-summer (approximately late July to August) in all sites. Other cultivars are either not infested or only lightly infested. The heavy infestation can result in a rapid increase in exposure of clusters to intense sunlight leading to sunburn, particularly in 'Lake Emerald'.

While controlling this pest can be easily accomplished with insecticides such *Bacillus thuringiensis* (Bt) and carbaryl (Sevin), this may be disadvantageous for the average homeowner.

To date, black rot infections have been observed in 'Lomanto' at two sites. In highly susceptible grape cultivars, black rot can result in total crop loss if left uncontrolled, but the loss observed in Lomanto has been less than 20%. In 2017, anthracnose was observed in Champanel at one site in East Texas. However, fruit loss was estimated to be less than 10%.

Figure 7. Grape leaffolder moth infestation of 'Champanel'.

Figure 8. Black rot infection in 'Lomanto'.

Figure 9. Anthracnose infection in 'Champanel'.

Literature Cited

Goldy, R. G., and Nesbitt, W. B. (1985). 'Nesbitt' muscadine grape. HortScience *20*:777.

Harp, D.A., D.C. Zlesak, G. Hammond, S. George, and W. Mackay. (2009). Earth-Kind rose trials— Identifying the world's strongest, most beautiful landscape roses. In: Zlesak, D.C. (ed.). Roses. Floriculture and Ornamental Biotech. *3*:166–175 (Special Issue 1).

Lane, R.P. (1989). 'Triumph' muscadine grape. HortScience *15*:322.

Lane, R.P. (1993). 'Tara' muscadine grape. HortScience *28*:232.

Mortensen, J. A., Harris, J. W., and Hopkins, D. L. (1994). 'SouthernHome': An interspecific hybrid grape with ornamental value. HortScience *29*: 1371–1372.

Mortensen, J. A., Harris, J. W., and Hopkins, D. L. (1994). 'SouthernHome': An interspecific hybrid grape with ornamental value. HortScience *29*: 1371–1372.

Overcash, J.P., C.P. Hegwood, and B.J. Stojanovic. (1982). 'Miss Blanc' A New Bunch Grape Cultivar. Mississippi Agricultural & Forestry Experiment Station Bulletin 909.

Scheiner, J. (2019). Grape Variety Profile: Champanel. Texas A&M AgriLife Extension Service EHT-120.

Scheiner, J. (2019). Grape Variety Profile: Lomanto. Texas A&M AgriLife Extension Service EHT-112.

Stover, L.H. (1954). The Lake Emerald Grape. University of Florida Agricultural Experiment Station Circular S-68.

How Do you Say That in Swahili? Meeting the Challenges of an International Workforce

Kata Kreß Wallace

Hoffman Nursery, 5520 Bahama Road, Rougemont, North Carolina 27572, USA

katakress@hoffmannursery.com

Keywords: Acclimating personnel, communication, cultural differences, images, learning the system, key translators, language differences, visuals

Summary

Over half the employees at Hoffman Nursery come from five different countries and combined, speak seven different languages. We have worked through Church World Services (CWS), which helps refugees and immigrants find security and opportunity in the U.S. With our international workforce, challenges center around communication and cultural differences. Our employees range from those who speak basic English to those who understand almost none. Most of our international employees work in our production department. Improving communication includes identifying key translators, focusing on basic work language, and using images and visuals. Apart from language, cultural differences can lead to miscommunications and poor performance. Fortunately, we have found ways to manage and work with those differences. Three approaches that have been successful include: 1) support and accommodation, 2) empathy: understanding their perspective, and 3) helping them acclimate and learn the system. Our employees and company benefit by helping them adjust to our workplace and understand our culture.

INTRODUCTION

Before Hoffman Nursery opened in 1986, the owners focused on landscaping. They had very little outside help until 1995 when propagating and growing became their priority. During that early era, the labor pool was made up mostly of Latinos and local workers. Currently, more than half the people who work at Hoffman Nursery were born outside the U.S. We have workers from five foreign countries who speak seven different languages.

Our international worker pool expanded in the early 2000s when the nursery's growth demanded more help than local labor could supply. We contacted Church World Services (CWS), which helps refugees and immigrants find security and opportunity in their new countries. Through the CWS program, the first group of Montagnards—indigenous peoples from the Central Highlands of Vietnam—came to work at the nursery. We also hired a few workers from Burma, but they only stayed a season or two. Around 2010, CWS connected us with refugees from the Republic of Rwanda and the Democratic Republic of the Congo who started working in our Production Department.

With our international workforce, it is no surprise our challenges center around communication and cultural differences. Having worked with the production team for almost 18 months, I would like to share what we've learned. I'll discuss practices that have helped improve communication and touch on areas where we've found management strategies that better fit our groups' cultural contexts.

MAIN CHALLENGES

Most of us have experienced the frustration of giving directions or explaining a concept to an employee only to find they did not understand. It happens when we speak a common language, and when working with non-native English speakers, it is even more likely to happen. Our employees range from those who speak basic English to those who understand almost none. Opportunities for miscommunications happen daily. In addition, each group in our workforce comes from a distinct cultural context. Some are from rural areas with an agrarian lifestyle, while others were professionals or office workers in their home countries. Almost all have adopted the U.S. as their new country - and a way of life that is profoundly different from what they knew.

Most of our international employees work in our production department. Their main jobs involve splitting and potting plants, with some helping with plant retrieval and delivery. In general, their tasks remain the same from day to day. Their jobs are critical for the nursery, so we must ensure they know how to do their jobs accurately and efficiently.

IMPROVING COMMUNICATION

Identify key translators

As we began working with each new group, we relied on hired translators. They came to the nursery and explained the job to the new hires. In 2015 we hired the first people from the Democratic Republic of the Congo. The production supervisor at that time selected one new person for intense training so that the new person could train other employees who spoke the same language (*Kinyarwanda*). With the help of his son, who is fluent in English, and the supervisor, the new employee learned the fine points of splitting and potting. We now have more people who speak *Kinyarwanda*, and they can help train new hires. That first Congolese worker was key to establishing a new cultural group in our workforce.

I have work plans for differing proficiency levels that include a timeline for what they learn and when. Regardless of their English proficiency, we partner all new hires with an experienced team member, especially for the first couple of weeks. Over time I have established a small team of people who help me translate and train new hires. With some languages, I can choose between a couple of people who are proficient in English and very good at their jobs. Sometimes I must work with translators whose skills aren't as sharp but who speak English and their native tongue well. I use that opportunity to help the translator brush up on the basics as well.

Focus on basic work language

During initial training and throughout their early months, we teach new hires key words and phrases. That set includes individual terms for equipment and short sentences that are essential for our workplace. For example, "This needs to be bigger, smaller…", "Can you please go help this person?", "Can you move to the potting line…" We use these words and phrases early and often, repeating them and asking the employee to practice saying them back to us. Repetition and active use facilitate the learning process. This practice also signals they'll be expected to understand and respond to these workplace communications.

Many factors influence language learning, including early experiences, age, frequency of contact with new language, and health. Motivation, economic status, and other circumstances play a role as well. Our groups differ widely in their situations. Many of the Congolese refugees speak multiple languages, while most of the other groups are fluent in one only. Furthermore, the different languages spoken at Hoffman Nursery are from disparate language families. Some spoken languages at the nursery have no written language, making translation even more difficult. We cannot use tools like Google Translate and must use an in-person translator. It is much easier for some employees than others to master the basics. I learned quickly that different methods work with our different groups here at the nursery.

Use images and visuals

In addition to using repetition, we work a lot with images and visuals. Demonstration is a universal language, so one of my earliest strategies was to print pictures, laminate them, and use them to demonstrate concepts. Sometimes images are enough; in other cases, we found a different solution. Here are some examples:

Work basket: When someone new starts in the Production Department, we give them a basket with all the equipment they need: scissors, pruners, apron, gloves, and a laundry bag They also get a sheet of paper that has the words and pictures for those five items. The same information is posted in the production break room. In addition, I ask them to name those items whenever I interact with them. They are responsible for their equipment, and if they need something new, they must ask for it with the English word.

Workgroup assignments: Our Production Team is divided into workgroups that rotate through various departments. They all have a written copy of the weekly schedule, and we also have a display board that uses pictures to show who works where on which day. Pictures of individual employees are arranged on the board so everyone can see with whom to go and where they will work.

Quality control with the board: When we're doing divisions, we need the splits to be consistent and appropriately sized. We routinely produce three liner sizes, and with custom projects we have the capacity to produce up to six different sizes. Changing sizes and switching to different plants can be confusing for the team. Our challenge then, is to communicate quality and consistency. Finding the best solution took time and trying several methods. When the nursery adopted "lean" strategies, they began using color-coded trays to indicate different sizes. They showed the team a sample division every time they switched a plant, so the team would know what size to make. However, the size of the divisions began to deviate over the course of a production run. It might move up a little larger or down a little smaller.

They figured having pictures of the target size would provide a reference and help keep the divisions consistent. That sounded like a great idea but was very time consuming.

Every time production switched to a different plant, someone had to print a picture, cut it out, and distribute it to everyone. That took too much time and still didn't achieve the desired quality.

When I started as Production Supervisor, solving this dilemma was a top priority, and we found a solution. Each team member has a board with a wire strung across it and three hanger clips hung on the wire. The wire clips match the colors of the destination trays for the divisions (Fig. 1). Before we switch production runs to a different plant, we quickly make sample divisions for the boards. We clip the samples on the board and then verbally explain what we'll be making. This has two main advantages: (1) they can see the sample throughout the production run, and (2) it is very easy to communicate which size goes into which colored tray and in which flat size it will be planted.

Figure 1. To keep quality consistent, visual boards show team members the proper size and corresponding tray color.

ADAPTING TO CULTURAL DIFFERENCES

Apart from language, cultural differences can lead to miscommunications and even poor performance. Fortunately, we've found ways to manage and work with those differences. Here are three approaches that have been successful.

Support and accommodate

Many of our team members are refugees and have experienced extreme hardships and traumatic life events. Some have relatives who are still in their home countries where it is not safe, and most cannot return to visit. They deal with situations and pressures that can be extreme. Below is an example of a particularly difficult day and how we dealt with it.

Last April, I came into work early to prepare and be ahead of my team. Several employees were already in the break room eating breakfast. As I passed the room, I saw two women crying, so I stopped to check in with them. On that single day, one's mother had died in Rwanda, the other's nephew had been killed in a war in the Democratic Republic of the Congo, and another man's grandson had been killed in a motorcycle accident in Vietnam. It suddenly became a very different day than I had planned. Unfortunately, days like this are not rare for us.

In these situations, I listen, and if necessary, ask someone to translate to tell me what has happened. It's important to me to treat them with compassion and caring, but I must balance that with getting the work done. One tactic that's helped me keep that balance is to adopt informal "office hours" in the morning before the workday starts. My team knows I'm there from 6:45am to 7:30am, so that's when they come talk to me about personal matters. We've developed the understanding that personal business happens then, and when we start work, their job becomes the priority.

Hoffman Nursery offers funeral leave, but most production team members choose not to take it. There's not usually a funeral in the U.S., and they prefer coming to work to staying at home. As one would expect, they tend to be distracted, and it's not business as usual. I try to accommodate these special situations by giving them choices. Usually, employees are either in a workgroup rotating through the nursery or with me splitting on a single table. When someone on the team has a death in the family or traumatic event, I let them choose where they want to work and whether they prefer to work alone. I believe if we let them choose the work environment that's most comfortable for them, they can still work efficiently given the situation.

Understand their perspective

Those who grew up in the U.S. and similar Western cultures understand deadlines, schedules, and time pressures. That mindset is very foreign to some of our team. Most of them aren't familiar with our traditional job structure of supervisors and employees—they've mainly worked for themselves or lived in a community where they shared the work. Everyone pitches in, and when it's your turn, everyone comes to your place to work. No supervisors, no compensation, no set schedule.

Some of our African employees had jobs more like ours, but they have a hard time with certain aspects of our work culture. They see Americans as always rushing around, being serious, and not having fun at work. My approach is to explain and help them understand our way of working—why we have deadlines, what our goals are, and how we structure our day. I find it easier to explain our approach if I learn more about their culture and their perspective.

My training has included numerous videos, books, and articles on how to manage people. Many advise managing by the numbers and using performance figures for mini-games and motivation. At the nursery, we collect employees' splitting and potting records daily, which I use to monitor individual performance and the overall production process. When I started as Production Supervisor, my great idea was to announce the best splitter and the best potter each morning. It took me about three months to realize two things: (1) my best splitter and potter is so good that the announcement never changed, and (2) it was frustrating and boring for everyone else to hear the same name every day.

To get more variety, I changed to announcing the top three performers. About three months later, I realized numbers for the two best splitters were decreasing. During an already scheduled performance review, which included a translator, I figured out what was happening. The best splitter did not understand the purpose of the announcement and did not like being called out every day. I realized she was changing her split rates so she wouldn't be the fastest. After reading and learning more, I discovered that many Southeast Asian cultures view competition negatively, and individuals prefer not to stand out among their peers. I still collect numbers—it is the best way to chart progress—but if their numbers change, I talk with employees individually.

Help them acclimate and learn the system

Many of our Production Team members struggle with life in the United States. Many experiences are new and unfamiliar, they don't yet speak our language, and tasks we find simple are often complicated for them. Thus, we try to simplify as much as possible.

Performance reviews: We do reviews with every new hire after their first three months, even with seasonal workers. We use simple illustrations and smiley faces to indicate the level of satisfaction with performance areas. Simplifying the form and using the format repeatedly (their first week, first month, three months in, and yearly thereafter) helps employees understand performance reviews and gets the message across.

Workplace changes: When there's a change that affects the Production Team, I go by a "3-time Rule." For example, we used to clean the production workspace at the end of the workday. We switched to cleaning in the morning so the crew could work up to closing time, and the team leaders and I could prep during the morning cleaning. With the 3-time Rule, I first communicate the change in our daily team meeting. Second, I touch base with "translators" from each cultural group immediately afterward. I make sure they understand enough to communicate it to the rest of the team speaking the same language. Third, if I believe it's necessary, I write a short, informal note that explains the change and send it home with all team members. I make sure during the next few days that everyone grasps the new procedure. As soon as I feel confident, they understand the change, they are responsible for following the new instructions.

Requesting time off: We use an online human resources tool for tracking employee paid and unpaid time off. Each full-time employee can request time off through the system using their individual login information. However, no one on my team is computer literate, and many do not read or write English. My solution combines expediency with responsibility. If they bring their login credentials and time off requests to me during my unofficial office hours (6:45am - 7:30am), I help them submit the request online. I help them navigate the system but still hold them responsible for gathering the information and requesting in an appropriate way.

FINDING WHAT WORKS

Even though we simplify our processes and language and give them extra help, we treat our international team members as adults. Our employees and our company benefit by helping them adjust to our workplace and understand our culture. I view working with them as an investment. Time spent teaching will get them up to speed and working independently. There are fewer misunderstandings, even if they don't learn English. That translates to better performance and a happier workplace. The strategies and practices we've shared will not work for everyone, nor will hiring international workers. But with our labor market and our business, we've found an approach that works. We hope some of our ideas and concepts can transfer to your workforce and help improve your operation.

Optimizing N Input Rate for Selected Asian Vegetable Production (luffa and long bean) in Florida

Yanlin Wang[1], Gabriel Maltais-Landry[2], Bala Rathinasabapathi[1], Steven A. Sargent[1] and Guodong Liu[1]

[1]Horticultural Sciences Department, 2550 Hull Road, PO Box 110690, University of Florida, Gainesville, Florida 32611, USA; [2]Soil and Water Sciences Department, McCarty Hall A, University of Florida, Gainesville, Florida 32611, USA

guodong@ufl.edu

Keywords: Asian vegetables, environment, health, *Luffa acutangular,* N input, *Vigna unguiculata* , yield

Abstract

There are more than 40 types of Asian vegetable crops grown in Florida. Recommendations for nitrogen (N) fertilization are being developed by UF/IFAS for Florida growers. Nitrate leaching is common in Florida due to the combination of sandy soils and high precipitation. Growers commonly over-fertilize with N as an insurance for high yield, which can negatively affect the environment and production costs. The objective of this study was to determine the optimal N fertilizer input rates for luffa (*Luffa acutangula*) and long bean (*Vigna unguiculata*) under conventional production in Florida. Our research shows that maximum yield occurred at N rates of 171 and 227 $kg.ha^{-1}$ (150 and 200 $lb.ac^{-1}$), respectively, for luffa and long bean.

INTRODUCTION

Nitrogen (N) is an essential element that is critical for amino acid synthesis and is taken up by plants as ammonium or nitrate. Applying N at the optimal rate is critical to maximize yield and profit. Insufficient N inputs will negatively affect plant growth and production (Wang, Li, and Bai, 2017). Over fertilization with N also inhibits infection and enzyme activity in N fixation (Fahraeus and Ljunggren, 1959; Sloger, 1976). Nitrogen inputs above or below the optimal rate decrease yield (Wang, 2018), and N fertilization can also affect fruit quality and nutritional value. High N inputs decrease the ascorbic acid concentration of fruit (Rajasree & Pillai, 2012), and can increase concentrations of carcinogenic N-nitroso from nitrate, as vitamin C is an inhibitor of this process (Mozafar, 1993). Since most soils in Florida are sandy, N fertilizers are prone to leaching due to low soil organic matter content, low cation exchange capacity, and high infiltration rate (Wetselaar 1962). This can be exacerbated by high summer precipitation.

Luffa (*L. acutangula*) is an annual vine in the cucurbit family that is adapted to hot and humid environments (Herklot, 1972; Purseglove, 1968). Long bean (*Vigna unguiculata* subs. *sesquipedalis*) is an annual vine in the legume family that is a good N-fixer, especially during blooming (Huang, Cai, Lv, Wu, & Cai, 1983). It is resistant to hot, dry, and infertile soils, which makes it more adaptive than most beans (Wang et al., 2013).

Both of these Asian vegetables are increasingly popular in Florida (Liu et al., 2015; Xie, Liu, Li, and Migliaccio, 2016). These new vegetables, along with 40 other species, enhance the diversity of vegetable markets. They are popular in Asian grocery stores in the U.S especially in California, New York and Florida (Xie, Liu, Li, & Migiaccio, 2016). In addition to nutritional benefits for customers, Asian vegetables are also more profitable for growers than traditional commodities. However, there is no official recommendation for N fertilization for Florida growers, even though UF/IFAS publishes a handbook (Agehara et al., 2019) every year on dozens of other vegetables. The objective of this study is to determine the optimal nitrogen fertilizer input rate for luffa and long bean under conventional production in Florida.

MATERIALS AND METHODS

Experimental Design. A field trial was conducted from March to July 2019 at the University of Florida Plant Science Research and Education Unit (PSREU) in Citra, Florida. Four treatments were established in a randomized block design: [0, 114, 171 and 227 $kgN.ha^{-1}$ (0, 100, 150 and 200 $lbs.N.ac^{-1}$)], with four replicate plots per treatment, ten plants per plot. Plants were fertigated with NH_4NO_3, along with 171 $kgP.ha^{-1}$ (150 $lbs.P.ac^{-1}$), applied twice a week via drip irrigation during a 10-week fertigation schedule.

In each plot, luffa (*Luffa acutangula* 'Jiao Gua') and yardlong bean (*Vigna unguiculata* subs. *sesquipedalis* 'Bai-lung') (Tainong Seed Co.) seeds were sown in the field on 19 March 2019. Seedlings were thinned to 10 plants per plot on 5 April 2019. Plants were irrigated twice per day throughout the trial: two hours in the morning and one hour in the afternoon.

Data collection

Three soil samples [0-30.5 cm (0-12 in.)] were collected for nitrate analysis before planting, 63 days and 110 days after planting and analyzed for NO_3-N concentration (Waters Agriculture Laboratory, GA). Thirty leaf samples per plot were collected at three growth stages: 7.6 cm (3-in.) long fruit, first harvest, and mid-harvest to measure leaf greenness (SPAD 502) and petiole sap NO_3 content (Horiba LAQUA NO_3 Meter). From 16 May 2019 until 7 July 2019, crops were harvested every 4-7 days. Long bean pods were weighed, and luffa fruit were counted and weighed for yield. At the end of the season, thirty plants per plot were sampled and separated into blade, petiole and stem, dried at 65 °C for 21 days and analyzed for total N (Waters Agriculture Laboratory, GA).

Data analysis

The data were calculated for treatment means and standard errors in Microsoft Excel. All quantitative data except for yield were analyzed using analysis of variance using SAS software (9.4 TS Level 1M6 X64-DSRV16 platform) and significant differences between means were determined using Tukey's test at $P < 0.05$.

RESULTS

SPAD readings

When luffa fruit were 7.6 cm (3-in.) long, there were no significant differences in SPAD readings among the four N treatments during the three measurement periods (Table 1). With long bean, only on 21 May 2019 were there significant differences using the Tukey Test. The 114 $kgN.ha^{-1}$ (100 lb. $N.ac^{-1}$) had a higher SPAD reading than the 0 or 171 $kgN.ha^{-1}$ (0 or 150 lb. $N.ac^{-1}$).

Table 1. SPAD reading. Data collected on the same day of each crop were analyzed as one set. Data marked by the same letters are not significantly different from each other using Tukey's test at $P < 0.05$.

	Luffa			Long bean		
	05/04/2019	05/21/2019	06/03/2019	05/04/2019	05/21/2019	06/03/2019
0 lb. N/ac	36.6±1.6 a	35.8±0.3 a	42.5±0.7 a	47.4±0.4 a	52.9±0.2 b	55.8±0.6 a
100 lb. N/ac	38.8±0.7 a	38.2±0.2 a	42.8±0.5 a	48.1±0.8 a	57.3±0.5 a	58.4±0.7 a
150 lb. N/ac	37.2±0.5 a	37.8±0.6 a	42.8±0.4 a	50.5±1.7 a	53.1±0.9 b	58.7±1.0 a
200 lb. N/ac	37.3±0.1 a	38.8±0.5 a	39.9±0.7 a	52.2±2.1 a	54.5±1.0 ab	60.4±1.5 a

Petiole sap NO_3

Luffa treated with 171 $kgN.ha^{-1}$ (150 lb. $N.ac^{-1}$) has the highest petiole sap NO_3 throughout the reproductive stage, except for the final sampling period (17 June 2019), when there were no significant treatment differences with the Tukey test (Table 2). With Luffa, there were no significant differences in petiole sap NO_3 among 0, 114 or 227 $kgN.ha^{-1}$ (0, 100 or 200 lb. $N.ac^{-1}$). There was trend in petiole sap NO_3 being lowest during the last measurement period: 17 June 2019. With long bean, 171 $kgN.ha^{-1}$ (150 lb. $N.ac^{-1}$) had the highest petiole sap NO_3 during the first sampling period, compared to the nonfertilized control (Table 2). There were no significant differences among treatments during the middle and final sampling period. As with Luffa, there was trend in declining petiole sap NO_3 during the final sampling period (Table 2). For long bean, there was a trend with plants at 171 $kgN.ha^{-1}$ (150 lb. $N.ac^{-1}$) having the highest petiole sap NO_3 - when fruit was 7.6 cm (3-in.) long and at first harvest. There was a trend with long bean plants grown in control plots having the lowest petiole sap NO_3 value.

Table 2. Petiole sap NO_3 content (ppm). Data collected on the same day of each crop were analyzed as one set. Data marked by the same letters are not significantly different from each other using Tukey's test at $P < 0.05$.

	Luffa			Long bean		
	05/21/2019	06/06/2019	06/17/2019	05/21/2019	06/06/2019	06/17/2019
0 lb. N/ac	1025±103 b	1035±54 b	543±36 a	725±26 b	805±36 a	625±48 a
100 lb. N/ac	733±73 b	733±72 b	637±12 a	1663±210 ab	933±37 a	988±91 a
150lb. N/ac	1800±114 a	2300±239 a	760±99 a	1833±284 a	1220±112 a	823±32 a
200 lb. N/ac	797±25 b	910±85 b	673±48 a	1433±227 ab	950±47 a	1093±92 a

Soil sampling

Background soil NO_3-N content was low in all four blocks (data not shown). After 63 days (2nd soil sample), long bean fertilized with 171 $kgN.ha^{-1}$ (150 lb. $N.ac^{-1}$) had the highest soil NO_3-N; there was a similar trend with Luffa (Table 3). After 110 days (3rd soil sample) long bean receiving 227 $kgN.ha^{-1}$ (200 lb. $N.ac^{-1}$) had the highest soil NO_3-N content, while there was a similar trend in Luffa at the highest N-fertility level (Table 3).

Table 3. NO_3-N (ppm) content for the two in-season soil sampling 63 days and 110 days after planting. The 2nd sampling was during the fertigation treatment, and the 3rd sampling was after the fertigation. Soil NO_3-N content before the experiment began, averaged across four blocks, was 0.1 ppm. Data collected on the same time of each crop were analyzed as one set. Data marked by the same letters are not significantly different from each other using Tukey's test at $P < 0.05$.

	2nd soil sample		3rd soil sample	
	Long bean	Luffa	Long bean	Luffa
0 lb. N/ac	0.24±0.01 c	0.31±0.11 a	1.01±0.35 b	0.46±0.03 a
100 lb. N/ac	1.01±0.18 b	0.68±0.05 a	1.40±0.15 b	0.32±0.07 a
150 lb. N/ac	2.65±0.30 a	1.40±0.30 a	1.74±0.16 ab	0.32±0.06 a
200 lb. N/ac	1.44±0.18 b	0.32±0.08 a	3.09±0.43 a	2.26±0.46 a

Daily fruit number and yield

The highest luffa yields were recorded with 227 $kgN.ha^{-1}$ (200 lb. $N.ac^{-1}$) during first harvest, 114 $kgN.ha^{-1}$ (100 lb. $N.ac^{-1}$) at mid-harvest or 171 $kgN.ha^{-1}$ (150 lb. $N.ac^{-1}$) at the mid to final harvest (Fig. 1a,b). Control plots of both species always had the lowest yields (Fig. 1a, b, c). Before May 28, long bean yields were highest with 171 $kgN.ha^{-1}$ (150 lb.

N.ac^{-1}), whereas after June 11, plants treated with 227 kgN.ha^{-1} (200 lb. N.ac^{-1}), had the highest yield (Fig. 1c). Yields decreased with lower N inputs (Fig. 1c).

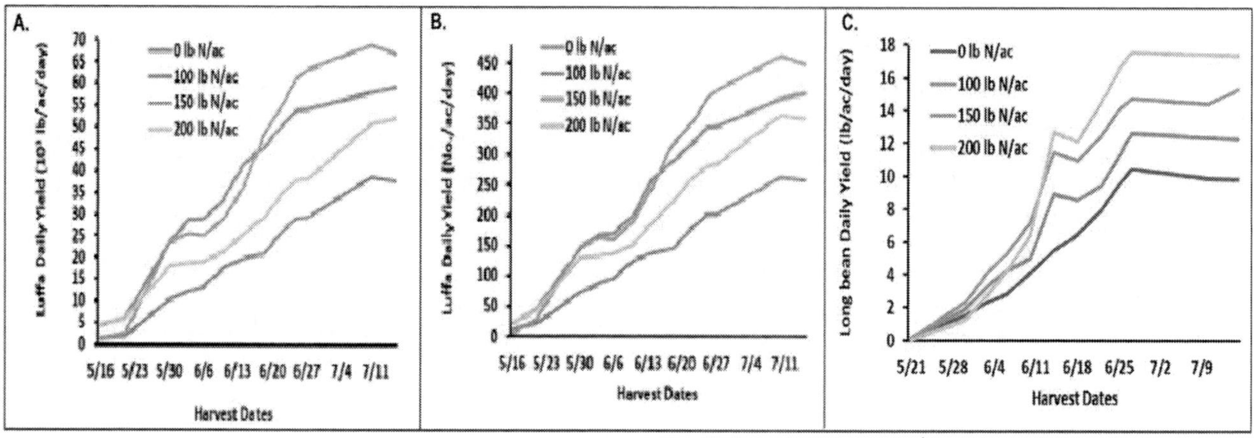

Figure 1. (a) Daily yield (10^3 lb./ac/day) of luffa fruits, (b) number of luffa fruits, and (c) daily yield of long bean pods (lb./ac/day).

Plant tissue N

For stem and petiole tissue, there was no difference in total N concentration among different N input treatments for both luffa and long bean (Fig.2 a, b). There was a trend in highest total N concentration in luffa leaf blades at 171 kgN.ha^{-1} (150 lb. N.ac^{-1}); however there were no significant differences in leaf blade total N among the four N treatments. With long bean, leaf blade N was highest with 171 kgN.ha^{-1} (150 lb. N.ac^{-1}) and lowest at the control / 0 N treatment (Fig. 2b).

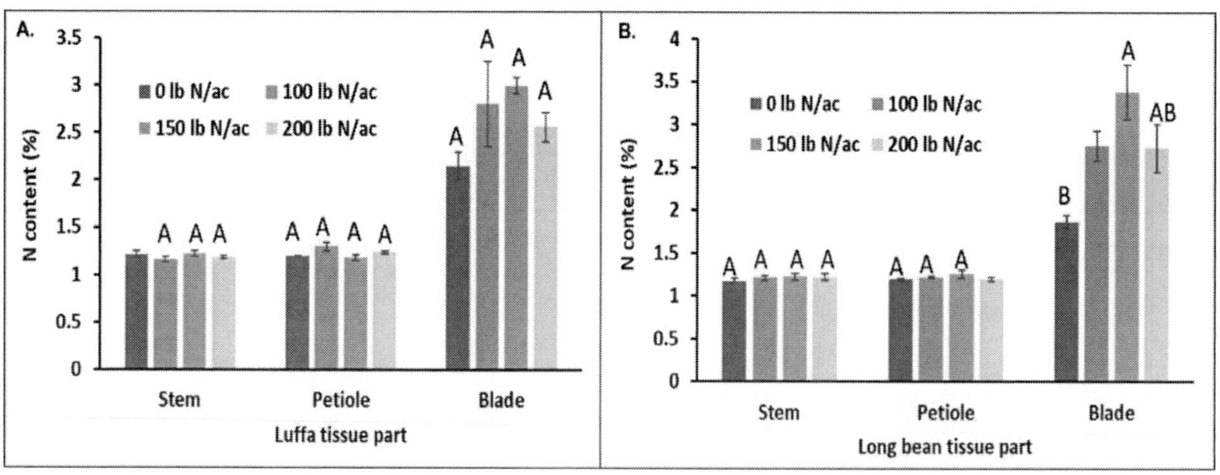

Figure 2. Plant tissue (stem, petiole, blade) analysis of total N content (%) of (a) luffa and (b) long bean plants. The bars represent the mean and standard error from four values each from individual plot. The data are compared within each tissue group. Data marked by the same letters are not significantly different from each other using Tukey's test at $P < 0.05$.

DISCUSSION

In this study we measured the effect of nitrogen fertilization levels by monitoring the leaf chlorophyll content via SPAD readings, tissue N content and soil N levels along with yield of vegetables. Although luffa and yard long bean share similar growth habits, long bean is a known nitrogen fixer while luffa is not. Hence, we hypothesized that these two vegetable crops could require different levels of N input for optimal yields. Luffa had the highest yield and similar N content in the leaf blade with 171 $kgN.ha^{-1}$ (150 lb. $N.ac^{-1}$) compared to 227 $kgN.ha^{-1}$ (200 lb. $N.ac^{-1}$). This is consistent with delayed flowering and yield reported by Dai et al. (2011). Nitrogen fertilizers can alter plant hormones and promote female flowers and reduce male flowers (Omini and Hossan, 1987). Furthermore, N greatly affects dry mater partitioning (Dai et al., 2011), which may reduce fruit production - given the extended co-existence of vegetative and reproductive development in cucurbits (Zhang et al., 2018). Thus, keeping a good balance of N sources and sinks via adequate fertilization is important to reach maximum yield (Dai et al., 2011; Tanemura, Kurashima, Ohtake, Sueyoshi, & Ohyama, 2008).

For long bean, 227 $kgN.ha^{-1}$ (200 lb. $N.ac^{-1}$) had the highest yield, and trends in greater petiole sap NO_3-N, leaf greenness, and soil NO_3 in the mid-late season. This is consistent with research by Wang, Li, and Bai (2017) who reported that in long bean as N fertilizer decreased - leaves were light green and stunted, as were vines; with increased N inputs, leaves were dark green and growth of vines and leaves were vigorous. The 171 $kgN.ha^{-1}$ (150 lb. $N.ac^{-1}$) treatment had the highest petiole sap NO_3 at first leaf sampling, which could be explained by long bean's greater N-fixation capacity during blooming (Huang, Cai, Lv, Wu, and Cai, 1983). Since NO_3 inhibits infection and enzyme activity in N fixation (Fahraeus and Ljunggren, 1959) and NH_4 decreases nodulation and N fixation by inhibiting Rhizobium (Sloger, 1976) -this could account for the lower yield of long bean with 227 $kgN.ha^{-1}$ (200 lb. $N.ac^{-1}$) during early harvests. However, we did not check for root nodules in long bean plants in the field.

The SPAD readings did not show any significant difference between treatments within the measuring stages except for long bean on 21 May 2019. These may due to sampling errors. In our experiment, the SPAD readings were not relevant for identifying N deficiency.

This study provides preliminary guidance to establish N input recommendations for Florida Asian vegetable growers. Greatest yield occurred at 171 and 227 $kgN.ha^{-1}$ (150 and 200 lb. $N.ac^{-1}$), respectively, for luffa and long bean. Future field trials will be conducted during the fall 2019. This is the first study focusing on N inputs in luffa and long bean production in Florida, which enhances our knowledge of best management practice for selected Asian vegetables in Florida. Future steps of this research will focus on the effect of N inputs on fruit quality, nitrogen use efficiency, and rhizobium of long bean.

Literature Cited

Agehara, S., Beuzelin, J., Boyd, N. S., Desaeger, J., Dittmar, P. J., Dufault, N. S., and Zotarelli, L. (2019). Vegetable Production Handbook of Florida, 2019–2020 edition. Gainesville, FL.

Dai, J., Liu, S., Zhang, W., Xu, R., Luo, W., Zhang, S., and Chen, W. (2011). Quantifying the effects of nitrogen on fruit growth and yield of cucumber crop in greenhouses. Scientia Hort, *130*: 551-561.

Fahraeus, G. and Ljunggren, H. (1959). The possible significance of pectic enzymes in root hair infection by nodule bacteria. Physiol. Plant. *12*:145-154.

Herklots G.A.C. (1972). Vegetables in southeast Asia. London: Allen and Unwin, pp. 326- 329, 333-338.

Huang, W., Cai, K., Lv, R., Wu, Y., and Cai, N. (1983). The intercropping green manure in citrus orchard—the evaluation of nitrogen fixation activity and amount of Indian long bean nodule. Journal of Fujian Agriculture and Forestry University (Natural Science Edition), 4.

Liu, G., Wang, Q., Li, Y., Dinkins, D., Wells, B., and Khatri, K. (2015). Long Bean—an Asian vegetable emerging in Florida. EDIS. Retrieved from http://edis.ifas.ufl.edu/hs1268

Mozafar, A. (1993). Nitrogen fertilizers and the amount of vitamins in plants: a review. J. Plant Nutrit. *16*:2479-2506.

Omini M.E., Hossain, M.G. (1987). Modification of sex expression in sponge gourd (*Luffa cylindrica* L.) Roem by mineral nutrient treatments. Genetica *74*: 203-209.

Purseglove, J.W. (1968). Tropical crops – dicotyledons. London: Longman. Pp. 128- 134.

Rajasree, G. and Pillai, G. R. (2012). Effect of nitrogen nutrition on fruit quality and shelf life of cucurbitaceous vegetable bitter gourd. J. Plant Nutrit. *35*: 1139-1153.

Sloger, C. (1976). Biochemistry of N2 fixation. World soybean research. L. D. Hill (Ed.). Danville: The Interstate printers and publishers Inc., Il1. Pp. 125-135.

Tanemura, R., Kurashima, H., Ohtake, N., Sueyoshi, K., and Ohyama, T. (2008). Absorption and translocation of nitrogen in cucumber (*Cucumis sativus* L.) plants using the 15N tracer technique. Soil Sci. Plant Nutrit. *54*: 108-117.

Wang, T., Li, C., and Bai, Z. (2017). The report of nitrogen (N) fertilization for yard long bean trial in Dengjiang County in 2013. Nong Ming Zhi Fu Zhi You, *4*:131-136.

Wetselaar, R. (1962). Nitrate distribution in tropical soils. III. Downward movement and accumulation of nitrate in the subsoil. Plant Soil *16*: 19-31.

Xie, Y., Liu, G., Li, Y., and Migliaccio, K. (2016). Luffa—an Asian Vegetable Emerging in Florida. EDIS. Retrieved from http://edis.ifas.ufl.edu/hs1285

Zhang, B., Li, M., Li, Q., Cao, J., Zhang, C., Zhang, F., and Chen, X. (2018). Accumulation and distribution characteristics of biomass and nitrogen in bitter gourd (*Momordica charantia* L.) under different fertilization strategies. J. Sci Food *98*: 2681-2688.

A Local Native Nursery Perspective on Propagation: One Way

Rick Webb

Louisiana Growers, 63279 Lowery Road, Amite, Louisiana 70422, USA

rick@lanativeplants.com; www.lanativeplants.com

Keywords: Constructed wetlands, ecological services, functional plants, green infrastructure, green industry, native plants, storm water management, sustainability

Summary

There are opportunities for the green industry to produce native plants that perform functional ecological services. Native plant species can be produced for building constructed wetlands for stormwater management, loosely termed "Green Infrastructure". Millennials are looking for functional, sustainably produced landscape plants – not just traditional, containerized shrubs and trees that look "polished" in the landscape. We include a list of suggested plant for green infrastructure management in Louisiana and environs.

INTRODUCTION

This paper will attempt to address how a small wholesale nursery in the U.S. Central Gulf South utilizes the owners' interest and education in the propagation and production of native plants. Louisiana Growers is a wholesale nursery farm specializing in plants native to the inland U.S. Gulf South, from grasses and wildflowers to canopy shade trees. Our products are utilized in enhancing constructed environments. Our customers are now demanding nursery products and services that perform functional ecological services, for example, the supporting of pollinators and birds, the restoration of disturbed lands, the reducing of garden maintenance and the production of edibles.

One involvement in ecological functional service is in the current movement to use nursery crops in the building of constructed wetlands for stormwater management loosely termed "Green Infrastructure".

New? I know it is not. Green Infrastructure has been around since "Adam and Eve". It has existed from gardening for food, to the shading and decorating of our homes - and even to managing storm water. Propagating, growing and functional use of plants is embedded in our civilization. The green industry has traditionally focused on the latest trends - from the latest variegated plant variety to the most floriferous forms as ornamental landscape plants to satisfy consumer demand. Yet we all know our green products and services are much more important than that.

The Green Infrastructure emphasis provides an avenue for our industry to enhance and broaden our status as the Green Industry. Please do not avoid the environmental opportunities of our future. Green is who we are. Green is what we do. And Green is how we should do it.

Stormwater management entails managing surface water runoff, mainly from rainfall or snowmelt. It embodies techniques, structures and land features used to slow, capture, reduce pollution, and to reuse or to timely release runoff water. This is most important in developed areas where impervious surfaces (roads, pavement, buildings, roofs, housing, industrial sites) limit direct absorption of water into the land.

We now think differently from previous designs of channeling and rapid disposal of water and run-off. We need to refocus water capture systems to manage rainfall: slow it, hold it, let it be absorbed or used. The instruments of our Green Industry trade in design, material supply, installation and maintenance are the perfect land shaping tools for addressing Green Infrastructure opportunities.

The following examples are some of the "G.I. Jane's and Joe's" species we produce for wetland construction.

Rhododendron viscosum. Swamp azalea is a deciduous native azalea that has white flowers and reaches a height of 1.5 m (5 ft). Found near, but not in, swamps. It is fragrant and grows well in some shade. Easy to grow, this native azalea blooms in late spring or early summer when the pollinator populations are high and active. It is a butterfly magnet.

Cyrilla racemiflora. Swamp Titi is a small, native semi-evergreen shrub that reaches 3-4.6 m (10-15 ft). It features summer flowers that are fragrant, creamy-white racemes that are 8-15 cm (3-6 in.) long and give the appearance of downward pointed fingers. Summer fruits are yellow-brown. It is suited best in full sun or partial shade and is perfect for naturalized plantings. In cold winters, interior leaves turn yellow and red before falling. Grows naturally on acidic, wetland soils, but can tolerate dry soils once it has become well established. It is highly attractive to bees and birds.

Hymenocallis sp. Spider Lily is a perennial that grows 0.3 – 0.6 m (1-2 ft) tall. Grows in high, sun or part-sun, with showy, white, fragrant flowers in spring and early summer. It grows in poorly drained sites such as ditches, marshes, and bottomlands. It is useful in bog gardens and along the side of ponds and lakes in the landscape.

Schizachyrium scoparium. Little Bluestem is a native ornamental grass with fine-textured foliage that forms dense mounds 46-61 cm (18-24 in.) tall. This grass produces slender, blue-green stems in August and reaches 0.9 m (3-ft) by September. It becomes mottled with mahogany-red with white, shining seed tufts in the fall. Grow well in full sun, also likes sun, part-shade in well-drained soil.

***Juncus effusus* 'Elfin Green'.** Soft Rush is a smaller growing selection of *Juncus* that grows in moist/wet soils, tolerating standing water to 10 cm (4-in.) depth. Adapted to full sun, but tolerates light shade. This plant will perform well in average garden soils as long as it receives consistent irrigation. It may be grown at the edge of a pond or water garden, in boggy areas, among wet gravel, rocks or in several inches of standing water. It delivers an evergreen component to a rain garden.

***Panicum virgatum* 'Short Sassy'.** Switchgrass is a more compact habit and fuller growing variety of the species. It reaches 0.9-

1.5 m (3-5 ft), has fall-flowering spikes with soft golden seed heads, which remain attractive into winter. It tolerates poor soil, from dry conditions to frequently inundated wetlands and full sun for strong growth. You should simply cut it back in early spring.

Callicarpa americana **'Bonner Creek'.** American Beautyberry is a more compact selection of the large loose deciduous shrub to 1.8 m (6-ft). It has medium foliage and small pinkish flowers lead to the eye-catching clusters of bright purple fruits that circle the stems. It has light green foliage that turns yellow in fall. It does well in full sun or under a pine canopy. Very tolerant of dry conditions and the fruit serve as a wildlife moisture source during drought. As with many fire-resistant (pyrogenic) shrubs, it should be severely cut-back every few years.

SOME ADVICE

Based on our experience, we suggest to others interested in Green Infrastructure is to take time to get out in the natural areas of your regional zones. Take note of the plants in their native spaces. Observe the hydrology, geology, light and ecology. Learn from these cues how select species might work in your production system and in your client's projects. These observational spaces start from roadsides to U.S. National Forests. So, get the right pair of water-proof boots - and Get out there!

Addendum

Selected text from the Fall 2019 newsletter of the Louisiana Nursery and Landscape Association reprinted with the author's permission: For those businesses already established in the horticulture industry, millennials have a different consumer mentality. As we progress and gain more purchasing power, we are doing our research and looking for responsible businesses that provide plants that do more than look 'polished' in a landscape. We want plants that have function. We're truly searching for ways we can improve our environment with each purchase. For growers and retailers, that means incorporating more native plants and sustainably grown merchandise into your business model. Native plants are in high demand due to this new generation and the growing force of green infrastructure. The horticulture industry should take note!

By Felice Lavergne. An urban planner specializing in stormwater management, a Master Naturalist, Native Plant Initiative member, and a Millennial!

An Exotic Insect Pest, Crapemyrtle Bark Scale (*Acanthococcus lagerstroemiae*) (Hemiptera: Eriococcidae): Host Range and Acceptance Among 19 Plant Species

Bin Wu[1], Runshi Xie[1], Gary W. Knox[2], Madison Dinkins[3], Hongmin Qin[3], Mengmeng Gu[4]

[1]Department of Horticultural Sciences, Texas A&M University, College Station TX 77843, USA; [2]Department of Environmental Horticulture, University of Florida/IFAS North Florida Research and Education Center, Quincy, FL 32351; [3]Department of Biology, Texas A&M University, College Station 77840, USA; [4]Department of Horticultural Sciences, Texas A&M AgriLife Extension Service, College Station, TX 77843, USA

bin.wu@tamu.edu; mgu@tamu.edu

Keywords: Lagerstroemia, Callicarpa, IPM, pest management

Abstract

Crapemyrtle bark scale (CMBS), *Acanthococcus lagerstroemimae* (Hemiptera: Eriococcidae), is an introduced, sucking pest mainly found on crapemyrtle (*Lagerstroemia* spp. L.). CMBS has been reported in 14 U.S. states. Confirming the range of CMBS's acceptance of different plant species is necessary to estimate its potential in aggravating risks in ecology and losses for the ornamental industry. Hence, in this study, a multiple-choice test was conducted in a greenhouse for 3 months to investigate the host range of CMBS as well as its acceptance among 19 plant species. Based on the current observation record, CMBS's host plants included six *Lagerstroemia* species (*L. caudata*, *L. fauriei* 'Kiowa', *L. indica* 'Dynamite', *L. limii*, *L. speciosa*, and *L. subcostata*) and nine *Callicarpa* species (*C. acuminata*, *C. americana* 'Bok Tower', *C. bodinieri* 'Profusion', *C. dichotoma* 'Issai', *C. japonica* var. *luxurians*, *C. longissima* 'Alba' *C. pilosissima*, *C. randaiensis* and *C. salidifolia*). Evaluation with a one-way ANOVA ($P<0.01$) indicated that CMBS showed significant difference in accepting 19 plant species.

INTRODUCTION

Crapemyrtle bark scale (CMBS), *Acanthococcus* (syn. *Eriococcus*) *lagerstroemiae* (*Kuwana*), is an exotic felt scale in the family Eriococcidae (Wang et al., 2016). *Acanthococcus. lagerstroemiae* seriously threatens growth and development of certain ornamental plants and reduces aesthetic quality because it sucks phloem sap and secrets honeydew. This consequently leads to declining plant health and black sooty mold (Gu et al., 2014). In the summer of 2014, CMBS was initially discovered in Richardson, TX (Merchant et al., 2014). Currently, CMBS has rapidly spread into 14 states including: Alabama, Arkansas, Georgia, Kansas, Louisiana, Mississippi, New Mexico, North Carolina, Oklahoma, South Carolina, Tennessee, Texas, Virginia (EDDMapS, 2019 and Washington (personal communication).

From the perspective of an herbivorous insect, "acceptance of a plant" means continuous food intake and oviposition occurs on the plant after the insect's central nervous system positively evaluates the plant (Schoonhoven et al., 2005). Schoonhoven et al. (2005) viewed "acceptance" as the pivotal decision taken during host-plant selection inasmuch as it directly makes the acquisition of nutrients and oviposition happen consequently. Thus, firstly confirming CMBS's acceptance of certain ornamental and crop plants is of utmost importance to assess potential ecological risks and economic losses Venette et al. (2010). Wang et al. (2019) reported that CMBS infestation was not only confined to crapemyrtle 'Natchez' (*Lagerstroemia indica* × *fauriei* L.) but also occurred on *Lawsonia inermis* L., *Heimia salicifolia* Link, *Punica granatum* L., *Lythrum alatum* Pursh, and *Callicarpa americana* L. 12 weeks after CMBS inoculation. Our previous observations on CMBS infestation test in a greenhouse also indicated that *Ficus carica* L., *Malus pumila* and *Glycine max* could support growth and development of CMBS. However, no systematic study was found investigating CMBS's acceptance of diverse *Lagerstroemia* species, assorted *Callicarpa* species and several *Ficus* species simultaneously.

MATERIALS AND METHODS

Host Plants

Six *Lagerstroemia* species (*L. caudata*, *L. fauriei* 'Kiowa', *L. indica* 'Dynamite', *L. limii*, *L. speciosa*, *L. subcostata*), nine *Callicarpa* species (*C. acuminata*, *C. americana* 'Bok Tower', *C. bodinieri* 'Profusion', *C. dichotoma* 'Issai', *C. japonica* var. *luxurians*, *C. longissima* 'Alba', *C. pilosissima*, *C. randaiensis*, *C. salicifolia*), three *Ficus* species (*F. pumila*, *F. roxburghii*, *F. tikoua*) and *Lythrum californicum*, as shown in Table 1, were tested and compared in the study. All plants were potted up in one-gal pots containing Jolly Gardener Pro-Line C/25 growing mixture (Oldcastle Lawn & Garden Inc, Poland Spring, ME)

Insects

CMBS-infected branches were collected from the nursery pad in the Department of Horticultural Sciences, Texas A&M University in May 2019. White ovisacs in good condition were selected to infest the 19 plant species.

Table 1. A multiple-choice test was used to evaluate 19 plant species as CMBS host candidates.

Species	Series	Section	Family	Native origin
Callicarpa pilosissima	*Callicarpae*	*Callicarpa*	Verbenaceae	China
Callicarpa acuminata	*Callicarpae*	*Callicarpa*	Verbenaceae	Mexico, USA
Callicarpa americana 'Bok Tower'	*Callicarpae*	*Callicarpa*	Verbenaceae	USA
Callicarpa bodinieri 'Profusion'	*Callicarpae*	*Callicarpa*	Verbenaceae	China, Vietnam
Callicarpa dichotoma 'Issai'	*Callicarpae*	*Callicarpa*	Verbenaceae	China, Japan, Korea
Callicarpa japonica var. *Luxurians*	*Verticirimae*	*Callicarpa*	Verbenaceae	China, Japan, Korea
Callicarpa longissima 'Alba'	*Callicarpae*	*Callicarpa*	Verbenaceae	China, Japan, Vietnam
Callicarpa randaiensis	*Verticirimae*	*Callicarpa*	Verbenaceae	China
Callicarpa salicifolia	*Callicarpae*	*Callicarpa*	Verbenaceae	China
Ficus pumila	--	*Sect. Rhizocladus*	Moraceae	East Asia
Ficus roxburghii (auriculata)	--	*Sect. Neomorphe*	Moraceae	Asia
Ficus tikoua	--	*Sect. Ficus*	Moraceae	Asia
Lagerstroemia caudata	--	--	Lythraceae	China
Lagerstroemia fauriei 'Kiowa'	--	--	Lythraceae	Japan
Lagerstroemia indica 'Dynamite'	--	--	Lythraceae	--
Lagerstroemia limii	--	--	Lythraceae	China
Lagerstroemia speciosa	--	--	Lythraceae	China
Lagerstroemia subcostata	--	--	Lythraceae	Japan, China, Philippines
Lythrum californicum	--	--	Lythraceae	Mexico, USA

Greenhouse Experiment

The multiple-choice experiment was conducted in Texas A&M University (TAMU) Department of Horticultural Sciences Greenhouse. Each plant was tied with CMBS-infested branches containing five fresh white ovisacs, and then one set of the 19 species was arranged in a cage with no canopy overlap. The cage (75cm × 50cm × 40cm) was made of PVC pipe frames, covered and enclosed with handmade mesh netting, and a 30cm-long zipper was added to the front mesh panel to water and observe plants easily. There were a total three sets of 19 species in three cages (three replicates) and the cages were placed separately on different benches in the greenhouse at 25 ± 5°C and 50 ± 10%RH.

Statistical Analysis

The plants in cages were examined biweekly starting three weeks after the CMBS inoculation. To confirm host range and evaluate CMBS acceptance among these plant species, the number of CMBS pupae (recognized by white narrowly spindle-shaped cocoons) and gravid females (recognized by white oval ovisacs) on each species as key parameters were counted and compared for 3 months. Plant species that supported several generations of CMBS were defined as host plants (Sands and Van Driesche, 1999), and variation in average amount of pupae and gravid females on each species under the same family was relevant to the insect's different feeding and oviposition acceptance. Data on average amount of pupae as well as gravid females were analyzed separately by One-way ANOVA using JMP Pro 14 (SAS Institute, Cary, NC).

RESULTS

Host Range Confirmation

Based on the current observation record (Figure 1), an increasing number of white pupae and white ovisacs were largely seen on all of the six *Lagerstroemia* species (*L. caudata*, *L. fauriei* 'Kiowa', *L. indica* 'Dynamite', *L. limii*, *L. speciosa*, and *L. subcostata*) as well as seven of the nine *Callicarpa* species (*C. americana* 'Bok Tower', *C. dichotoma* 'Issai', *C. japonica* var. *luxurians*, *C. longissima* 'Alba', *C. pilosissima*, *C. randaiensis*, and *C. salicifolia*). The average amount of pupae in *C. dichotoma* 'Issai', for instance, increased to 217 on August 22nd, while average gravid females increased to 124.

There was a remarkable increase in *L. limii* with 756 pupae and 377 gravid females. In marked contrast, only 9 pupae and 4 gravid females emerged on *C. acuminata*; 5 pupae and 2 gravid females were observed on *C. bodinieri* 'Profusion'. Just one pupa formed on each *F. tikoua* plant 3 months after CMBS inoculation. The result indicated that the six *Lagerstroemia* species as well as the nine *Callicarpa* species mentioned above were CMBS host plants.

CMBS Feeding and Oviposition Acceptance among 19 Plant Species

Variation in average number of pupae and gravid females on each species was relevant to the insect's different acceptance - including feeding and oviposition. On 22 August 2019, the data was taken and analyzed with a one-way ANOVA running JMP Pro 14 - to determine the plant species effect on number of pupae and gravid females. The One-way analysis result demonstrated significant difference in the average amount of pupae (F=2.9606, DF=18, 37; P<0.01) and gravid females (F=6.1820, DF=18, 37; P < 0.001) among the 19 species. Thus, CMBS was accepted differently among the 19 plant species. The largest number of pupae was observed on *L. subcostata* (826), followed by *L. limii* (735), *C. dichotoma* 'Issai' (217) and *L. fauriei* 'Kiowa' (184) (Fig. 1). The largest number of gravid females occurred on *L. limii* (377), followed by *L. subcostata* (302), *L. fauriei* 'Kiowa' (200) and *C. dichotoma* 'Issai' (124).

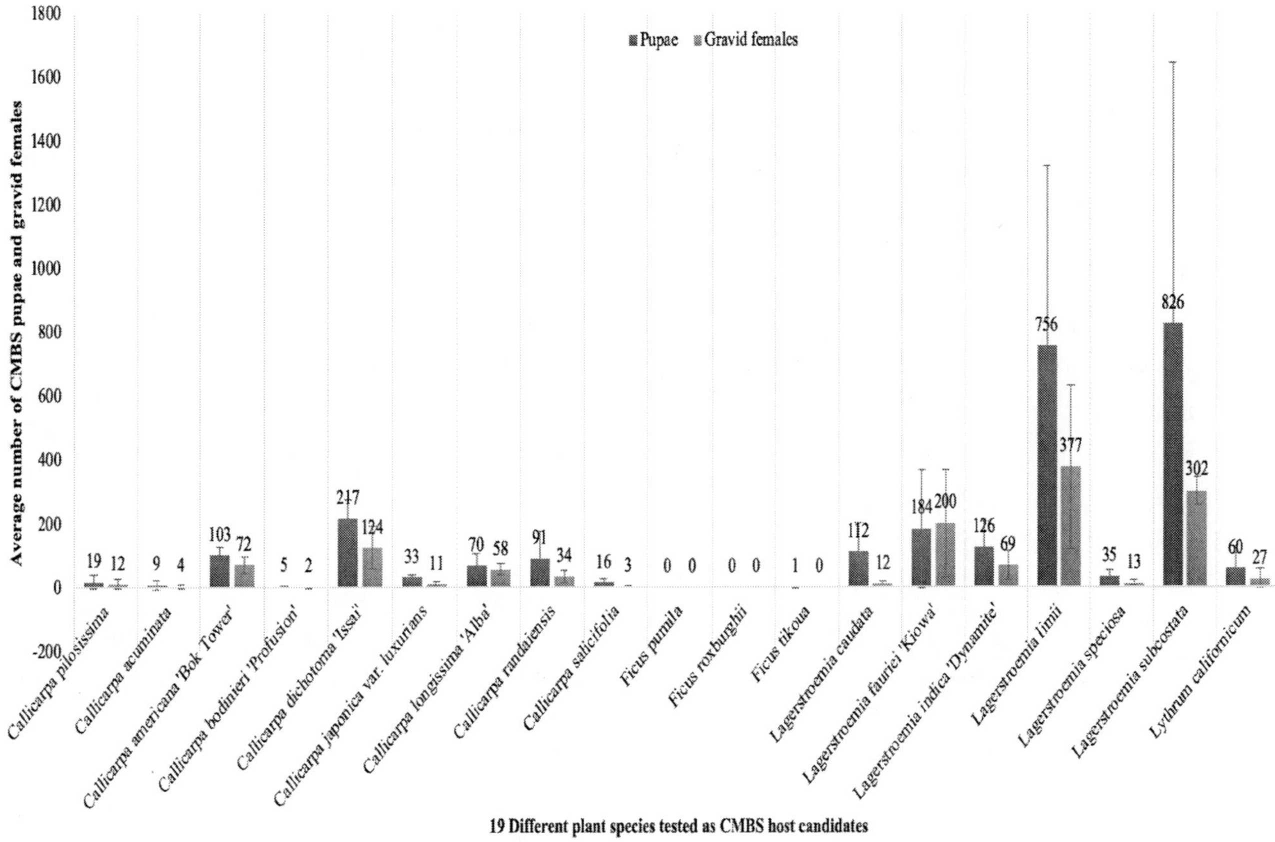

Figure 1. Average amount of pupae and gravid females on *Callicarpa*, *Ficus* and *Lagerstroemia* and *Lythrum* plant species (22 August 2019).

DISCUSSION

Our study shows that CMBS was able to complete its life cycle and largely proliferate on all of the six *Lagerstroemia* species and the nine *Callicarpa* species. There were noted differences in CMBS accepting the evaluated 19 plant species. The underlying cause of CMBS's differential acceptance of these plant species is worthy of further study. Varying acceptance could be caused by differences in physical features (trichomes, tissue thickness and wax microstructure) and/or chemical compounds (volatiles, carbohydrates, amino acids and some plant secondary metabolites) among the different species and cultivars (Schoonhoven et al., 2005). For example, CMBS fed and laid eggs on *F. carica* but cocoons or white ovisacs were hardly seen on *F. pumila* and *F. roxburghii*. The reason why *F. roxburghii* did not support growth and development of CMBS was probably because its surface texture was too thick for CMBS to probe. Future research will evaluate and compare differences in surface morphology and select primary plant metabolites among these three species. Determining parameters affecting plant acceptance can provide insight for improved pest management.

Literature Cited

EDDMapS. (2019). Early detection and distribution mapping system. The University of Georgia - Center for Invasive Species and Ecosystem Health. http://www.eddmaps.org last accessed August 31, 2019.

Gu, M., Merchant, M., Robbins, J., and Hopkins, J. (2014). Crape myrtle bark scale: A new exotic pest. Texas A&M AgriLife Ext. Service. EHT 49.

Merchant, M.E., Gu, M., Robbins, J., Vafaie, E., Barr, N., Tripodi, A.D., and Szalanski, A.L. (2014). Discovery and spread of *Eriococcus lagerstroemiae* Kuwana (Hemiptera: Eriococcidae), a new invasive pest of crape myrtle, *Lagerstroemia* spp. *https://bugwoodcloud.org › pdf › ESAPosterDiscovAndSpread2014*

Sands, D., and Van Driesche, R. (1999). Evaluating the host range of agents for biological control of arthropods: rationale, methodology and interpretation, Host Specificity Testing of Exotic Arthropod Biological Control Agents: The Biological Basis for Improvement in Safety. Proceedings of the Xth International Symposium on Biological Control of Weeds. pp. 69-83.

Schoonhoven L.M., Van Loon, B., van Loon, J.J., and Dicke, M. (2006). Insect-Plant Biology. Oxford University Press. 2nd Ed. ISBN: 9780198525950

Venette, R.C., Kriticos, D.J., Magarey, R.D., Koch, F.H., Baker, R.H., Worner, S.P., Gómez Raboteaux, N.N., McKenney, D.W., Dobesberger, E.J., and Yemshanov D. (2010). Pest risk maps for invasive alien species: a roadmap for improvement. BioScience *60*:349-362.

Wang, Z., Chen, Y., Gu, M., Vafaie, E., Merchant, M., and Diaz, R. (2016). Crapemyrtle Bark scale: a new threat for crapemyrtles, a popular landscape plant in the U.S. Insects *7*: 78.

Wang, Z., Chen, Y., and Diaz, R. (2019). Temperature-dependent development and host range of crapemyrtle bark scale, *Acanthococcus lagerstroemiae* (Kuwana) (Hemiptera: Eriococcidae). Florida Entomologist *102*:181-186.

Developmental Stages and Host Range Confirmation of Crapemyrtle Bark Scale (*Acanthococcus lagerstromiae*)

Runshi Xie[1], Bin Wu[1], Gary W. Knox[2], Hongmin Qin[3], Mengmeng Gu[4]

[1]Department of Horticultural Sciences, Texas A&M University, College Station, Texas 77843, USA; [2]Department of Environmental Horticulture, University of Florida/IFAS North Florida Research and Education Center, Quincy, Florida 32351; [3]Department of Biological Sciences, Texas A&M University, College Station, Texas 77840, USA [4]Department of Horticultural Sciences, Texas A&M AgriLife Extension Service, College Station, Texas 77843, USA

fushe001@tamu.edu; mgu@tamu.edu

Keywords: Crapemyrtle, crapemyrtle bark scale, developmental stages, host range

Abstract

Crapemyrtle Bark Scale [(CMBS) *Acanthococcus lagerstroemiae*] is an exotic pest species that is causing aesthetic and economic damages to crapemyrtles and posing potential threats to other horticultural crops. Although previous studies had reported on the life history of this insect pest, much information is still unclear or missing in terms of the insect developmental process and its ability to infest alternative hosts. In this study, two different insect rearing methods were utilized to obtain a detailed documentation of each developmental stages of CMBS. Infestations of CMBS were confirmed on apple (*Malus domestica*), *Malus angustifolia*, *Chaenomeles speciosa*, *Disopyros rhombifolia*, *Heimia salicifolia* and *Lagerstroemia* 'Spiced Plum'. No sign of CMBS infestation was observed on *Rubus* 'Arapaho', *Rubus* 'Navaho', *Rubus idaeus* 'Dorman Red', *Rubus fruticosus*, *Buxus microphylla* var. *koreana x B. sempervirens*, *Buxus harlandii*, *and Diospyros virginiana* during a 14-week experiment period.

INTRODUCTION

Crapemyrtle Bark Scale [(CMBS) *Acanthococcus lagerstroemiae*] originates from East Asia and has been reported to infest important horticultural crops, such as crapemyrtles and pomegranate, in Japan, Korea, India, and China (Egolf and Andrick, 1978; Gu et al., 2014a). Since its first discovery in 2004 in Richardson, TX, the infestation of this scale insect has been confirmed in New Mexico, Oklahoma, Texas, Arkansas, Louisiana, Tennessee, Mississippi, Alabama, Georgia, North Carolina, Virginia (EDDMapS, 2019), and Washington (Wang, et al. 2016). As an exotic pest species, its wide spread is also posing threats to other potential alternative host crops, especially native plant species in the U.S. The most common host of CMBS, crapemyrtles (*Lagerstroemia spp.*), have been utilized as a very important landscape tree in the southern U.S., which generates around $66 million of wholesale market value per year (Wang et al., 2016). The wide use of crapemyrtle could be partly due to its ease to propagate, grow and manage. However, infestation of CMBS can greatly affect performance of crapemyrtle, causing aesthetic and economic damages.

According to previous studies and observations, the development and biology of crapemyrtle bark scale is categorized as incomplete metamorphosis, with a life cycle of 56 to 83 days (Jiang and Xu, 1998). The number of generations of the scale ranges from two to four depending on the climate zones (Gu et al., 2014b; He et al., 2008; Jiang and Xu, 1998). In the U.S., up to four generations were observed in the field in Dallas, TX (Gu, et al., 2014b). Under greenhouse condition, the development of males involves five nymph stages, and three nymph stages were found in females (Wang, et al., 2015). However, the underlying mechanism of the insect development, including the defining differentiation and the lengths (the numbers of days within a life cycle) of each developmental stage of CMBS - remains elusive. The difficult in defining the developmental stages is largely due to the small size (ranging from 0.3 mm to 3 mm depending on the age and sex) of the nymph, which constitutes a large portion of its entire life cycle. The better understanding of CMBS life cycle is important for understanding its entire developmental process. This is crucial for investigating the factors that affect insect mortality in terms of effectively controlling this pest.

In this study, two experiments have been designed to investigate the developmental stages of CMBS, and its host ranges - in order to develop effective Integrated Pest Management (IPM) programs for controlling this pest.

MATERIALS AND METHODS

Experiment 1: Developmental stages of crapemyrtle bark scale:

Insect source and plant material

Branches/twigs infested with CMBS were collected from crapemyrtle trees on campus (Texas A&M University, College Station, TX), and stored in zip-lock bags under constant temperature (25°C). The gravid females collected from infested plants were used for the experiments immediately, or within one or two days after the collection.

Construction of feeding chamber

Feeding chambers were constructed with small petri dishes and clear plastic food wrap. Around half of petri dish was wrapped by clear plastic to create space for medium. Water agar was poured into the bottom of the petri dish at a 70-degree angle to fill around one third portion of the petri dish. One piece of detached stem from *Lagerstroemia fauriei* 'Kiowa' (food source) was stuck in agar medium to stay turgid (Figure 1).

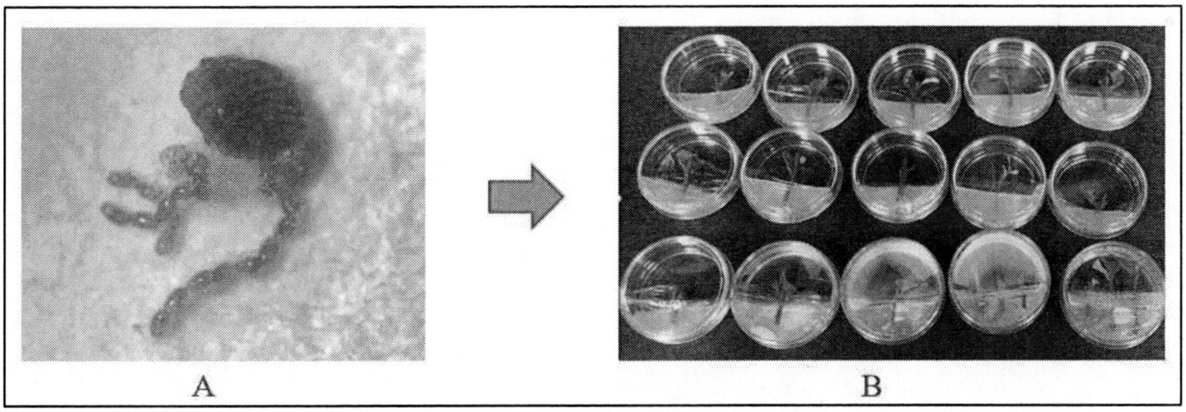

Figure 1. (A) Female CMBS producing eggs within 24 hr; (B) CMBS feeding chambers containing detached plant stems (from *L. fauriei* 'Kiowa') as food source and agar medium.

Collecting CMBS eggs

Gravid female was obtained by gently opening white ovisac on infested branches using a fine pin. All existing eggs were removed, and the gravid female were transferred onto a wet filter paper placed in a petri dish. All eggs laid by gravid female were collected in the next day (after 24 hours) and used as day-1 eggs in the rearing experiment (Fig. 1). Day-1 eggs were collected and kept under 25°C until hatched.

Insect rearing experiment

Detached crapemyrtle stems with leaves and bud nodes was used as food sources and placed in the feeding chamber as described above. One or two newly hatched (within one day) crawlers/nymphs were transferred by using a fine brush on detached stem with separate feeding chambers. Daily observation was made to record the settling status for nymphs. Feeding chambers with nymphs that failed to settle on plant were discarded when the mortality or escaping were confirmed.

Data collection

Egg incubation time and the number of settling (feeding) nymphs were first recorded to evaluate to mortality rate of insects at the early nymph stages. Daily observation continued as nymphs start feeding, and the duration of each developmental stage (including nymph stages, pupa, and adult stages) were recorded. When a male reached the adult stage, it will be transferred to pair with a female for mating in order to complete the life cycle of the female. Fecundity data (the number of eggs that an adult female produces), and longevity (the number of days a female lives) were recorded as gravid females completed their life cycles. The developmental stages of both male and female CMBS were determined by the number of times the nymphs molted, which was determined by keeping track of the exuviates.

Experiment 2: Host Range Confirmation of crapemyrtle bark scale:

Insect source and plant material

The branches/twigs infested with CMBS were collected from crapemyrtle trees on campus (Texas A&M University, College Station, TX), and stored in zip-lock bags under constant temperature (25° C). The CMBS collected from infested plants were used immediately for experiments, or within one or two days after the collection. Fourteen plant species and cultivars from seven genera, including *Malus domestica* 'Fuji', *M. domestica* 'Red Delicious', *M. angustifolia*, *Rubus* 'Arapaho', *R.* 'Navaho', *R. idaeus*, 'Dorman Red', *R. fruticosus*, *Buxus harlandii*, *B. microphylla* var. speciosa, *Chaenomeles speciosa*, *Diospyros rhombifolia*, *D. virginiana*, *Heimia salicifolia*, and *Lagerstroemia* 'Spiced Plum', planted in 1-gal pots were used in the experiment.

Construction of plant chamber

Plant chambers [1.3 m x 1.3 m x 1.2 m (4.2 ft × 4.2 ft × 3.8 ft)] were constructed using PVC pipes as frame and enclosed by mesh fabric sown together with a sewing machine. Zippers were sown onto mesh fabric to form an opening for transporting the plants (Fig. 2).

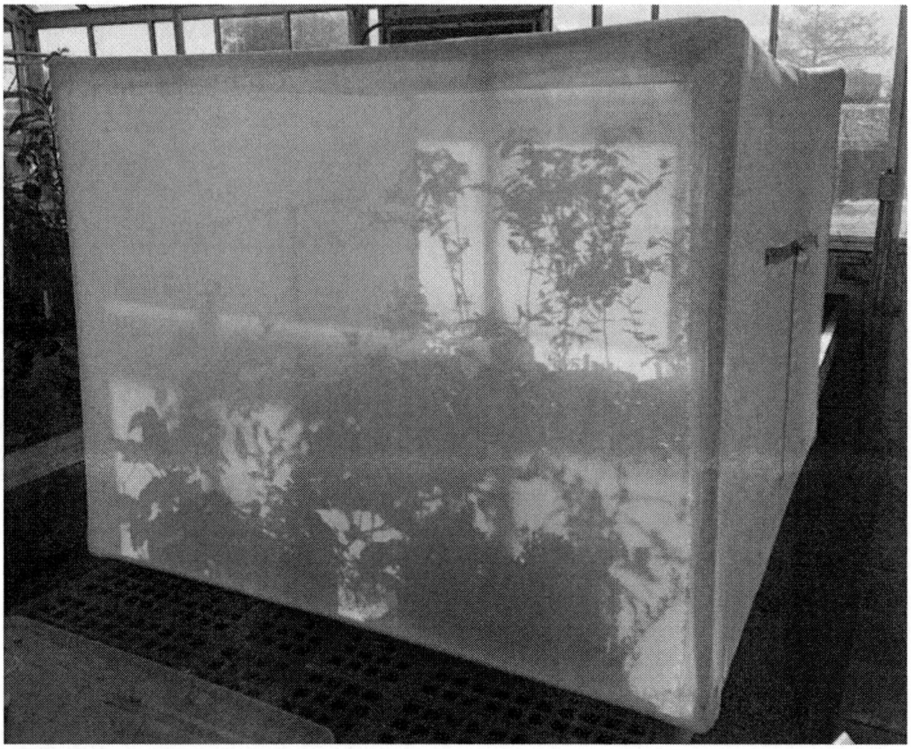

Figure 2. Plant chambers [1.3 m x 1.3 m x 1.2 m (4.2 ft × 4.2 ft × 3.8 ft)] containing fourteen plant species and cultivars.

Host range confirmation

All plants were placed inside the plant chambers before inoculating with CMBS with one set of 14 plant species and cultivars per chamber. One infested crapemyrtle branch with 5 fresh ovisacs was attached each tested plant. To ensure successful inoculation of CMBS, plants were inoculated twice on 13 May and 15 June 2019, respectively. The plants were observed weekly from May to August 2019, and the numbers of male pupae and gravid females were recorded biweekly.

RESULTS AND DISCUSSION

Developmental stages of crapemyrtle bark scale

Observations performed in this study confirmed that the developmental stages of adult males consisted of an egg, two nymphal stages (1st and 2nd instar), three different stages of pupa, and the winged adult stage. The development of a female entails four major stages: egg, two nymph stages (1st and 2nd instar) and an adult stage. The egg incubation time for both males and females is a little over 12 days under 25 °C. For males, the average duration for 1st instar, 2nd instar, pupa1, pupa2, pupa3, and adult period is around 14, 9, 3, 2, 4, and 4 days, respectively. For females, the average duration for 1st instar, 2nd instar, and adult period is around 13, 13, and 31 days, respectively. The males had shorter life span of around one month compared to the females, which lived much longer and produced eggs for up to 57 days. This might be due to the sole purpose of males for mating and indicating their specific role within the sexual reproduction of CMBS.

The rearing experiment was successful in supporting the development CMBS. Insects were able to successfully complete their life cycle under the experimental conditions of the study. The feeding chamber constructed for rearing CMBS allowed for direct monitoring of the entire developmental process of CMBS, which would normally be difficult to achieve under field conditions. Hence, this novel rearing method can be used to further compare the effects of different plant hosts on CMBS. Experiments can be set up to extract data for life table analysis, which is one fundamental tool in the field of entomology that builds up the knowledge foundation and provides insight of the population dynamics of arthropods of interest.

Host Range Confirmation of Crapemyrtle Bark Scale

During the 14 weeks (May to August 2019) of the experiment, 7 out of 14 tested plant species and cultivars (*Malus domestica* 'Fuji', *Malus domestica* 'Red Delicious', *Malus angustifolia*, *Chaenomeles speciosa*, *Diospyros rhombifolia*, *Heimia salicifolia*, and *Lagerstroemia* 'Spiced Plum') were confirmed to have different levels of CMBS infestation (Figs. 3, 4 and 5). The control plant *Lagerstroemia* 'Spiced Plum' had the highest number of male pupae and gravid females throughout the 14-week period. Population numbers were highest durng week 14, with 414.7 pupae and 100.3 gravid females (Fig. 4 and 5).

The number of CMBS pupa and gravid females were much lower on *M. angustifolia*, and *Diospyros rhombifolia,* respectively. Gravid females were observed on 'Fuji' (2 gravid females), 'Red Delicious' (0.7 gravid females), *C. speciosa* (0.7 gravid females), and *H. salicifolia* (9.7 gravid females) (Fig. 4 and 5). The lower number on other plant species, indicates that CMBS showed greater feeding preference toward *Lagerstroemia*. However, our study confirmed the ability of CMBS to infest alternative plant species, especially plants within genera of *Malus*, *Chaenomeles*, and *Heimia*.

Figure 3. Infestation of (A-B) *A. lagerstromiae* (red arrows) found on *Malus domestica* 'Fuji', (C), *Malus domestica* 'Red Delicious', (D), *Malus angustifolia,* (E-G), *Chaenomeles speciose,* (H-J) *Heimia salicifoliam,* and (K-L) *Lagerstroemia* 'Spiced Plum' recorded from May to August 2019.

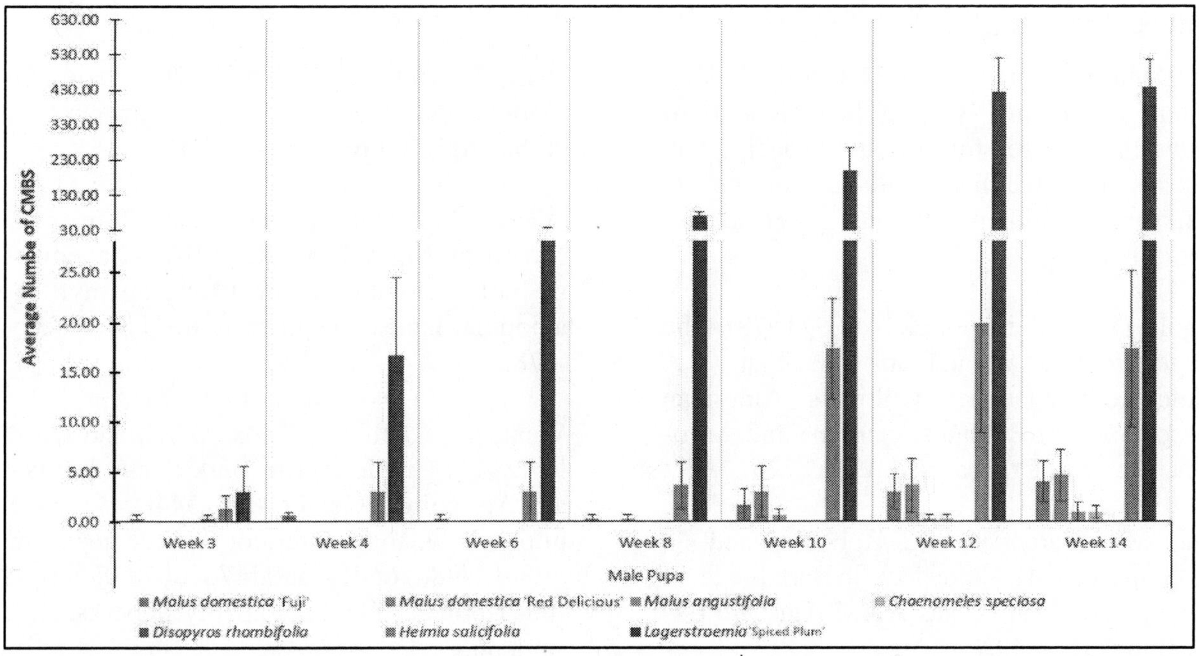

Figure 4. Average number of male pupae of *A. lagerstromiae* on *Malus domestica* 'Fuji', *Malus domestica* 'Red Delicious', *Malus angustifolia*, *Chaenomeles speciosa*, *Diospyros rhombifolia*, *Heimia salicifoliam*, and *Lagerstroemia* 'Spiced Plum' recorded from May to August 2019.

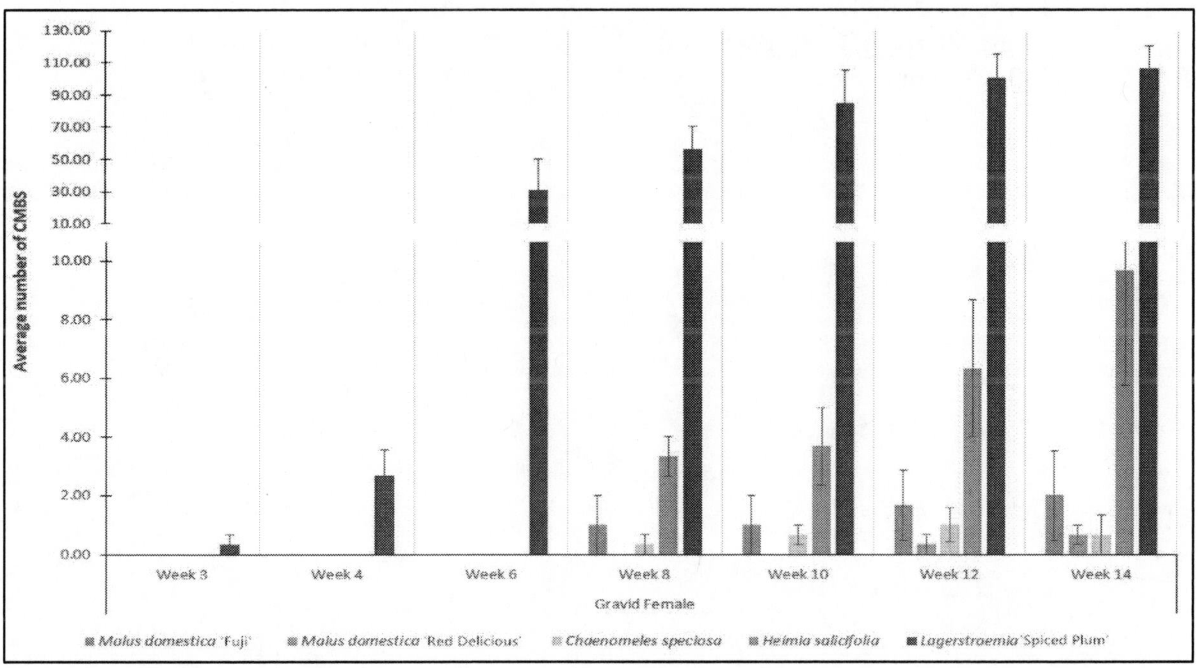

Figure 5. Average number of gravid females of *A. lagerstromiae* on *Malus domestica* 'Fuji', *Malus domestica* 'Red Delicious', *Malus angustifolia*, *Chaenomeles speciosa*, *Diospyros rhombifolia*, *Heimia salicifoliam*, and *Lagerstroemia* 'Spiced Plum' recorded from May to August 2019.

Literature Cited

EDDMapS. (2019). Early Detection & Distribution Mapping System. The University of Georgia - Center for Invasive Species and Ecosystem Health. Available online at http://www.eddmaps.org/; last accessed August 31, 2019.

Egolf, D. R., and Andrick, A. O. (1978). The Lagerstroemia Handbook-Checklist: A Guide to Crapemyrtle Cultivars. American Association of Botanical Gardens and Arboreta, Inc., Las Cruces, NM, USA. 72p.

Gu, M., Merchant, M., Robbins, and J., Hopkins. (2014). Crape Myrtle Bark Scale: A New Exotic Pest. Texas A&M AgriLife Extension. EHT-049.

He, D. Y., Cheng, J., Zhao, H., and Chen, S. S. (2008). Biological characteristic and control efficacy of *Eriococcus lagerstroemiae*. Chinese Bul. Entomol. *5*: 034. http://en.cnki.com.cn/Article_en/CJFDTOTAL-KCZS200805034.htm

Jiang, N., and Xu, H. (1998). Observation on *Eriococcus lagerstroemiae* Kuwana. J. Anhui Agri. Univ. *25*: 142-144.

Wang, Z., Chen, Y., Gu, M., Vafaie, E., Merchant, M., & Diaz, R. (2016). Crapemyrtle bark scale: a new threat for crapemyrtles, a popular landscape plant in the U.S. Insects *7*:78.

Wang, Z., Chen, Y., Knox, G.W., and Diaz, R. (2015) Crape Myrtle Bark Scale. Louisiana Agriculture Center. Pub. 3440. http://www.lsuagcenter.com/~/media/system/7/8/d/1/78d165df43ac0d4767607d88dadfb841/pub3440bugbizcrapemyrtlebarkscale_final.pdf

Hardwood and Sugarcane Biochar Can Replace Bark-Based Substrate for Container Production of Tomato (*Solanum lycopersicum*) and Basil (*Ocimum basilicum*) Plants

Ping Yu[1a], Mengmeng Gu[2], Qiansheng Li[2], Lan Huang[3]
[1]Department of Horticulture Sciences, Texas A&M University, College Station, TX, 77843, USA ; [2]Department of Horticulture Sciences, Texas A&M AgriLife Extension Services, College Station, TX, 77843, USA; [3]Institute of Urban Agriculture, Chinese Academy of Agricultural Sciences, Chengdu, 610000, China

yuping520@tamu.edu

[a]Second Place- Charlie Parkerson Graduate Student Research Paper Competition

Keywords: Container media, substrate physical properties.

Abstract

Biochar (BC) has potential as a supplement for more expensive peat and bark media components in container production of plants. This research demonstrates that mixed hardwood biochar (HB) can replace 50% of bark-based substrate, while sugarcane biochar (SBB) can replace 70% of bark-based substrate in container mixes for tomato and basil production. There was no adverse effect on plant growth. Tomato plants grown in SBB amended substrates had lower total dry weight, but similar or higher fruit dry weight in comparison to the control. The suitable rates of SBB and HB to replace bark-based substrate for container production of other crops are worthy of further investigation.

INTRODUCTION

Biochar (BC) is rich-carbon material with porous structure produced by the thermo-chemical decomposition of biomass in an oxygen depleted or oxygen-limited atmosphere (Demirbas and Arin, 2002; Lehmann, 2007; Nartey and Zhao, 2014). Research has shown that BC from certain raw materials and conditions can be a potential alternative to commonly used substrates (Gu et al., 2013; Guo et al., 2018a; Guo et al., 2018b;

Huang and Gu, 2019; Huang et al., 2019) due to its suitable properties for plants growth. BC can increase water and nutrient holding capacity, ameliorate acidity and provide a suitable environment for plants (Dumroese et al., 2011; Vaughn et al., 2013; Zhang et al., 2014). Under certain conditions, BC can increase greenhouse crop growth, yield and quality (Tian et al., 2012; Headlee et al., 2014; Zhang et al., 2014; Nieto et al., 2016; Méndez et al., 2017).

There are huge substrate demands for greenhouse plants production (Gu et al., 2013). According to the USDA, around 5.4 million (M) ft^3 substrate was used for potted plants production in 2017. The substrate use is considerably greater, since only entities with over $1 million in sales from 15 states were included in the survey (USDA-NASS, 2018). The substrate commonly used for potted plants production in greenhouse is mainly peat moss-based. However, there are negative environmental impacts with peat moss extraction, such as destroying rare habitats and cultural heritage, and adversely affecting water management and climate change (Alexander et al., 2008). Thus, the United Kingdom and other countries have environmental policies to restrain unnecessary peat extraction and to encourage use of peat alternatives. The price of peat and bark is constantly increasing, especially when transportation cost were taken into consideration (USGS, 2016). This directly affects growers profitability (Gu et al., 2013). Bark is a peat alternative. While it is less expensive than peat moss, the supply of bark has decreased due to fluctuation in housing demand, lumber and paper supplies (Wright and Browder, 2005).

Research has focused on finding commonly used container substrate alternatives from industrial and agricultural waste - such as switchgrass and miscanthus straw (Altland and Krause, 2009; Altland and Locke, 2011), clean chip residuals (Boyer et al., 2008) and animal manures. Although some of these materials have potential to be good substrate components, the lack of reliable supplies limited their use. As a novel material, which has potential to be widely used as substrate component, BC has also attracted researchers' attention. There is no universal standard for BC addition to plant production. The effects of BC on container substrates depend on many factors including BC feedstock, production conditions and the percentage of BC. Our previous study showed that mixed hardwood biochar (HB) performed well as greenhouse media amendments (Huang et al., 2019). This research was conducted to determine the effects of different BC as bark-based substrate amendments on container plants growth.

MATERIALS AND METHODS

Plant material and experimental design

Tomato (*Solanum lycopersicum* 'Red Robin™') (Fred C. Gloeckner, Harrison, NY, USA) and basil (*Ocimum basilicum*) (Johnny's Selected Seeds, Winslow, ME, USA) seeds were sown in plug trays (cell depth: 5-cm; cell top length and width: 4-cm; volume: 55ml) with commercial mix (BM2 Berger, Saint-Modeste, Quebec, Canada), one seed per cell on 26 February 2019. After the first pair of true leaves expanded, uniform seedlings were transplanted into 6-in. azalea pots (depth: 10.8-cm; top diameter: 15.5-cm; bottom diameter: 11.3-cm; volume: 1,330 ml) with commercial potting mix (Jolly Gardener, Oldcastle Lawn & Garden Inc. Atlanta, GA) incorporated with either sugarcane biochar (SBB) at two rates (50%, 70%, by vol.) or mixed hardwood biochar (HB) at 50%. A commercial potting mix was used as control. The SBB was produced by American Biocarbon LLC (White Castle, LA) using proprietary methods, and the HB was the by-product of fast pyrolysis of mixed hardwood produced by Proton Power Inc. (Lenoir City, TN, USA). During transplanting, slow-release fertilizer Osmocote Plus (15-9-12, Scotts-Sierra Horticultural Products Company, Marysville, Ohio) were applied as surface dressing at the rate of 4.8 g/pot for basil and 7.7g/pot for tomato. This experiment was designed as random complete block design with six replications per treatment. Pots were placed in a greenhouse at Texas A&M University, College Station, TX. The average greenhouse temperature, relative humidity and dew point were 23.7°C, 81.8% and 19.6°C, respectively.

Measurements

Potting mix physio-chemical properties

Physical properties of all the potting mixes were measured using North Carolina State University Horticultural Substrates Laboratory Porometer (Fonteno et al., 1995). The leachate electrical conductivity (EC) and pH were measured biweekly starting at one week after transplanting (1WAT) with a portable EC/pH meter by pour-through method (LeBude and Bilderback, 2009).

Plant growth

Plant growth index was calculated at 1, 3, 5 and 7WAT using the formula-- Plant growth index=Plant height/2+(Plant width 1+Plant width 2)/4. Plant stem, leaf, fruit (tomato) and flower (basil) were harvested separately and their dry weight (SDW, LDW, FDW) were weighed after being oven-dried to a constant weight at 80 °C. Total dry weight (TDW) of above-ground part were calculated by adding SDW, LDW, and/or FDW. Plant roots were washed under running water after harvest, and root length, root surface area, root diameter and the number of root tips were measured by scanning under a root scanner (WinRHIZO, Regent Instruments Canada Inc., Canada).

RESULTS

Potting mix physio-chemical properties

The HB is alkaline while SBB is acidic. Most of the mixes' physical properties were within the recommended range even though for SBB mix (Table 1). Their TP and CC were slightly higher than the recommended ones and 50% SBB mix had slightly lower than recommended. For both tomato and basil (Fig.1), the EC of all treatments decreased during the experiment.

Table 1. The physio-chemical properties of biochar, commercial substrate and their mixes.

Composition	pH	EC µS/cm	TP[W] %	CC %	AS %	BD g/cm^3
SBB[X]	5.9	753	74	71	3	0.11
HB[Y]	10.1	1,058	87	66	20	0.13
50%SBB+50%CS	6.3	2,073	81	75	7	0.13
50%HB+50%CS	7.5	1,370	78	62	17	0.13
70%SBB+30%CS	6.4	1,830	89	76	13	0.14
Commercial Substrate[Z]	6.5	1,819	97	85	12	0.15
Suitable range [R]	-	-	50-80	45-65	10-30	0.19-0.7

[X] SBB =Sugarcane Bagasse Biochar produced by American Biocarbon LLC.
[Y]HB = Mixed Hardwood Biochar provided by Texas A&M and produced by Proton Power, Inc.
[Z] Commercial bark-based substrate, Jolly Gardener, Oldcastle Lawn & Garden Inc. Atlanta, GA, USA. [W] TP=Total porosity, CC=container capacity, AS=Air space, BD= Bulk density. [R] Recommended physical properties of container substrate by (YEAGER et al., 1997).

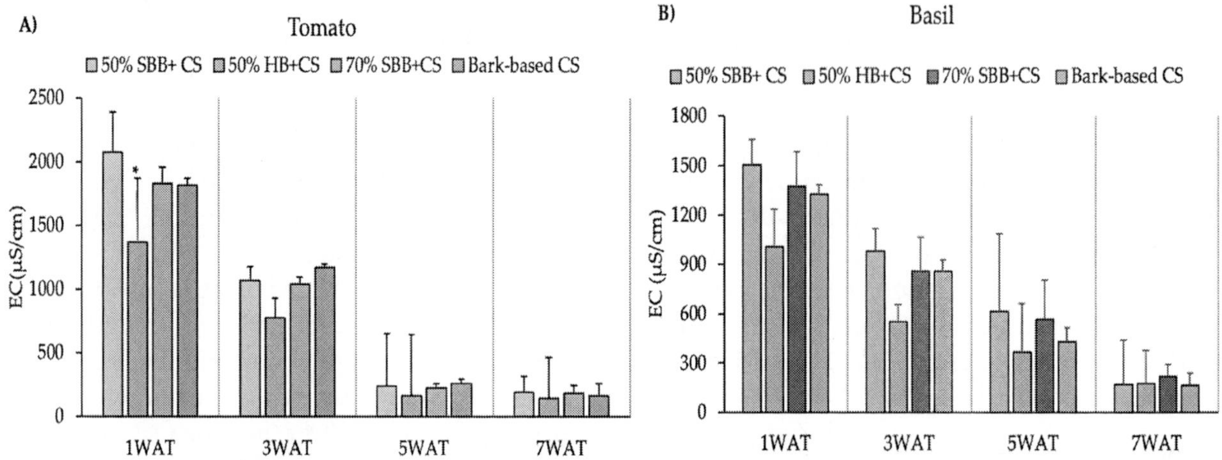

Figure 1. EC (mean ± standard error) of containers media with 50% SBB, 50%HB, 70%SBB or 0% BC (by vol.) mixed with bark-based commercial substrate in the growing period, grown with tomato A) and basil B) plants. * indicated significant difference from the control using Dunnett's test at P<=0.05.

For tomato, (Fig. 2A), treatments with 50%HB had significantly higher pH than the control at 1, 3 and 7WAT. At 1WAT, 50% SBB treatment had significantly lower pH than the control, while at 7WAT, 70%SBB had significantly lower pH than the control. For basil plants (Fig. 2B), treatment with 50%HB had significantly higher pH in comparison to the control for all the weeks, and SBB treatments (both 50% and 70%) had significantly lower pH compared to the control at 5WAT and 7WAT.

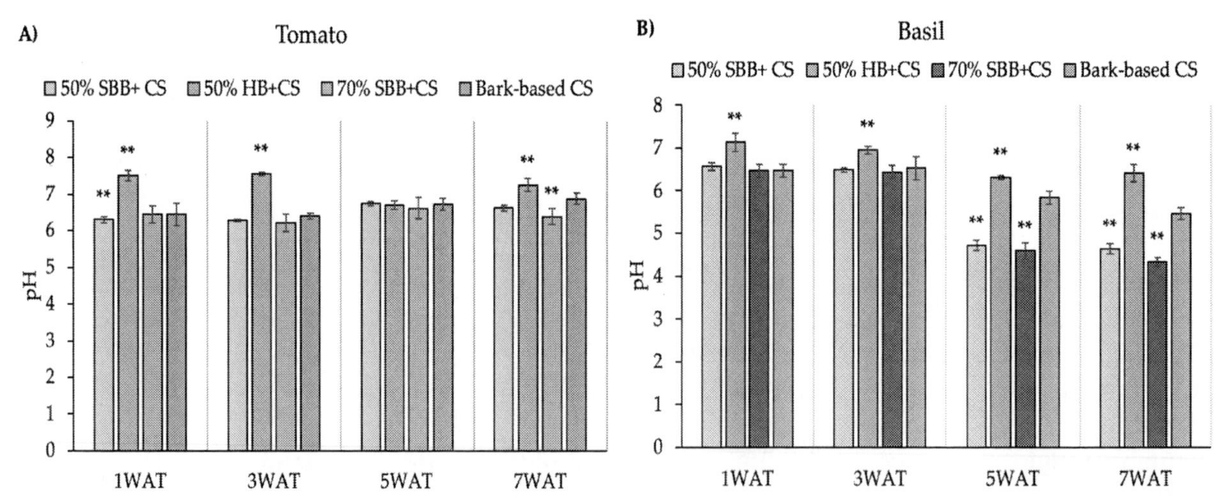

Figure 2. The pH (mean ± standard error) of containers media with 50% SBB, 50%HB, 70%SBB or 0% BC (by vol.) mixed with bark-based commercial substrate in the growing period, grown with tomato A) and basil B) plants. * indicated significant difference from the control using Dunnett's test at P<=0.05(*), P<=0.01(**).

Plant growth

For tomato plants (Fig. 3A), the 70%SBB had significantly higher GI than the control at 5WAT. There we no other significant differences among the four treatments. For basil plants (Fig. 3B), there were no significant differences among the four treatments.

Figure 3. Tomato A) and basil B) plants after grown in containers with 50% SBB, 50%HB, 70%SBB or 0% BC (by vol.) mixed with bark-based commercial substrate for 7 weeks.

For tomato plants, there were no significant differences among the four treatments on root length and average diameter (Fig. 4A). The SBB treatments had a significantly smaller root surface area than the control. Treatments with 50% SBB had significantly less root tips compared to the control while other treatments had similar or more tips than the control.

For basil plants (Fig. 4B), there were no significant differences among the treatments in root surface area. All the BC treatments had significantly shorter root length than the control, but a significantly larger average diameter. Treatments with 50% of BC had significantly less root tips compared to the control.

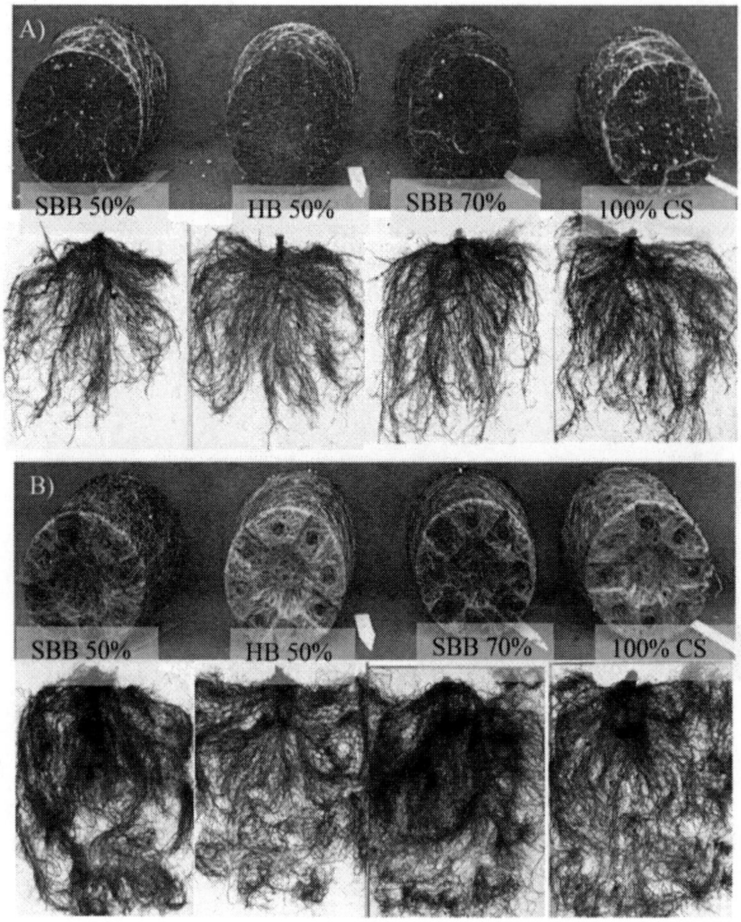

Figure 4. Root development of tomato A) and basil B) plants after grown in containers with 50% SBB, 50%HB, 70%SBB or 0% BC (by vol.) mixed with bark-based commercial substrate for 7 weeks.

CONCLUSION

Mixed Hardwood biochar can replace 50% of bark-based substrate and sugarcane biochar can replace 70% of bark-based substrate in the potting mixes for tomato and basil production, without affecting the plant growth in this experiment. Tomato plants grown in SBB amended substrates had lower TDW, but had similar or higher FDW compared to the commercial control. The suitable rates of SBB and HB to replace bark-based substrate for container production of other crops merits further investigation.

Literature Cited

Alexander, P., N. Bragg, R. Meade, Padelopoulos, G., and Watts, O. (2008). Peat in horticulture and conservation: the UK response to a changing world. Mires & Peat 3.

Altland, J.E. and Krause, C. (2009). Use of switchgrass as a nursery container substrate. HortScience *44*:1861-1865.

Altland, J.E. and Locke, J.C. (2011). Use of ground miscanthus straw in container nursery substrates. J.Environ. Hort. *29*:114-118.

Barassi, C., Creus, C., Casanovas, E., and Sueldo, R. (2000). Could Azospirillum mitigate abiotic stress effects in plants. Auburn University. http://www. ag. auburn. edu/argentina/pdfmanuscripts/brassi. pdf.

Boyer, C.R., Fain, G.B., Gilliam, C.H., Gallagher, T.V., Torbert, H.A., and Sibley, J.L. (2008). Clean chip residual: A substrate component for growing annuals. HortTech.*18*:423-432.

Demirbas, A. and Arin, G. (2002). An overview of biomass pyrolysis. Energy Sources *24*:471-482.

Dumroese, R.K., Heiskanen, J., Englund, K., and Tervahauta, A. (2011). Pelleted biochar: Chemical and physical properties show potential use as a substrate in container nurseries. Biomass Bioener. *35*:2018-2027.

Fonteno, W., Hardin, C. and Brewster, J. (1995). Procedures for determining physical properties of horticultural substrates using the NCSU Porometer. Horticultural Substrates Laboratory, North Carolina State University, Ralegh, N.C.

Gu, M., Li, Q., Steele, P.H., Niu, G., and Yu, F. (2013). Growth of 'Fireworks' gomphrena grown in substrates amended with biochar. J. Food, Agri. Environ. *11*:819-821.

Guo, Y., Niu, G., Starman, T., Gu, M. (2018a). Growth and development of Easter lily in response to container substrate with biochar. J. Hort. Sci. Biotech. *94*:1-7.

Guo, Y., Niu, G., Starman, T., Volder, A., and Gu, M. (2018b). Poinsettia growth and development response to container root substrate with biochar. Horticulturae 4:1.

Headlee, W.L., Brewer, C.E., and Hall, R.B. (2014). Biochar as a substitute for vermiculite in potting mix for hybrid poplar. Bioenergy Res. *7*:120-131.

Huang, L. and Gu, M. (2019). Effects of Biochar on container substrate properties and growth of plants—A review. Horticulturae 5:14.

Huang, L., Niu, G., Feagley, S.E., and Gu, M. (2019). Evaluation of a hardwood biochar and two composts mixes as replacements for a peat-based commercial substrate. Industrial Crops Prod. *129*:549-560.

LeBude, A. and Bilderback, T. (2009). Pour-through extraction procedure: A nutrient management tool for nursery crops. North Carolina Cooperative Extension:1-8.

Lehmann, J., (2007). A handful of carbon. Nature *447*:143-144.

Méndez, A., Cárdenas-Aguiar, E., Paz-Ferreiro, J., Plaza, C., and Gascó, G. (2017). The effect of sewage sludge biochar on peat-based growing media. Biolog. Agri. Horticul. *33*:40-51.

Nartey, O.D. and Zhao, B. (2014). Biochar preparation, characterization, and adsorptive capacity and its effect on bioavailability of contaminants: an overview. Advances in Materials Science and Engineering 2014.

Nieto, A., Gascó, G., Paz-Ferreiro, J., Fernández, J., Plaza, C., and Méndez, A. (2016). The effect of pruning waste and biochar addition on brown peat based growing media properties. Scient. Horticul. *199*:142-148.

Tian, Y., Sun, X., Li, S., Wang, H., Wang, L., Cao, J., and Zhang, L. (2012). Biochar made from green waste as peat substitute in growth media for Calathea rotundifola cv. Fasciata. Scientia Hort. *143*:15-18.

USDA-NASS. (2018). Agricultural Statistics, p. 202-210. In: USDA (ed.), United States Government Printing Office Washington.

USGS, 2016. (2016). Minerals YearBook-Peat.

Vaughn, S.F., Kenar, J.A., Thompson, A.R., and Peterson, S.C. (2013). Comparison of biochars derived from wood pellets and pelletized wheat straw as replacements for peat in potting substrates. Indust. Crops Prod. *51*:437-443.

Wright, R.D. and Browder, J.F. (2005). Chipped pine logs: A potential substrate for greenhouse and nursery crops. HortScience *40*:1513-1515.

Yeager, T., Gilliam, C., Bilderback, T., Fare, D., Niemiera, A., and Tilt, K. (1997). Best management practices: Guide for producing container-grown plants. South. Nurser. Assoc. Atlanta, GA.

Zhang, L., Sun, X.Y., Tian, Y., and Gong, X.Q. (2014). Biochar and humic acid amendments improve the quality of composted green waste as a growth medium for the ornamental plant Calathea insignis. Scientia Hort. *176*:70-78.

The Effects of Biochar Incorporation on Plant Growth in Container Production

Ping Yu[1], Lan Huang[2], Yanjun Cecilia Guo[3], Mengmeng Gu[4]

[1]Department of Horticultural Sciences, Texas A&M University, College Station Texas 77843, USA
[2]Institute of Urban Agriculture, Chinese Academy of Agricultural Sciences, Chengdu 610000, China
[3]Syngenta Flowers, LLC, Gilroy, California 95020, USA
[4]Department of Horticultural Sciences, Texas A&M AgriLife Extension Service, College Station, Texas 77843, USA

mgu@tamu.edu

Keywords: Substrate, physical properties

Abstract

Biochar is the solid product of pyrolysis of various biomasses, including a lot of agricultural and forestry waste products. The physical and chemical properties of biochars vary significantly due to the differences in feedstock, pre- and post-treatment, and pyrolysis conditions. Based on our ten years of research on incorporating different types of biochar in container substrate, we are confident that biochar made of locally available materials, such as mixed hardwood or sugarcane bagasse, could replace significant amount (50%) of peat or bark in container mix, without negatively affecting plant growth - and in many cases could be beneficial.

INTRODUCTION

In the absence of oxygen or with limited oxygen, biomass heated at high temperature under high pressure will yield biogas, bio-oil and biochar. Depending on many factors, such as source of feedstock, temperatures and pressure, and pre-and post-treatment, biochar could have different properties, pH, electric conductivity (EC), bulk density, particle size distribution, and surface area. The biochar's physical properties as container substrate such as air space, container capacity and total porosity, will be different too.

For example, biochar from fast pyrolysis of pine wood (PB) at 450 °C has an EC of 0.15 mS·cm^{-1} and pH of 5.3 (Table 1). The pH and EC of Sugarcan bagasse biochar (SBB; American Biocarbon LLC White Castle, LA, USA) are 5.9and 0.75 mS·cm^{-1}, respectively, and the pH and EC of mixed hardwood biochar (HB; Proton Power Inc. Lenouir City, TN, USA) were 10.1and 1.06 mS·cm^{-1}, respectively. Similarly, the total porosity (TP), container capacity (CC), air space (AS), and bulk density (BD) of these biochars are also different.

The chemical and physical properties of substrate components are not as important as the properties of the container substrate as a whole. Based on the chemical and physical properties of different types of biochars, appropriate substrate components could be mixed at certain ratios to make the final mix with chemical and physical properties within or close to the suitable ranges (Table 1).

Table 1. The pH, EC, total porosity (TP), container capacity (CC), air space (AS), and bulk density (BD) of pine wood biochar (PBC), sugarcane bagasse biochar (SBB), mixed hardwood biochar (HB), perlite (P), peatmoss (PM), P:PM mix (70:30, by vol.) and a commercial substrate (CS), and suitable ranges.

Substrate Component[z]	pH	EC (mS/cm)	TP (%)	CC (%)	AS (%)	BD (g/cm^3)
PBC	5.3	0.15	83	49	34	0.17
SBB	5.9	0.75	74	71	3	0.11
HB	10.1	1.06	87	66	20	0.13
P	7.3	0.57	92	59	34	0.05
PM	5.03	0.18	69	58	11	0.11
PM:P=70:30	5.6	0.16	79	62	16	0.09
CS	6.5	1.82	97	85	12	0.15
Suitable range	6.2-6.8	--	50-80	45-65	10-30	0.19-0.7

[z] SBB=Sugarcane bagasse biochar produced by American Biocarbon LLC, LA, USA; HB=Mixed Hardwood biochar produced by Proton Power, Inc, TN, USA; CS=commercial substrate, Jolly Gardener, Oldcastle Lawn & Garden Inc., Atlanta, USA; P=Perlite (Kinney Bonded Warehouse, Tyler, TX, USA; PM=Peat moss, Voluntary purchasing Group Inc., Bonham, TX, USA.

PINE WOOD BIOCHAR (PBC)

Preliminary experiments found the growth of 'Fireworks' gomphrena was improved when grown in peat-based commercial substrate (CS) mixed with 10%, 20% and 30% (by volume; Gu et al., 2013) PBC. Then PBC was used in container substrate to replace pine bark (PB) or CS at 20%, 40%, 60%, 80% or 100% (by volume). Tomato, lettuce (Fig. 1), chrysanthemum and basil (Fig. 2) were grown in these mixes and had similar growth in mixes with up to 60% or 80% PBC compared to PB and CS, respectively (Choi et al., 2018; Peng et al., 2018).

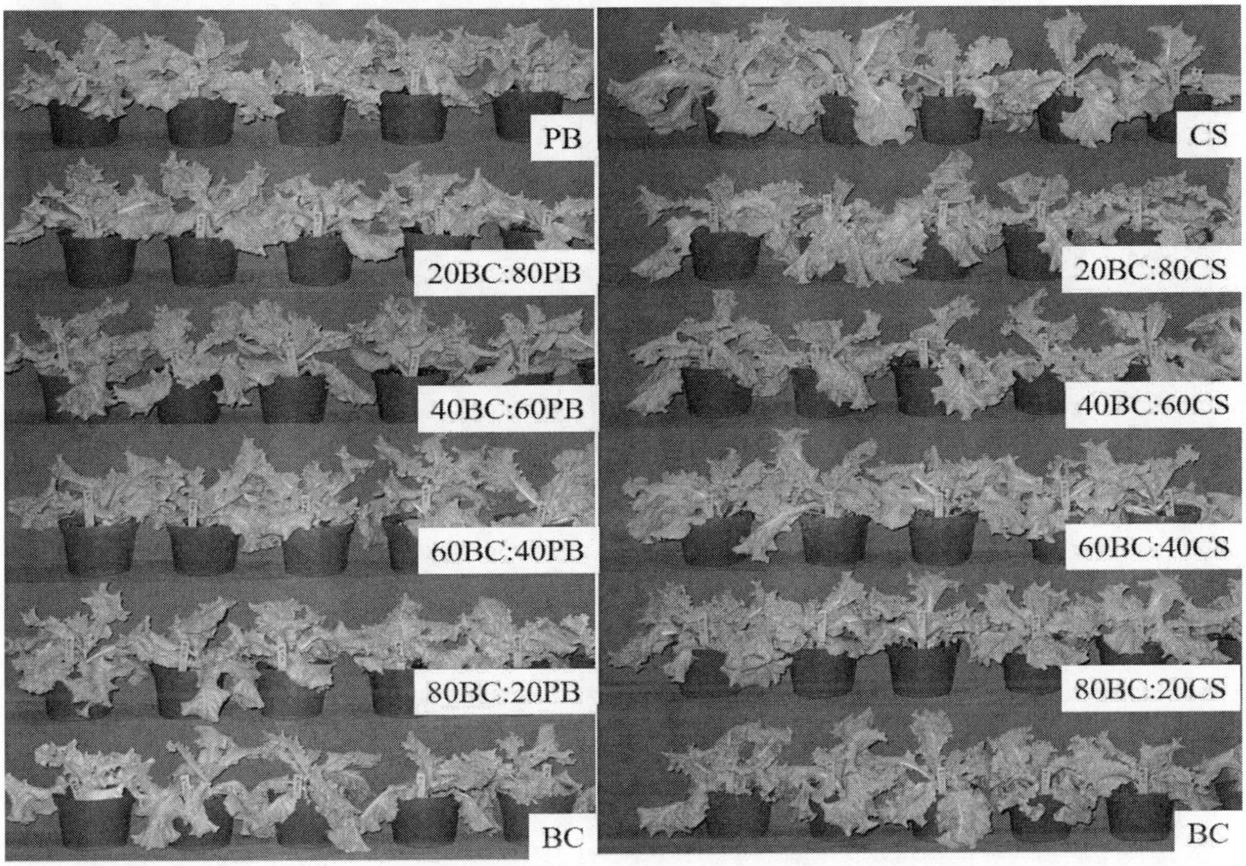

Figure 1. 'Simpson' lettuce grown in pine bark (PB), pine wood biochar (BC):PB mixes (numbers indicate the percentage by volume), peat-based commercial substrate (CS), BC:CS mixes, and BC.

Figure 2. The root balls of 'Genovese' basil grown in pine bark (PB), pine wood biochar (BC):PB mixes (numbers indicate the percentage by volume), peat-based commercial substrate (CS), BC:CS mixes, and BC.

Two relatively long-growth (3-4 months) greenhouse crops, poinsettia (Fig 3) and Easter lily (Fig. 4), were tested in CS replaced by PBC. Poinsettia plants grown in up to 60% PBC were similar to plants in CS and Easter lily up to 80%. The plants had similar visual rating, respectively.

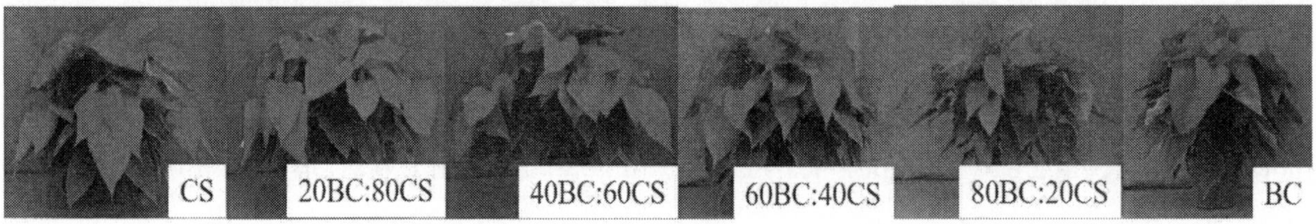

Figure 3. 'Prestige Red' poinsettia grown in peat-based commercial substrate (CS), pine wood biochar (BC):CS mixes (numbers indicate the percentage by volume), and BC.

With the type of plants and duration of production tested, the PBC could be potentially used at high incorporation (up to 80%) in growing greenhouse crops. During these experiments, plants were maintained as the plants grown in CS. If production practices, such as fertilization and watering schedule, were adjusted to meet the plants needs in mixes with high rates of PBC, plants in PBC mixes may have grown even better. And it may be possible to use 80%-100% PBC as container substrate.

Figure 4. Easter lily grown in peat-based commercial substrate (CS), and pine wood biochar (BC):PB mixes (numbers indicate the percentage by volume).

SUGARCANE BAGASSE BIOCHAR (SBB)

Sugarcane bagasse biochar has similar low pH as PBC (Table 1), however, the SBB particles are much smaller than PBC, resulting in much higher container capacity and lower air space. So trials with SBB did not include peat-based commercial substrate, but instead, bark-based commercial substrate was used. There's still significant portion of peat in the bark-based substrate, but bark-based substrates generally have higher air space. In a trial including 50% SBB, 70% SBB and 50% HB, both tomato and basil plants in biochar mixes had similar growth index and yield, compared to bark-based CS. It is possible that plants in SBB mixes might perform better than bark-based CS if the biochar mixes were mixed with bark alone, as high air space of bark may be complementary to the SBB with low air space.

MIXED HARDWOOD BIOCHAR (HB)

Challenges of using HB in container substrate are the high pH and EC (Table 1). Its container capacity and air space are within or close to the recommended range. Substrate components of finer particle sizes, including vermicompost and chicken manure compost, were used in HB trials.

In one trial, 5%, 10%, 15% and 20% vermicompost were mixed with 20%, 40%, 60% and 80% HB with the rest being commercial peat-based substrate. Basil plants grown in any of the 16 HB mixes performed equally to or better than the commercial substrate. All the tomato plants grown in HB mixes had higher or similar growth index and total dry weight compared to the control.

Chicken manure compost is a much cheaper resource than vermicompost. In another trial, 5% vermicompost or chicken manure compost was mixed with 60%, 70%, 80%, or 90% of HB with the rest being peat-based commercial substrate. Basil plants in 60% and 70% HB:vermicompost mixes had similar total dry weight compared to control. Tomato plants in all HB mixes (with vermicompost or chicken manure compost) had similar growth index compared to control. Chicken manure compost had high salt level and EC. Basil plants had low salt tolerance while tomato plants had high tolerance. This may explain why tomato plants performed well in more HB mixes (especially those with chicken manure compost), compared to basil plants. So chicken manure compost would not be recommended for salt sensitive plants.

DISCUSSION

Biochar is not for all growers. But for growers with close and cheap access to biochar, it is worth of a trial to include biochar in their growing media. Although two of the three biochars in our trials had low pH, most biochars have high pH. Addition of high-pH biochars could reduce or eliminate lime to increase pH of bark or peat. The huge surface area of biochar may provide greater habitat for microbes, which deserve further investigation.

Literature Cited

Choi, HS, Zhao, Y., Dou, H., Cai, X., Gu, M. and Yu, F. (2018). Effects of biochar mixes with pine bark-based substrates on growth and development of horticultural crops. Hort. Environ. Biotech. *59*:345-354. https://doi.org/10.1007/s13580-018-0035-x. https://rdcu.be/MaPY.

Gu, M., Li, Q., Steele, P.H., Niu, G., and Yu, F. (2013). Growth of 'Fireworks' gomphrena grown in substrates amended with biochar. J. Food, Ag. Environ. *11*:819-821.

Peng D., Gu, M., Zhao, Y., Yu, F., and Choi, H.S. (2018). Effects of biochar mixes with peat-moss based substrates on growth and development of horticultural crops. Hort. Sci. Tech. *36*:501-512.

PROCEEDING'S PAPERS

EUROPEAN REGION OF

NORTH AMERICA

Tim Lawrance-Owen, Regional editor

Fifty-second Annual Meeting - 2019

Stratford-Upon-Avon, Great Britain

Ornamental Production at Bordon Hill Nurseries

Geoff Caesar

Bordon Hill, Stratford-upon-Avon CV37 9RY, UK

sales@bordonhill.com

Keywords: Management, crop scheduling, WhatsApp, Plant nutrition

INTRODUCTION

Having spent over 30 years in the nursery stock sector growing shrubs, trees, climbers and perennials for the retail market I joined Bordon Hill Nurseries growing seed and cutting raised plugs for the grower market. Whilst similar in some ways, we grow an ornamental plant destined for the retail market. The sector and the product are very different. My presentation explores the challenges the sector and in particular the grower faces and how Bordon Hill works to meet these challenges.

One enviable advantage we have as a grower of young plants that are quick to finish is that whilst we do produce some speculative product the vast majority of what we grow is not stuck or sown until we have an order for it. Our customers place firm orders for specific delivery weeks. This helps ensure we have minimal wastage and keeps our risk to the fluctuations in the market place far lower than growers who are continually speculating for the majority of their sales.

As a result of placing firm orders for specific delivery weeks our customers expect us to supply what they ordered when they ordered them for, and this is where our challenges begin. We supply both the commercial and mail order markets and offer 35 tray formats across a broad range of varieties resulting in over 4,100 different products grown across 22,500 batches being produced at the three Bordon Hill sites. When we grow a product for a specific week number in most cases it will only remain saleable for that one week, when we sow or stick, we are sowing or sticking for supply in only one week.

I recall the challenges of having to get a number of different varieties of shrub or perennial ready for the first week of a sales month. In week 14 of this year Bordon Hill supplied 151,975 trays across 1,255 different products or batches, each product having a different length of time in production, a different nutrient requirement and a different climate requirement to ensure it was in ideal condition for sale in week 14. Keeping track of all of these batches and making sure they get just the right growing requirements is incredibly complicated from the management of orders to the ordering of seed through sowing, gapping, transplanting, climate control and nutrition there are endless opportunities for it to go wrong. It may be attractive to have short production cycle but if the plant is sown late, gapped late or transplanted late we will miss our sales week. If there is a P+D or nutrition problem with a product that has a 4-week growing period identifying the problem and putting it right is time sensitive to say the least.

Management of Information

Accurate management of information and making this available to our staff is critical from ensuring the availability of seed, materials and staff to planning work flows and recording outputs. Whilst off the shelf planning packages do exist, we have not taken this route but have a dedicated IT team who have written a bespoke planning and information management system. Initially this was written as an access database system but more recently the team has been writing and migrating our systems over to a web-based system improving security, functionality and access from multiple locations.

Being able to track progress of any batch through the system is critical from a production management perspective but also from the perspective of being able to provide information to our customers. The system enables us to be able to see progress with a customer's order from order placement through all production processes through transport to the point of delivery.

Growing profiles

Every product has a pre-determined growing period or profile which changes as we go from one season to another. A plant with a 5-week profile for sowings in week 50 to week 10 may then have a 4 week profile for later sowings. The growing profile determines the sow week, gapping week and transplanting week if applicable to meet the required sales week. Running trial production batches of new product introductions enables us to establish accurate profiles for new introductions.

Growing protocols

Growing protocols determine temperature, shading requirement, lighting, particular pests or diseases to monitor for and details of nutritional requirements. We are working to ensure all of our products have growing profiles that guide the growing team in making decisions on how to treat the crop. They are not hard and fast blueprints as the season and general climate have such an impact on growth. Profiles and protocols are reviewed regularly as improved breeding can result in changes beings required to both.

WhatsApp

With such short growing profiles, If there is an issue or problem we need inform our technical team in as timely a manor as we can and arrive at a diagnosis as quickly as possible. We use Whatsapp to help achieve this forming a technical group who post images, batch details and information about the issue. The tech team will then go on site to assess the batch, collect samples, diagnose the problem and post a recommendation for treatment. Squeezing water out of the plug and analyzing for EC and PH is relatively straight forward whereas plating samples and using the microscopes to diagnose soil borne diseases is a little more involved but it is essential that we diagnose diseases like Pythium or Rhizoctonia within 24 hours. We can lose a whole crop if delayed for more than this.

Nutrition

We can't use control release fertilizers in tiny cells, the distribution is not accurate enough and we also need far great control to ensure we hit the correct week numbers for every batch. We use a series of liquid fertilizers starting with a base level feed that most crops receive then more specific feeds to suite the particular crop and the period of its growing cycle. We use the growing protocols to guide our growers but they will also make changes and alterations to manipulate growth and hit week numbers.

Irrigation control

Getting irrigation right is critical with such short growing periods. Too wet and root growth is poor and disease can be a problem, too dry and the plant is dead within hours. The temptation is almost always to over water as too dry is obvious and devastating whilst too wet means the plant is still alive, all be it compromised. Sensors can be used for larger cells but don't currently exist to use in the tiny cells we are working with. We have been working with a system of tray weighing to help turn something like irrigation that can be so subjective into a more objective science. In reality it is our growing team who are establishing and working to wetting up and drying down targets and so improving growth control through the control of irrigation.

Growth control

Whilst good irrigation management can help growth control, we still need plant growth regulators to help us produce the plants our customers require. Our growing supervisors walk the crop on a daily basis to agree PGR requirements using hand held devices to set up PGR application plans for the growing team. Weather and time of year have a huge impact on PGR requirement so rather than work to pre-determined programs we have to have to continually assess requirements. Reducing our dependence on PGR's is an area we are working on over the next 12 months, climate control, nutrition, irrigation and improved breeding should all help reduce the need for growth regulators.

Vapour pressure deficit or VPD

Vapour pressure deficit is the difference between the amount of moisture in the air and the amount of moisture the air can hold when it is saturated. The higher the VPD the greater the stress the plant is under. Some VPD is essential in pushing the plant to make roots and put on growth but too high a VPD can result in the plant being stressed and suffering growth abnormalities. During this season we identified tip abortion in Petunia during a period of very low VPD. We have been using accurate VPD monitoring on susceptible crops to help identify periods of stress and using shade screens, irrigation and climate control to reduce stress and maximize growing conditions. Whilst VPD is a relatively new technology for us it is proving to be another powerful tool to help improve plant quality.

Supply

We are constantly assessing crops through their production cycle and making changes in feeding, climate control, PGR application and irrigation to ensure they meet their supply date. We don't get it right 100% of the time but do inform customers if changes need to be made, offer alternate delivery dates or substitute varieties should we need to.

Tech transfer

There may be techniques used by seed raised bedding growers that other growers could utilize to help in production. On site disease diagnosis, objective irrigation analysis (sensors can be used in containers larger than tiny plugs), Vapour pressure deficit, WhatsApp groups and supplementary liquid feeding to name but a few. Many growers will be using these techniques but if you are not there are opportunities you could consider.

Crop Production Horticulture, The Next Generation

Mark Diggines

Trainer and Assessor in Plant Production

Pershore College of Horticulture, Warwickshire College Group

madiggines@warwickshire.ac.uk

Keywords: Education, apprenticeship, internship

INTRODUCTION

I will be looking at the perceived perception of horticulture from the younger generation despite the value that the industry has within our economy. This will then be followed by looking at the objectives horticultural groups such as the RHS and the All-Party Parliamentary Gardening and Horticulture Group have put in place as a result of this.

After this, a brief look into whether these objectives are working or are even realistic will be discussed and what you, as employers, can do to promote the industry we work in.

The paper will be finished with a look into what Pershore College is doing to help create the next generation of crop/plant production employees.

A report compiled by Oxford Economics, commissioned by Ornamentals Round Table, looking into the value of the ornamental horticulture industry. https://www.hortweek.com/

The numbers in the report are quite staggering considering the perception of horticulture amongst the younger generation in our country.

These figures are from 2013/14, however, a survey, which is co-funded by AHDB Horticulture, Royal Horticultural Society, Arboriculture Association, British Association of Landscape Industries, Chartered Institute of Horticulture and Land Based Colleges Aspiring to Excellence following a need identified by the Ornamental Horticulture Roundtable Group, will see 1,000 ornamental businesses contacted in a bid to understand the skills gap in the ornamental industry.

A number of objectives were then set on the back of this looking at how the Government can promote the industry within education. In addition, the All-Party Parliamentary Gardening and Horticulture Group produced a report 12 months ago looking into the big issues around the horticultural industry at this time, specifically looking at:

- Incentivizing UK production, biosecurity and trade.
- Nurturing innovation to support health and the environment.
- Training the future workforce and seasonal worker availability.
- Incentivizing UK production, biosecurity and trade.
- Nurturing innovation to support health and the environment.
- Training the future workforce and seasonal worker availability.

Apprenticeships

One sentence from the Securing-the-future-of-the-gardening-and-horticulture-sector-report really frustrated me. Here we are complaining that we can't get the next generation into our industry and yet a Horticultural based group says that students will struggle to find opportunities for work placements and hands-on experience.

What is clear to see is that we can't simply wait until Government initiatives take hold. What are we/you as employers doing? If colleges are struggling to find willing employers to support work placements and an individual's education, then we are always going to have the same problems.

The IPPS motto is "To seek and to share" and has a tagline under the logo, "Sharing plant production knowledge". Seek out those that might be interested in horticulture but have little guidance from their school/college and then share your plant knowledge with them.

In my experience, those employers who have invested time in the local communities and schools are those that regularly have very few issues with recruiting apprentices and then seeing them progress through their business. However, we tend to see the same employers year after year. Therefore, I just want to show you a little bit about what we do at Pershore College alongside members of the Midland Regional Growers group with our apprenticeship programme.

Trailblazers apprenticeships were first available to us from 2017 with the old-style frameworks being fully withdrawn in July 2020, with the Crop Technician standard being launched in September 2018. These standards allow us to formulate more bespoke training programmes and work alongside employers to deliver exactly what they need.

Our first group of students have just finished the first year of their 2-year course and are progressing well. Current numbers for enrollment for this year look like doubling that of our first intake. In the last 12 months the students have been to:

- Frank Matthews trees to do some grafting
- Wyevale Nurseries
- Melcourt
- Ball Colegrave summer showcase
- HTA National Plant Show
- IPPS conference
- ICL Hort science live day next week
- Trips to Bordon Hill arranged
- IOH young hort of the year competition

Plastics in Horticulture

Alex Everett[1] and David Chilvers[2]

[1]Aeroplas, Great Western Way, Tipton DY4 7AB UK and [2]Bransford Webbs Plant Company, New House Farm, Bransford WR6 5JB UK

david@bransfordwebbs.co.uk

Keywords: Containers, recycle, waste, environment.

INTRODUCTION

At the end of 2017 David Attenborough and the team at Blue Planet 2 brought the startling truth about our use of plastics into every living room in the country. The nation sat up and realized that we are dumping huge amounts of plastics into our rivers and oceans every day, and that we needed to act before it was too late. Alex Everett (Aeroplas) and David Chilvers (The Bransford Webbs Plant Company) highlight one way in which the Ornamentals Horticulture industry has led the way in a collaboration to develop one solution to help alleviate the problem.

The Problem

For many years the horticulture industry has used black polypropylene and expanded polystyrene as its main material to hold compost for growing. These materials have served the industry well and provide a versatile and robust product that can stand up to the rigours of plant production through all seasons and fit

with increasingly mechanised systems. Black plastic (usually polypropylene in the case of pots) is cheap because it can contain many different colours or recycled material that are then blended with carbon black to produce a uniform colour. The sorting automation within UK recycling facilities use Near InfraRed scanners to grade out plastic containing carbon black so the pots are generally discarded to landfill or energy recovery. This means that although they have been made from recycled material and are potentially recyclable, in reality they become single use as the consumer cannot recycle them easily.

The Challenges

The industry needed to find a solution that would meet all the demands of growing plants over very different growing periods in a multitude of environments and weather conditions, whilst being able to be used through mechanised potting methods and remaining economically viable.

The Solution

Experts in the pot manufacturing industry had already seen the need to change. Alex and Aeroplas had been working on solutions to the problem as far back as 2014, but it wasn't as simple as it sounds. Sourcing recycled carbon free, detectable polymer wasn't easy. Fortunately, new sources were discovered, and volumes grew. To produce consistent colours you need to add masterbatch and this also contained carbon pigment, so Alex had to work with these suppliers to provide a solution. By early 2018 Aeroplas had a product that was ready to show customers.

Collaboration

Once the nursery industry woke up and realised that we needed to change our use of plastic, a group of some of the leading growers to the retail market came together to develop a joint approach. In what is seen as an industry leading collaboration these businesses put aside commercial advantage and worked together for the greater good of the industry and environment. This led to the birth of the taupe pot and a push to move all plant production to be in a kerbside recyclable solution as soon as possible.

Combined Proceedings IPPS 69:372-378. 2019.

Managing Hydration of Herbaceous Cuttings – From Harvest to Stick

William Healy

Senior Technical and Research Manager, Ball Horticultural Company, West Chicago Illinois 60185, USA

Keywords: Cutting propagation, perennials, vegetative annuals, cutting storage, hydration

Abstract

During the last 20 years, the Floriculture industry has seen an exponential growth of vegetatively propagated plants originating from offshore production locations. With the growth of offshore production, the need to manage URC post-harvest life is critical to insure uniform establishment of root liners. We have investigated the role of relative humidity (RH) on post-harvest performance of a wide range of species currently available for sale. Uniform and rapid root development occurs when unrooted cuttings (URC) are fully hydrated prior to sticking, regardless of hydration status upon arrival at the rooting location. The best hydration treatment was removal of URC from all packaging and placing in a RH controlled (90-99%) cooler at 8C. Leaving URC in the plastic bag, but out of the shipping containers, was acceptable if there was enough time (12-24 hours) in a RH controlled chamber to fully hydrate the URC. The worse treatment that resulted in poor, uneven and reduced rooting was the current practice of placing URC in a non-RH controlled chamber at 8C without packaging. Once URC were stuck, the first 5-7 days required high RH to maintain URC turgidity. Rooting speed was optimized when the duration of the on-cycle was controlled to maintain foliage moisture without increasing the soil moisture to saturation. Optimizing the dry and wet soil moisture target weights were critical factors for promoting callus formation with lower dry target weights needed to promote uniform root growth throughout the soil profile.

INTRODUCTION

Prior to the 1990's the vegetatively propagated products used in Floriculture consisted of a narrow range of species that were first introduced as early as the beginning of the 20th century. The number of genera was limited and the varieties within each genus consisting of 1-100s of unique varieties. Growers would produce their own stock plants each season and harvest then root the unrooted cuttings (URC) as needed for their own or local requirements.

During the early 2000's, commercial breeding companies began to aggressively introduce new genera and varieties with improved floriferousness, basal branching and consumer appeal. Concurrent with the introduction of the new varieties, large offshore producers began supplying growers around the world with URC which eliminated the need for onshore self-propagation stock plants. The first challenge the offshore producers addressed was the need to manage the cold chain from harvest to arrival at the rooting location. Managing the cold chain has improved over the last 20 years but still is a significant challenge due to gaps in the transport chain where temperature control is lost, and URC are exposed to fluctuating temperatures.

The off-shore URC supply chain produces a more uniform performing URC than did self-propagators. Unfortunately, the off-shore producers periodically ship URC that won't root uniformly. There is no apparent reason why the URC failed to perform. The poor performance is unpredictable depending on various conditions, but ultimately resulted in reduced liner yield compared to the URC shipped. Rooting trials that evaluated chemical treatments, rooting media and misting intervals periodically reduced losses, but did not eliminate the inconsistent rooting performance. We have followed 'inconsistent rooted' liners though the production supply chain from rooting station to retail outlets. These 'inconsistent rooted' liners usually resulted in substandard plants at retail with a significant loss in retail value. Recognizing the lost sales opportunities, our research group of interns and grower cooperators asked the question "What factors can we operationally manipulate to improve not only speed of rooting but also uniformity?" We know from work on seed raised plugs that improving the uniformity of the population significantly improves final sales value at retail. Our current research focused on the role of hydration on URC performance from the time of receipt until roots began to form.

Results

Trial 1. Our initial trials assumed that offshore URC could be re-hydrated like cut flowers where cut stem bases are placed in saturated soil conditions to facilitate water uptake to maintain hydration. Several trials were conducted to look at different media and saturation levels. No trial resulted in consistent rooting improvements or re-hydration of the URC. We determined that although placing URC directly in water would result in re-hydration, it was not possible to maintain saturated conditions in soil at the base of the URC to facilitate hydration.

Trial 2. We evaluated the relative humidity (RH) in the holding chamber (8C) to determine if we were hydrating or de-hydrating the URC when we had to hold the URC overnight (Fig. 1). The RH varied between 80% to 100% which resulted in an oscillating vapor pressure deficit (VPD) of 0 to -0.2. Due to the continued accumulation of the -0.2 VPD there is a slow but steady drying out of the URC. This can be verified by looking at the condensation on the inside of the bag in the picture to the left. This moisture is moving from the URC to the air and then due to cool temperatures the water vapor condenses on the bag surface.

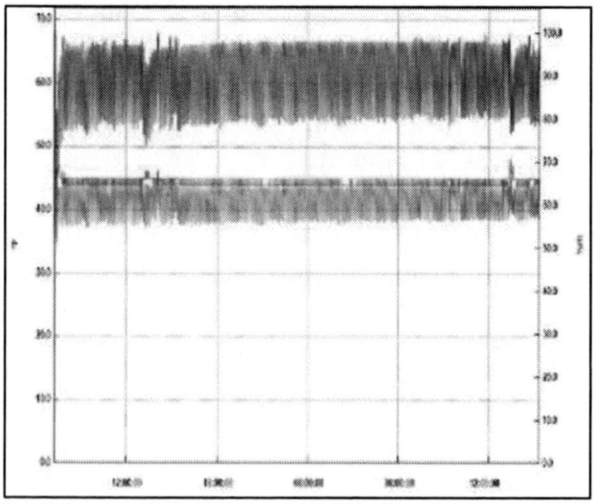

Figure 1. A 13-hour oscillation of the temperature (green line), RH (blue line) and dew point (red line) in a chamber with no RH control.

When we created a RH controlled chamber in the cooler, we observed a more consistent RH and temperature (Fig. 2) compared to the non-RH controlled chamber. Due to the lower VPD accumulation, the URC were able to absorb moisture from the air and became fully hydrated. We concluded that the cooler could create an elevated VPD which could cause dehydration of the URC. The longer the URC were held in the non-RH chamber the greater the dehydration of the URC.

<u>Trial 3</u>. To test the effect of chamber RH, we created a 2x2 trial of +/- RH versus URC in/out of bag. The '-RH+Bag' (ambient RH + URC left in bag) is the current process while '+RH No Bag' (added RH + URC removed from bag and exposed to elevated RH conditions) would be the treatment that creates the best hydration condition. The most stressful condition should be the '-RH no bag'.

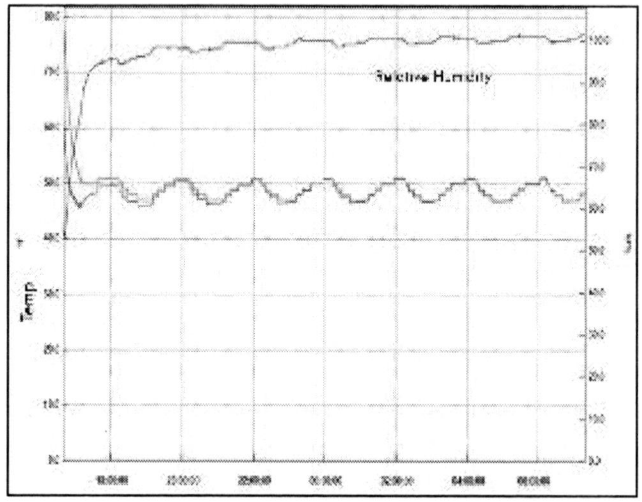

Figure 2. Temperature (green line), RH (blue line) and dew point (red line) in a chamber with RH control.

The degree of hydration can be observed by looking at the turgidity of the leaves and stems (Fig. 3).

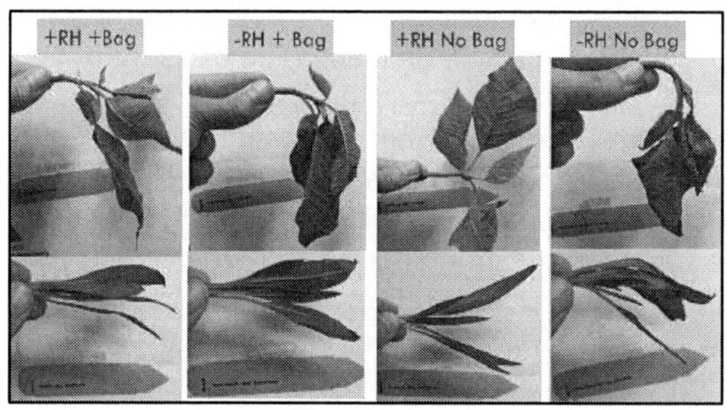

Figure 3. Turgidity after 24-hr in a chamber "limp test".

In a separate trial we quantified the amount of dehydration and leaf turgidity. The '-RH No Bag' would be <70% full hydration while the '+RH No Bag' would be at full hydration. The other treatments would be at various % of full hydration. It is interesting to note that the '+Bag' treatments never reached full hydration or were <70% hydration after 24 hours in the chamber due to the bag providing a barrier to minimize the effect of the accumulated VPD in the chamber. Therefore, to maximize hydration there needs to be air exchange when bags are used to encourage re-hydration. An alternative we have used when RH controlled chambers are not available is to wrap the URC in wet paper towels and then place in bags. This increases the RH in the bag, and we observe fewer dehydrated URC at the time of sticking.

Trial 4. We were interested in how the URC in trial 3 rooted. We stuck the URC from trial 3 and evaluated them 21 days later for root development and uniformity. Multiple species were trialed with similar results to what is shown in Fig 4. The best hydration treatment (B) clearly had more roots and greater uniformity than either the current process (C) or the most stressful treatment (D). It is interesting to note that the best treatment (B) and second most uniform (A) are slightly different in the root number and root length. Based on this trial we are beginning to understand that hydration is the primary factor which influences uniformity of rooting in herbaceous URC.

Figure 4. Effect of supplemental +/- RH and removing the unrooted cuttings from the bag on rooting after 21-days.

Trial 5. Since hydration is a critical factor in predicting rooting performance, we investigated the effect of time between removal from the +RH chamber and sticking on URC hydration. We previously determined that the high essential oil crops, which have significant rooting and shoot breakdown problems, arrived at 24-60% dehydration. We observed that once the URC reached 75+% hydration rooting was impacted and <75% hydration levels resulted in marginal leaf collapse. Table 1 shows what happens when the URC are moved from a RH chamber to the sticking area and held for different durations before moving to a high RH environment. After 1 hour the URC drops to ~80% hydration and by 2.5 hours this continues to drop to almost 70%.

Table 1. The effect of handling time on rooting percentages in three herbs.

Variety	Hours Prior to Sticking		
	0	1	2.5
Lavander Hidcote Blue	100%	80%	74%
Rosemary Tuscan	100%	84%	81%
Thyme Golden Lemon	100%	81%	76%

When we look at maximum root development (Fig 5) we can see that % rooting uniformity drops from 75% to 50% as the time out of RH chamber increases. Therefore it is crtical that the URC are held in the RH chamber for as long as possible or held at the sticking area under high RH to prevent a decline in rooting uniformity and speed. When setting up a rooting area it is better to have the area on the right vs the area on the left which is way too dry and will cause hydrated URC to decline in performance. The grower on the right is able to hold stuck URC overnight with an observed improvement in performance. During the day, the URC are held for several hours in the high RH before moving into the rooting area. By holding at high RH prior to moving to the rooting area they are able to assure hydrated URC that insure uniform rooting and faster root development.

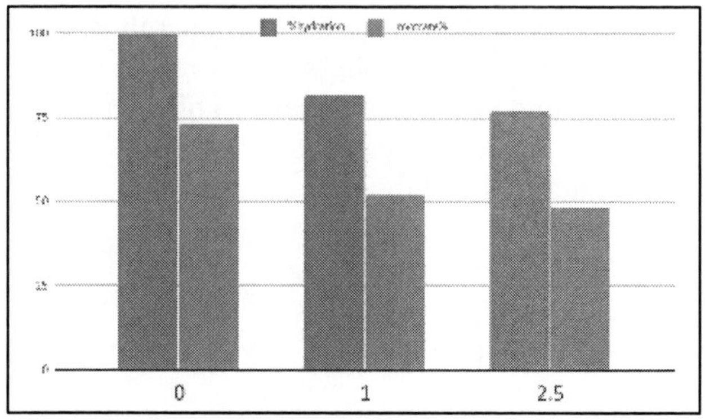

Figure 5. Dry down hydration percentage (blue bar) and rooting uniformity (orange bar) after 21-days.

Trial 6. Previous trials showed that rooting was directly related to the soil moisture. If the soil at the time of sticking was too dry or too wet then rooting was delayed. We have worked with growers to develop tray weight targets that will optimize callus formation and subsequent root development. A trial was done to determine the optimum belt speed for the watering tunnel to insure optimum moisture levels. Fig 6. Shows the effect of belt speed and the tray moisture levels.

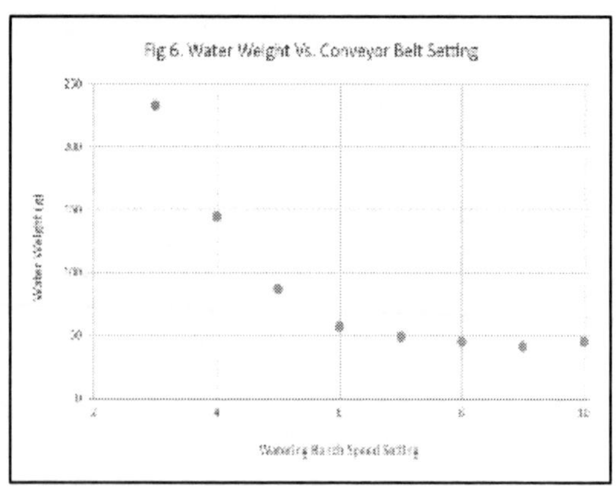

Figure 6. The relationship between tray weight and belt speed.

This intuitive observation is critical for growers to understand. The importance of setting up the sticking line to prevent saturated trays can't be over stressed since saturated trays delay rooting. Each grower needs to determine the optimum tray weight to achieve a level 4 moisture which appears to be optimum for callus formation.

Trial 7. We have observed different dry down rates of the media used in propagation systems as measured by how rapidly they reach different target weights (% full saturation). This is a critical piece of information to determine the correct mist settings. We believe that callus formation requires the soil to stay >80% saturation (level 4) but root development is best at 50% (level 2) and rewetting to 80% (Fig. 7). Note that Hydrafiber drops to <80% saturation within 16 hours which would restrict callus formation if not misted aggressively.

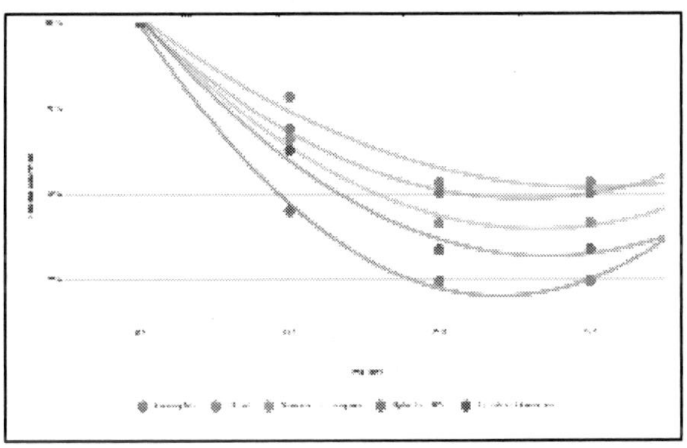

Figure 7. Dry down curves of the industry standard (Ellepots-orange) and several other media including the newest entry in the market 'Hydrafiber' (pink).

Conversely Ellepots remain >80% for up to 40 hours so managing the misting duration would be critical to prevent saturating the soil. To maximize root formation and development growers need to understand the dry down curves of their media and adjust the mist cycles to provide the correct targett moisture.

Trial 8. The final trial was designed to evaluate the effect of misting duration and frequency during the first 72 hours on re-hydration and rooting. To determine the optimum response we used 2 different VPD mist frequency settings (O.5 and 1.5) when the mist 'on' cycle would be triggered. We then used different 'on' durations (2-14 sec) to provide different quantities of water every 24 hours (Fig. 8).

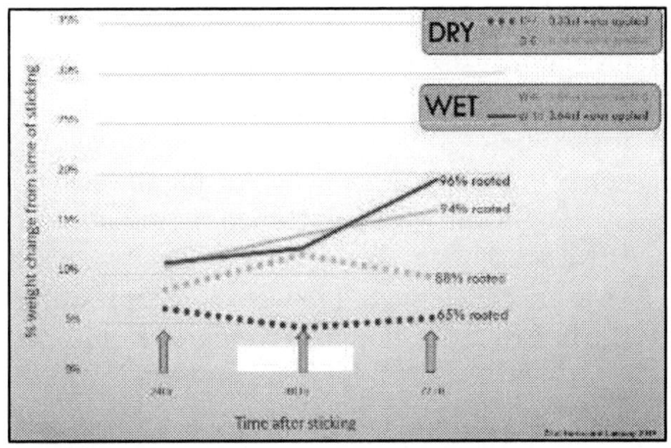

Figure 8. Impact of mist frequency and duration on cutting hydration after 72-hr.

The maximum rooting occurred at 21 days when the wet VPD frequency (0.5) was applied with eith 6 or 10 seconds of mist. When <0.6 ml or >1.2 ml of water per 24 hours was applied daily during the first 72 hours, rooting was affected. The data showed that there is an optimum frequency and duration that growers must determine to optimize root development in the shortest period possible. There are a number of factors that influence the mist settings and each grower must determine the optimum settings for their media and species.

DISCUSSION

Based on our research trials and grower validation trials, we feel confident to recommend growers install VPD controlled RH systems in their holding coolers and sticking operations. We find that growers who try to use low pressure mist systems or wet floors to raise the RH, never achieve the same results that growers who install a high-pressure system. An effective high-pressure system can be purchased for USD$2-5,000 depending on size of cooler and rooting areas. Considering the weekly value of the URC in the cooler and rooting area, an improvement of 5% in root development and uniformity can provide a rapid ROI that continues for years to come.

Managing the soil moisture after sticking and during the first 72 hours in the greenhouse, is critical to insure continued hydration of the URC and promotion of callus formation. Growers can rapidly establish the correct targets for growth by using the 'water by weight' process developed by the author. By aggressively managing the target weights growers can achieve 1-2 weeks faster rooting (level 5 root development) and more uniform rooting within the production block. Disease issues were also reduced when growers used water by weight to target the optimum moisture which naturally reduced excess wet soil and foliage.

We have seen significant improvements in order fulfillment and plant quality when growers implement a robust URC husbandry program. Many of the 'recommendations' that are used in the industry become irrelevant when the URC hydration is correctly managed. Historically we had sticking sequence with many varieties on a 'stick immediate' recommendation. When the hydration is managed these 'stick immediate' URC can be held 1-2 days in high RH conditions with improved performance. URC husbandry is one of the last procedures that rooting operations must implement to improve their overall performance.

Advances in Our Understanding of the Disease Biocontrol Potential and Enhancement of *Trichoderma harzianum* Strain T22 (Trianum®)

Adrian Jackson, Consultant

Koppert UK

Keywords: Disease management, fungus, *Tagetes, Lavandula, Ceanothus, Deutzia*

Abstract

The commercial product Trianum-P is a registered plant protection product for the control of plant diseases in growing media. It contains the *Trichoderma harzianum* strain T22 which is arguably the most widely applied beneficial fungus in commercial horticulture today (Harman et al., 2004a). As part of this year's Trianum-P development work, the author carried out a series of small-scale tests and trials to evaluate the potential benefits of combining Trianum-P with various commercial biostimulants in the production of ornamental plants. One of the challenges of using disease biocontrol agents in crop production is an apparent inconsistency in field performance. It is incumbent on us to find ways to enhance the disease biocontrol activities of T22 and other disease antagonistic micro-organisms. The results obtained in this preliminary work have shown that the combined treatments of Trianum-P and various biostimulants produced various plant growth effects and reduced disease incidence. Fungal biomass development of *T. harzianum* (T22) was favoured by 0.05% Vidi Parva and the pelleted plant-based product Vidi Funda. The biostimulant Fortafol-D inhibited *T. harzianum* conidiospore germination at the commercial rate (0.1%v/v). The significance of this is discussed. Early plant growth stimulation effects by Trianum-P were observed in *Tagetes*. *Lavandula* plugs treated with Trianum-P + Vidi Parva developed less root disease incidence and more shoot extension growth compared to untreated controls. In *Ceanothus* the Trianum-P treatment produced the best overall plant shape and more spring growth than all other treatments. A reduced number of flower heads in the Trianum-P treated *Ceanothus* may have been attributed to stress reduction effects and growth hormone induction resulting from *T.harzianum*'s symbiotic associations with the plant's roots. In *Hebe* the greatest root biomass was produced by the Trianum-P + Vidi Parva and M149 + Vidi Parva treatments. Trianum-P + Vidi Parva gave the best results overall, scoring highly in root biomass production, strong lateral root production and achieving the greatest reduction in disease incidence. In *Deutzia* more root biomass and the best root quality was produced in the Trianum-P and Trianum-P + Vidi Parva treatments.

INTRODUCTION

Trichoderma harzianum strain T22 was generated by combining protoplast tissue from *Trichoderma harzianum* strains T95 and T12 (Stasz *et al.*, 1988) and is now sold commercially as the product Trianum. The development of T22 marked the beginning of a new commercial era in disease biocontrol around the World. Although there are two commercial formulations available, this paper will focus on the powdered formulation Trianum-P. This has several label and off-label approvals for use in ornamentals, soft fruit and salads (Anon, 2019a). Additional product literature is available at www.Koppert.co.uk (Anon, 2019b, 2019c).

In early studies both of T22's parent isolates T95 and T12 demonstrated strong disease biocontrol activity and rhizosphere competence (Ahmad & Baker, 1988). The commercial production of Trianum is carried out under clinical conditions in computer controlled stainless steel fermenter vats. The computerized system ensures that the precise biotic and abiotic requirements are provided so that optimal conidiospore to biomass ratios are achieved. As with plants, the precise nutritional needs of T22 must be met in order to generate the necessary high-quality yields in a cost-effective manner.

Mechanisms of action

Gradually scientific research has unraveled the complex methods employed by T22 to gain a competitive advantage over disease-causing plant pathogenic fungi. Different effects and observations pointed to multiple mechanisms of action including mycoparasitism, antibiosis, a high competitive saprophytic ability, the production of pathogen cell wall degrading enzymes, ability to colonise the roots of plants (rhizosphere competence) and the ability to induce systemic resistance in plants. When a pathogen is under attack, the T22 hyphae coil around its host, and produce chitinase and β-1,3 glucanase enzymes which produce scars, pits and perforations in the pathogen hyphal wall causing leakage of cellular contents (Harman *et al.*, 2004a; Harman, 2006). T22 produces secondary metabolites in artificial culture and under natural conditions in the plant's rhizosphere at concentrations which are fungitoxic to pathogens (Hanson, 2005; Vinale *et al.*, 2006; Harman, 2006). T22 is rhizosphere competent and occupies the root surface preventing fungal plant pathogens access to the root surface or infection court.

The biocontrol agent is also able to penetrate root epidermal cells causing the production and release of a variety of compounds that lead to induced systemic resistance and pathogen suppression on the shoot remote from T22's original point of entry on the root (Heil & Bostock, 2002). Root application of *T. harzianum* strain T22 reduces foliar infections in tomato plants against cucumber mosaic virus (Vitti *et al.*, 2016) and *Alternaria solani* (Harman *et al.*, 2004a). Maize seeds treated with T22 increased plant resistance against foliar anthracnose caused by *Colletotrichum graminicola* (Harmen *et al*., 2004b).

Nursery trials and experiments

Vidi Parva and T22 spore germination

Trianum-P can be applied with Vidi Parva to provide root growth benefits and a disease prevention measure at the same time. Vidi Parva is a 100% natural plant-based biostimulant which is manufactured using cold compressed seaweed extracts from *Ascophyllum nodosum*. The L-tryptophan content of Vidi Parva is a pre-cursor amino acid involved in the production of plant auxins (Zhao, 2012). Specific auxins stimulate the formation of root branches and root hairs and supports the development of a strong plug plant with a well-developed root system (Nibau, 2008).

A static culture experiment was carried out on three separate occasions to determine if Vidi Parva can act as a source of nutrition and directly benefit the growth of T22 mycelium. The assessment was based on a visual estimate of the percentage volume of solution that was occupied by T22 mycelium. The most concentrated solution was prepared by adding 5 ml of Vidi Parva to 95 ml of tap water in 150 ml clear plastic bottles, and this was used to perform a 100-fold serial dilution to produce seven Vidi Parva dilutions in total. There was 100 ml of solution per container and each solution was inoculated with 0.05 g of Trianum-P. The containers were sealed with a lid and maintained at room temperature (18-21°C) for 7 days. The 0.05% test dilution was the closest concentration that matched the recommended application rate for Vidi Parva (1-2ml/m^2 in 1-2 litres of water) in commercial plant propagation situations.

T22 spore germination was evident after 2 days in all solutions except the most concentrated solution (5%). The 0.05% solution produced more biomass within the first two days than any other solution. In the weakest solutions ($5x10^{-10}$ and $5x10^{-12}$) the small quantity of mycelium was similar to the water control. The results suggest that the commercial rate used for root development purposes could feasibly promote the development of T22 when used together in a plant production environment.

Trianum-P + Vidi Funda

The activity of resident disease antagonistic micro-organisms in soils can be stimulated by the addition of organic amendments such as farmyard manure, green manures and straw. These materials are freely available but vary in composition. In early product development work, the high quality manufactured pelleted organic amendment Vidi Funda also increased the growth of various beneficial fungi. Therefore, the product was tested as a potential 'solid-substrate partner' for Trianum-P. The product is sold as a soil conditioner and organic fertiliser delivering N, P and K (7:2:4) over 4-12 weeks.

An experiment was set up to compare six treatments. There were three replicates per treatment. Each container (11x11x5.5 cm) consisted of 20 g of Klasmann bedding compost and either 2 g of Vidi Funda (low rate) or 10 g of Vidi Funda (high rate). Containers were either treated with 2 g Trianum-P, 2 g M149 (a coded trial product containing *Trichoderma spp*) or remained untreated. All containers received 15 ml tap water and were lightly mixed with a spatula. The lidded containers were incubated at room temperature and checked for fungal colonisation over 28 days. The biomass produced by M149 and Trianum-P was increased at both rates of Vidi Funda, compared to untreated controls, but more biomass was produced at the high rate. This experiment demonstrated that the colonisation of non-sterile growing media by Trianum-P was increased by the addition of Vidi Funda.

Trianum-P + Fortafol

Fortafol is a biostimulant which is formulated as a water-soluble emulsion and contains humic and fulvic acids, and plant extracts from *Thymus* and *Mentha*.

Since it is feasible that Fortafol and Trianum-P could be used as part of the same integrated crop management programme, it was necessary to test the compatibility between the two products. Previously it has been shown that the viability of fungal spores of various plant pathogens are rendered non-viable following exposure to a Fortafol solution. Further proof and clarity regarding the compatibility between Fortafol and strain T22 was required.

The effects of Fortafol, Switch and Cercobin on T22 spore germination were tested at five concentrations ranging from 0 to 1% v/v in a 100-fold serial dilution. To each dilution 0.05 g Trianum-P was added to 100 ml of the diluted solution in 150 ml clear plastic bottles with sealable caps. The pots were placed in the dark in a box at room temperature (16-20^{0}C).

In the 1% and 0.01% solutions of Fortafol, Switch and Cercobin, the spores of T22 failed to germinate after 14 days, but spore germination and mycelium development did occur in the weak solutions ($1x10^{-4}$ and $1x10^{-6}$) and water control. The recommended application rate for Fortafol as a substrate treatment is 1ml/l (0.1%v/v). Since no germination occurred at 1% and 0.01%, it is reasonable to assume that germination would not occur at the commercial rate either. This experiment confirms that Fortafol-D must not be applied as a tank mix. The current advice for a growing media drench or drip line treatment is to apply Trianum-P first followed by Fortafol-D three days later.

Trianum-P & plant growth promotion in *Tagetes*

The plant growth promoting effects of *Trichoderma* are well documented (Baker, 1988; Ousley, *et al.*, 1994). An experiment was set up to compare two Trichoderma species as growth promotors of marigold (*Tagetes sp*).

A total of 60 marigold seeds were sown in Klasmann bedding plant substrate in polystyrene trays in early March. There were two trays per treatment and 10 seeds per tray. The trays were watered, covered with a plastic film until seedling emergence and placed in a cold glasshouse with frost protection (minimum temperature of 5°C). After 7 days all seeds had germinated and had reached the cotyledon stage. Trianum-P and M149 were applied as a compost drench after 4 days. The untreated control seedlings were treated with water. Trianum-P and M149 were applied at 1.5 g/l of water and each cell received 5 ml of solution. A second treatment was applied 10 days later at a rate of 0.5 g per cell followed by irrigation to move the dry product into the compost profile.

Assessments were made 15 days and 24 days after the first treatment. Growth differences between treatments became evident from the cotyledon stage. After 15 days, the mean stem height, mean cotyledon length and mean length of the first true leaves were determined. The largest growth differences occurred in the Trianum-treated seedlings, followed by the water-treated control. The smallest seedlings were produced in the M149 treatment. After 24 days the stem height difference between water and Trianum-treated seedlings had narrowed and was no longer significantly different. The most significant difference occurred in the first true leaves, measuring mean lengths of 47.1 mm (18.5% increase), 40.4 mm and 36.2 mm (10.4% decrease) for Trianum-P, untreated and M149 respectively.

Lavandula hidcote propagation trial

Very few studies have been carried out to evaluate the combination of Trianum-P with the rooting stimulant Vidi Parva, in the propagation of hardy ornamental nursery stock.

A *Lavandula* propagation trial compared the treatment combinations of Trianum-P + Vidi Parva (TP) and M149 + Vidi Parva (MP) against an untreated control (U). Vidi Parva was included due to its positive effect on T22 spore germination and growth as reported earlier in this paper. Due to the limited number of *Lavandula* cutting trays available for this trial, it was not possible to include solo treatments of Trianum-P, M149 or Vidi Parva.

The nursery had previously experienced a high disease incidence in *Lavandula* leading to significant losses, and there were also three-week-old diseased *Lavandula* cuttings in an adjacent area when the trial was set up. There was also an unprotected period of 7 days between the sticking date and treatment date.

There were 12 trays of cuttings with four trays per treatment. There were 60 cells per tray and each tray measured 45cm x 27 cm. The cuttings were prepared on 26th April and the treatments were applied on 3rd May. Trianum-P and M149 were applied at the rate of 2.5 g/m^2 and Vidi Parva at 5 ml/m^2 in a water volume of 2 l/m^2. The cuttings were maintained on a sand bed and misted frequently to maintain humidity. After six weeks there was significantly ($P=0.05$) less disease incidence in the TP treated plugs compared to untreated ones. Other treatment mean comparisons yielded no significant differences.

Trianum-P growth effects in nursery stock

A commercial trial was set up to study the growth effects of Trianum-P in *Deutzia nikko*, *Hebe* x *andersonii* 'Variegata' and *Ceanothus repens*. On close examination some of the *Deutzia* roots were discoloured and collapsed caused by soil borne disease infections. The *Hebe* roots were also discoloured but this was considered normal for this plant species. The *Ceanothus* roots were white and healthy. A total of 126 plugs were potted-on into 9 cm pots (volume 570 cm^3) in Klasmann container substrate on 26th March. The trial consisted of seven treatments and 18 plants per treatment (three plant species x six plant replicates per treatment). All pots received two treatments, the first on the 1st April and the second on the 2nd May. After potting-on all 126 pots received 1 g of Vidi Funda (7:2:4) which was applied to each pot individually to provide plants with equal levels of fertiliser for the three-month trial period. The treatments were:

1) Untreated
2) M149 (2g/m^2)
3) Trianum-P (2g/m^2)
4) Vidi Parva (2.8 ml/m^2)
5) M149 (2g/m^2) + Vidi Parva (2.8 ml/m^2)
6) Trianum-P (2g/m^2) + Vidi Parva (2.8 ml/m^2)
7) Fortafol (2.8 ml/m^2)

Treatment solutions were applied at a rate 100 ml per pot when the growing substrate was beginning to dry. The untreated pots received an equivalent volume of plain water. The target rates were in line with the recommended rates: Trianum-P – 15 g/1000 pots, Parva – 3 l/ha, Fortafol – 3 l/ha. The trial product M149 was also applied at the rate of 15 g/1000 pots.

A. *Ceanothus repens*: The assessments carried out on the 8th May (37 days after the first treatment) identified a number of growth differences. There was a statistically significant ($P=0.05$) increase in the mean lateral length produced in Fortafol-treated *Ceanothus* plants, when compared with the Vidi Parva and water treatments, but not Trianum-P solo, or M149 solo, or combination treatments. There was no difference (P=0.05) between the number of lateral shoots produced in all treatments. On the 11th May (41 days after the first treatment) there were 72%, 44.4%, 33.3%, 27.8%, 27.8% and 27.8% fewer open flower heads in the Trianum-P, Fortafol, M149, Vidi Parva, Trianum-P+ Vidi Parva and M149 + Vidi Parva treatments respectively, compared to the control. On 7th June the Trianum-P treated plants were rated as being the best quality (based on leaf colour, number and leaf area), having the best shape (good branching; moderate gap between leaves) and having the greatest amount of spring growth overall (plant height, vegetative growth).

B. *Hebe* x *andersonii* 'Variegata': There were significantly more shoots produced in the Trianum-treated *Hebe* plants compared with the untreated control at the 5% probability level. There were no significant differences between all other paired treatment means. An assessment of root growth was carried out on the 12th May (41 days after the first treatment). Compared to the untreated plants, the greatest root biomass was produced in the Trianum-P + Vidi Parva and M149 + Vidi Parva treated plants. These treatments also produced the healthiest looking roots. Blackened disease-infected roots were present in all treatments with the lowest incidence in Trianum-P + Vidi Parva and Trianum-P solo treated plants. Root lateral branching occurred in all Vidi Parva treatments (combinations and solo) and Trianum-P solo treated roots, but lateral branching was low in Fortafol, M149 and untreated plants.

C. *Deutzia nikko*: An assessment of root growth was carried out on the 9th June (69 days after the first treatment). Compared to the untreated plants, the greatest root biomass and best root quality was produced by the Trianum-P + Vidi Parva and Trianum-P treatments.

Conclusion and Discussion

The trials demonstrated that Trianum-P can stimulate the growth of different plant species raised from seeds (*Tagetes*), cuttings (*Lavandula*) and plugs (*Deutzia, Ceanothus and Hebe*). The application of Vidi Parva as a tank mix partner can also contribute to growth and health improvement in *Hebe, Deutzia* and *Ceanothus*. This work provided some important evidence to support the use of Trianum-P and Vidi Parva in future nursery stock production for plant growth and development and disease prevention purposes. Vidi Parva yielded an increase in T22 growth in artificial culture suggesting that Trianum-P was able to utilise Vidi Parva as a source of nutrition. The provision of high (specific) levels of nutrition under natural growing conditions could be important for the biocontrol activities of mycoparasitism (enzyme production) and antibiosis (secondary metabolite production) (Jackson *et al.*, 1991a). The growth of T22 was also supported by the plant-based pelleted fertiliser Vidi Funda. It is therefore feasible that this food source could also enhance T22's disease biocontrol activities. It might also favour T22's longer-term survival prospects and greater potential for self-perpetuation (Jackson *et al.*, 1991b). It was confirmed that Fortafol is fungitoxic to T22 spores at the commercial rate and must not be applied as a tank mix. This preliminary work has given an insight into how biostimulants and Trianum-P could be used together in future crop management matrices.

Literature Cited

Ahmad, S.J. and Baker, R. (1988). Implications of rhizosphere competence of *Trichoderma harzianum*. Canadian J. Microbiology *34*: 229-234.

Anon. (2019a). Authorised pesticide products and extension of authorization for minor use. www.pesticides.gov.uk.

Anon. (2019b). How to use Trianum-P. www.koppert.co.uk.

Anon. (2019c). The Trianum and Vidi Parva foundation. www.koppert.co.uk

Baker, R. (1988). *Trichoderma spp*. as plant growth stimulants. CRC Critical Reviews Biotechnol. *7*: 97 106.

Harman, G. (2006). Overview of mechanisms and uses of *Trichoderma spp*. Symposium: The Nature and Application of Biocontrol Microbes II: *Trichoderma spp*. *96*:190-194.

Harman, G.E, Howell, C.R., Viterbo, A., Chet, I. and Loriti, M. (2004a). *Trichoderma* species — opportunistic, avirulent plant symbionts. Nature Rev. Microbio. *2*:43–56.

Harman, G. E., Petzoldt, R., Comis, A. and Chen, J. (2004b). Interactions between *Trichoderma harzianum* strain T22 and maize inbred line Mo17 and effects of these interactions on diseases caused by *Pythium ultimum* and *Colletotrichum graminicola*. Phytopathology *94*:147-153.

Heil, M. and Bostock, R. M. (2002). Induced systemic resistance (ISR) against pathogens in the context of induced plant defences. Annals Bot.*89*:503-512.

Jackson, A.M., Whipps, J.M., Lynch, J.M. and Bazin, M.J. (1991a). Effects of some carbon and nitrogen sources on spore germination, production of biomass and antifungal metabolites by species of *Trichoderma* and *Gliocladium virens* antagonistic to *Sclerotium cepivorum*. Biocontrol Sci. Technol.*1*:43-51.

Jackson, A.M., Whipps, J.M. and Lynch, J.M. (199b). Production, delivery systems and survival in soil of four fungi with disease biocontrol potential. Enzyme Microb. Technol. *13*:636-642.

Nibau, C., Gibbs, D. J. and Coates, J.C. (2012). Branching out in new directions: the control of root architecture by lateral root formation. New Phytologist, *179*:595–614.

Ousley, M.A., Lynch, J.A. and Whipps, J.M. (1994). Potential of *Trichoderma spp*. as consistent plant growth stimulators. Biol. Fert. Soils *17*:85-90.

Stasz, T.E, Harman, G & Weeden, N.F. (1988). Protoplast preparation and fusion in two biocontrol strains of *Trichoderma harzianum*. Mycologia *80*:141.

Vitti, A., Pellegrini, E., Nali, C., Lovelli, S., Sofo, A., Valerio, M., Scopa, A. and Nuzzaci, M. (2016) *Trichoderma harzianum* T22 induces systemic resistance in tomato infected by cucumber mosaic virus. Frontiers Plant Sci.*7*:1-11.

Zhao, Y. (2012). Auxin biosynthesis: A simple two-step pathway converts tryptophan to indole-3-acetic acid in plants. Molecular Plant *5*:334–338.

Efficient Water Use in Ornamental Production

Georgina Key

Resource Management Scientist, Agriculture and Horticulture Development Board

Stoneleigh Park, Kenilworth Warwickshire CV8 2TL UK

Keywords: Crop management, fertilizer, fertigation, irrigation

Abstract

Growers in the north west of Europe tend to be more technologically savvy and innovative compared with Mediterranean or eastern European regions. However, there is a need, and a desire, to apply basic practices well, and learn more about newer technologies, and the Fertigation Bible is an importance resource to help growers do this.

INTRODUCTION

FERTINNOWA was a European project that collected information on water use from several EU countries. The project concentrated on horticulture across Europe. As a result of the project, a "Fertigation Bible" was created, which brings together all current information on ways to collect, store, clean and recycle, for farmers to use, for free. https://www.fertinnowa.com/the-fertigation-bible/.

In the talk I give a couple of examples from each section of the Fertigation Bible. It is a huge document, so it is worth focusing in on areas that are relevant for you and your business.

Water provision

Water can come from multiple sources (mains, borehole, river etc.), and each can have issues associated with them. The best quality water is rainwater, which you do not have to pay for if you collect it yourself before it hits the ground. However, it is a misconception that rainwater is free, as investment is required for efficient collection and storage. Water silos of varying sizes can be used – it depends on the amount of capital you have available to invest. Alternatively, ponds and reservoirs to hold water can be anything from a small clay-lined affair to a huge, properly lined reservoir. The Fertigation Bible gives average costs to install both silos and reservoirs of varying size. It also gives examples of easy wins to improve water quality – such as adding a float to your water pump to avoid sucking up sediment – through to more expensive options such as climate models. Improving the quality of stored water can be done using four different approaches: altering chemical composition, particle removal, algal removal, and disinfection. Again, the Fertigation Bible gives information about different methods to use all of these approaches. For example, a relatively cheap method to treat algae is to use blue food dye, which cuts out the wavelength of light algae need to photosynthesise (e.g. Dyofix Pond Blue). A more expensive option is to use ultrasound pulses to disrupt algal movement.

Drip systems

There is a section on fertigation equipment, and in the talk I use drip systems as an example of the Fertigation Bible giving different cost options. Drip systems can be split into three cost levels. The cheapest lies flat on the soil surface. For a better water flow, drip emitters costing 2-3 p each can be added. Drip tape with pressure-compensating drip emitters, for use on slopes, costs just under £4 per metre, but allows more controlled irrigation over problematic contours. The most expensive form of drip irrigation is tape buried beneath the soil surface to deliver water directly to the roots. This can cost between £900-2000/ha, and can have a shorter lifespan, but can reduce evaporation and water loss.

Fertigation management

There are a host of different ways to manage fertigation in your system. The range of technology and products to help you do this is increasing all the time. It can involve whole system approaches, such as water deficit irrigation (keeping plants dry almost to the point of wilting before watering), or specific technology to measure soil water, crop water or plant tissue nutrient status. The project found differences between countries in their approaches – in the UK growers tend to send leaf samples off for analysis very infrequently, usually when a problem starts appearing, whereas growers in Germany pay for analysis on a very regular basis so that they can respond to changes in the plant before any problems occur.

Preventing pollution

Finally, there are multiple options for cleaning wastewater before it goes back into the environment. If your site is on a slope (even a slight one), it may be worth digging a ditch or pond to collect runoff or water percolating through the soil. It will collect and allow excess nutrients to settle out of the water. Duckweed can be grown to take up these nutrients. Digging a ditch or trench with reeds grown in clay pellets requires slightly more investment. Water enters one end of the ditch, the reeds take up the nutrients, and as the water progresses through the channel, the nutrients are taken up, and the much-improved water can either be reused or allowed to flow into a river. The reeds require cutting once or twice a year. Constructed wetlands are the most expensive option for collecting water runoff in an agricultural setting. They can be vast, or a long, relatively narrow series of ponds containing reeds and other marginal plant species.

These ponds slow down the flow of water, allowing the nutrients to settle out of the water. The plants then take up the nutrients. This is a slower process, but can be relatively cheap once in place, and can have the added benefit of attracting wildlife.

There are also more experimental options being developed, for instance iron oxide-coated sand. This sand is placed in a filter, and as water runs through it, the sand extracts any remaining phosphorus from it. Researchers have recently figured out how to remove the phosphorus from the sand so that it can be used again. It is currently being tested in the field, on the continent.

Nursery Production Efficiency – The Three W's: Wastage, Weeds and Water

Kyle Ross

Production manager at Wyevale Nurseries, Wyevale Way, Kings Acre, Hereford HR4 7AY UK

Keywords: Nursery production, pest management, IPM, WhatsApp.

INTRODUCTION

Wyevale Nurseries in Hereford is a wholesale nursery of hardy nursery stock including field grown trees and transplants and containerised shrubs, ferns, herbaceous and grasses. I have been in my current position for a year and a half, before that I was a management trainee for 2 years after graduating from Pershore College with a degree in Horticulture. My responsibilities include overseeing all aspects of growing, crop protection, stock control and product development for the container side of our business. I am also responsible for biosecurity and plant health for the entire business. This paper will be centred around efficiencies that I try to improve within my job role. For me the three greatest challengers as a grower are the three W's: wastage, weeds and water.

We have a dedicated product development and trial area where we record every detail we possibly can about new products, this helps us decide whether to take the crop on commercially and if so, to come up with correct growing plans, environmental aspects and watering conditions etc. We try to choose varieties that will consistently perform well and can survive in all UK climates.

Wastage

Our business supplies plants to garden centres and retail nurseries as well as to the amenity landscape market. Pictured here is a scheme we supplied at Bath University. Choosing plants that both markets can use is crucial to reducing wastage.

Shrubs are found to have the most dual marketability and are what we focus most of our attention on. Selecting and growing crops that can be sold in both markets means that we overwinter less and have the maximum sell through possible. Seen is another image of the sort of schemes we supply, Jellicoe Water Gardens.

An example of a new crop for both markets this year are Euonymus White Spire, for which we have the UK propagation license. This crop maintains its variegation as has been used for promotional garden centre sales and has been planted on new housing schemes.

Other examples of dual market crops we have recently introduced is the *Dianella* 'Blue Stream', an attractive strappy plant with glaucus blue foliage and an upright habit. *Dryopteris* 'Jurassic Gold' is a new fern which produces shoots of golden new foliage in the spring.

As all growers will know, pest and disease can be a huge thorn in their side. We have a dedicated team of four crop protection specialists onsite operating daily. We ensure the team are trained to use any equipment necessary and in the identification of pests and diseases on the huge range of crops we do grow.

Utilising the resources available from ADAS and Hutchinsons we have regular consultant visits from experts in horticultural pest and disease management. These visits really help us tailor our plan of attack on more unusual issues that may arise and provide crucial information on changes to chemical products and legislations.

Integrated Pest Management systems are in place to reduce our dependency on chemicals and therefore minimize the chance of chemical resistance within the pest population. We have a solid routine of using nematodes to combat vine weevil and *Atheta* rove beetles for *Sciarid* fly at the propagation stage. We are known for growing *Phormium* and for that reason we use *Phytoseiulus* mites to control RSM, switching to *Amblyseius californicus* during hot temperatures, these specialist predators were invaluable in last year's heatwave. *Aphidius colemani* wasps are also used to control aphid populations.

The team applies predators on crops like *Euonymus* weekly from the end of March to the end of September. These predators are dependent on daylight hours, so careful monitoring of this will make the purchasing of the natural predators more efficient.

Communication is key and were not always the best at it. Having clear lines of conversation between departments builds relationships and strengthens teams. As simple as it may seem, WhatsApp has completely changed the way we work. We all know how long it can take to get around to replying to the countless emails we receive on a daily basis. Having WhatsApp groups for production, crop protection, sales and marketing has shown to be incredibly beneficial. For example:

- All team leaders, supervisors and managers are on the groups, if the growing team is working in a crop and sees the beginning stages of a red spider mite infestation, a picture goes straight to the crop protection team and the problem is handled quickly before it is able to become a major issue.
- From another wastage point of view, we can of course lose crops to the elements, we are in the process currently of preparing for winter moving. This is a labour-intensive job, but these costs can be reduced by the moving team speaking with the sales teams on which crops need to be sold first, and they are also able to tell them which crops are at risk of going over soonest.
- The quick image sharing means that we can fire off pictures of poor stock received at the potting stage back to suppliers and request credit notes.

Weeds

Every nurseryman's most persistent enemy. Over the past few years we have noticed some species of weeds are building up a tolerance and in some cases resistance to herbicides. We have had a full review of how we use these products and have come up with a succinct plan of action. The 75-acre nursery has been split into areas, with each area being allocated a different herbicide product to use in the hope of breaking this resistance. It does appear to be working as we have had our cleanest year in the past decade.

As before mentioned we have many different teams on the ground, each of which carry out their own crop walks to generate work lists, moving space, pest and disease identification and weekly 'looking good' lists for sales. This means we have the maximum number of eyes on the ground. A weekly weed list is created which helps us to channel resources into particularly bad areas or crops that need weeding out.

Weeds on young plants or compost is of course an issue, whilst these issues have improved dramatically over the years can still be an issue. If *Oxalis* comes in on young plants, we reject the crop. Zero tolerance, the same is said for mealy bug. We do also carry out regular weed identification tests much to our team's delight.

We are ISO14001 Environmental accredited, so water usage is a key monthly measure for us. We use 3 boreholes primarily and occasionally top up with mains when necessary. Of our 70-acre site a good 30-acre area of that uses purely collected and recycled water. These beds are designed to collect and filter water through to a central sand filter and then through an ozone tank to kill of any microorganisms.

By conforming to ISO14001 standards we have also successfully reduced the amount of chemicals used on crops and increased green products like bio stimulants and natural predators by 40%.

Water

Drip irrigation is used on larger pot sized shrubs. This reduces the need for overhead watering. Understanding your crops watering requirements plays a key role in grouping plants together. Correct bed planning can not only increase water efficiency but also reduce crop wastage as you don't run the risk of over or underwatering.

Our old reservoir has been converted into a natural habitat to promote biodiversity on-site, we have seen nesting birds of various species as well as countless beneficial insects including the ever-important bees.

Smart IPM – Data Driven Decisions

Ant Surrage

Technical Development Director, Fargro,

Vinery Fields, Arundel Road, Poling, Arundel, West Sussex, BN18 9PY UK

Keywords: Integrated pest management, disease control, computing systems, biocontrol

INTRODUCTION

Integrated pest management (IPM) adoption has increased considerably within the horticultural sector across an array of crops and situations. The driving forces behind this are the increased resistance to and the loss of synthetic chemistry and increasing legislative and societal pressures. Couple this with an increase in investment in the development of alternative control methods and the increase in adoption becomes clear.

However, there are barriers for a grower to the adoption of IPM. Those principally being, control methods being less efficacious, less predictable and more complex. Many industries have grasped the concept of and seen the value in the data they collect, within horticulture and IPM, data offers significant value and potential to optimise programmes and reduce barriers to entry.

Growers primarily are interested in an "output" that is in this case improved efficacy from an IPM programme. Improved sensor technology can enhance the quality of the data set we put into a process by making it more representative, granular and consistent. Cloud-based computing systems then allow that higher quality data set to be better analysed, presented and collaborated upon which in turn leads to the insight from the data being of more value to the grower.

Controls within an IPM system, be they macrobiological or biopesticides, in a majority of cases, are living organism. As living organisms, they have adapted to perform optimally in certain environmental conditions. It is therefore extremely important that when applying controls a grower thinks about the environmental conditions and the impact this has on the pest or disease and the control method.

For example, *Beauveria bassiana,* as either Naturalis - L or BotaniGard. When *B. bassiana* is used in the correct environmental conditions, is capable of extremely efficient control of aphids, whitefly and a range of other pests. However, when used in cool, low humidity environments the efficacy is considerably lower. The ideal conditions for the fungus is between 20 - 30°C and above 60% relative humidity (RH), with sporulation taking place above 80% RH.

Dashboards like the example shown in Fig. 1 translate this information in the form of gauges and graphs giving a grower insight into how well the fungus will perform at that moment, and also the environmental patterns give an indication of how well *B. bassiana* will perform over time. This concept can be applied to most if not all the control methods in an IPM programme, from macrobiologicals to conventional chemistry.

We can start to build on this foundation by using environmental data to help predict disease and pest outbreaks and help refine the application timing of control methods.

For example, the fungus *ampelomyces quisqualis* is hyperparasite of powdery mildew. The fungus offers little curative action but as a hyperparasite requires powdery mildew to be present to survive. Therefore, there is a small window of opportunity to optimally apply the control to achieve maximum efficacy. By monitoring environmental conditions and programming alerts for risk of powdery mildew development growers can more confidently apply the product and can expect to see improve efficacy.

In summary, within an IPM system, there is a multitude of variables that are affected by environmental conditions, from lifecycle speeds of pests and diseases to feeding of macrobiologicals. It is therefore critical to the success of a programme that these conditions are properly monitored and insight from that data is easily transformed into actions that can help optimise the performance of an IPM programme.

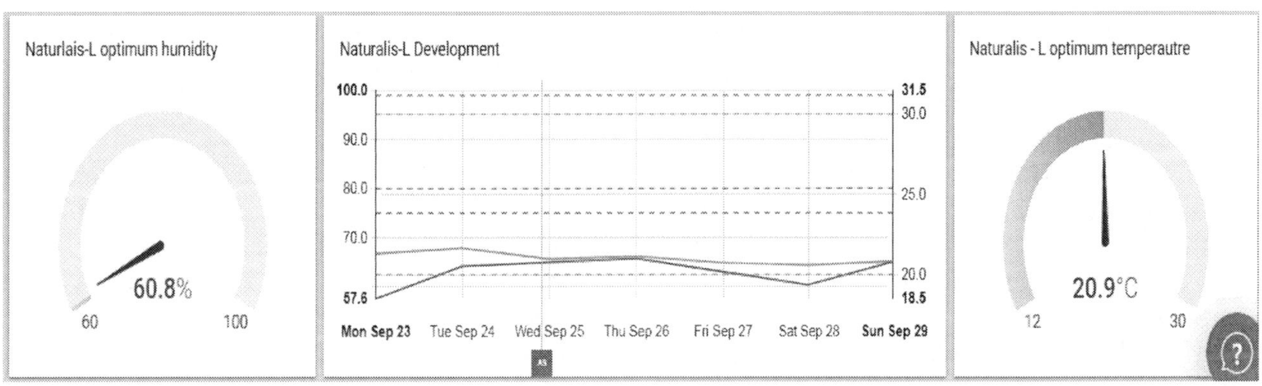

Figure 1. Image showing the environmental conditions required for Naturalis – L, showing current conditions in gauges that change colour based on conditions. With an additional line graph showing conditions over the last week with bands programmed in for the development conditions for *B. bassiana*. (Construct on ZENSIE).

Commercial Production and Showing Daffodils

Johnny Walkers

Walkers Bulbs, Washway House Farm Holbeach, Spalding Lincs PE12 7PP UK

walkers@taylors-bulbs.com

Keywords: Narcissus, forcing, Chelsea flower show, bulbs.

INTRODUCTION

I am often quoted as saying that the UK grows more daffodils than the rest of the world put together. This was certainly true 20 years ago and probably still is today as the UK area has increased slightly and the other major country has decreased its daffodil area. The crop in the UK is grown for both bulb and flower production for home and export markets.

There are 3 main areas of production in the UK. Traditionally growers in the Cornwall (the south-west) were small family units and grew daffodils mainly for flowers with the bulb very much a secondary consideration while in Lincolnshire (the eastern counties) the bulb was all important and the flower a secondary consideration as the bulbs were used for producing early flowers in glasshouses. That has all changed with the flowers being important to all areas as glasshouse production has declined.

There has been a decline in glasshouse forcing of daffodils. Bulbs are sold for pre-packing and sold in garden centres and multiples and for amenity use. Since the late 60's a substantial export market has been established for bulbs and flowers. There is an estimated 7000t of bulbs exported to countries as far apart as North America and Russia and a large tonnage to The Netherlands. Actual figures for number of bunches of flowers exported is difficult to obtain but I estimate it to be in the region of 50 - 60 million bunches.

Traditionally flower sales were through regional wholesale markets taking flowers from individual growers on a commission basis and prices dictated daily by supply and demand. Today's market is very much led by the multiples who want a consistent product at a fixed price.

Consequently, a large proportion of flowers now go through packers that have contracts with the supermarkets and buy from the growers on a fixed price. However, being a field crop the vagaries of the weather can mean shortages of flowers at critical times such as Mothering Sunday and Easter which are important to the daffodil market. This partly offset by the fact that the earliest flowers come from Cornwall, followed by Lincolnshire and Scotland bringing up the rear. In years past it was possible for a particular area to have a bonanza due to the effects of supply and demand but with fixed price contracts that is less common.

Fifty years ago, much of the labour to pick flowers was family members and local casual labour looking to earn some pocket money but today most of it is East European supplied by Agencies and will often follow the crop from Cornwall – Lincolnshire – Scotland. Flowers are picked on a piece work basis with pickers earning 10.5 pence per bunch (£10.50 per 100) and a good picker will be on 200 bunches per hour, even when they are picking in the rain with a gale up their backsides.

Once the flowers have been picked, they will be transported to a temperature-controlled store for storage at 2°C prior to transport by refrigerated trailer to the flower packer.

The daffodil crop is unusual when compared to many other agri/horticultural crops in that there is the possibility of streams of income from flowers or bulbs or both. Whereas with most crops one plants the seed and harvests the flowers or food with daffodils there is an opportunity to take flowers and bulbs and if the weather comes wrong and there is a glut of flowers and they are not picked there is an initial loss of income but there is the bulb crop to mitigate that loss. In many cases the bulb crop will be 10-15% better. One might therefore ask why not leave the flowers, but they come on stream at a time of year when cash flow is under severe strain, so it is better to take the hit further down the line.

Daffodils are usually a 2- or 3-year crop with an 8-year gap between planting on the same field. On occasion bulbs maybe left in the ground for 4 years. For maximum bulb yield a 2-year crop is best but for maximum flower yield a 3 year crop is best. A 4-year crop will give an excellent flower crop, but the bulbs harvested will be very small, take a long time to get back to flowering size and are at a high risk for pests and diseases. Bulbs are planted in August/ September after undergoing what is known as HWT (hot water treatment) or sterilizing

This is a process of immersing the bulbs in hot water at 44.4°C for 3 hours. The temperature and duration is critical. The temperature is sufficient to control daffodil eelworm (*Ditylenchus dipsaci*) and large narcissus fly (*Meredon Equestris*) and bulb scale mite (*Tarsonemus*) without damaging the bulb providing it is done at the right time. Once HWT has been completed the bulbs are dried back and planted mechanically. Whereas 50 years ago we would plant bulbs in 6 row beds with every bulb sitting on its bottom (very labour intensive) today the machine scatters the bulb in the bottom of the ridge at a depth of 13cm. Planting rate is normally 15-20 tonne per hectare which should yield 36-42 tonnes after 2 years. Lower planting densities will give a better % weight increase and bigger bulbs however this requires more land and the market for large bulbs is very limited. Daffodil bulbs when sold to pre-packers and in the UK and abroad are sold by weight and obviously there are not so many big bulbs in a tonne as there are medium size bulbs making the price per bulb higher.

Two-year schedule

After 2 years the bulbs are lifted with modified potato harvesters. In Lincolnshire, lifted bulbs go straight into trailers, while in Cornwall and Scotland bulbs are windrowed to allow Mother Nature to part dry bulbs prior to storage. Bulbs are placed onto forced air drying systems to dry prior to grading. Grading is mechanical and will separate the bulbs into sizes according to circumference, the biggest and smallest bulbs destined for the planter and the mid-size bulbs for market. Most of the large growers (one or two will be growing 1000 hectares) will be growing large quantities of a few varieties but there is also quite a vibrant market for unusual and new varieties.

Although the basic growing techniques are the same the marketing is completely different. There are several growers specializing in this market. They do not market the flowers and bulb sales are through mail order and extensive advertising either through the press or attending shows. Whereas the large grower might have 15-20 varieties I have been growing 400 varieties on 2 hectares. On some of these varieties there may only be 50 bulbs available and price per bulb will reflect this. There are two important shows as far as I am concerned but many others as well.

The most important show is Chelsea Flower Show at the end of May after the natural flowering season is finished so we must manipulate flowering time. When I first started showing in 1986, we managed to use field grown flowers but climate change and the loss of some fungicides means that is no longer possible.

Following several experiments, we have devised a programme to produce flowers when we want. As soon as I get back from Chelsea we start planning for the following year. When we start harvesting bulbs in July whilst grading the 150 bulbs of 150 varieties are pulled out (22,500 bulbs) These are placed in a temperature-controlled store at 22°C where they stay until January 2 when planting commences.

Forced daffodils start with bulbs stored until January. Bulbs planted into good quality soil-less compost approximately 6-cm deep and covered with 1-cm compost. For the post planting programme bulbs go into temperature-controlled storage for 10 days at 10°C followed by 13 weeks at 2°C. Three weeks before the Chelsea show bulbs go into the glasshouse. Although, the initial display only uses 70 varieties we know that some will not make it for one reason, or another so like to have plenty in reserve.

Until we start picking flowers we can water overhead. Hopefully, ten days after housing we will start picking the first flowers. We are however still at the mercy of Mother Nature if we have dull cold days the flowers will not grow even with heat on, but more worrying is if it very sunny and hot the flowers grow too quick and face upwards instead of sideways. We have a selection of early, mid-season and late varieties and they will flower in that order. As a variety starts to flower, we may only get five flowers the first day and then 10 the next day and so on. As the flowers reach optimum size and colour they are cut and put into buckets of water in the store at 2°C.

Transition to responsibly sourced growing media in UK horticulture (CP 138)

Chloe Whiteside

Horticultural Consultant
RSK ADAS Ltd. Boxworth, Cambridgeshire CB23 4NN UK

Chloe.whiteside@adas.co.uk

Keywords: Integrated pest management, disease control, computing systems, biocontrol

INTRODUCTION

CP138 "*Transition to responsibly sourced growing media use within UK Horticulture*" is a five year project funded by Defra, AHDB Horticulture, Growing Media Manufacturers and Growers. The project is led by RSK ADAS Ltd with project partners Quadram Institute Bioscience and Stockbridge Technology Centre. The project will develop confidence in the use of alternative growing media materials to diversify a market that has been dominated by high performing peat products for many years. The pressure to seek other materials has come from a combination of government environmental policy and consumer preference for plant products produced in "peat alternatives".

Commercially available growing media, other than peat, is grouped into four main raw material types: coir, wood fibre, bark and green compost. Over the last 20 years much progress has been made by the growing media manufacturers in the reliable sourcing and conditioning in sufficient quantities of each material. In some sectors such as soft and cane fruit there has been a successful switch to coir from peat based growing media. Coir alone, however, is not suitable for all plant types and production systems and sufficient, high quality amounts at an affordable price could not be sourced to replace peat; furthermore, it would also again mean dependence on a single raw material type. It is appropriate on a sustainable availability, supply, performance and cost basis to mix up to four raw materials in a "blend", to produce commercially acceptable "peat alternative" plant products in containers and blocks.

In sectors which are the largest users by volume of growing media and where peat dominates (hardy nursery stock and bedding), growers have found that peat-reduced growing media, typically 25% other materials, can produce reliable and consistent results. Beyond this and towards 40-50% reduction can be described as “super reduced” and at this level and up to 100% peat free, then results have been variable, or just not suitable from a practical mechanisation and growing system perspective.

As an industry, to make the cross-sector leap beyond an average inclusion rate of 25% for materials other than peat then there has to be a reliable way to predict the performance of “peat alternative” blends. To date the only way to test 100% peat-free blends has been to conduct stand-alone trials. If, however, the raw materials change between testing and manufacture for supply then there can be some discrepancy between expected and actual commercial plant performance.

To develop sufficient experience, knowledge and confidence in alternative material blends, can be time consuming. There is a need therefore to short-circuit this process and be able to reliably predict the performance of blends at the point of manufacture; this is the main deliverable of CP138. If this can be achieved then it will not only increase the range of materials that can be sourced and used by the horticulture industry but expedite the uptake of alternative materials that can perform as well as, or better than, the industry standard, peat.

To achieve this, a programme has been developed, covering research and development, knowledge exchange, demonstration trials and dissemination of best practice throughout all the relevant horticulture sectors; vegetable and salad propagation, protected edible crop production, mushroom production, soft fruit propagation and production, top fruit propagation and production and ornamentals propagation and production (including container-grown plants).

All data generated from the trials work goes into developing a final model, which can be used to create growing media blends with particular characteristics to produce plants of a certain specification. The model itself is based on three key physical properties of growing media; Air-filled porosity (AFP), Available water (AW) and dry bulk density (D_b) (presentation slides 5, 10, 13, 20). Prototypes used in the trial were initially created on the basis that their physical properties would be similar to peat (slide 10). Prototypes created later in the project were then used to explore a wider range of the parameters offered by the materials available, and so were deemed more ‘extreme’ blends.

The final year of the project (2019) has been designed to test the model, by taking new materials, characterising them for AFP, AW and D_b, and creating blends based purely on those characteristics.

For further information please refer to the annual reports which can be found at: https://horticulture.ahdb.org.uk/project/transition-responsibly-sourced-growing-media-use-within-uk-horticulture

PROCEEDING'S PAPERS

IPPS Singapore Symposium - 2019

Hayden Foulds, Special Editor

Singapore, Malaysia

IPPS Singapore Symposium 2019

Hayden Foulds

40 Gordon St, Woodville, New Zealand 4920

haydenfoulds@gmail.com

Keywords: Parks, nurseries, greening

INTRODUCTION

At first glance, Singapore might seem to be an odd choice to hold an IPPS symposium. With a population density of 7800 people per square kilometre (New Zealand has 18), one would wonder if there was any room to grow plants. Add in the tropical climate of being located one degree off the equator, one would wonder what would actually grow in such a climate?

But horticulturally, there is a bit happening in Singapore despite the challenge of being a city state. In 1967, the vision of a city in a garden was introduced putting greening initiatives at the forefront of future development. Almost 10% of the land area has been set aside for parks and reserves. Recent developments like Gardens by the Bay have also lead to Singapore becoming a horticultural hotspot in Asia.

So after months of planning by Eddie Welsh and Philip & Janine Smith, Paul Fisher from the University of Florida and Gardens by the Bay staff in Singapore, the IPPS Singapore Symposium was underway. 34 members of IPPS New Zealand joined others from Australia, Thailand and India for five days of visits, lectures and sightseeing around Singapore.

The first day was an all-day field trip around Singapore. The first visit was to Hua Hng Trading Co Ltd, one of the largest general nurseries in Singapore which has an extensive selection of plants sourced from all over the world. The idea that plants were brought in from overseas and sold was of course interesting for us from New Zealand but then I guess with so little room to grow things in Singapore, it helps to widen what is available. Plants come from Malaysia, Thailand, China and Australia with a small percentage grown on site. Hua Hng sells direct to the public as well as garden centres, landscapers and other nurseries. Being in the tropics, there were many plants found you would not see in New Zealand on display and there tended to be large numbers of each line, rather than the usual 10 or less you typically find in New Zealand. The most expensive lines were bonsai large grade specimens of Casuarinas at over S$13000! It was also interesting to see potted watermelon plants for sale complete with ripening fruit!

Figure 1. Potted watermelon plants for sale, complete with ripening fruit at Hua Hng Trading Co Ltd.

Our next visit was to Woon Leng Nursery, a specialist orchid nursery which specialises in the production of Phalaenopsis orchids. Woog Leng were the first to do these in Singapore and buy the plants in via tissue culture from Taiwan, then grow them on in climate controlled greenhouses before being sold. Producing them this way keeps the cost down and also reduces the damage that can be done to flower spikes during transport. It is a very labour intensive operation with 40 staff employed in the business. The orchids are sold all around Singapore to private customers as well as for displays in public areas. They looked stunning in bloom in the nursery.

Figure 2. Phalaenopsis Orchids, a specialist but labour intensive crop for Woon Leng Nursery.

After lunch, we visited the Kampung Admiralty. This is a unique development which integrates housing for the elderly with a range of social and commercial services including childcare, so there is somewhere for the grandkids to go. On the roof of the 11 story complex is a community garden which is terraced. This provides space for people to go, meet up with others and also to tend the community garden. At times it was hard to believe you were on the top of a building in one of the most densely populated cities on earth. The development will be the prototype for similar developments in the future and has already won several international architecture awards.

Figure 3. Part of the rooftop garden of the Kampung Admiralty development.

We then visited the green wall at the Heeren Shopping Centre just off Orchard Road. The six-story high green wall was developed by Greenology and is a substrate panel system with an aluminium backing. The panels were planted and grown in a nursery before being installed on the wall. Once up, it basically looks after itself with minimal maintenance required.

Figure 4. The six-story high green wall at the Heeren Shopping Centre.

The next day was spent at Gardens by the Bay where we were very kindly hosted by CEO Felix Loh and his team. Opened in 2012, the gardens had 12 million visitors in 2018 and is one of the most popular tourist attractions in Singapore.

The first part of our visit was a guided tour around the gardens, fortunately we were driven around the 101-hectare site. We were able to take in the Skyway, a 128 metre suspended walk between two of the large super trees which are another feature of the gardens. At 22 metres high, it gave users great views over the park and surrounds, including the magnificent Marina Bay Sands Hotel with its three columns, topped with a large platform. We also visited the Floral Fantasy, the latest attraction opened at Gardens by the Bay. It has been designed to cater for those that don't have time to visit the larger conservatories but still want to see a floral display on a smaller scale. There are four separate themed areas which you walk through and then you can end your time with a 4D ride 'Flight of the dragonfly' through the gardens.

We also had time to explore both the Flower Dome and the Cloud Forest conservatory and marvel at the displays inside both. The Flower dome has nine gardens inside ranging from the stunning Baobab trees from Africa through to an Australia garden and an olive grove with centuries old trees growing. At the centre is the flower field which is changed on a regular basis. For our visit, there was a Christmas themed display of Poinsettias on show – I never knew how many different types there are, and it certainly looked magical for Christmas.

Figure 5. The spectacular display of Poinsettia in the Flower Dome

The Cloud Forest conservatory features a 35-metre-tall 'mountain' which you take a lift to the top and then walk down via a series of elevated walkways and internal stair. The inside of the mountain also has things to see and do including a display of crystals and interactive displays. At the top of the mountain is a display of carnivorous plants (complete with Lego® versions) and then as you wander down, you get to fully appreciate the complexity of the planting on the side of the mountain and what it must take to keep it looking good.

Figure 6. The 'mountain' inside the Cloud Forest Conservatory

After lunch, it was the start of the lecture program which were all grateful it was being held in an air-conditioned room. Mr Felix Loh, CEO of Gardens by the Bay welcomed everyone and spoke fondly of his time studying in New Zealand at Massey University. He then spoke about the challenges of maintaining Singapore's reputation as a garden city and how Gardens by the Bay helps with this. Philip Smith on behalf of the IPPS New Zealand Region presented Felix with an IPPS plaque

Dr Puay Yok Tan from the National University of Singapore spoke about the motivations, benefits and challenges of greening Singapore. There have been remarkable results from greening Singapore over the last five decades and one quote he used from the founding Prime Minister of Singapore Mr Lee Kuan Yew "Greening is the most cost effective project I have launched". He spoke of the environmental, social and cultural benefits that greening has brought but also of some of the challenges ahead, especially the loss of secondary forests in Singapore.

Chris Dalzell, originally from South Africa, but now with Gardens by the Bay wowed us with beautiful photos of many South African plants suitable for the tropics. He spoke of the diverse and unique flora of South Africa which makes up 9% of the worlds plant species but also of the threats including the unlimited ability for people to collect plants from the wild for traditional medicine.

Following dinner hosted by Felix and his team, we enjoyed the light and sound show under the Supertrees – an amazing experience and not to be missed as a part of a visit to Gardens by the Bay.

A visit to the Singapore Botanic Gardens started the next day off. Celebrating its 160th year, the 82-hectare gardens are the only tropical garden to be designated a UNESCO World Heritage Site. The gardens started off as a public park developed by the Singapore Agri-Horticultural Society and in the late 1800's, played an important role in research into rubber production. In recent times, the garden has played an important role in the development of Singapore as a garden city. In 2009, another 18 hectares was added to the gardens and this has been developed into a number of different attractions.

During our visit, we saw the giant palm *Lodoicea maldivica* which blew us away with its giant fronds. The curtain of roots from the Curtain Ivy *Cissus verticillata* were also very impressive to see as were some of the bonsai on display. An area of tropical rainforest older than the gardens themselves also gave an insight into what Singapore looked prior to human settlement.

Figure 7. The giant palm *Lodoicea maldivica*.

We were also able to look around the National Orchid collection which contains over 1000 species and 2000 hybrids. Part of the collection is the VIP Orchid collection where orchids have been named for famous visitors to the gardens which range from Princess Diana to Nelson Mandela. It was surprising to see orchids named for two former Governor Generals of New Zealand – Sir Anand Satyand and Sir Jerry Mataparae on display. A very impressive collection of orchids and well presented in a garden situation.

We left the gardens just as the heavens opened to head back to Gardens by the Bay and an afternoon of presentations. Thomas Seo spoke about the three main areas of research and development at the gardens: Orchids including the breeding program at Gardens by the Bay, Flower Trials and production which involves controlling of flowering and longevity; and sustainability which is becoming an important aspect of how the gardens are operated.

Mei Leng Lim spoke on the orchid program at the gardens which includes managing them in displays, trialing them for flower longevity and breeding new hybrids. Breeding focuses on creating hybrids between highland and lowland species but also those with better colours, more flowers and compact habits.

Carly Anderson spoke on growing spectacular hanging baskets. A display of 100 baskets is planned for the Flower dome in September and since established baskets are difficult and costly to freight, they are being grown in house. Carly ran through some of the considerations for baskets from what substrate to use to plant material and managing pests and diseases.

Dex Chen spoke about sustainability at the gardens including making better use of resources and minimising waste. One example was the use of old Christmas trees (real) which were sent to Singapore Zoo and used in enclosures as a tool to keep the animals amused.

Dr Sladana Bec then presented on Disease Diagnosis in Plants. This included the interaction between the host, pathogen and environment in diagnosing diseases and differences between fungal, bacterial and viral infections. It was also good to have a refresh course on differences between signs and symptoms of disease and that not all symptoms are caused by disease. This was followed by a hand on workshop where various options were used for diagnosing diseases.

The final day of the Symposium started with the final session of presentations. Dr Nura Abdul spoke about Mr. Lee Kuan Yew's vision of a city in a garden and how this has been achieved. Dermot Molloy from the Royal Botanic Gardens in Melbourne spoke on the development of the Oman Botanic Gardens, an exciting development underway in the Middle East. The gardens will cover 423 hectares and will be one of the largest in the world. It will also house almost all of Omans flora from deserts to monsoon forests. Dermot spoke about some of the unique plants found in Oman and of the challenges in setting up an onsite nursery at the Botanic Gardens.

Fellow Australian David Hancock from Western Australia spoke on developing a nursery for a wetland revegetation project in Oman. Large quantities of water are produced from the oil fields of Oman which is separated out but still contaminated with residues. Previous practice has been to pump this back underground, but an engineering solution was sought to develop a wetland to treat the water. 250 hectares was required to be planted in four species so a nursery was set up. This presented numerous challenges which David and his team had to overcome in the successful delivery of the project. One of the takeaway points David made was that horticultural skills can make the difference between success and failure.

Paul Fisher then spoke on Phosphorus and the ins and outs of how it affects plant growth including what deficiencies look like and its use in nurseries vs landscape production. Yours truly then spoke on Roses of New Zealand featuring some of the more prominent breeders and their creations. Eddie Welsh brought the presentations to a close by speaking on the 'Spirit of IPPS' and why he thinks it has been so successful around the world.

The afternoon saw more visits, first up to Khoo Teck Puat Hospital. No, no one needed to see a doctor from the group but rather we were here to see this green hospital. As many were to remark during our time here, it did not feel like we were in one. The hospital has extensive plantings including a large area between two of the buildings landscaped like a park, complete with water features. The roof of one of the buildings features a roof top garden where vegetables and fruiting plants are grown with produce given to volunteers and staff. There was also a medicinal garden with plants that "heal, thrill and kill", coloured coded of course by the end outcome.

Figure 8. The rooftop garden at Khoo Teck, Puat Hospital

It was then onto Changi Airport to visit Jewel Changi. This is a nature themed entertainment and shopping complex located between three of the passenger terminals. Starting at the top, there are numerous activities to do including a maze, slides and a glass floored walkway out over the edge. The centre is dominated by a large waterfall cascading from the roof which is surrounded on the outside by plantings of thousands of trees and shrubs making it feel like a tropical rainforest. We were treated to a light and sound show complete with falling fake snow. This certainly makes visiting an airport and enjoyable experience – take note Auckland, this is how it can be done.

Figure 9. Changi Jewel.

With that, the rest of the tour boarded the bus back to the hotel while I went and checked in for my flight home to New Zealand. Five days went by very quick but visiting Singapore was a fantastic experience, even if the heat made things hard going at times (thank goodness for air conditioning and a hotel swimming pool).

A big thank you must go to all of those who hosted us on visits in Singapore, you all made us feel very welcome and nothing was ever a problem. Thank you also to the Gardens by the Bay staff and especially Chad Davis and Mihkaali Ng for ensuring everything went smoothly for our visits.

Also, a big thanks to Eddie Welsh who did a lot of behind the scenes work that made everything come together so well for the symposium.

And lastly a big thank you to IPPS New Zealand whose assistance through a Plant Production Scholarship enabled me to travel to Singapore for this event.

Growing Spectacular Hanging Baskets and Showcasing Horticultural Excellence

Carly Anderson

Gardens by the Bay, 18 Marina Gardens Drive, Singapore 018953

carly.anderson@gardensbythebay.com.sg

Keywords: Gardens by the Bay, hanging baskets, container production

INTRODUCTION

Under the Research and Horticulture Department at Gardens by the Bay, a team of growers produce novel and spectacular plants and flowers to be showcased in cooled conservatories. Singapore, a non-agricultural county, has limited options for purchasing finished flowering plant material. To mitigate supply and delivery risks, and to showcase our horticultural excellence, the team of growers produce plants in glasshouses throughout the year for breath taking floral displays.

Growing Facilities

Located at the Gardens, 9 state-of-the-art glasshouses equipped with air conditioning, shade systems and photoperiod lighting enhance 3,000 m^2 of premier growing space. Singapore's equatorial location provides outdoor conditions that consist of high daily temperatures, high humidity and day neutral photoperiod.

Our air-conditioned glasshouses with photoperiod lighting provide the team with cooler temperatures that are necessary for our herbaceous flowering plant material. These glasshouses are also used for orchid collection and breeding, plant production, flower trials, quarantine, and imported plants holding areas.

Figure 1. Part of the growing facilities at Gardens by the Bay.

Producing Spectacular Hanging Baskets

Logistically, importing hanging baskets can be costly and often results negatively on over all plant quality. Producing hanging baskets within our growing facilities allows for greater control on quality and helps the growers to achieve the necessary specs for our in-house end users and customers.

Production Goals

In order to achieve a desirable final product, we strive to reach goals that are driven by our carefully curated standard operating procedures. Having a structured process in place ensures uniformity and quality. Schedules are created with the production tasks and growing times in mind to finish the hanging baskets for our internal customers and floral designers. With the partnership and collaboration of international breeders we are able to produce high quality with the newest genetics available on the market. The use of numerous plant growth habits, forms and textures gives us the freedom to create combinations that are difficult to procure and not available in the marketplace.

Keys to Growing Spectacular Hanging Baskets

Before a production begins the Research and Horticulture growers will carefully plan what is required to successfully grow hanging baskets. Spacing, grow time calendars, and culture guidelines are carefully drafted and implemented prior to the beginning of production. The cultural guidelines include substrate needs, pot size, plant material and quantity, moisture management, nutritional programs, photoperiod requirements and pest and disease programs. Having a detailed plan in place before production begins allows for a smooth growing season.

Figure 2. An example of a finished hanging basket

The Hanging Gardens

The Flower Dome, a glass biome that replicates the cool-dry climate of the Mediterranean, hosts 7 annual changing floral displays each year at Gardens by the bay. In September 2020 the 1st ever hanging basket display, "The Hanging Gardens", will be showcased at the Gardens. The Research and Horticulture growers will be supporting the show by producing 1,000 – 40 cm baskets of spectacular flowering material for the floral designer. This opportunity to showcase our horticultural excellence will be on display for all of our local and international visitors to admire.

Sustainability by the Bay

Dex Chen

Gardens by the Bay, 18 Marina Gardens Drive, Singapore 018953

dex.chen@gardensbythebay.com.sg

Keywords: environment, energy, water, heating, cooling

INTRODUCTION

Underlying the concept of Gardens by the Bay are the principles of environmental sustainability. For Bay South, the first phase of the Gardens' development, a concerted effort was made to plan and design for sustainable cycles in energy and water.

Cooled Conservatories

Comprising two glass biomes, the cooled conservatories – Flower Dome and Cloud Forest - replicate the cool-dry climate of the Mediterranean and semi-arid sub-tropical regions and the cool-moist climate of the Tropical Montane region respectively. They house a diverse collection of plants that are not commonly seen in this part of the world.

The energy used for powering the chillers is produced by a biomass furnace that burns horticultural waste collected from parks and gardens island wide.

The cooled conservatories are a statement in sustainable engineering and apply a suite of cutting-edge technologies that provide energy-efficient solutions in cooling:

The two structures are fitted with a glass material that allows optimal light in but cuts out a substantial amount of heat.
They also apply the strategy of cooling only the occupied zones, thus reducing the volume of air to be cooled.

The air in the conservatories is first de-humidified by liquid desiccant, which reduces the moisture content of the air. This cuts down the amount of energy required in the cooling process.

This suite of technologies can help to reduce energy consumption, compared to conventional cooling technologies.

Supertrees

The Supertrees reflect the form and function of mature trees to illustrate the emergent giants of the rainforest.

Ranging in height from 25 to 50 metres, the 18 tree-like structures serve as unique vertical gardens showcasing a diverse variety of bromeliads, ferns, orchids and tropical flowering climbers, on a scale never before presented in a garden.

Eleven of the Supertrees are embedded with environmentally sustainable features. Some have photovoltaic cells on their canopies to harvest solar energy to offset the energy for lighting up the Supertrees at night. Others are integrated with the de-humification process and serve as ventilation that expels hot, moist air.

Lake System

The Gardens' lake system acts as a catchment and supplies all the water required for irrigating the outdoor gardens. In addition, it also functions as a living system, providing aquatic habitats for biodiversity such as fishes and dragonflies.

Encompassing two main lakes – the Dragonfly Lake and Kingfisher Lake, the lake system is designed to be an extension of the Marina Reservoir. Water run-off from within the Gardens is captured by the lake system and channelled through filter beds, comprising aquatic reeds, and wetlands, before being discharged into the reservoir.

The lake system exemplifies the role and importance of plants in the healthy functioning of our ecosystem. It raises awareness of the value that aquatic plants play in nature and highlights the significance of water in sustaining biodiversity.

The Journey Continues

Moving forward, the Gardens will go beyond current sustainability practices and venture into new, sustainable ways to optimise operational efficiency. We are exploring long-term approaches in charting the sustainability journey, particularly in our efforts to protect the environment, a duty that is expected of a world-class garden attraction.

Extra Phosphorus for Flowering and Other Myths

Paul R. Fisher[1] and Jacob H. Shreckhise[2]

[1] University of Florida, PO Box 110670, Environmental Horticulture Dept., Gainesville FL 32611-0670, USA; [2] United States Department of Agriculture - Agriculture Research Service, Agricultural Engr. Bldg., 1680 Madison Ave, Wooster, OH 44691, USA

pfisher@ufl.edu

Keywords: Nursery production, Landscape Management, Fertilizer, Flowering, Controlled Release, Best Management Practices

INTRODUCTION

Do you use a high-phosphorus fertilizer such as 10-10-10 or 14-14-14 to promote roots or flowers? If so, you are following an outdated recipe for nursery production and landscape management. Hopefully, this article will convince you to avoid that practice. In the process, you might save on fertilizer costs, protect natural water resources, and avoid an environmental damage lawsuit!

Phosphorus: A key plant element

Phosphorus is an essential plant nutrient with an interesting mythical origin. This 13th element in the periodic table was discovered by the alchemist Hennig Brand in Germany in 1669 in his quest for the Philosopher's Stone which would turn other metals into gold.

Hennig experimented with thousands of liters of human urine, figuring that it had a promising golden color. He discovered an interesting residue which would not turn lead into gold but did burn and glow in the presence of oxygen (hence its modern use in match heads). The element was named phosphorus, meaning "bearer of light". Appropriately, given the raw material from which it was first derived, its chemical symbol is P.

Phosphorus is one of 12 essential fertilizer nutrients (Figure 1). The ability of P to add and drop electrons leads to a major role in plant metabolism as the plant's battery, allowing energy from photosynthesis (the sun) to be used for many processes in the plant. Like other essential plant elements and environmental conditions, lack of P can limit plant growth and flowering. However, it is not specifically a root- or flower-promoting nutrient.

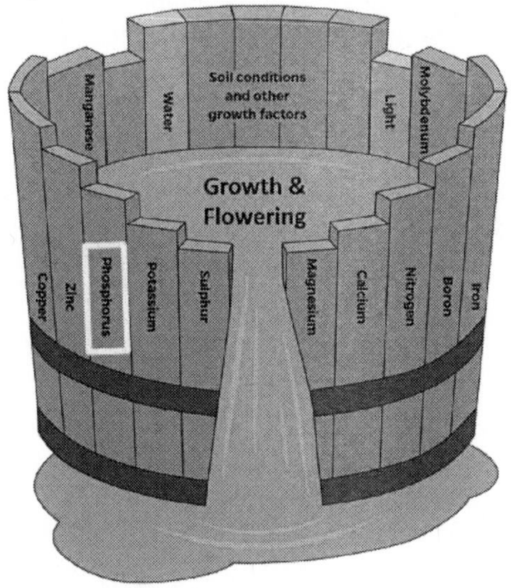

Figure 1. Phosphorus is one of the essential requirements for plant growth.

Phosphorus is the "P" in NPK fertilizers (along with nitrogen, N, and potassium, K). The traditional way to write fertilizer phosphate is P_2O_5, and K_2O is commonly called potash. Depending on your location and supplier, blended fertilizers are often expressed in two ways. Fertilizer labels can list N–P_2O_5–K_2O (which we will use in this article), or as elemental NPK. This can cause challenges when interpreting fertilizer recommendations:

- Nitrogen (N) is always described on a fertilizer label as elemental N.
- Elemental phosphorus (P) = $P_2O_5 \times 0.4364$; P_2O_5 = elemental P $\times$ 2.2915.
- Elemental potassium (K) = K_2O (potash) $\times$ 0.8301; K_2O (potash) = elemental K $\times$ 1.2047.

A common symptom of P deficiency is purple or red leaves (Figure 2).

Figure 2. Purpling of leaves from phosphorus deficiency in verbena and kale.

However, avoid diagnosing P deficiency based on physical appearance alone. As shown in Figure 3, purpling can result from several other causes, such as cold temperature, nitrogen deficiency, excess light level, and pesticide phytotoxicity.

Figure 3. Red or purple coloration can be caused by several factors, not just P deficiency.

Some species show other P deficiency symptoms, such as tip burn in azalea or decreased growth (Figure 4). To confirm a P deficiency, it is therefore necessary to send soil and leaf tissue samples to an analytical laboratory.

You can ensure adequate P in container crops by providing a moderate fertilizer level and substrate-pH. Phosphorus deficiency may result from lack of fertilizer. When growing in container substrates, P will remain available if the substrate pH is less than 6.5. Insoluble calcium phosphate can form at high pH (above 6.5) in both field soil or container substrates. Phosphorus can also be tied up in acidic field soils, because high levels of iron and aluminum in mineral soils reduce P solubility at low pH.

Figure 4. Evergreen azaleas show tip burn (top) and hydrangea ‘Limelight’ has decreased growth (bottom) in response to phosphorus deficiency. Research by Virginia Tech University.

Excess P fertilizer can have negative environmental impacts. Because the soluble, plant-available form of P (phosphate, PO_4^{3-}) has a negative charge in the substrate solution, it can easily drain from the substrate (just like nitrate, NO_3^-) and subsequently leach into groundwater or runoff to nearby streams and lakes. Around 40 to 70% of P fertilizer typically leaches from nursery containers. This is a wasted expense, which is important because and the cost of P fertilizer is rising quickly and profit margins are often tight for horticulture businesses.

Phosphorus and N contamination are the main algae-promoting nutrients in natural waterways, and a very low concentration (0.1 ppm P) can be enough to trigger eutrophication. Mining P from calcium phosphate in the ground can also cause a significant waste issue, which has been experienced in phosphate mines in Florida, US. As the "green industry", we want to both stay out of the spotlight and be good stewards of the environment.

Figure 5. Phosphorus is easily leached from container substrates and causes eutrophication of water resources.

Do high P fertilizers promote flowering?

Blooming annuals: Many fertilizers sold to consumers as flowering fertilizers contain high levels of P. For example, the N-P_2O_5-K_2O ratio of one retail "Bloom Plus" product is 10-54-10. However, how much P from a water-soluble fertilizer (WSF) do plants need to flower? Henry and Whipker (2015) from North Carolina State University showed that 5 to 10 parts per million (ppm) of P provided with each irrigation using a WSF is adequate for most flowering and foliage annuals. Additional P did not increase blooms or growth. Another research team from North Carolina State University (Kraus et al., 2011) reported similar findings for herbaceous perennials, citing that growth and flowering of *Rudbeckia* and hibiscus fertilized with 3 ppm P were similar to those given 50 ppm P. In fact, 100 ppm P (with 100 ppm N) severely decreased growth of 'Luna Blush' hibiscus. This is just another reason why it's better "to aim low" when selecting your P fertilizer levels.

Table 1 will help you calculate how much P is in a WSF. The concentration of blended WSF is usually described in terms of ppm of N. Typical constant WSF concentration for annuals is based on 100 ppm N or up to 200 ppm N for heavy-feeding crops such as petunias. With different N-P_2O_5-K_2O ratios, this means the ppm of P increases if the overall fertilizer concentration is increased, or if we choose a fertilizer with a high P_2O_5 content relative to N. If we used the Bloom Plus fertilizer, even at a low N level (100 ppm N), we would be applying 236 ppm P (which is 20 times the amount that a plant would need for flowering). Instead we could apply a fertilizer such as 15-5-15 at 100 ppm N, and not waste all that P fertilizer.

Table 1. The concentration of elemental P in a water-soluble fertilizer with different N-P_2O_5-K_2O ratios when applied at two nitrogen concentrations (ppm N). Levels in bold italics are close to recommended 6 to 13 ppm of P for maximum flowering (for example, 100 ppm N from 15-5-15 provides 11 ppm P).

	ppm of elemental phosphorus (P)	
Fertilizer N-P_2O_5-K_2O	**At 100 ppm N**	**At 200 ppm N**
10-54-10	236	472
20-20-20	43	86
20-10-20	***22***	43
15-5-15	***11***	***22***
13-2-13	4	***9***
15-0-15	0	0

Flowering shrubs: Perhaps you grow flowering shrubs and therefore think your plants are different and need higher P. In fact, research has found that only 5 to 10 ppm P on a constant feed basis (similar levels to annual plants) is needed for maximum growth and flowering. Best management practices (BMPs) adopted around the US for nursery growers to provide healthy growth (and to avoid environmental law suits) is to only use 5 to 15 ppm P when applying constant WSF.

Table 2. Research showing the minimum ppm P in constant water-soluble fertilizer required to maintain maximal growth in flowering shrubs.

Study	**Plant Taxa**	**ppm P**
Graca and Hamilton, 1981	*Cotoneaster divaricatus*	5
Yeager and Wright, 1982; Wright and Niemiera, 1985	*Ilex crenata* 'Helleri'	5
Havis and Baker, 1985	*Rhododendron* 'Victor'	2.5
Havis and Baker, 1985	*Cotoneaster adpressus* var. *praecox*	10
Shreckhise, Owen & Niemiera, 2018	*Hydrangea paniculata* 'Limelight'	5
Shreckhise, Owen & Niemiera, 2018	*Ilex crenata* 'Helleri'	1
Shreckhise, Owen & Niemiera, 2018	Azalea hybrid 'Karen'	3

Figure 6. Growth and flowering response of hydrangea and holly to several CRF products. Research by Virginia Tech University.

Controlled release fertilizer: You may use controlled release fertilizer (CRF) and it is hard to relate these WSF concentrations to your nursery. Research at Virginia Tech found that 0.3 to 0.6 grams P per 1-gallon (3.8L) pot provided maximum growth of holly and hydrangea. To achieve that level of P depends on how much fertilizer you apply to a container and the N–P_2O_5–K_2O ratio. For example, plants grown with 18-3-12 or 18-4-12 CRF applied at the medium recommended label rate had as much growth and flowering as a 15-6-12 CRF that contained up to twice the P level (Figure 6).

Table 3 shows examples of the P contribution from three common CRF products. A high-P fertilizer (14-14-14) greatly over-applies P if it is top-dressed at a high rate of 24 grams of fertilizer per container.

Table 3. The grams of elemental P from controlled release fertilizers with different N-P_2O_5-K_2O ratios when top-dressed at two weights of total fertilizer per container. Levels in bold italics are close to recommended 0.3 to 0.6 grams of P for maximum growth (for example, 12 grams of fertilizer per 3.8L container from 15-9-12 provides 0.5 grams P).

	Grams of elemental phosphorus (P) per 1 gallon (3.8 liter) nursery container	
Fertilizer N-P_2O_5-K_2O ratio	**Top-dress 12 grams fertilizer per container**	**Top-dress 24 grams fertilizer per container**
14-14-14	***0.7***	1.5
15-9-12	***0.5***	0.9
13-3-13	0.2	***0.3***

Landscapes: Are you still not convinced? Perhaps your business is in landscape maintenance. Many landscapes where manures, composts, and general fertilizers have been applied in the past are already high in P. You might be able to save money without decreasing plant performance by not applying any P fertilizer. Best management practices for P fertilization in landscapes is to take a soil test first (this is actually a requirement in some areas of the US). If the soil analysis shows high P, no fertilizer is needed (or even permitted in some areas). For example, Florida has porous sandy soils, high rainfall, a high-water table, and a subtropical climate, and is therefore very sensitive to algal blooms. Landscapers in Florida are required to have BMP training and follow several guidelines, such as:

- No more than 0.25 kg P_2O_5/100 m^2 per year may be applied to urban turf without a soil test. A one-time application of up to 0.50 kg P_2O_5/100 m^2 is permitted for establishment of new turf.
- Annual landscape rates for established Florida garden beds (kg per 100 m^2 per year) are up to 1.0 kg N, 0.5 kg P_2O_5 = 0.2 kg P, and 1 kg K_2O = 0.8 kg K, using slow-release fertilizer or compost in order to reduce rapid leaching.

Recommended rates vary depending on the plant, soil, and location, but we recommend reviewing BMP documents provided by university extension services. Free fact sheets are available from University of Florida IFAS Extension (see https://ffl.ifas.ufl.edu). Note that when interpreting US units, 1 lb/1000 ft^2 = 0.5 kg/100 m^2.

Do high P fertilizers promote rooting?

Here is the kind of misinformation that permeates the internet: "[I]f you want a fertilizer that supports root growth, ensure the second and third numbers are larger than the first. For example, a 3-20-20 fertilizer that contains 3 percent nitrogen, 20 percent phosphorus and 20 percent potassium encourages roots to grow strong and healthy."

It **is** true that high levels of nitrogen encourage excess shoot growth. Therefore, do not over-apply nitrogen. However, does high P increase rooting? No.

The billions of commercial transplants grown each year provide evidence for avoiding high P. Seedling plug growers often purposely create a slight P deficiency by limiting phosphorus fertilizer supply in order to produce transplants with compact shoot growth, strong roots, and dark green leaves. That is why a low P fertilizer (13-2-13 N-P_2O_5-K_2O) is widely used in seedling plug production in the US. Increasing P does not increase root/shoot ratio, and these growers limit both P and N (especially the use of ammonium-N, favoring instead nitrate-N) to avoid leggy shoot growth.

As an essential element, P is of course required by plants to grow. That is true of growing both roots and shoots. However, P is not a root-promoting nutrient. Research at Virginia Tech found that increasing ppm P when fertilizing hydrangea and holly had a big effect on increasing shoot growth, but had less effect on increasing root growth. In other words, as P was increased from 0.5 to 6 ppm P, there was less allocation by the plant to growth of roots relative to the growth of shoots (the "root-to-shoot ratio" decreased). Phosphorus does not specifically target root growth.

In Conclusion

Phosphorus has mythical origins. Unfortunately, our industry also hangs on to a persistent myth about the need for high P fertilizer. This negatively affects both our wallet and our environment. Why is it so hard to change our beliefs, and for the nursery and landscape industry to reduce P application? We can put it down to the psychology of bloody-mindedness:

- We would rather deny new, uncomfortable information than reshape our worldview
- When doubts do creep in, we dig in our heels
- There is a grief process to change: denial - anger - bargaining - depression - acceptance
- We love myths!

Don't just take our word for it, but look at the research that this article is based on and take the following steps:

- Run a soil test in the landscape to see if any P fertilizer is needed. Remember that N, not P, is the most common production and landscape deficiency.
- Use a tissue analysis to diagnose a P deficiency (not just red or purple leaves).
- Use slow release forms, including CRF or compost in the landscape.
- Provide nutrients in the ratio that plants can use: A 4-1-4 N-P_2O_5-K_2O is always adequate for vegetative and flowering growth.
- See for yourself by running trials.

Using best management practices can help improve plant quality, reduce production cost, and is our responsibility as stewards of the environment.

Literature Cited:

Havis, J.R. and Baker, J.H. (1985). Phosphorus requirement of *Rhododendron* 'Victor' and *Cotoneaster adpressa praecox* grown in a perlite-peat medium. J. Environ. Hort. *3*:63–64.

Henry, J. and Whipker, B.E. (2015). Revising Your Phosphorus Fertilization Strategy. E-Gro Research Update 2015.10 http://e-gro.org/pdf/2015-06%20NCSU%20PGR%20Research.pdf accessed April 2020.

Kraus, H.T., Warren, S.L., Bjorkquist, G.J., Lowder, A.W., Tchir, C.M., and Walton, K.N. (2011). Nitrogen:phosphorus:potassium ratios affect production of two herbaceous perennials. HortScience *46*:776–783.

Shreckhise, J., Owen, J. and Niemiera, A. (2018). http://magazine.nurserymag.com/article/july-2018/understanding-phosphorus-for-containerized-nursery-crops.aspx accessed April 2020.

Wright, R.D. and Niemiera, A.X. (1985). Influence of N, P and K fertilizer interactions on growth of *Ilex crenata* Thunb. 'Helleri'. J. Environ. Hort. *3*:8–10.

Yeager, T.H., and Wright, R.D. (1982). Phosphorus requirement of *Ilex crenata* thunb. Cv. Helleri grown in a pine bark medium. J. Amer. Soc. Hort. Sci. *107*:558–562.

Acknowledgements:

We thank USDA-ARS Floriculture and Nursery Research Initiative #58-3607-8-725, National Institute of Food and Agriculture, USDA "Clean WateR3 - Reduce, Remediate, Recycle", #2014-51181-22372, Horticultural Research Institute, Virginia Agricultural Council, Virginia Nursery and Landscape Association, and industry partners of the Floriculture Research Alliance at the University of Florida (floriculturealliance.org) for supporting this research.

Roses of New Zealand

Hayden Foulds

• 40 Gordon St, Woodville 4920, New Zealand

haydenfoulds@gmail.com

Keywords: *Rosa*, breeding

INTRODUCTION

New Zealand has a very strong gardening culture based on its English heritage, mild climate and fertile soils. One of the most popular cultivated plants are cultivars and species of the genus *Rosa* which are grown for their wide variety of flower colours and forms, plant types and that many are also fragrant.

New Zealand has no native *Rosa* species; rather all species and cultivars present have either been introduced to New Zealand or have been bred here. The first recorded introduction of roses to New Zealand were by the missionaries in the early 1800's. *Rosa chinensis var semperflorens* also known as 'Slaters Crimson China' was introduced in 1814 with Rosa rubiginosa 'Sweet Briar' introduced between 1815 and 1820. The former is still grown in gardens today; the latter unfortunately is now a weed in parts of New Zealand.

While many private residences had roses in their gardens from early settlement, the first public rose garden was established in the Christchurch Botanic Gardens in 1910 and still exists today. Most towns and cities around New Zealand today have public plantings of roses.

Nurseries grew roses from the middle of the 1800's to meet the demand for plants for private gardens. It was easy to import new varieties from overseas breeders in this period and with the country still in the early stages of its development, there was little need to develop new cultivars in New Zealand. In fact, very little happened in the way of rose breeding up until the middle of the 20th century.

The first rose breeder of note was William Lippiatt (1863 – 1941) who ran a large rose nursery in Auckland and imported new varieties from leading European breeders. Sadly, it seems none of his varieties exist today.

With the rose becoming popular, the National Rose Society of New Zealand (now known as the New Zealand Rose Society) was founded in 1931. Its motto is still the same today as it was then 'To implant the rose in the hearts and gardens of the people'.

Like many organisations in New Zealand to-day, the membership of the rose society is much smaller than in its heyday due to changes in society, but the passion remains as strong as ever. A separate group, Heritage Roses New Zealand was formed in 1980 and is mostly dedicated to conserving heritage roses so they are not lost to New Zealand.

Early rose trials were established in Auckland, Morrinsville and at Massey Agricultural College in Palmerston North but for various reasons, none of these survived more than a few years. In 1969, the New Zealand Rose Society established International Rose Trials in the newly developed Dugald Mackenzie Rose Garden in Palmerston North with strong support given from the Palmerston North City Council. This was the first international rose trials established in the southern hemisphere and tests un-released roses from New Zealand and international breeders over a two-year period.

Figure 1. The New Zealand Rose Society Trials, Palmerston North.

In 1990, the Auckland Rose of the Year trial was established at the Auckland Botanic Garden. This focused on varieties already in commerce and was judged on one day a year. In 2001, the trials moved to the Rogers Rose Garden in Hamilton and were rebranded as the New Zealand Rose of the Year.

ROSE BREEDING IN NEW ZEALAND

There have been many roses bred in New Zealand, the large majority by amateur breeders who have created them as a hobby rather than as a commercial operation. The domestic market is seen as too small as to make a living from breeding roses so breeders need to crack overseas markets to have any chance of making a reasonable income from rose breeding – a difficult task with strong competition from other breeders, cost and plant import rules among the barriers. But it is not impossible, and some breeders have managed to have varieties commercialised in other parts of the world. The following are some of the breeders who are currently active along with some significant breeders who have made a contribution to rose breeding in New Zealand.

Sam McGredy

Arguably, New Zealand's greatest rose breeder, Sam McGredy immigrated to New Zealand in 1972 from Northern Ireland. He brought to New Zealand the reputation as one of the best rose breeders in the world and only enhanced this once here. He also freely shared his knowledge with other rose breeders and brought international exposure to roses in New Zealand. Sam also lobbied successful for the introduction of Plant Variety Rights (PVR) legislation which was passed into law in 1975. It was only fitting that a McGredy variety 'Matangi' (Macman) was the first plant to be protected under the legislation.

Figure 2. Sam McGredy.

Sam was a very innovative breeder and among his roses created were the 'creepy crawly' groundcovers such as 'Snow Carpet' (Maccarpe), 'hand-painted' roses such as 'Picasso' (Macpic), 'Old Master' (Macesp) and 'Rock N Roll' (Macfirwal); and striped roses such as 'Oranges N Lemons' (Macoranlem), 'Michelangelo' (Mactemaik) and 'Hurdy Gurdy (Macpluto).

Figure 3. Rosa 'Oranges N Lemons' (Macoranlem).

He also bred the traditional Hybrid Tea's, Floribunda's and Climber's which sold well including 'Aotearoa' (Macgenev), 'Paddy Stephens' (Macclack), 'Solitaire' (Macyefre), 'Sexy Rexy' (Macrexy), 'Trumpeter' (Mactru), 'Bantry Bay', 'Dublin Bay' (Macdub) and 'Schoolgirl'. A range of compact free flowering miniature roses were often given New Zealand place names such as 'Kaikoura' (Macwalla), 'Wanaka' (Macinca) and 'Kapiti (Macglemil).

Figure 4. Rosa 'Aotearoa' (Macgenev).

Sam was also a very astute businessman and often roses were named for a particular person or product. The most successful of Sam's roses in this aspect was 'Olympiad' (Macauck) which was the official rose of the 1984 Olympic Games in Los Angeles. 750000 plants were propagated in that year alone.

Figure 5. Rosa 'Olympiad' (Macauck).

Among the many roses named for products or causes were 'Auckland Metro' (Macbucpal), 'Arthur Bell', 'Mulland Jubilee', 'Matawhero Magic' (Macamster) and 'Massey University' (Macwhitba). Well known people who Sam named rose for were 'Picasso' (Macpic), 'Violet Carson' (Macio), 'Elizabeth of Glamis' (Macel) and 'Ginger Rogers' while those for New Zealand celebrities included 'Maggie Barry' (Macoborn), 'Susan Devoy' (Macreno) and 'Paul Holmes' (Macwyom).

Sam McGredy retired from rose breeding in 1991 but continued to support and encourage other New Zealand rose breeders right up until his passing on August 25th, 2019 aged 87.

Nola Simpson & John Ford

A computer scientist by profession, Nola had a lifelong interest in roses. She took up rose breeding in the 1970's after marrying her husband John Simpson, a well-known amateur rose breeder whose rose 'Velvet Lustre' was always Nola's favourite. Upon his passing in 1977, she decided to carry on his work.

Figure 6. Nola Simpson.

Nola loved breeding roses with good shape to the blooms so they could be used for exhibition in rose shows in New Zealand. Among those created for this purpose were 'Silky Mist' (Simsilko), 'Reflections' (Simref) and 'Snow Queen' (Simseen). Nola also loved breeding roses that had novel coloured blooms and became world renown for her world with roses in the brown shades. Her most well known rose was 'Hot Chocolate' (Simchoco) while 'Chocolate Prince' (Simchoka) and 'Chocolate Ripples (Simstripe), the latter two having been released by Meilland International in France.

Figure 7. Rosa 'Hot Chocolate' (Simchoco).

Nola passed away in 2011 but her nephew John Ford, a well-known rosarian in his own right, is continuing Nolas work. Among the roses he has released so far include 'Southern Beauty' (Forauty), 'Caramel Swirl' (Forusty) and 'Simply Gorgeous' (Formaui).

Figure 8. John Ford.

Figure 9. Rosa 'Caramel Swirl' (Forrusty).

Rob Somerfield

Rob has had a lifelong interest in roses and started breeding them in 1983. In the mid 1990's, he along with wife Linda started Glenavon Roses near Tauranga which supplies rose plants around New Zealand.

Figure 10. Rob Somerfield.

Rob's first commercial release was 'Blackberry Nip' in 1998, a rose that created much interest in New Zealand. Rob aims to breed very healthy plants that also have masses of flower and have a compact habit. Some of Rob's most well-known roses are 'Sky Tower' (Somskywer), 'White Romance' (Somgle01) 'Lemon N Lime' (Sombeethree), 'Little Miss Perfect' (Somgle07), 'Pacific Glory' (Somlinone), 'Looking Good' (Somannmac) and 'Wild Cherry' (Somredsky). Rob has an interest in breeding green roses and his 2020 release 'Green with Envy' (Sompounamu) is the culmination of his work so far.

Figure 11. *Rosa* 'Green with Envy' (Sompounamu).

Bob Matthews

Bob was born into roses with his parents establishing Matthews Nurseries near Wanganui in 1947. Bob and his wife Cath took over in 1978 and it continues to be one of the leading rose nurseries in New Zealand. Bob began breeding in the 1990's with his first commercial release 'Anniversary' (Mattlace) in 1997 celebrating the nurseries 50th anniversary.

Figure 12. Bob Matthews.

He is committed to producing roses that can be grown without spraying for pests and diseases. Some of Bob's well known roses include 'Akaroa' (Mattanstill), 'Catherine' (Mattlaslo), 'Diamonds Forever' (Mattdiafor), 'Lest we Forget' (Mattzac), 'Platinum' (Mattmilano) and 'St Pauls Cathedral' (Mattcan). He has also developed 'My Mum' (Mattmum) which was named for his mother which has now lead to the 'My Rose' collection of roses which now includes 'My Dad' (Mattmaf), 'My Grandma' (Mattlewanna) and 'My Treasure' (Mattamb).

Figure 13. Rosa 'My Mum' (Mattmum).

Mike Athy

Mike began breeding roses in the mid 1990's near Gisborne and aims to create free flowering, healthy garden plants.

His roses are probably more well-known overseas than in New Zealand with agents in Africa, South America and Europe. One of his varieties 'The Daisy Rose' (Athyfaala) won five awards at the Biltmore Rose Trials in North Carolina in 2013. Among Mikes other creations are 'My Love (Athypacross), 'Thank You' (Athyou), 'Hi Ho Silver' (Athysumo) and 'Caroline Bay' (Athybay).

Figure 14. Rosa "Hi Ho Silver' (Athysumo).

Doug Grant

Doug is a plant breeder specialising in breeding Pumpkins and Onions based at Pukekohe. He also breeds roses and aims to breed healthy roses that can handle the humidity that the Upper North Island experiences. His first commercial release was 'Millennium' (Gralove) named for the Year 2000 celebrations. He has also bred a series of compact climbers including 'All My Love' (Grakita), 'Candy Kisses' (Graemma) and 'Cherry Kisses' (Grachloe) which are suited for growing in smaller gardens.

Figure 15. Doug Grant with Rosa 'Cherry Kisses' (Grachloe).

SUMMARY

Since the mid-20th century, many fine roses have been created by New Zealand rose breeders and this paper has just given a short preview of some of the breeders and their creations. With roses continuing to be one of the most popular cultivated ornamental plants in New Zealand, there will be many more new varieties developed by breeders for years to come. We look forward to their future creations.

Development of a Nursery for a Wetland Project in Oman

David Hancock

Natural Area Environment Consulting Services
233c Drumpellier Drive, Whiteman Western Australia, Australia 6068

david@naturalarea.com.au

Keywords: Middle East, water purity, ecology

INTRODUCTION

The oilfields of Oman, along with others in the Middle East and elsewhere in the world, produce large amounts of ground water as a by-product of the oil extraction process. This oil contaminated water is often dealt with by return to the deep oil field aquifers. Petroleum Development Oman has established the largest constructed wetland in the world by engaging a German environmental consultant to design and construct a vegetated wetland to purify the oil-laden water to avoid the use of fossil fuel power sources to pump the contaminated water deep underground. The volumes of water are enormous, and the success of the project is demonstrated not only by the water purity of the wetland outfall but also by the wetland supporting extensive wildlife that otherwise would not be present and to offset habitat loss in other migratory zones.

This presentation will demonstrate the science and engineering bought together by specialists from around the world to develop an environmentally responsible solution and provide an example of the power of constructed natural processes to address water pollution challenges. The presentation will include the latest staged addition to the wetland, involving the propagation of wetland species endemic to Oman in a collaboration between international wetland scientists and plant propagation specialists.

Gardens by the Bay Orchid Programme

Mei Leng Lim, Andrea Kee, Jassy Phua, Siti Nurziana Yacob, Yu Qin Ng, Shawn Tay

Research & Horticulture Department, Gardens by the Bay, Singapore

lim.mei.leng@gardensbythebay.com.sg

Keywords: Trials, shows, hybridization, flowering

INTRODUCTION

Gardens by the Bay (GB) – a garden of 101 hectares in the new downtown area of Singapore, is a national garden and a premier horticulture attraction for local and international visitors. The Gardens comprise Bay South, Bay East and Bay Central Gardens. The Bay South Garden, largest of the three gardens is 54 hectares and opened its doors to the public in 2012. Bay South Garden consists of a series of thematic outdoor gardens, the Supertrees and cooled conservatories. The outdoor thematic gardens consist of 2 series of gardens; The Heritage Gardens and World of Plants are collections of landscape gardens to highlight the history and culture of Singapore's ethic groups and colonial past. The World of Plants, on the other hand, depicts the system of how plants adapted to adverse environments. The Supertrees are giant tree-like structures measuring 25 to 50 meters in height and planted with vertical greenery. Conservatories are climate-controlled glasshouses where sub-tropical and temperature plants are grown and featured in landscape display.

As a horticultural attraction in the tropics, GB showcases an array of tropical plants. However, with the cooled conservatories, GB can also grow and showcase plants from the sub-tropical regions that required a lower temperature for growth. With this unique feature of having diverse growing environments under one roof, the GB orchid team is able to embark on something that other gardens might not have the opportunity to do. This is the capability to grow and flower tropical and sub-tropical orchids together. Thus, in 2015, GB Orchid Programme was officially launched.

There are four components in GB Orchid Programme. They are the orchid thematic displays, participation in orchid shows and competitions, orchid trials and orchid breeding. The orchid thematic display is a bi-annual landscape display designed and installed by the orchid team in the Cloud Forest Conservatory. This landscape display is curated to feature a specific theme or orchid genera for a period of six months. A total of five displays, featuring orchids such as *Cattleya*, *Dendrobium* and scented orchids, had been completed (Tay, 2019).

Besides curating the orchid display in Cloud Forest Conservatory, the orchid team also participated in the landscape category of 2017 World Orchid Conference held in Ecuador. The orchid landscape display won a gold medal for the Best In-show Category (Channel News Asia, 2017).

Figure 1. Dendrobium display in Cloud Forest Conservatory, Gardens by the Bay.

Figure 2. Gardens by the Bay display at the 22nd World Orchid Conference at Ecuador 2017.

The second component in GB Orchid Programme - participation in orchid shows and winning awards in competitions, is another important goal that the orchid team strives hard to achieve. Participation in orchid shows and competitions encourage the orchid team staff to learn and hone their horticultural skills at growing orchids. These shows, and competitions, provide a platform for the staff to interact with local and international orchid nurseries and specialists, as well as to engage in and develop mutual co-operation with them. These learning opportunities are essential elements in ensuring good quality orchids are grown and used in landscape displays.

Conducting of orchid trials, which is the third component in the orchid programme, focuses on two areas in the flowering morphology of selected orchids. The two areas of focus are: initiation of flower buds in orchids and extension of flower longevity. Trials were carried out on selected orchid genera that possess good floral values that can be used in landscape displays or entries into show competitions. Manipulation of temperature was used to initiate flower buds in the trials of selected orchid species and hybrids. The aim of the trials was to acquire the ability to control flowering in selected orchids so that these orchids can be used at desired times for display or entry into show competitions. Extension of flower longevity was aimed at extending the shelf life of orchids so that the orchids can have a longer display duration before the need to be replaced arises. With a longer display duration, the frequency for orchid replacements can be reduced and cost savings can be achieved.

Equipped with the unique characteristic of having both tropical and sub-tropical growing environments under one roof, GB can grow orchids from two very diverse groups – the lowland and highland orchids together. The capability to grow these orchids together presents a golden opportunity to create orchid hybrids that can thrive in the two different climatic zones through hybridisation (Hong, 2017). Thus, orchid breeding is the final component that completes the GB orchid programme.

Conclusion

The GB Orchid Programme is in its fifth year, much of the work such as orchid trials and breeding are still on-going. Future work would include incorporating more literature review on the developments in orchid breeding, trials to improve our cultivation techniques of the orchids and producing better quality of orchids for display and competition, as well as following up on what is current in the displays of orchids. Much learning and discovery await us.

Literature Cited

Channel News Asia, (2017). Gardens by the Bay clinches gold at World Orchid Conference. Singapore: Mediacorp Pte Ltd.

Hong, J. (2017). New orchid blooms in hot or cool climates. Singapore: SPH Digital News.

Tay, S. C. (2019). Scent of an orchid. Singapore: SPH Digital News.

Oman Botanic Gardens

Dermot Molloy

Royal Botanic Gardens Victoria, Birdwood Ave, Melbourne, Victoria Australia 3004

Dermot.Molloy@rbg.vic.gov.au

Keywords: Propagation, biodiversity, conservation, research

INTRODUCTION

The Oman Botanic Gardens is in the Sultanate of Oman near the country's capital Muscat. The reason for the visit began at the 2013 IPPS conference in Melbourne Australia when two staff members from OBG Bhuthainia and Hunan attended. They both work in the nursery at OBG and were very keen to learn about all new propagation methods. David Hancock met them at the conference and asked if they would like to visit his nursery Natural Area Nursery in Perth. After the visit David sent out an invitation to all Australian and New Zealand IPPS members to help OBG with difficult species propagation. David and I visited in in November 2015.

Upon arrival David and I had a couple of free days to get to know the beautiful city of Muscat. We attended the Sultan Qaboos Grand Mosque, the Mutrah Souk and the Corniche in the port of Muscat.

The late Sultan Qaboos Bin Said Al Said did a wonderful job of modernising his country over the last 50 years. The Sultan decreed that he wanted a Royal Grand mosque, a Royal Opera House and a Royal Botanic Garden for Oman. The mosque and the opera house are completed, and construction is underway at OBG. The Oman Botanic Garden will be a world class facility to conserve, display and research the plants and ethnobotany of Oman. The gardens are led by Scientific Director Dr. Annette Patzelt and the nursery manager Ian Oliver who has since moved back to South Africa. Dr Khalid Al Farsi now manages the nursery.

Dr Darach Lupton is a Horticultural Botanist at OBG, and he and his team collects plants and seeds from the field to then be grown at OBG, Darach is also a member of IPPS. Bhuthainia and Hunan lead the propagation team in the nursery. The garden currently has 80 staff that work in the field and the nursery.

The vision of OBG is to inspire people, to cherish and conserve the biodiversity and botanical heritage of Oman for a sustainable future. The gardens mission is to discover, cultivate and protect Oman's unique plant diversity and ethnobotany through innovative research, exciting displays and engaging communication.

OBG is being constructed on a site that is 420 Hectares and will contain two glasshouses called the Northern Biome and the Southern Biome. The Northern Biome will display flora from the mountainous regions of Oman where the temperature is a lot cooler.

The Southern Biome will showcase plant species and habitats prevalent in the Dhofar region of the Sultanate. There will also be a visitor centre with interpretive exhibits, cable car, café and gift shop.

Figure 1. Overview of Oman Botanic Gardens. Photo credit: OBG Landscape Designers.

The first day of our visit we had a tour around the large nursery which has the largest documented collection of Arabian plants in the world. The number of plants in cultivation is 70000 with 14000 in propagation and 46000 in the living collection. The total taxa (seed bank and living collections) is 927 of 1407 and plant families 101 of 117.

We then went on a field trip to Jebel al Akhdhar or green mountain to visit OBG's Northern Biome nursery. The temperature on the day was 24C while in Muscat it was 44°C. In the afternoon we did some seed and cuttings collection of species like *Berberis*, *Lonicera* and *Daphne*. We stored our seed and cuttings back at the main nursery for the night before processing the following day. The next day begun by running workshops on seed and cutting processing as well as formulation of new propagation mixes and treatments for the material we had collected.

David and I thoroughly enjoyed our time at Oman Botanic Gardens. The people were very friendly and welcoming, and we believe they learnt a lot from our experience. David and I have since visited Oman on other projects including Nimr Wastewater Treatment Plant.

I would recommend visiting this wonderful country and its people to anyone wanting a unique Omani experience.

Figure 2. Pomegranate terraces Jebel Al Akhdar, Oman. Our field trip visited these mountains.

Spirit of IPPS

T. Eddie Welsh

27 Bay Rd., Ostend, Waiheke Island 1081, New Zealand

starter.plants@xtra.co.nz

Keywords: Sharing, knowledge, plant propagation, membership

INTRODUCTION

In the mid-20th Century, Jim Wells an English nurseryman transplanted to the East Coast of the USA got together with other likeminded propagators and suggested setting up a Society to promote the craft of propagating plants that is starting new plants from seed, cuttings, grafting and tissue culture. The idea was to hold annual conferences where members gave talks on their special techniques for propagation and growing plants. These papers are edited and published as Combined Proceedings known as the Black Book. Regions also hold field days to visit nurseries and gardens where ideas flourish and friendships are made.

This Society spread amongst Northern American propagators and three regions were subsequently formed: Eastern, Southern and Western Regions which span Canada and the USA. Jim Wells was also instrumental in forming a Region in his mother country and the GB&I Region (now known as the European Region) was established. A trip was made to Australia and New Zealand with Regions also forming in those two areas. In the early 1990's it was decided to expand into new Regions with the New Zealand Region cultivating interest in Japan and Australia spreading the word to South Africa. Those two areas are now full IPPS Regions.

Once again, the Society is looking at expanding into new areas which includes India and China. This meeting in Singapore is to test interest in tropical areas of Southern Asia.

So, one might ask why a person in the business of propagating would and selling plants want to give away "trade" secrets. One propagator put it this way: "You go to a meeting and give away one secret and you come home with a bag full." The Motto of the society is "Seek and Share". To become a member, you must be willing to do so. To join the society, you must be nominated by one member and be involved in plant propagation, growing plants, maintaining plants (Botanic Gardens) or supply of growing materials. Most importantly you must be willing to share. This can be demonstrated by giving a paper at conference, opening your nursery for visits, attending conferences and sharing your knowledge even if it is in the hotel bar. Some go onto serve on the Regional committee with the possibility of going onto represent the Region on an International level which involves attending International Board meetings and pre-conference tours. Regions take in turn the responsibility of hosting the International Board's meeting and tours.

Why do I think IPPS has been so successful?

- Membership is on an individual basis not as a business or organization. You might have up to ten members from one nursery and all are members on an equal basis. It is not a Society just run by the "Bosses"
- In my 40 years as a member it has been a great way to see most parts of New Zealand attending Conferences and field days in various locations. I have also met a wide range of propagators. I now have over 200 "mates" I can call on for technical advice.
- The Executive that governs the Region cycles through executive positions. Those members who want to serve in a governance role go onto to Vice President, President and International Director. After giving that service to the society you then you bow out and let others take the helm. However, for efficient administration Treasurer, Secretary and Editors serve longer terms.
- Conferences and Field Days are educational but also fun and a chance to catch up with mates. Some late nights, music and dancing, dress ups and always too much food and drink make it the highlight of the year for most propagators.
- We are a family. Even though many are competitors in business we leave business grudges and ambitions behind for a weekend to gain from each other. It is frowned on to promote your business or products directly as part of a presentation. However indirect promotion does occur with sponsorship of Conferences given by allied businesses or member nurseries which helps keep the costs of Conference affordable particularly for new members.
- We encourage new blood. Young propagators can apply for scholarships and might be nominated to attend conference at no cost to assist with conference tasks. Known as the "six pack" in Australia and "four pack "in New Zealand.
- In my career as a teacher and nurseryman IPPS has been my source of professional development. When I was a young Tutor at Massey University, I attended my first International Board meeting and tour at GB&I. I regularly made use of three textbooks while teaching practical's on plant propagation. All three authors of those text books were on the tour and became friends.
- Finally, it can be lonely propagating plants in isolation. You need to share your experience with likeminded people. Friends and partners don't really want to know about your latest tricks to get roots on your new and wonderful plant discoveries. Most plant propagators I know can't keep secrets anyway and they need to brag and show off their achievements.

Here is a message from our current International Chairperson Alan Jones who resides in the Eastern Region.

On behalf of the IPPS International Board of Directors, I would like to congratulate the New Zealand IPPS Region for organizing a spectacular program in Singapore. The Singapore meeting truly encompasses the IPPS motto "To Seek and Share" and fully embraces the international component of IPPS. Singapore is an ideal location to hold a meeting and generate interest in IPPS as in recent years IPPS members have participated in conferences in

China and India with the hope of establishing IPPS regions in those countries.

The IPPS website www.IPPS.org offers members unique access to over 68 volumes of regional conference papers containing over 32,000 edited pages of industry information. IPPS can also be followed on Facebook, LinkedIn and Instagram.

We are also excited by our rapidly growing student membership. IPPS Regions around the world now offers a free membership to over 800 student members who have an interest in the Horticulture industry.

Please consider joining the 2021 IPPS International Tour hosted by the Western Region of North America. The tour starts in British Columbia, Canada and ends in Washington State, USA. The 2022 International Tour October 16th – 24th is hosted by the Japan Region.

IPPS celebrates its 70th year in 2021. The Society is always open for new membership and new ideas.

CONFERENCE ABSTRACTS

Next Generation Plant Production and Utilization Technologies

Dr. Rakhi Chaturvedi - Convener
Indian Institute of Technology Guwahati, Assam, India

International Conference on NGPPBUT- 2019

Guwahati, Assam, India

Message from Convener

Rakhi Chaturvedi

Department of Biosciences & Bioengineering Indian Institute of Technology Guwahati, Assam, India

rakhi_chaturvedi@iitg.ac.in

Abstract

Namaskar! On behalf of the Organizing Committee of "Next Generation Plant Production and Utilization Technologies (NGPPBUT-2019)", it is my privilege and honor to welcome you all to Indian Institute of Technology Guwahati (IITG), the most dynamic and beautiful educational campuses in the country . This year IIT Guwahati has completed it 25 years of journey and this conference is being organized to celebrate its silver jubilee.

NGPPBUT-2019 also marks the first meeting of the International Plant Propagators' Society in India. It endeavors to provide a unique forum to more than 250 Indian/International attendees for three days of intensive presentations, beneficial discussions and field tours. It will bring together leading academic institutions and plant biotechnology industries, making the conference a platform for aspiring academicians, plant biotechnologists, leading industrialists, technocrats, students, entrepreneurs and plant growers to promote information exchange, networking and strategic alliances. The conference will establish the best possible model for livelihood transformation, plant production technologies, bioresources management and utilization in the changing world. The scope of the conference will include all aspects that are pertinent to successful development and commercialization of quality plants and their products.

We once again welcome you to International Conference on NGPPBUT 2019 and are confident that your engagement, scientific contribution and ideas will pave a way to provide solutions to challenges in the field of Plant production technology. IIT Guwahati, with its idyllic setting in a region tremendously rich in flora and fauna, provides the perfect background to the engagements and possible solutions that we seek to have.

Advances in Nursery Production Technology

Alan M. Jones

Manor View Farm, 15601 Manor Road,
Monkton, Maryland, USA

ajones@manorview.com

Keywords: Nursery, Industry

Plenary speaker

Abstract

This talk will review a new, unique, and different technology that is being utilized in the nursery industry in the USA and around the world. The talk will also look at research that will inspire new and different uses for the technology within the industry.

Designing Crop Plants for the Future: Molecular Genetic Enhancement of Rice Yield and Resilience

Prakash P. Kumar

Department of Biological Sciences,
National University of Singapore,
Singapore

prakash.kumar@nus.edu.sg

Keynote speaker

Keywords: Avicennia officinalis, Casparian bands and suberin, lamellae, rice plants, salinity tolerance

Abstract

Salinity stress affects growth and yield of crops. Genetic variations cause different plant species to exhibit different degrees of stress tolerance. Hence, we need to understand how plants respond to stresses in order to design plants with improved tolerance to salinity without compromising yield. We will discuss our attempts to apply findings from *Arabidopsis* and the mangrove *Avicennia officinalis* to improve rice plant. A study of *OsTPS8* of the trehalose-phosphate-synthase (*TPS*) gene family showed that *OsTPS8* controls salinity-stress tolerance and key agronomic traits in rice. *OsTPS8* belongs to class II TPS, whose involvement in salinity tolerance was not studied previously. Loss-of-function *ostps8* mutant was salt sensitive, and ectopic expression of *OsTPS8* confers salt tolerance in rice without any undesirable effects. Another aspect to be discussed is application of findings from *Avicennia* that grows in saline environment with the help of specific adaptations, including NaCl exclusion (ultrafiltration) at roots. *Avicennia* roots exclude ~95% salt by developing enhanced hydrophobic barriers (Casparian bands and suberin lamellae in the endodermis and exodermis). Cytochrome P450s play a key role in biosynthesis of suberin precursors. We identified several *Cytochrome P450* (*CYP*) genes that were differentially expressed upon salt treatment in *Avicennia* roots. Using an *Arabidopsis* mutant, *atcyp86b1*, we characterized the function of *CYP86B1* in regulating suberin biosynthesis. When treated with salt, the mutant plant roots exhibited reduction in growth, suberin lamellae, and Casparian bands. We identified specific WRKY transcription factors as the upstream regulators of *CYP* genes.

We exploited this molecular regulatory mechanism, and by heterologous expression of the *Avicennia CYP* genes in rice, higher salinity tolerance was conferred to transgenic rice plants. Hence, besides contributing to basic knowledge of the underlying molecular regulatory mechanism, our findings provide fresh approaches for generating abiotic-stress-tolerant crop plants in the future.

Plant Nursery of the Future

Dharam P. Sharma

Dry Creek Labs, Duarte Nursery,
Hughson, California 95326

dharam@drycreeklab.com

Keynote speaker

Keywords: Aquaponics, carbon mitigation, environmental controls, hydroponics, plant breeding, plant nursery

Abstract

Plant propagation has come a long way since the times when man learnt how to sprout a seed or stick a shoot in soil and make it grow into an independent tree. In 2018, the total turnover of global seed market was US \$52.69 billion with 35% contribution from U.S. alone (www.mordorintelligence.com). Nursery and garden stores in the same year in the United States had a market share of US\$ 48 billion (www.ibisworld.com), and the sales are projected to grow more than 7% every year until 2025. However, the constantly increasing population pressures are limiting the availability of land, water, and other inputs besides adversely impacting the environment. New innovations and technologies are driving agriculture to adopt cultural practices to conserve precious natural resources and nursery industry is no exception. Carbon mitigation, hydroponics, aquaponics, vertical farming, biocontrol and automated gadgets to control environmental variables such as light, temperature, humidity, and so forth are increasingly being used to address such concerns and may see a further surge in the coming decades. The changing consumer demands, political and regulatory pressures, emerging marketing, and advertising strategies and availability of labor shall further shape the nursery industry in future. The product development and plant breeding strategies will be dictated by efficiency, quality, and consumer preferences. The online shopping is gradually taking over visiting garden centers or nurseries and the trend may increase in future. Overall, we shall witness micro-managed, more automated nursery industry with increased environmental controls.

Application of *Vitis* Regeneration System Production and Genetic Improvement

Sadanand A. Dhekney[1] and Dennis J. Gray[2]

[1]Department of Agriculture, Food and Resource Sciences, University of Maryland, Eastern Shore, Princess Anne, Maryland, USA
[2]Mid-Florida Research and Education Center, University of Florida, Apopka, Florida

sdhekney@umes.edu

Keynote speaker

Keywords: Grapevine, *in vitro* regeneration, micropropagation

Abstract

Grape is among the most ancient crops cultivated worldwide and is highly valued for its value-added products such as wine, juice, jelly, and jam. Fungal and viral pathogens affecting grapevine can result in significant decrease in productivity of vineyards worldwide. Availability of clean, disease-free planting material is an important prerequisite for ensuring sustainable productivity and long-term vineyard health. Micropropagation serves as an alternative to traditional propagation methods for rapidly increasing the amount of clean, disease-free planting material of elite grape cultivars. Cultures initiated from shoot apical meristems proliferate on a medium containing cytokinins and can be maintained by transfer to another fresh medium at regular intervals. Rapidly proliferating shoots can be regenerated following growth on a culture medium containing auxins. Grapevine somatic embryogenesis, which utilizes asexual tissues for plant regeneration, has wide applications for clonal propagation and genetic improvement. Grapevine somatic embryos are ideal target tissues for insertion of desired traits of interest and recovery of modified grapevines. Grape species and cultivars widely vary in their embryogenic response, which necessitates the optimization of protocols for individual cultivars. Factors influencing the production of embryogenic cultures including explant type, development stage, growth media, and culture conditions have been optimized for a large number of grape species and cultivars. The development of efficient grapevine micropropagation and somatic embryogenesis systems can enable large-scale production of clean planting material while advancing efforts for grape genetic improvement using precision breeding.

The Enhancement of Functional Ingredient in the Vegetables and Future Issues

Kenta Nakanishi[1] and Emiko Yanase[1, 2]
[1]Graduate School of Natural Science and Technology, Gifu University, Gifu, Japan
[2]Faculty of Applied Biological Sciences, Gifu University, Gifu, Japan

e-yanase@gifu-u.ac.jp

Keynote speaker

Keywords: Brassicaceae, glucosinolate, isoflavone, legumes, secondary metabolites, sprouts

Abstract

Vegetables are important sources of many nutrients, especially secondary metabolites. Some of secondary metabolites in vegetables, such as polyphenols, are known for improving our health and have attracted the attention as the functional components. "Sprouts" are one of the vegetables, artificially germinated by seeds and then eaten by consumers. Sprouts are said to be rich source of vitamins, minerals, and phytochemicals due to the metabolism with the germination. Legumes and Brassicaceae are two well-known sprout types in Japan; isoflavone and glucosinolate are their known as functional compounds.

The productivity of these secondary metabolites is known to be easily affected by growing conditions, such as drought, temperature, and elicitor. Conversely, by increasing the productivity of secondary metabolites, there is a possibility that functionality of vegetables can be enhanced. In this research, to develop the high-value-added vegetables by enhancement of functional ingredients, we investigated effects of plant-growing conditions on the content of phytochemicals, focusing on the soybean and broccoli sprouts. Furthermore, as various metabolites are contained in vegetables, we also investigated how plant-growing conditions affect the production of other metabolites.

Propagating Indigenous Flora for Eco-Restoration

Todd Jones

Fourth Corner Nurseries, Sand Road,
Bellingham, Washington, USA

toddjones57@hotmail.com

Keynote speaker

Keywords: Anthropocentric, native plants, next-generation restoration, propagation

Abstract

Choosing indigenous plant material for eco-restoration is an evolving science with several schools of thought. This can often make the propagation of plant material for restoration a challenge, requiring completely different methods than those used in ornamental landscape plant production. Understanding the purpose of the installation, and the goals the restoration hopes to achieve, is a critical part of the propagation process. A biocentric approach to habitat restoration considers the importance of biodiversity in the landscape and recognizes that plants do not grow in isolation from other living things around them. With the rapid reduction of wild, undisturbed ecosystems the need to preserve genetic diversity is of primary importance. This type of restoration requires the use of carefully sourced native flora; yet, the definition of what is native is not universal and can add an additional level of complexity to the propagation process. An anthropocentric approach to restoration utilizes both natural and social sciences and may not be as concerned with what is and is not native flora, its objectives may be quite different than the latter approach. Mitigating effects of human disruption to offset health dangers and concerns related to climate change are two common focus areas that utilize native plantings to achieve their objective. This type of installation may require a different plant production approach. As a long-time propagator and grower of native flora, I will outline protocols I have developed to address these differences and share some of the important elements I believe a successful next-generation restoration nursery needs in order to address this rapidly changing industry.

Soil-Centric Approach to Advancing Food Security and Improving Environment

Rattan Lal

Carbon Management and Sequestration Center, The Ohio State University, Columbus, OH 43210, USA

lal.1@osu.edu

Keynote speaker

Keywords: Climate change adaptation and mitigation, conservation agriculture, eco-intensification, soil degradation, soil organic carbon sequestration, yield gap

Abstract

India has made a commendable progress in increasing food production since 1960s. However, there are second-generation problems related to degradation of soil, contamination and depletion of water resources, pollution of air, and extinction of biodiversity. These problems are exacerbated by increase in population, rapid urbanization, and growing demands of the ever-affluent population. Despite the increase in yield of food crops and other agronomic commodities, there exists a large scope and the necessity for improvement. Important among options to narrow the yield gap is the strategy of restoring soil health. Soils of agroecosystems are prone to degradation by accelerated erosion, depletion of soil organic carbon (SOC) and plant nutrients, secondary salinization, and decline in aggregation amount and strength leading to crusting, compaction, and hard setting. Concentration of SOC in soils of some agro-ecoregions is as low as 0.05% compared with the desired range of 1.1–1.5% or more. Thus, the use efficiency of inputs (e.g., improved varieties, fertilizers and other amendments, and irrigation) is low and soils are a major source of greenhouse gases including carbon dioxide, methane, and nitrous oxide .Yet, these degradation trends must be reversed through restoration of degraded soils and ecosystems by conversion to a restorative land use and adoption of recommended practices of soil, water, crop/trees, and livestock management. A system-based conservation agriculture (CA), involving a judicious combination of no-till, retention of crop residue mulch, incorporation of a cover crop in the rotation cycle, and integration of crops with trees and livestock, can create a positive soil/ecosystem carbon budget, making soils a sink of atmospheric carbon dioxide with an attendant increase in soil organic carbon concentration and stock along with gradual improvement in soil health (physical, chemical, biological, and ecological). Adoption of CA can be promoted by payments to farmers and land managers for provisioning of ecosystem services and advancing sustainable development goals of the United Nations.

Engineering Optimization Protocols for Plant Cell/Hairy Root Cultivations

Ashok Kumar Srivastava

Department of Biochemical Engineering and Biotechnology, Indian Institute of Technology Delhi, Hauz Khas, New Delhi, India

ashokks@dbeb.iitd.ac.in

Keynote speaker

Keywords: Azadirachtin production, biopesticide, bioreactor cultivation, engineering optimization protocols, hairy root cultivation, plant cell cultivation

Abstract

Plants are considered as the versatile source of strategic secondary metabolites (flavors, pesticides, drugs, etc.) for society. Presently, these compounds are produced by large-scale cultivation of natural plants, wherein plants are uprooted, which is followed by (solvent) extraction of bioactive compounds from different plant parts. This traditional production process features several disadvantages, for example, inadequate, sustained availability of the source plant part throughout the year, low product yield arising from the slow growth of natural plant, and so forth. Sometimes the overexploitation of plant parts (particularly roots) for recovery of secondary metabolite leads to extinction of plant. Therefore, there is a desperate need to establish alternate *in vitro* plant cell/hairy root bioreactor cultivation technologies and to develop innovative bioreactor designs to not only eliminate dependence on natural plants but also significantly increase the availability of bioactive compounds for growing demands of society.

It is invariably observed that plant cell cultivation is a relatively new and faster production protocol to cultivate specialized plant cells (developed from different plant parts) and mass produce bioactive compounds but yet there is a huge scope of use of engineering optimization protocols and development of innovative bioreactor designs to significantly increase the concentration, yield, and productivity of the desired secondary metabolite not only in minimum time but also with least experimental trials. The above-mentioned methodology(ies) and some newer bioreactor designs will be presented for the mass production of biopesticide (azadirachtin) using plant cell/hairy root cultivations in a bioreactor.

Advances in Nursery Production

Robert Geneve

Department of Horticulture, University of Kentucky, Lexington, Kentucky, USA

rgeneve@uky.edu

Keynote speaker

Keywords: Automation, cuttings, plant propagation, robotics, seeds

Abstract

Nurseries produce high-value crops for the ornamental industry. They rely on seeds and vegetative propagation to provide the initial liners or plugs that initiate a production cycle. Propagation has traditionally involved a significant amount of hand labor. Recent efficiencies in work flow have reduced labor costs throughout the industry. Additional efficiencies in production have relied on computer-assisted production practices. Many of these practices reduce or eliminate traditional hand labor operations. These include environmental control systems, computer-aided irrigation control, mechanical conveyance of plant material, and robotic systems for cutting and seed propagation. This presentation will endeavor to illustrate some of the recent advances in nursery production technology with an emphasis on innovations related to plant propagation.

Designer Plants to Survive Global Warming and Climate Change

Puthiyaparambil Josekutty

Skybury Coffee Pty Ltd., Mareeba, Queensland 4880, Australia

josekutty964@gmail.com

Keywords: Breeding, climate change, gene editing, genetic diversity, genetic engineering, global warming

Abstract

Global warming and subsequent climate change are impacting agriculture and horticulture across the world. Impact of climate change affects plants to a greater extent than animals because of their confinement to the environment they are born for their entire life. Natural adaptations do occur in the plant kingdom and the fittest species and individuals survive adversities, changes to the environment arising from climate change.

Abiotic stresses such as extreme heat, cold, drought, flood, sea level rise, and biotic stresses such as new diseases and pests as well as weeds/super weeds are major impacts of the climate change. Various methods "conventional and modern" for developing "Designer Plants" with capabilities to survive adversities caused by the climate change will be discussed in this paper.

Subcellular Bioengineering of *Artemisia annua* L. for Enhanced Biosynthesis and Accumulation of Artemisinin

M. Z. Abdin

Department of Biotechnology, School of Chemical and Life Sciences, Jamia Hamdard, Hamdard Nagar, 110062, New Delhi, India

mzabdin@jamiahamdard.ac.in

Keywords: Active pharmaceutical ingredient, *Artemisia annua* L., Artemisnin biosynthesis, Secondary plant metabolism

Abstract

Plants synthesize a vast array of secondary plant metabolites through a network of complex metabolic pathways regulated by endogenous and environmental factors. Although these compounds are believed to be non-essential for plants to live, they play an important role in interaction of plants with the environment ensuring their survival in the ecosystem. Some of these secondary metabolites are also of immense medicinal importance because of their therapeutic value. These are referred as active pharmaceutical ingredients (APIs). The concentrations of these compounds in medicinal plants are, however, very low, limiting their commercial exploitation. Artemisinin is one of these APIs isolated from aerial parts of *Artemisia annua* L. It is a potent antimalarial drug against drug-resistant malaria. In recent times, the demand for artemisinin is exponentially increasing with the increased incidence of drug-resistant malaria throughout the world, especially African and Asian continents. However, the commercial production of artemisinin-based combination therapies has limitation due to the presence of low concentration of artemisinin in plants. Therefore, we employed bioengineering approach to develop transgenic lines of *A. annua* L., overexpressing HMG-Co A reductase (hmgr), amorpha-4, 11-diene synthase (ads), and cytochrome P450 monooxygenase (cyp71av1) genes to enhance artemisinin content. The selected transgenic lines were found to accumulate 1.29% to 1.44% artemisinin. Thus, results obtained in these studies, clearly indicate that the synthesis of APIs in medicinal plants is tightly regulated, and bioengineering approach can be used in modulating plant metabolism to improve their biosynthesis, so that drugs manufactured from these APIs could be available at cheaper rates to the public.

Containerisation of a Plant Production Nursery Affords a Wide Range of Benefits in Propagation and Disease Management

David O. Cliffe

Narromine Transplants Pty Ltd., Narromine, NSW, Australia 2821

david@transplants.com.au

Keywords: Irrigation, nursery, plant production, terracotta containers

Abstract

During the last 60 years there has been a transformation in many plant production nurseries worldwide as they took advantage of the availability of new plastics that would partly replace the practices of traditional in-ground growing and the use of terracotta containers. Coupled with a proliferation of container shapes and sizes and the advent of more modern substrates that better suit the use of containers, a revolution in the production and management of plants has taken place. These changes have also allowed closer management of nutrition, irrigation application, closer attention to nursery hygiene, and significant changes in nursery infrastructure. This paper attempts to describe the evolution that has occurred over this period and the ongoing benefits that have accrued to those that have commenced or continued to practice best management principles.

In vitro plant propagation and commercial cultivation in the Micronesian region: challenges and measures for sustainable commercial pepper production

Virendra M. Verma

Micronesia Plant Propagation Research Center, Kosrae Agricultural Experiment Station (USDA Landgrant Program), College of Micronesia-FSM, Kosrae, Micronesia, USA

vmv_vmv@hotmail.com

Keywords: Best management practices, commercial propagation, organic fertilizers, pepper, soil amendments, sustainable cultivation

Abstract

Black pepper (*Piper nigrum* L.), a flowering vine of the Piperaceae family, is valued for its dried berries called peppercorns, which are known for their health benefits and used as a spice and seasoning. Native to the humid jungles of the Malabar Coast of Southwestern India, the plant is cultivated in the tropics worldwide. In the Micronesian region, it is gaining commercial importance as an important cash crop because of the premium price of peppercorns. However, the limited availability of disease-free black pepper seedlings and the trunks of the native tree fern (*Cyathea nigricans*), which are used as supports for black pepper vines, are becoming limitations for sustainable commercial black pepper cultivation in the region. Therefore, to ensure the year-round availability of uniform, disease-free, and high-quality planting material in Micronesia, an efficient micropropagation and acclimatization protocol was developed for a local commercially important black pepper cultivar (*P. nigrum* cv. Srilanka). Shoot apical meristems were used as explants for culture establishment. Best culture initiation was observed on Murashige and Skoog medium augmented with 5 μM 6-benzylaminopurine (BAP). For further growth and subsequent multiplication, the established cultures were transferred on 7.5 μM BAP and 5 μM indole-3-acetic acid (IAA). The number of multiple shoots produced from each explant after two subcultures varied from eight to 20. Best rooting was observed on 2 μM indole-3-butyric acid (IBA). Plantlets were acclimatized with 68% survival rate in 10 weeks. Research trials for sustainable commercial black pepper cultivation were designed, implemented, and vigorous vegetative growth was observed. To overcome the limitations of live tree-fern supports, nonliving supports such as reinforced cement–concrete standards were specifically designed and used to support the vines for commercial cultivation. First harvesting was done after 12 months of planting and data collection and analysis are being continued. Outcomes of this analysis would be used to provide assistance to the regional farming communities to promote sustainable commercial cultivation of black pepper in the region.

Commercial Aspects of Micropropagation and Hydroponics, The Future of Farming

Shweta Narasimhan
Plant Tissue Culture, HiMedia Laboratories Pvt. Ltd., Mumbai, Maharashtra, India

nshweta@himedialabs.com

Keywords: Conservation, conventional propagation techniques, hydroponics, plant tissue culture

Abstract

Food and Agriculture Organization's vision is of a "world free from hunger and malnutrition, where food and agriculture contributes for improving the living standards of all, especially in an economically, socially, and environmentally sustainable manner." Gains in productivity and technological advances have contributed to more efficient resource use and improved food safety. Around 795 million people still suffer from hunger, and more than two billion from micronutrient deficiencies or forms of undernourishment. Plant biotechnology holds a promise to resolve the problem and is indeed a blessing to achieve global prosperity. Understanding basic biology of plants is a prerequisite for proper utilization of the plant system or parts thereof. Plant tissue culture and hydroponics-growing plants without soil, have emerged as promising tools to increase farming outputs and grow plants in artificial habitats. As emerging technologies, both have a great impact on agriculture and industry by enriching plant population, needed to meet the ever- increasing world demand. It has made significant contributions to the advancement of agricultural sciences in recent times and today they constitute an indispensable tool in modern agriculture. With modernization in technology, currently several engineering techniques (robust, automated, and computerized) such as hydroponics have been applied to micropropagation with the objective of providing optimum environmental conditions to *in vitro* plant stock at a larger level. With the increasing demand in the technology, the demand for supply of resources has increased over a period of time allowing us to develop a wide range of products for both the techniques of farming helping growers to meet their needs.

Sustainable Farming Technologies Focused on Micro-Irrigation to Double the Irrigable Command Area and Farm Income

M. S. Ramesh

Jain Irrigation Systems Ltd., New No. 15, Old No. 10, Lakshmi Nivas, West Avenue Road, Kodambakkam, Chennai, Tamil Nadu, India

ramesh.ms@jains.com

Keywords: Crop and farm productivity improvement, drip irrigation, doubling the irrigable area with a given volume of water, integrated irrigation systems, sustainable farming practices

Abstract

In the Asian region, the irrigation is done through flow irrigation thro' canals and wells. In the canal system of irrigation, problems include (i) wastage of water thro' percolation (up to 40%) and evaporation (up to 20%); (ii) excessive use of water especially at the head end of the delivery point; (iii) unreliable water supply due to poor control over the entire system; and (iv) non uniformity of water application due to flow irrigation, rotational water supply (once in a few days), and seasonal water supply (either for a season or for 4 months in a year). Hence, the utilization is less than the irrigation potential created and there is an increasing gap between these two.

"Resource to Root™" (meaning delivering water directly from water resource to roots of individual plants), a revolutionary concept that provides effective solutions to above-mentioned challenges. While it provides an easy, precise, reliable control for the distribution of water, it also delivers water straight to roots of a plant, thus increasing the "Water Use Efficiency" by nearly 50%.

As water is distributed under pressure through closed pipe network, it results in uniform and equitable distribution to all stakeholders. Water delivered directly at root zone of plants thro' drip irrigation system ensures optimal utilization of water by crops, thus reducing water losses thro' percolation and evaporation during water conveyance from the source and also during farm-level irrigation. The saving of water in the whole project is up to 50% compared with traditional ways, thus doubling the irrigable command area. The proven benefits of "Resource to Root™" with very high overall project efficiency and water factor productivity result in doubling the irrigated area, increase crop yields, and ultimately generate more income and profits for farmers. Deliverables make these projects economically viable, thus contributing to increased share of agriculture sector in overall GDP of the nation.

Nanocarriers-Mediated Smart Delivery Applications for the Next-Generation Seed and Horticultural Products

J. Pon Samuel

Single Cell Technologies, ETT Core Technology, SIBS (Systems Innovations for Breeding Seeds product) R&D, Corteva agriscience™, Agricultural Division of DowDuPont™ (ADOD), Johnston, Iowa, USA

pon.samuel@corteva.com

Keywords: Cell-penetrating peptides (CPPs), nano-conjugates, nanoparticles, quantum dots, smart delivery, subcellular internalization

Abstract

Producing food for a healthy tomorrow calls for key enabling transformational technologies for cell manipulation by defending the genome integrity and at the same time to develop innovative future product concepts. The recent development of smart delivery applications across the intact cell walls using nanocarriers including quantum dots (QDs) and plant-derived cell-penetrating peptides (CPPs) have been facilitating the development of sensitive fluorescence biosensors due to their high quantum yield, narrow and tunable emission spectrum, and good photostability. We first introduced the use of QDs and plant-derived CPPs, their preparation and functionalization approaches for smart delivery into intact plant cells, tissues, and organs. In this presentation, we describe methods for introducing a molecule of interest into plant cells with intact cell walls by using QD-peptide linked to CPPs. The use of QDs or polystyrene nanoparticles with improved biocompatibility further promotes biological applications and we summarize QDs-based fluorescent biosensing for proteins and nucleic acids, and QDs-based applications in cellular and *in vitro* and *in vivo* subcellular targeting and imaging. Last but not the least, we envision the potential of such smart non-invasive delivery and tracking technologies for future seed products.

Iron Biofortification in Bananas by Expression of *Oryza sativa* Nicotianamine Synthase Genes

M. Mayil Vaganan, C. Sivagandhi, S. Ganesan, M. Kumaravel, I. Ravi, K. J. Jeyabaskaran, S. Backiyarani, and S. Uma

ICAR National Research Centre for Banana, Thayanur Post, Thogamalai Road
Tiruchirappalli 620102, Tamil Nadu, India
mmayilv@gmail.com
Keywords: Grand naine, iron deficiency, OsNAS1, OsNAS2, rasthali, transgenics

Abstract

Iron is an essential micro nutrient for plant growth and human health. Iron deficiency is one of the most prevalent micro nutrient deficiencies in the world, causing anemia to several billions people especially in developing countries. The production of iron-biofortified staple crops will help to ameliorate iron-deficiency anemia. Banana is one of the staple crops of Asian and African population. Two commercial banana genotypes, Rasthali and Grand Naine, are biofortified by introduction of two rice (Oryza sativa) nicotianamine synthase genes, OsNAS1 and OsNAS2, driven by Zea mays ubiquitin promoter to increase the iron content in fruit pulp. Agrobacterium-mediated transformation of the constructs carrying these genes and selectable marker, nptII, was carried out with embryogenic cell suspension (ECS) of the bananas. One hundred independent transgenic events of each banana genotype for each gene construct were generated and being field grown along with untransformed control plants in iron-sufficient soils under confinement in transgenic net houses equipped with all biosafety standards as per guidelines of Department of Biotechnology, Government of India. The transformants, before planting, were confirmed for transgene presence and selectable marker with specific primers using polymerase chain reaction (PCR) and checked for Agrobacterium contamination using Vir-C gene-specific primers to eliminate false positives. Presently, transgenic lines are 9 months old and in shooting and fruit-bearing stages. Estimation by Inductively coupled plasma atomic emission spectroscopy (ICP-OES) of iron element content of 25 representative lines of each genotype at vegetative stage is promising with seven Rasthali lines and six Grand Naine lines having average iron content of 24.5 mg/100 g dry wt. as compared with non transgenic lines with maximum of 11.2 mg /100 g. As the increase in iron concentration was 2.2 times in the transgenic lines, the results offer promise to effectively increase the level of iron in the fruit. Transgene integration and the copy number of the transformed events are being performed.

Biochar Utilization in Benefiting Plant Productivity

Chien-Teh Chen, [1] Irene Kosasih, [2] San-Gwang Hwang,[3] Ying Yong,[3]
Yi-Jun Chen,[3] and Keng-Tung Wu[4]

[1]Department of Agronomy, National Chung Hsing University, Taiwan
[2]International Master Program of Agriculture, National Chung Hsing University, Taiwan
[3]Department of Horticulture, National Chung Hsing University, Taiwan
[4]Department of Forestry, National Chung Hsing University, Taiwan

ctchen41@nchu.edu.tw

Keywords: Agricultural waste, biomass seed shell, *Caliandra*, sweet potato vines, tomato stem, and leaves (RTS)

Abstract

Normally, the agriculture waste is mixed with soil as fertilizer. However, some of agro-wastes are not suitable to be dumped into soil. For example, seed shells of oil tea (SST) are not allowed abandoned openly because of having considerable amount of alkaloid; sweet potato vines (SPV) if abandoned on the field may cause the root rot disease in the next season; residual of tomato stems and leaves (RTS) are unfavorable for continuous tomato cultivation because of having high content of phenolic compounds. The main objective of this study is to convert the eco-unfriendly agricultural waste into biochar for further utilization. Our results showed that the biochar from SST inhibited growth of rice seedling even as low as 1% in soil. SPV can be added into soil for 10% without inhibition to rice. RTS can be added into soil for 50% without inhibition to rice. It was noticed that the biochar of RTS showed significant benefits on the growth of tomato seedlings.

Biochar could be the byproduct of biomass gasification power plant and has been widely known for benefiting plant productivity. *Caliandra calothyrsus* recently utilized in biomass power plant in Indonesia, the effect of biochar on nursery of *C. calothyrsus* seedlings was conducted. Experiments were processed on the Island of Kundur, Riau Islands Province, Indonesia. A half-year of nursery planting for *C. calothyrsus* was performed. Seedlings that meet the qualification for field planting were counted for the number, height, and diameter. Results showed biochar amendment increased the qualified seedling by 18%. In 1-year trial of field planting, biochar amendment increased wood and leaf yield of *C. calothyrsus*. It is concluded that the application of biochar in *C. calothyrsus* nursery planting could be included in circulating system for biomass power plant.

Biomass Utilization for Development of Value-Added Products for Food Packaging Applications

Vimal Katiyar

Department of Chemical Engineering, Indian Institute of Technology Guwahati, Guwahati 781039, Assam

vkatiyar@iitg.ac.in

Keywords: Biodegradable polymers, biofiller, cellulose nanocrystal, packaging, silk nanocrystal

Abstract

The utilization of biomass for the development of various tailor-made sustainable polymers including poly (lactic acid) (PLA), poly (ε-caprolactone) (PCL), polyhydroxyalkanoates (PHA), and so forth targeting food packaging application is the main focus of the presentation. It is noteworthy to mention that both plant- and marine-based biomass can be transformed into value-added chemicals including nanocellulose (NC), silk nano crystals (SNCs), functionalized nanochitosan (FNC), nano-gums (NG), and so forth, which are components of both edible and no edible food packaging materials. Further, the application of proper strategies for fabricating the mentioned biomass-derived materials can help to deliver tuned properties in the concerned application. Thus, the formulation of biocomposites by incorporating the addressed biomass-derived materials (NC, SNCs, FNC, and NG) into the above-mentioned biodegradable polymers (PLA, PCL, and PHA) is a way to deliver tuned properties as food packaging materials. Further, the development of bio-composites of sustainable polymers has found to offer improved properties in terms of thermal, mechanical, and barrier properties, which are an important criterion for its uses in food packaging. In addition to above, the addressed chemicals are also considered a promising candidate in developing edible food packaging with enhanced thermo mechanical properties and health benefits. The application of herbal bioactive agents to biopolymers for acting as edible food packaging materials will further improve shelf life of perishable fruits, which allow consumers a value-added food product with reduced plastic-based waste. In this regard, chitosan has the properties of nontoxicity, biocompatibility, biodegradability, and antimicrobial activity. Further, chitosan provides many health beneficial properties including antioxidant property, antidiabetic property, weight-reducing activity, anticancer activity, cholesterol-lowering activity, and others, which make it a promising agent for edible films and coatings with added benefits. Moreover, residual biomass of green algae is another excellent source of various bioactive compounds such as protein, lipids, antioxidant, vitamins, minerals, and so forth. Thus, the utilization of algae for the extraction of nutraceuticals would be a great source of value addition of this class of biomass.

Further, it can be utilized for the production of bio-energy through microbial fuel cell technology, which will also be discussed during the talk. These sustainable materials can further be utilized as emerging candidates as a catalyst for various processes, additives for tailoring of polymers properties, efficient energy storage and power-generation medium, bioactive medium in biomedical applications, and so forth.

Biotechnological Intervention for Boosting *Citrus* Industry

Ashok K. Sharma

Tissue Culture Laboratory, CSIR National Botanical Research Institute, Lucknow, Uttar Pradesh, India

ashok_nbri@hotmail.com

Keywords: Citrus fruit, cloning, germplasm preservation, tissue culture

Abstract

Citrus may be considered as the number one fruit of the world and number three fruit of India after mango and banana in view of its nutritional and medicinal value as well as its production. Despite Northeastern India being the native place of *Citrus* and also having a huge orchard area, rich germplasm, and favorable agroclimate, the *Citrus* fruit production in India has been declining at an alarming rate during the past several decades mainly due to the destruction of orchards by diseases (particularly viral) and general neglect toward replenishing the declining orchards with certified healthy and high productive varieties of *Citrus*. The poor state of *Citrus* industry obtained in the country can greatly be alleviated or even changed to a situation in which India becomes a major *Citrus*-producing country in the world with the application of biotechnology, precisely plant tissue culture. At National Botanical Research Institute (NBRI), Lucknow, the first *Citrus* tree in the world through *in vitro* culture has been produced in 1972 and later remarkable success has also been achieved in various important aspects of *Citrus* tissue culture research, such as shoot meristem culture, unpollinated pistil culture, and micrografting for virus elimination; haploid production through androgeneis for genetic improvement; production of cloned plants of several commercially important scion species through nodal stem segments; and production of cloned as well as disease-free rootstocks by exploiting nucellar polyembryony. Under multilocational field trials, the shoot meristem-regenerated plants of *C. aurantifolia* exhibited better performance. Besides, efficient *in vitro* processes for germplasm preservation of several *Citrus* spp., including *C. indica* (an endemic threatened wild relative of *Citrus*), have been developed for establishing "Germplasm Repositories" of *Citrus* spp. growing in diverse agroclimates.

Integrating biotechnology and ecology for threatened plant conservation of Sikkim Himalaya

Mithilesh Singh, and Devendra Kumar

G.B. Pant National Institute of Himalayan Environment and Sustainable Development, Sikkim Regional Centre, Pangthang, Gangtok, 737101, Sikkim, India

singmithilesh@gmail.com

Keywords: Himalaya, MaxEnt modeling, propagation, threatened plants

Abstract

Sikkim Himalaya that forms a part of the "Himalayan Biodiversity Hotspot" is one of the biologically richest areas in the Eastern Himalayas. The region encompasses more than 4400 species of plants including those of medicinal value and otherwise economically important ones. However, rise in human population with demand on land for farming, construction of roadways, hydropower projects and allied works, and of late the tourist influx have collectively resulted in building up of considerable pressure on the survival of important plant species in Himalayan region. Declining in the number of species and significant changes in their natural habitat, as well as in some cases, complete population annihilation in the wild have become strong issues of concern, and a compelling reason to start work on the conservation of threatened plants of the region. In this study, very strong efforts have been made to improve conservation status of two threatened plans of Sikkim Himalayan Region, namely Rhododendron leptocarpum and Phoenix rupicola. Maximum entropy (MaxEnt) based distribution modelling algorithm was used to identify suitable habitats for plant reintroduction. Biotechnological interventions have been made for producing large number plants of R. leptocarpum and P. rupicola. For reintroduction, about 1000 plants of R. leptocarpum and 5000 plants of P. rupicola were produced through micropropagation and macropropagation techniques, respectively. Combined approach of ecological niche modeling and biotechnological techniques used in this study is ideal for the conservation of endangered plant species of Himalaya.

The Majesty of Plant Secondary Metabolites: Unlocking the Futuristic Trends of Bioresource Augmentation

Suchitra Banerjee

Plant Biotechnology Division, CSIR-CIMAP, Lucknow, Uttar Pradesh, India

suchitrabanerjee07@yahoo.com

Keywords: Bioresource augmentation, disease-pest infestations, MAPs, nutraceuticals, plant secondary metabolites

Abstract

Plants have been used for centuries to cure diseases throughout history owing to their ability to synthesize a fascinating class of phytochemicals, called "plant secondary metabolites (PSMs)." The vast array of chemically distinct PSMs of "Medicinal and Aromatic Plants' (MAPs)" origin represent diverse/novel molecular scaffolds, which are unachievable through synthetic chemistry leads. Their huge ranges of pharmacological activities have revolutionized the global history of blockbuster drug-development process by their direct use or as templates for semisynthetic derivatization. Recent advancements in the area of combinatorial chemistry and computational drug designing processes have accelerated their demand in drug discovery through identification of novel drug targets. Apart from pharmaceuticals, PSMs find multifaceted uses in nutraceuticals, fragrance / perfumery, cosmaceuticals, herbal remedies, and bio-insecticides. Additionally, current insights into the multifunctionality of plant volatiles in raising healthy livestock and aquaculture have further propelled their rapid commercial need. In light of the escalating global demand for PSMs, pharmaceutical companies are finding it difficult to comply with the year-round supply of biochemically consistent raw materials of MAPs because of their chronic limited supplies owing to multiple impediments. Indiscriminate wild harvesting without strategic cultivation practices, habitat loss due to rapid urbanization/industrialization, climate change, disease–pest infestations, and so forth not only radically jeopardized the sustainable use of MAPs but also imposed an endlessly mounting threat of extinction on a huge list of already endangered natural MAP resources. Creating the situation of assured affordability of commercially desirable MAPs' raw materials epitomizes the supreme challenge of this century. The concept of growing MAPs along with conventional crops for sustenance of their adequate resources is slowly changing the value of agriculture to the rapidly developing perception of entrepreneurial opportunities. Moreover, combining and refining the abilities of plant biotechnology with the major regulatory challenges of the pharmaceutical industry are also progressively drawing global attention by complying with the mandatory promise of more environmentally sound, economical, and effective PSM-delivery processes under controlled environment. Credible examples illustrating such strategies will be discussed to unlock the futuristic trends of PSMs' bioresource augmentation.

Role of Secondary Metabolites in Crop Protection

Ajeet Singh Tomar

International Marketing Development Department, Gharda Chemicals Ltd., Mumbai, Maharashtra, India

astomar@gharda.com

Keywords: Crop protection, microbial secondary metabolites

Abstract

Globally there is an increasing concern regarding the food production to sustain the exponentially increasing human population under a frequently changing environment (global warming). To meet the growing food demand, without affecting the human health and damaging the environment, compounds derived from microbes "secondary metabolites" can play an important role in crop protection as alternatives to chemical pesticides. Numerous bacterial and fungal secondary metabolites have been isolated and evaluated for fungicidal, insecticidal, and herbicidal activity. Of these, 33,500 microbial metabolites, about 12.5%, are metabolites of unicellular bacteria and cyanobacteria, 41% are products of Actinomycetes fermentations, and about 47% are of fungal origin. most effectively commercialized bacterial metabolites as bio-insecticides are endotoxins produced by *Bacillus thuringensis*, among soil microbes, actinomycetes produce wide-spectrum biologically active substances already commercialized as kusagamycin (bio-fungicide), abamectin (bio-insecticide), spinosad (bio-insecticide), and streptomycin (antibiotics). Thus, there is a huge potential for screening of new secondary metabolites that can be applied in crop protection, which would have least impact on food and environment and would also be benign to other non-target organisms.

Identification and Characterization of Known and Novel Cyclotides in the Indian Medicinal Plant *Viola odorata* and Its *In Vitro* Cultures

M. Narayani, Anju Chadha, and Smita Srivastava

Department of Biotechnology, Indian Institute of Technology Madras, Chennai, Tamil Nadu

smita@iitm.ac.in

Keywords: Cyclotides, cytotoxic, hemolytic, *in vitro* cultures, mass spectrometry, *Viola odorata*

Abstract

Cyclotides are a class of cyclic plant proteins with a unique topology that is responsible for their exceptional chemical, thermal, and enzymatic stability. This makes them a potential candidate for diverse commercial applications such as agrochemicals, pharmaceutical scaffolds for drug delivery, and therapeutic agents. Currently, cyclotides are obtained only via direct extraction from limited plants. In this study, known and novel cyclotides were identified for the first time in the Indian variety of the medicinal plant *Viola odorata* using liquid chromatography and Fourier transform mass spectrometry. Specific protocols were developed for successful identification and characterization of cyclotides in the plant that included confirmation based on their mass (2.5–4 kDa), hydrophobic nature, disulfide bonds, circular structure, and amino acid sequence. A total of 71 known and 98 putative new cyclotides were identified in the Indian varieties of *V. odorata*. Among the 98 putative new cyclotides, amino acid sequences of eight cyclotides have been established using *de novo* sequencing approach. This study revealed that the production of cyclotides in plants varies with geographical location and the type of the plant tissue, hence cannot serve as a reliable source for the production of cyclotides. Moreover, owing to extensive wild crafting of *V. odorata* for several commercial applications, it is categorized as an endangered species in parts of India. Hence, to establish an alternative and sustainable production platform for cyclotides, cell (callus, cell suspension) and organ cultures (somatic embryo and shoot cultures) of *V. odorata* were developed to investigate the production of known and new cyclotides. Furthermore, the somatic embryos (rich in cyclotides) demonstrated equivalent and in some cases superior biological activities (cytotoxic, hemolytic, and antimicrobial) than the natural plant, suggesting it as an alternative source for several therapeutic applications.

Commercial Micropropagation—It's About Time!

Gayle R.L. Suttle

Microplant Nurseries, Inc., Gervais, Oregon, US

gayle@microplantnurseries.com

Keywords: Tissue culture, Industry impact

Abstract

The speaker will discuss her personal journey over the past four decades growing plants in culture for profit. She will highlight some of the technical challenges her company has overcome, discuss the impact micropropagation has had on our horticultural industry, and share her perspective of the opportunities ahead.

Advances in Agricultural Tools and Technology: A Transition from Traditional to Conventional Agriculture Practices in India

Pankaj Rawat

Bhanu Biotech Pvt. Limited, Bank Enclave, Laxmi Nagar, New Delhi 110092, India

bhanubiotech1@gmail.com

Keywords: Automation, farming, high tech green house, IoT

Abstract

Greenhouse technology and protected cultivation have proved to be an advantage for researchers/farmers by providing favorable climatic conditions in agricultural sector to harness improvements in marketable yields, significant reductions in the control of pests and weeds, increased density of crops, efficient utilization of water, and the accommodation of the use of different cultivation methods. This type of farming system has long been in the background and has not yet explored its maximum capability. However, recently observations have been made about the use of greenhouse technology in the agriculture sector. Plant growth modules (high tech green houses, tissue culture laboratories, cold rooms, and plant growth chambers) and allied scientific equipment cater the needs of Science community at a large in agriculture from automation to technology directed with the Internet of Things (IoT). The company design and implement environment monitoring system along with plant tissue culture laboratory establishment with various wireless sensor to control and monitor the plant propagation. Our latest innovative technology includes front-end data acquisition, data processing, data transmission, and data reception and at the same time, researchers may view, analyze, and store data that provide real-time statistics for agricultural greenhouse/cold room/plant growth chamber/clean room facilities, and other weather control equipment, thus achieving real-time weather updates. The company provides latest drone-based agriculture solutions for pesticide spraying and crop monitoring. Additionally, observations made indicate that a lack of knowledge and experience in these new technologies are the most pressing challenges faced by Indian researchers/greenhouse producers and most traditional farmers. Thus, latest agricultural tools/farming techniques and their systems are applicable apparatus for developing and improving the agricultural produce of India.

Ex Situ Conservation of *Coelogyne ovalis* Lindl. Through Asymbiotic Seed Germination and Assessment of Genetic Variation for Reintroduction

Nutan Singh and Suman Kumaria

Department of Botany, Plant Biotechnology Laboratory, North-Eastern Hill University, Shillong-793022, Meghalaya, India

sumankhatrikumaria@gmail.com

Keywords: Coelogyne ovalis, conservation, genetic fidelity, reintroduction

Abstract

For conservation of an endangered, ornamental, and medicinally important orchid *Coelogyne ovalis* Lindl., an efficient protocol was developed via asymbiotic seed germination. Eight months old capsules of *C. ovalis* were inoculated aseptically on different media viz., Gamborg (B5), Knudson C (KC), and Mitra, with and without supplementation of plant growth regulators (PGRs) to check their effect on the growth of seedlings. Among the studied media, the best seed germination response was found in KC medium. The highest germination percentage was found in KC medium supplemented with 6-benzyl-aminopurine (BAP). For shoot induction, seedlings were further cultured in KC medium supplemented with PGRs viz., BAP and 1-naphthaleneacetic acid (NAA). The best shooting was observed in KC medium supplemented with 15 μM of BAP and 5 μM of NAA. Optimum rooting frequency of regenerated shoots was achieved in KC medium augmented with 15 μM NAA and 30 μM phloroglucinol. Well-developed plantlets were acclimatized in a compost mixture. Genetic variation of *in vitro* raised regenerated plantlets was ascertained using start codon targeted polymorphism (SCoT) and inter simple sequence repeat (ISSR) markers. The present report on *in vitro* generated *C. ovalis* insures rapid propagation of plantlets for conservation purposes and then their reintroduction in fields.

An Efficient Method of *In Vitro* Propagation and Hardening of Plum (*Prunus salicina*) Cultivar Santa Rosa

Vishal Sharma, Niharika, Manu Vivek, and Manisha Thakur

Department of Biotechnology, Dr Y. S. Parmar University of Horticulture and Forestry, Nauni, Solan, Himachal Pradesh, India

vishalshrm073@gmail.com

Keywords: Plum, *In vitro* propagation, hardening

Abstract

In the present investigation, a technique for *in vitro* propagation and hardening of plum (*Prunus salicina* L.) cultivar Santa Rosa has been developed. Treatment of explants with 1% sodium hypochlorite for 10 min was found to be the best as it gave the maximum number of uncontaminated buds and buds' survival. Maximum *in vitro* establishment (75%) of explants was achieved on Murashige and Skoog (MS) medium fortified with 0.5 mg/L 6-benzylaminopurine (BAP) and 0.05 mg/L indole-3-butyric acid (IBA), in the month of February and March. The highest multiplication rate (1:8) was obtained on MS medium fortified with 0.5 mg/L BAP, 0.1 mg/L gibberellic acid (GA_3), and 0.1 mg/L IBA. Shoot multiplication rate and shoot length showed an increase with the increase in passages, which increased to a maximal of 1:10 and 4 cm, respectively, and showed a decline in further passages after the fourth passage.

Two procedures for rooting, that is, single step and two steps were adopted, and the maximum rooting (70%) was observed on ½-strength MS medium supplemented with 0.5 mg/L IBA following the single-step approach. In two-step procedure, 30% rooting was observed after 48 h dark incubation in ½-strength liquid MS medium fortified with 0.5 mg/L IBA followed by transfer to semisolid ½-strength MS basal medium within 4 weeks of culture. *In vitro* rooting efficiency increased with the increase in passages, which increased to 70% during the ninth passage. Best hardening was observed by following hydroponic approach in which the plantlets were dipped in liquid MS medium without sucrose and myo-inositol for 15 days with 66.6% survival after 4 weeks of transfer to cocopeat. Drenching of potting mixture with 15 mL Jeevamrit (3%) showed the survival of 60.33%. *In vitro* regenerated plantlets showed no morphological variations when compared with mother trees.

In Vitro Propagation of Plum (*Prunus salicina* L.) Cultivar Frontier Through Control of Shoot Tip Necrosis

Manisha Thakur and Vishal Sharma

Department of Biotechnology, Dr Y. S. Parmar University of Horticulture and Forestry, Nauni, Solan, Himachal Pradesh

drmanisha72@yahoo.com

Keywords: Plum, shoot tip necrosis, tissue culture

Abstract

The present investigation was carried out to develop an efficient protocol for *in vitro* propagation of plum (*Prunus salicina* L.) cultivar (cv.) Frontier and to control shoot tip necrosis (STN). High yielding trees of cv. Frontier, indexed against *Prunus* necrotic ring spot virus, cherry leaf roll virus, and apple chlorotic leaf spot virus, were selected for the study. Shoot apices and axillary buds were collected from mother plants and cultured on Murashige and Skoog (MS) medium containing different concentration and combination of phytohormones. MS medium supplemented with 0.1 mg/L 6-benzylaminopurine (BAP), 0.1 mg/L kinetin, and 0.05 mg/L indole-3-butyric acid (IBA) was found to be the best and gave highest percentage (79.50%) of shoot bud establishment after 4 weeks of culture; however, all shoots showed necrosis and died after the first subculture. Necrosis appeared with yellowing of shoot tips within 7 days of the first subculture and gradually increased downward, resulting in yellowing of leaves followed by death of shoots. Among various reported methods tested to control STN, combination of fructose (0.55 mM) and calcium chloride (1.0 mM) proved 100% effective in STN control. Highest *in vitro* shoot multiplication of 1:5 was achieved on MS medium supplemented with 0.5 mg/L BAP, 0.1 mg/L gibberellic acid (GA_3), 0.05 mg/L IBA, 0.55 mM fructose, and 1.0 mM calcium chloride. For rooting, a two-step rooting procedure was followed, where microshoots were dipped in half-strength MS broth containing 0.5 mg/L IBA and incubated in dark for 48 h before transferring to the same strength basal MS medium, resulting in 85.00% rooting. Rooted plantlets were successfully hardened in cocopeat with 70–80% survival after 10 weeks and after 1 year, these plants were transferred to the field with 100% success. Regenerated plants showed no morphological variation when compared with mother trees.

Formulation of Agro-Waste-Based Bacterial Biofertilizer and Its Plant-Growth-Promoting Effects

Gargi Chakravarty
Department of Botany, D. K. College, Mirza, Assam, India

gargichakravarty2008@gmail.com

Keywords: Agro waste, *Bacillus subtilis*, biofertilizer, plant growth-promotion, *Pseudomons fluorescens*

Abstract

The main purpose of this research is to find a possible way for efficient utilization of agricultural wastes for sustainable plant production systems. Agro wastes were used as substrates for the growth of plant growth-promoting bacteria *Pseudomonas fluorescens* and *Bacillus subtilis* and production of biofertilizers. Altogether nine substrate combinations were studied, namely fruit wastes, vegetable wastes, fallen leaves, and rice straw each one inoculated with *Pseudomonas fluorescens* (C1–C4), again each of the four organic wastes inoculated with *Bacillus subtilis* (C5–C8), and the final substrate C9 containing a mixture of the four agro wastes inoculated with both strains of *P. fluorescens* and *B. subtilis*. Control (C10) consisted of a mixture of the four agro wastes inoculated with any of the two bacterial strains. Enzyme activities and plant growth-promoting traits of the selected bacterial strains were tested. Temperature and pH during the composting period were measured at regular intervals up to 28 days, and chemical analysis including organic carbon, nitrogen, phosphorus and potassium concentrations of the various inoculated substrates were carried out. The viable bacterial population of the decomposed substrates was evaluated at 15 days interval up to 90 days. The three best bioformulations depending on bacterial load were selected for study of their effects on shoot length, leaf surface area, total leaves and branches emerged in the plant, and fruit yield in the test plant *Solanum melongena*. The carbohydrate, protein, and chlorophyll content of the treated and control plants were also recorded. The formulation with a mixture of the four agro wastes and both bacterial strains (C9) showed the best result to be used as a potential and effective biofertilizer, meeting the need of environment-friendly food production systems and providing balanced nutrient supply and waste recycling.

Artificial Seed Technology for Short/Medium-Term Germplasm Storage in *Aquilaria malaccensis* Lam.; a Commercially Important Tree

Seram Devika Devi, Meera C. Das, and Suman Kumaria

Plant Biotechnology Laboratory, Centre for Advance Studies in Botany, North-Eastern Hill University, Shillong 793022, Meghalaya, India

seram.devika@yahoo.com

Keywords: *Aquilaria malaccensis,* artificial seed, recalcitrant, regeneration, response, storage

Abstract

Aquilaria malaccensis is a tropical tree of high commercial value due to the production of a non-timber forest product called "agarwood". Storage is highly problematic in the species due to the recalcitrant nature of seeds. Artificial seed technology is known to be the best alternative for short/medium-term storage of germplasm in such plant species that possesses seeds with very short shelf life. The study therefore explored the effect of two temperatures, that is, 4°C and 23± 2°C on storage potential of *A. malaccensis* zygotic embryos and *in vitro* derived nodal buds. Maximum responses were observed in liquid full-strength Murashige and Skoog's medium from both the explants, while regeneration was observed only from *in vitro* nodal buds in the same medium. Encapsulated zygotic embryos could be stored for only 20 days at both the studied temperatures; in case of nodal buds as explants, storage was possible for 60 and 50 days at 4°C and 23±2°C respectively. Meanwhile, encapsulated zygotic embryos failed to regenerate into shoots in addition to its inefficiency for storage while all survived encapsulated nodal buds regenerated into shoots. Shoot development from the encapsulated nodal buds stored at 4°C was found to be better than that of 23 ± 2°C.

A Comparative Study of Selective Macrophytes for *In Situ* Bioconcentration From Sediment–Water Continuum as Self-Remediative Lacustrine Function

Umar Nazir, A. Salim Khan, and Anisa B. Khan

Department of Ecology and Environmental Sciences, Pondicherry University, Puducherry 605014, India

salimamu@gmail.com

Keywords: Autotrophy, biofiltration, nitrate, phytofiltration, remediation, senescence

Abstract

Macrophytes perform optimization and self-purification functions; otherwise, the evident water quality could get worse. They behave as a living link between abiotic water–sediment interfaces as depicted by the seasonal nutrient regeneration fluctuations. Tissue concentrations of nutrients and trace elements differ species wise but well correlate with ambient water and sediment media. The peak biomass values (gm^{-2} dry weight basis) measured are 880.2 in case of *Myriophyllum aquaticum*, 678.4 in *Nelumbo nucifera*, 182.4 in *Ceratophyllum demersum*, and 45 in *Salvinia natans*. Biomass parameters such as dry weight, productivity, NPP and specific growth rate establish similar variations in the experimental species but species turnover is highest in case of *S. natans* and lowest for *C. demersum*. N and P are intimately related to biological productivity of aquatic ecosystems. Higher temperature favor bioproduction and elevate carbonate-driven pH significantly. The diminishing conductivity at improved temperature conditions suggest higher bioaccumulation rate of nutrients. The peak nutrient uptake and bioconcentration coincide with peak biomass in summer and autumn. BCF criterion indicates hyperaccumulation for most of the metals in case of *C. demersum* and *S. natans*. Although the quantum of removal potential for different elements in the analyzed species is divergent but the pattern is related, which suggests unselective absorption. The elemental turnover rates in selected macrophytes closer to the reference value of 1 has significance. Although the emergent macrophyte proved efficient in mineral and metal retention, but, submerged ones provide a better biofilters' option in terms of spontaneous occurrence and site aesthetics. Based on the outcome of the study, lake ecological restoration is possible by limiting human perturbations, practising periodic dredging and sediment trapping, scaled-cum-selective deweeding, and construction of vegetation buffer strips for decelerating cultural eutrophication. Therefore, internal structure, external inputs, compartment cycling, and resource exploitation together quantify changes in inland waters.

Recovery of Lignin from Agro-Waste

M. Arjun, A. K. Harigovind, R. Ananyanjali, Maryl Ann, N. Gazliya, and K. Aparna

Department of Chemical Engineering, National Institute of Technology, Calicut, India

gazliya.pn@gmail.com

Keywords: Agro waste, alkaline treatment, banana stalk, lignin, lignocellulose

Abstract

India contributes to 29.19% of banana production, being the world's leading producer. Banana occupies 20% of the total area under crop production in India, which covers almost the entire country. Lignocellulose biomass is gaining increased attention currently; because it is renewable and sustainable in the form of agro-waste. The agro-waste that is produced after harvest is plant biomass consisting of energy-rich lignocelluloses that can be efficiently used to make value-added products such as lignin, which in turn, can be used for the production of chemicals, fuel, and electricity. Banana stalk is an unused part of the plant, which is discarded in local markets and packing centres, where residues are dumped into the open and water resources. The stalk consists of ~16% lignin under dry biomass weight. The present work is undertaken to recover lignin from black liquor, produced by the alkaline treatment of banana stalk. Under peroxide-assisted, mild-alkaline, and ambient-temperature conditions, 6.6% lignin yield was obtained from 25 g of dry biomass, which accounts for 41.25% of the total lignin content. The lignin is recovered by acid precipitation and characterized by Fourier Transformed Infrared spectroscopy (FTIR), Differential scanning calorimetry (DSC), Thermo-gravimetric Analysis (TGA), X-ray diffraction (XRD), and molecular-weight analysis

Standardization of an *In Vitro* Regeneration Protocol in Khasi Mandarin

Panchashree Das

Department of Agricultural Biotechnology, Assam Agricultural University, Jorhat, India

panchashreedas@gmail.com

Keywords: Regeneration, root induction, shoot induction, zygotic seedlings

Abstract

Citrus species are the most widely grown fruit crops within the whole world. India is the fourth largest producer of orange in the world. North-Eastern India is considered as one of the centres of origin of many citrus species. Among them Khasi Mandarin is the most widely grown citrus species. According to Ministry of Agriculture and Irrigation, Govt. of India, the yield of Khasi Mandarin is declining day by day drastically due to biotic and abiotic stresses. Conventional breeding for overcoming this problem is restrictive due to non-availability of resistant sources. Recent advances in plant tissue culture have made it possible to develop abiotic and biotic stress-resistant cultivars. For developing such cultivars a suitable *in vitro* regeneration protocol is prerequisite. Citrus cultivars vary in their response to *in vitro* organogenesis. This results in the need for cultivar-specific optimization of *in vitro* regeneration protocol. The present study was conducted in 2015–2017 at Department of Biotechnology, Assam Agricultural University, Jorhat. In the present investigation, *in vitro* regeneration of Khasi Mandarin was optimized using zygotic seedlings as explants. Modified Murashige and Skoog (MS) medium containing 1 mg/L 6-benzylaminopurine (BAP), 0.5 mg/L 1-naphthaleneacetic acid (NAA), and 0.4 mg/L kinetin shows the best result for multiple shoot induction with an efficiency of 68%. The average number of multiple shoots developed was 5. The modified MS medium containing 0.25 mg/L BAP, 0.5 mg/L NAA, 0.5 mg/L indole-3-butyric acid shows the best result for rooting of explants with an efficiency of 82% and average root length of 4 cm. These results suggest that standardization of these factors can help in development of a commercially viable tissue culture system for Khasi Mandarin.

Bio-Oil Generation and Characterization from a Woody Biomass as Sustainable and Renewable Energy Source

Harshal Kawale and Nanda Kishore

Department of Chemical Engineering, Indian Institute of Technology Guwahati, Guwahati, India

nkishore@iitg.ac.in

Keywords: Biomass, bio-oil, gas chromatography–mass spectroscopy, ^{1}H-NMR, pyrolysis

Abstract

The petroleum depletion over a period of time and the need of value-added petrochemicals drives a force to develop fruitful technologies. The recent trends in renewable energy production are yet to qualify the energy needs. Thermal processes for biofuels from biomass results biofuels in the form of solid, liquid and gas, which can be utilized as fuel and value-added products. Among these technologies, pyrolysis and catalytic pyrolysis are cost effective technologies on pilot scale as well as these can be delivered to form a biorefinery. We have conducted pyrolysis experiments at variable-temperature ranges to get biofuels from a lignocellulosic biomass. The source of biomass is available at tremendous amount around the far region of land but scarcely researched. Main product obtained was pyrolytic bio-oil, which was characterized using Fourier transform infrared spectroscopy (FTIR), and the analysis showed the number of functional groups present in it. Gas chromatography – mass spectroscopy (GC-MS) results in large number of components present in the bio-oil. ^{1}H-NMR spectroscopy analyzed to show various functional groups present in the sample as well as the amount of aromatic carbon content in the bio-oil. Thus, these characterization techniques revealed the fuel potential of the biomass to be used as an energy crop for the production of energy in a sustainable manner.

Metal Uptake Potentiality of *Salvinia cucullata*, Roxb. to Paper Mill Effluent – A Phytoremediation Approach

Kisholay Mazumdar and Suchismita Das

Department of Life Science and Bioinformatics, Assam University, Silchar, Assam, India

drsuchismita9@gmail.com

Keywords: Bioconcentration, hyperaccumulation, paper mill effluent, phytoremediation, translocation

Abstract

In the human development, paper mill industry plays a significant role but simultaneously it causes health hazard by emitting various heavy elemental constitution in the form of effluent. A novel fern *Salvinia cucullata* has the potentiality to resolve the problem by hyperaccumulating trace elements from contaminated sites. To acknowledge the phytoremediation or metal uptake potentiality of the plant, experiments were begun by treating *S. cucullata* with different concentrations of paper mill effluent (25%, 50%, 75%, and 100% v/v) for 28 days. It revealed that the plant has the ability to accumulate 10 different heavy metals such as Cd, Cu, Cr, Ni, Pb, Mg, Mn, P, Fe, and Zn at different effluent concentration treatments. However, the plant scored both translocation factor and bioconcentration factor values >1 at all the four effluent concentration treatments, which reflects metal hyper-accumulation potentiality of the plant. *S. cucullata* however undoubtedly a suitable plant for phytoremediation of paper mill effluent.

Phytoremediation of Heavy-Metal-Contaminated Soil using *Pongamia pinnata*; a Biofuel Plant

Pallabi Borah,[1,2] Tanmoy Karak,[3] Latha Rangan,[4] and Sudip Mitra[2*]

[1]Department of Environmental Science, Tezpur University, Assam, India
[2]Centre for Rural Technology, Indian Institute of Technology, North Guwahati, Assam, India
[3]Upper Assam Advisory Centre, Tea Research Association, Dikom, Dibrugarh, Assam, India
[4]Department of Biosciences and Bioengineering, Indian Institute of Technology, North Guwahati, Assam, India

sudipmitra@iitg.ac.in/sudipmitra@yahoo.com

Keywords: Bioaccumulation factor, heavy metals, phytoremediation, *Pongamia pinnata*, translocation factor

Abstract

The aim of the study is to assess the phytoremediation potential of *Pongamia pinnata*, a biofuel plant. Both, pot and field experiments were conducted in soil contaminated with paper mill wastes and municipal wastes. The pot experiment was conducted with different combinations of contaminated and control (forest) soil @ 0%, 25%, 50%, 75%, and 100%. The pot study showed its effectiveness in removing heavy metals (HMs) (Cd, Cr, Cu, Fe, Mn, Ni, and Zn) from the contaminated soil. The plants grown in 25% contaminated soil exhibited highest uptake capacity of the studied HMs. High average concentration of HMs were observed in roots except for Cd and Mn, whose concentration were highest in stems. Cu, Cr, and Ni were effectively removed by *P. pinnata*, with more than 50% removal efficiency. The average bioaccumulation factor values were found to be <1 for all the selected HMs, thus belonging to the excluder category. The translocation factor values exceeded 1 for Cd, Cr, and Mn in paper mill soil combination but in municipal wastes soil it exceeded 1 only for Cd and Mn. In the field experiment, *P. pinnata* showed accumulation of Zn and exclusion of other heavy metals (Cd, Cr, Cu, Fe, Mn, and Ni). Thus, the present study suggests that combination of soil @ 25% polluted soil with 75% control increases the uptake potential of *P. pinnata*.

In Vitro Studies of Histological and Biochemical Induction of Rooting of *Bacopa monnieri* (L.) Wettst.

Aanchal Goel[1] and Anil Kumar[2]

[1]University Institute of Biotechnology, Chandigarh University, Gharuan, India
[2]Department of Biotechnology, TIFAC CORE, Thapar University, Patiala, India

anchalgoel.biotech@cumail.in

Keywords: Alzheimer's disease, meristemoids, polyphenol oxidase, primordia, rhizogenesis

Abstract

Bacopa monnieri (L.) or Brahmi has its importance in ayurvedic medicine for treating nervous system disorders such as Alzheimer's disease and anxiety and psychiatric disorders such as improving memory. The present study was carried to investigate histological and biochemical changes during adventitious rooting of microcuttings of *Bacopa monnieri* (L.) Wettst. Histological studies were conducted to identify different phases of rooting in these microcuttings. The root meristemoids with distinct polarity become visible after 3 days and mark the beginning of *in vitro* root initiation phase. Biochemical studies were also conducted from basal portions of microcuttings. Total carbohydrate content was lower during initial periods (up to day 1) and was found to increase during root initiation and primordia development, which reflects high energy demands for active cell divisions. A significantly higher level of phenols was observed in microcuttings on medium supplemented with indole butyric acid (IBA). Polyphenol oxidase, peroxidase (POX), and ascorbate peroxidase activities were also found to vary during different phases of rhizogenesis. Early phases were also marked with the lower activities of POX and indole acetic acid oxidase (IAAO). This study reveals significant roles of enzymes, sugars, and phenols during different phases of rooting.

Dedifferentiated *In Vitro* Cell Lines: A Bioresource Utilization Method for Enhanced Spilanthol Production by Optimizing Media Constituents using Response Surface Methodology from *Spilanthes paniculata*

Radhika Rajendran[1] and Rakhi Chaturvedi[1]

[1]Department of Biosciences and Bioengineering, Indian Institute of Technology Guwahati,Guwahati-781039, Assam, India

rakhi_chaturvedi@iitg.ac.in

Keywords: In vitro dedifferentiated cultures, response surface methodology, *S. paniculata*, spilanthol

Abstract

Spilanthes paniculata is a medicinal herb with rich source of therapeutic constituents. *In vitro* callus cultures were established from leaf-disc explants of *S. paniculata*, which is an alternative method of cell biomass utilization and conservation of natural plant resources. Additionally, response surface methodology (RSM) was performed to maximize the production of *N*-alkylamide, Deca-2*E*, 6Z, 8E-trienoic acid isobutylamide (spilanthol) from *in vitro* cell cultures. In the first step of optimization, with Plackett-Burman design (PB), Murashige and Skoog (MS) major salts, sucrose, 2,4-dichlorophenoxyacetic acid (2,4-D), N6-benzylaminopurine (BAP) were found to be the important factors affecting spilanthol production significantly. In the second step, a 2^4 full factorial central composite design was applied to determine the optimal concentration of each significant variable. A second-order polynomial was determined by the multiple regression analysis of the experimental data. Optimum values for the critical components were obtained as MS (1.5), sucrose (5%), 2,4 –D (1.8 μM) and BAP (4.82 μM), with a predicted value of maximum spilanthol production of 3.72 mg/g dry weight (DW). Under the optimal conditions, the experimental value of spilanthol production was 2.81 mg/g DW. The coefficient of determination (R^2) was 0.9922, which ensures adequate credibility of the model. Furthermore, the higher production of spilanthol was achieved by statistical model as compared with that of non-optimized media constituents. Before optimization, callus cultures and leaves from parental plant (control) yielded 1.75 and 0.26 mg/g DW spilanthol, respectively. Thus, RSM is an effective tool for optimizing the media combinations, which uses quantitative data from an appropriate experimental design to simultaneously solve and determine multivariate equations on maximizing the production of therapeutic compounds. Current work is unique information on statistical optimization and production of spilanthol from leaf-disc callus cultures of *S. paniculata*.

A Comparative Biochemical and Molecular Approach to Understand the Factor Determining Salt Tolerance in Rice Cultivars

Amber Gupta[1] and Birendra Prasad Shaw[1]

Plant and Microbial Biotechnology Unit, Institute of Life Sciences,
Nalco Square, Bhubaneswar, 751023, Odisha, India

bpshaw@ils.res.in

Keywords: Active transport, antiporters, energy gradient, H^++ ATPase, NHX, salt tolerance

Abstract

Among the abiotic stresses high soil salinity, contributed largely by Na^+ and often compounded with drought, is the main factor that adversely limits growth and productivity of major crop plants, including rice. Salt (NaCl) affects a plant in two ways. First, it lowers the water potential of the environment, resulting in reduction in turgor pressure. Second, its accumulation in cytoplasm has toxic effect on cellular processes, leading to inhibition of growth and development. As a counter measure, plant adapt a number of strategies. Low concentration of Na^+ ions in the cytoplasm is apparently achieved by regulation of Na^+/K^+ selectivity of antiporters/channels across the membranes. The functioning of secondary transporters is activated by proton pumps, which are the primary active transport systems in membranes. The free energy gradient ($\Delta\mu H^+$) produced by the plasma membrane and vacuolar H^+ ATPase is presumed to provide the driving force in regulating the functioning of secondary transporters (housed in respective membranes) for the maintenance of ionic balance. Majorly two secondary transporters SOS1 (plasma-membrane-bound Na^+/H^+) and NHX (vacuole-membrane-bound Na^+/H^+) have also been reported to play important role in salt tolerance in plants. However contribution of biochemical processes in salt tolerance is not well defined hence the current study was conducted to understand their functioning in rice cultivars; tolerant and sensitive one to salinity. The present study was planned to go for a comparative study and to find out few of the most potential and universal salinity-stress-responsive gene in two salt-tolerant (Nona Bokara and Pokkali) versus two salt-sensitive (IR29 and IR64) rice varieties, so as to get a holistic picture of their importance and involvement in salt tolerance.

Phytochemical Diversity and Micropropagation of *Paris Polyphylla* Sm. Rhizomes from Northeast India

Khumuckcham Sangeeta Devi,[1] Pramod Tandon,[2] Suman Kumaria,[2] Ashiho Asosii Mao,[3] Nripemo Odyuo,[1] Deepu Vijayan,[1] Ashutosh Pathak[1]

[1]Botanical Survey of India, Eastern Regional Centre, Shillong, Meghalaya, India
[2]North-Eastern Hill University, Mawlai, Shillong, Meghalaya, India
[3]Botanical Survey of India Headquarter, Kolkata, West Bengal, India

khsangbotany@gmail.com

Keywords: HPLC, khonoma, micropropagation. *Paris polyphylla*, saponins

Abstract

Identification of medicinal plant for elite genotypes requires the assessment of phytochemical diversity across different populations. Such study is fundamental for further scale of plant resources as well as subsequent drug development for the treatment of human ailments. The present study was taken up to assess the total steroidal saponins diversity in *Paris polyphylla* across the northeastern region of India. Nine populations from four northeastern states of India were assessed in the present study. Quantification of the steroidal saponins in the selected population was carried out by comparing against standard saponins using high-performance liquid chromatography (HPLC). From the study, it was found that *P. polyphylla* populations from Khonoma showed the highest total saponins content, recording an average of 32.06 mg/g dry weight in comparison with all other populations under study Micropropagation of Khonoma populations was carried out for large-scale propagation of this elite zonal chemotype. Efficiency of two cytokinins with different sucrose concentrations on minirhizome induction was studied, and it was found that 6-benzylaminopurine 0.5 mg/L + 6% sucrose and 2-isopentyladenine (2iP) 1.0 mg/L + 6% sucrose resulted in the best response giving 88.6% and 89.2% with 1.27 ± 0.02 g fresh weight and 1.36 ± 0.10 g fresh weight of minirhizome, respectively.

DNA Barcoding of Commercially Important *Vanda* Species (Orchidaceae) of Nagaland

Joyrison Kamba and Chitta Ranjan Deb

Department of Botany, Nagaland University, Lumami, Nagaland, India

debchitta@gmail.com

Keywords: DNA barcoding, molecular markers, orchids, recognition tag

Abstract

Orchids are known for their ornamental and medicinal importance, besides their beautiful looks and attractive color that has led to indiscriminate collection and smuggle extensively in the wild, leading to rare, endangered, and threatened many important species without proper documentation. Moreover, they can be crossed within the same genus or with different genera, leading to the production of various types of new hybrids, having similar morphological traits in nature. So, in the present work, we established barcodes for *Vanda* species, which are used for rapid identification at any stage of life and detection of species that relies on short DNA sequence variation and provides a unique recognition tag to a species. For these, six *Vanda* species, that is, *Vanda coerulea*, *V. bicolor*, *V. stangeana*, *V. ampullacea*, *V. testacea*, and *V. alpina* were tested with three molecular marker system (*matK*, *rbcL*, and ITS) from cytoplasmic and nuclear regions. The species identity generated from BLAST results obtained by these markers system correctly matched morphological traits, which indicate that DNA barcoding can be used for the rapid identification and detection of any unknown biological sample at the molecular level for species and for variety certification for protection and conservation purposes.

In Vitro Induction of Cormlet in *Crocus sativus* and its Future in Genetic Transformation

Mona Hamal Thakuri and N. K. Chrungoo

Centre for advance studies in Botany, North Eastern Hill University, Shillong 793022, Meghalaya, India

hamalthakurimona@gmail.com

Keywords: Agrobacterium, cormlet, genetic transformation, saffron, sterile triploid

Abstract

Crocus sativus (saffron) is a triploid sterile plant belonging to family Iridaceae. Its dry stigma is considered to be the world's most expensive spice, having diverse medicinal properties. The plant has an active growth period in the autumn and winter along with flower and leaf formation. Corms multiply very slowly producing only four to five daughter corms per mother corm. They survive for only one season, forming daughter corms and perish. Genetic improvement in saffron is not possible due to its triploid nature and male sterility. Crop improvement is confined to the evaluation and selection of naturally occurring clonal variants. To overcome these challenges, application of tissue culture methods offers great potential for mass multiplication and genetic improvement. In the present study, we have induced microcorms from different somatic tissues that were grown on Murashige and Skoog (MS) half-strength medium supplemented with 0.5 mg/L 6-benzyl amino purine (BAP) + 0.1 mg/L α-naphthalene acetic acid (NAA). Formations of cormlet-like structure from these embryos was observed in MS medium supplemented with 2 mg/L thidiazuron (TDZ) and 1 mg/L indole acetic acid (IAA). Germination of cormlets could be achieved on MS medium containing 1 mg/L BAP and 0.5 mg/L NAA. Alternately, transformants containing key regulatory genes involved in enhancing the phytochemical property of saffron or genes involved in inducing increased number of cormlets can be carried out. The need of the hour is to develop a transfer mechanism for the genetic transformation, for which calli and embryos were cocultivated with *Agrobacterium* and direct *in planta* cocultivation of mother corms were carried out using the native expression vector. Screening of the transformant was done using green fluorescent protein and β-glucuronidase (GUS) analysis. The development of a method for gene transfer will help in designing an effective method for the development of disease-resistant and high-yielding saffron plants.

Tissue-Culture-Mediated Propagation of Some Medicinal Plants for Conservation and Sustainable Utilization

Soumendra K. Naik
Department of Botany, Ravenshaw University, Cuttack, India

sknuu@yahoo.com

Keywords: ISSR, medicinal plants, micropropagation, RAPD, synthetic seeds

Abstract

Plants in general and multipurpose medicinal plants in particular are under severe stress in their wild habitat and depleting rapidly due to various anthropogenic activities. A number of such plants have limitations in propagation through conventional means. Thus, in our laboratory, we have developed various tissue-culture-mediated plant propagation systems for few such medicinal plants species, namely *Paederia foetida*, *Bacopa monnieri*, *Withania somnifera*, and *Operculina turpethum*. Protocols for the production of synthetic seeds using axenic nodal segments by encapsulation method have also been developed for plant propagation and germplasm exchange of *P. foetida* and *B. monnieri*. In some of these protocols, clonal fidelity of micropropagated plants with that of the mother plant was confirmed using various molecular markers including Random amplified polymorphic DNA (RAPD) and Inter simple sequence repeat (ISSR). Plant propagation methods of these plant species have the potential to provide constant supply of planting materials independent of season, thus useful for their reintroduction in wild and/or providing raw materials with the uniform quality for manufacturing therapeutics and other uses in a sustainable manner without disturbing wild populations.

Iron Biofortification In Bananas by Expression of *Oryza sativa* Nicotianamine Synthase Genes

M. Mayil Vaganan, C. Sivagandhi, S. Ganesan, M. Kumaravel, I. Ravi, K. J. Jeyabaskaran, S. Backiyarani and S. Uma

ICAR National Research Centre for Banana, Thayanur Post, Thogamalai Road Tiruchirappalli 620102, Tamil Nadu, India

mmayilv@gmail.com

Keywords: Grand Naine, iron deficiency, *OsNAS1*, *OsNAS2*, Rasthali, transgenics

Abstract

Iron is an essential micronutrient for plant growth and human health. Iron deficiency is one of the most prevalent micronutrient deficiencies in the world, causing anemia to several billion people especially in developing countries. The production of iron-biofortified staple crops will help to ameliorate iron-deficiency anemia. Banana is one of the staple crops of Asian and African population. Two commercial banana genotypes, Rasthali and Grand Naine, are biofortified by introduction of two rice (*Oryza sativa*) nicotianamine synthase genes, *OsNAS1* and *OsNAS2*, driven by *Zea mays* ubiquitin promoter to increase the iron content in fruit pulp. *Agrobacterium*-mediated transformation of the constructs carrying these genes and selectable marker, *nptII*, was carried out with embryogenic cell suspension (ECS) of the bananas. One hundred independent transgenic events of each banana genotype for each gene construct were generated and being field grown along with untransformed control plants in iron-sufficient soils under confinement in transgenic net houses equipped with all biosafety standards as per guidelines of Department of Biotechnology, Government of India. The transformants, before planting, were confirmed for transgene presence and selectable marker with specific primers using polymerase chain reaction (PCR) and checked for *Agrobacterium* contamination using *Vir-C* gene-specific primers to eliminate false positives. Presently, transgenic lines are 9 months old and in shooting and fruit-bearing stages. Estimation by Inductively coupled plasma atomic emission spectroscopy (ICP-OES) of iron element content of 25 representative lines of each genotype at vegetative stage is promising with seven Rasthali lines and six Grand Naine lines having average iron content of 24.5 mg/100 g dry wt. as compared with nontransgenic lines with maximum of 11.2 mg /100 g. As the increase in iron concentration was 2.2 times in the transgenic lines, the results offer promise to effectively increase the level of iron in the fruit. Transgene integration and the copy number of the transformed events are being performed.

Long-Term Seed Storage and Acclimatization of *In Vitro* Derived Plantlets of *Paphiopedilum villosum* – A Threatened Commercial Slipper Orchid

Reema Vareen Diengdoh, Meera Chettri Das, and Suman Kumaria

Plant Biotechnology Laboratory, Department of Botany, Centre for Advanced Studies, North-Eastern Hill University, Shillong 793022, Meghalaya, India

rimavareen@gmail.com

Keywords: Acclimatization, *Ex situ*, orchid seeds, SEM, storage, sustainable utilization

Abstract

In today's times, orchids are appraised for their beauty and discussed for their scanty population. Heterogeneous orchid seeds exhibit, by nature, features like minute seed size and ample availability per capsule. Cryopreservation is an efficient *ex situ* strategy, but orchid seeds conservation is hampered by poor storage conditions and *in vitro* regeneration with effective acclimatization protocols need to be standardized. The advantages of seed storage are indefinite storage period, genetic stability, reduced infrastructure, and the stored genetic material does not require manipulation. Stored seeds of *Paphiopedilum villosum* showed the best result on storage at −196°C with no significant variation in *in vitro* germination (81.5%) as well as viability analysis (80.1%) up to 360 days. Mature seeds stored at −196°C were appropriate for long-term storage, having germination of 81.3 ± 1.5% with similar viability percentage 80.1 ± 1.9%, over 360 days storage supported by scanning electron microscopic (SEM) studies. Plantlet growth *in vitro* was best recorded in Murashige and Skoog medium containing 20 μM 6-benzylaminopurine + 5 μM indole-3-acetic acid with high cumulative response of 84.5 ± 3.1%. While for acclimatization in two experimental locations, the combination of compost mixture comprising charcoal + brick stone + soil + layer of moss litter in the ratio 1:1:1:1 was found with good survival of 59.1 ± 3.1% with average shoot length of 11.8 ± 0.8 cm. *Ex situ* conservation offers safer security backup system for the conservation, allowing accessibility for research evaluation, commercial propagation, and ultimately lead to sustainable utilization for the future.

Efficient *In Vitro* Propagation of *Hedychium coronarium* J. Koenig using Rhizome Segment and Assessment of the Genetic and Biochemical Fidelity of Micropropagated Plants

Shashikanta Behera,[1] Pradeep K. Kamila,[2] Kedar K. Rout,[3] Durga P. Barik,[1] Pratap C. Panda,[2] and Soumendra K. Naik[1]

[1]Department of Botany, Ravenshaw University, Cuttack, Odisha, India
[2]Plant Taxonomy and Conservation Division, Regional Plant Resource Centre, Bhubaneswar, Odisha, India
[3]Siddhi Binayak Science College, Angul, Odisha,India

sknuu@yahoo.com

Keywords: Hedychium coronarium, HPTLC, ISSR, medicinal plant, micropropagation

Abstract

Hedychium coronarium J. Koenig (Zingiberaceae) is an important medicinal plant with ornamental values. The rhizome and flower of the plant also possess volatile compounds and thus useful in perfume industries. *H. coronarium* has already been enlisted as a medicinal plant of conservation concern in various states including Madhya Pradesh, Odisha, Kerala, and Karnataka due to unsustainable harvesting to meet the demand. Tissue-culture-mediated plant regeneration has the potential to produce large number of plants, overcoming limitations of conventional propagation methods, thus fulfilling the demand of the plant and plant materials in a sustainable manner. Therefore, the present study was undertaken to develop an efficient protocol using rhizome segments for clonal propagation of *H. coronarium*. Best shoot proliferation was recorded on Murashige and Skoog (MS) medium containing 0.8 mg/L thidiazuron (TDZ), followed by their subculture on MS augmented with 1.0 mg/L gibberellic acid (GA_3). Upscaling of shoots was carried out using axenic stem segments derived from primary *in vitro* shoots. Simultaneous root development from shoots was observed during shoot multiplication. This phenomenon eliminates the requirement of an additional step of rooting, thus reducing both the cost and time of the plant propagation. The present protocol is an efficient one and could produce ~540 plantlets starting from a single explant within 14 weeks. *In vitro* plantlets were successfully acclimatized and eventually established in the field. Monomorphic banding profile of micropropagated plants viz-à-viz mother plant obtained by inter simple sequence repeat (ISSR) confirmed the clonal fidelity of tissue-culture-raised plants while bio-chemical stability of micropropagated plants was also ascertained using various quantitative phytochemical analyses including high-performance thin-layer chromatography (HPTLC). This protocol could be useful for commercial-scale propagation and conservation of *H. coronarium*.

Evaluation of Nutritive Value, Dietary Antioxidants and *In Vitro* Antioxidant Activity of Some Edible Flowers from Ethnic Sources: Exploration of Lesser Known Food Sources of North East India

Phunumani Baruah, Kangkon Saikia, Khanjan Das, Karabee Dutta, A. K. Handique

Department of Biotechnology, Gauhati University, Guwahati, Assam, India.

phunumanibaruah@gmail.com

Keywords: Antioxidant activity, dietary antioxidants, edible flowers, HPLC-DAD, nutritive value, polyphenols

Abstract

Certain flowers are use as food supplement which is an old food culture for many ethnic communities in the world and is part of traditional knowledge system. However, information about their basic food chemistry is very limited and so there is a big knowledge gap between traditional knowledge, food culture, and scientific scrutiny to validate the traditional knowledge. The present study was focused on evaluation of nutritive and nutraceutical values of edible flowers that are traditionally consumed by various ethnic communities of North-East India. Seven flower samples were selected for study namely, *Moringa oleifera, Nyctanthes arbor-tristis, Musa balbisiana, Phlogacanthus thyrsiformis, Carica papaya, Adhatoda vasica* and *Curcuma angustifolia*. The objectives of the present study are: (1) To study food value in terms of major nutritional parameters; (2) To quantify major dietary antioxidants namely, total phenolics and flavonoids along with profiling of major polyphenols using high-performance liquid chromatography coupled with diode array detection (HPLC-DAD), (3) To evaluate antioxidant efficacy by *in vitro* antioxidant assays. The present study revealed that the flowers are nutritionally rich with protein content of 10-20%, carbohydrate 8-14%, dietary fibre 3-17% and lipid 1.5-4%. Total mineral content was 6-17% with major content of K, P, Na, Ca, Fe and Zn. Among dietary antioxidants, content of total phenolics was 0.3–3.4 mg GAE/g and total flavonoids were 1.9-57.1 mg RE/g. The major polyphenols found were quercetin, naringin, rutin, chlorogenic acid, vanillic acid, sinapic acid, *p*-coumaric acid and gallic acid. Significant variation in antioxidant activity was observed among the flowers. The findings will help to eliminate the misconception that non-conventional and ethnic food are "poor man's food". Being excellent source for nutritional components as well as dietary antioxidants edibles flowers can be considered as a non-conventional source to fight nutritional security.

Role of Elicitors in Eugenol Production in Hairy Root Cultures of *Ocimum tenuiflorum* L.

Swati Sharan,[1,2] Kunal Mukhopadhyay,[2] and Anjana Rustagi[3]*

[1]School of Life Sciences, Jawaharlal Nehru University, New Delhi, India
[2]Department of Bio-Engineering, Birla Institute of Technology, Mesra, Ranchi, India
[3]Department of Botany, Gargi College, University of Delhi, Delhi, India

anjana.chaurasia@gmail.com

Keywords: Eugenol, hairy root cultures, methyl jasmonate, Ocimum tenuiflorum, salicylic acid, yeast extract

Abstract

Ocimum tenuiflorum is an annual aromatic herb highly prized for its culinary and therapeutic values. Eugenol is one of the major active components of *O. tenuiflorum*, which has been recommended for the treatment of different ailments under different systems of medicines. In the present study, we report different tissue culture methodologies and synergistic effects of elicitors for enhanced biomass production and eugenol accumulation in hairy root cultures of *O. tenuiflorum* and their subsequent quantitative estimation using high-performance liquid chromatography. The concentration of elicitors, age of cultures, and exposure time were studied for optimization. Our investigations suggest that the determination of the right stage of culture for addition of elicitors is one of the important parameters for enhanced production of biomass and eugenol accumulation. A sigmoid growth curve was obtained for increase in dry weight (DW) and/or fresh weight (FW) versus accumulation of eugenol. Roots of *in vitro* cultured plants were inoculated in a liquid culture in flasks and multiplied to the maximum biomass, which was 6.63-fold higher than that of initial inoculum after 25 days. Eugenol accumulation in hairy roots versus time graph was a sigmoid curve. Different concentrations of various elicitors, yeast extract, methyl jasmonate, and salicylic acid, were added to 17-day-old (exponential phase) and 22-day-old (stationary phase) hairy root cultures for different exposure times (4, 8, and 12 days). Results suggest that 8 days of exposure to yeast extract is the optimum condition for the maximum biomass production and the accumulation of eugenol in 17-day-old hairy roots cultures. These optimum conditions led to a 6-fold increase in eugenol production. This study recognizes the potential of hairy roots and role of different elicitors for enhanced biomass production and the accumulation of secondary metabolites under *in vitro* conditions.

In Vitro Culture Establishment and Shoot Multiplication in *Buchanania lanzan* (Chironjii)

Fatima Shirin, Deepti Bhadrawale, Sonam Singh Baghel, Sushma Marawi, Trilok Gupta, and Mukesh Sonkar

Genetics and Tree Improvement Division, Tropical Forest Research Institute, P.O. RFRC, Mandla Road, Jabalpur 482 021, India

fatimashirin1997@gmail.com

Keywords: Benzyl adenine, *Buchanania lanzan*, indole acetic acid, *in vitro*

Abstract

Buchanania lanzan (family Anacardiaceae), known as char or chironji, is a commercially useful tree species of India. Its seeds are used as dry fruits and sold at the rate Rs. 600–2000/kg. Seeds are the major source of natural regeneration. The major problem in the reforestation of *B. lanzan* is the low-percentage germination of seeds due to fungal contamination associated with storage of seeds. Seeds exposed to sunlight fail to germinate and soon lose their viability. In forest biotechnology, micropropagation is a promising choice for mass propagation of superior forest tree genotypes. The present research work was undertaken with a view to solve the above-mentioned problem. Ripe fruits were collected from three different healthy trees of chironji located in Tropical Forest Research Institute campus. The hard seed coat was removed manually, and seeds were collected after depulping of fruits. They were surface sterilized by using 0.1% mercuric chloride solution ($HgCl_2$) for 4–5 min followed by rinsing with sterile distilled water. Nodal segments of germinated seedlings were used as explants. Axillary shoot proliferation through nodal segments was tried on Murashige and Skoog (MS) medium supplemented with different concentrations of benzyl adenine (BA; 1, 3, and 5 mg/L) and indole-3-acetic acid (IAA; 0, 0.1, and 0.5 mg/L). The maximum sprouting was obtained on MS medium supplemented with 3 mg/L BA and 0.5 mg/L IAA. Around three to four shoots of 2–3 cm shoot length were formed after 20–30 days of inoculation. These results will be very helpful in shoot multiplication of this commercially important tree species.

Flexibility in Genetic System Directing Diversification and Migration of Medicinally Important Genus *Artemisia* L. Inhabiting North West Himalayas, India

Uma Bharti, Eshan sharma, and Namrata Sharma

Department of Botany, University of Jammu, Jammu, Jammu and Kashmir, India

umabotany786@gmail.com

Keywords: *Artemisia*, chromosome, polyploid, population, rootstock, variability

Abstract

Cytological and morphological variations are usually found in taxa having a wide distribution range; it helps them to survive and adapt to different kinds of environmental conditions. Such variations occur both at inter- and intrapopulation levels and reflect the adaptability as well as future survival that can help the plant diverge its habitat in apparently random directions. *Artemisia* L., a taxonomically complex, medicinally important, and cytologically flexible genus of family Asteraceae, is an adequate example of the same. *Artemisia* L. has a long history of use in herbal medicine especially in matters connected to digestive system and treatment of worms. Present work is based on four species of this genus that are *Artemisia nilagirica* (C.B. Clarke) Pamp, *Artemisia scoparia* Waldst. & Kit, *Artemisia maritima* L., and *Artemisia tournefortiana* Rchb. forming populations at variable altitudinal regimes (332–3350 masl) of J & K state, India. Cytological details reveal that *A. nilagirica* is the most complex with four cytological races and chromosome numbers as $2n = 18, 32, 34$, and 54; others are cytologically stable. Abnormalities in male meiotic tract of *A. nilagirica* affect the genetic constitution and viability of male gametes. Such gametes result in to the origin of aneuploids, polyploids, and species complexes with derived/new chromosome numbers in this genus. Besides, these are also responsible for reduced reproductive success through seeds, that is, reduced percentage of healthy seeds. Although little amount of variability survives, it is maintained and ramified through alternate means using rootstock. The rootstock hence acts as the main organ for plant propagation and it can be used for large-scale multiplication of favorable genotypes, if the need arises in the near future, keeping in mind the economic value of various species of this genus.

Defense-Related Gene Expression in *Musa paradisiaca* var. Kachkal against *Odoiporous longicollis* Infestation

Sudarshana Borah,[1,2] Dipsikha Bora,[2] and Priyadarshini Bhorali[1]

[1]Department of Agricultural Biotechnology, Assam Agricultural University, Jorhat, Assam, India
[2]Department of Life Sciences, Dibrugarh University, Assam, India

sborah151@gmail.com

Keywords: Defense, pseudostem weevil, resistance, transcriptome

Abstract

Banana pseudostem weevil, *Odoiporus longicollis*, is an important pest of banana, which causes significant yield loss. In Assam, it is a serious constraint to banana production and management is highly dependent on the use of chemical pesticides. In the present study, we investigated the defense transcriptome of a banana cultivar resistant to *O. longicollis* infestation, Kachkal, in an attempt to monitor the defense-related transcriptomic changes taking place in the resistant plant at molecular level and to identify genes potentially involved in conferring resistance. We artificially introduced the pest to plants raised in controlled environment. The insect causes extensive tunneling to the core of pseudostem and leaf sheath forming a lethal condition for the host. Transcriptome sequencing of uninfested and infested plants depicted a differential pattern in terms of defense-related genes specifically modulated in response to the pest infestation. The transcriptome data show that important defense-regulated genes such as ones encoding pathogenesis-related proteins, chitinase, lipoxygenase, lectin, and so on, showed high fold change upon insect infestation. Important defense-related genes highly expressed in response to infestation were further validated through quantitative polymerase chain reaction (qPCR). In the present research, we are presenting the analysis of the transcriptomic changes taking place in Kachkal specifically against pseudostem weevil infestation, and validation of expression of important defense genes that could be potentially utilized for generating resistant banana lines against the pest.

Characteristics of Biomass Briquettes Prepared in a Low-Power Screw-Press Machine using Wild *Colocasia esculenta* Tuber as Binder

Anjali Narzary[1] and Amarendra Kumar Das[2]

[1]Centre for Rural Technology, IIT Guwahati, Guwahati 781039, Assam
[2]Department of Design, IIT Guwahati, Guwahati 781039, Assam

anjalinarzary@iitg.ac.in

Keywords: Binder, briquette, calorific value, *Colocasia esculenta* (taro)

Abstract

Biomass briquetting is a very promising technology to tackle problems such as agriculture and forest waste management and also generate high-density, good burning characteristics fuel energy that could be a great substitute for fire wood used for cooking purposes in rural areas, and this could in turn solve the problem related to deforestation. Biomass briquettes were made using dry leaves, wood charcoal, and straw. The binder used in this work was wild *Colocasia esculent* tuber (Taro). The following two methods were used for the preparation of the binder: gelatinization and starch extraction. The result shows that wild Taro can make a good substitute for other complicated binders used in different studies. Charcoal and dry leaves were mixed in three different ratios to find the best results. Among all the ratios, 3:1 dry leaves and charcoal briquettes showed the highest volatile matter content of 83.21%, low ash content of 8.45%, high bulk density of 0.605 g/cm^3, and high calorific value of 5414.5 MJ/kg. Binder was tested for two different percentages, 20% and 40%, for dry leaves briquette, and the 20% binder showed better results such as 82.03%, 7.63%, 0.532 g/cm^3, 4239.98 MJ/kg, and 9.03% of volatile matter content, moisture content, bulk density, calorific value, and ash content, respectively. The comparison was also done for gelatinization and starch extraction procedure, out of which the latter showed better results.

Evaluation of Nutritional Parameters, Antioxidant Potential and Polyphenol Profile of *Moringa oleifera* Lam. using HPLC/ESI MS-MS.

Bhanita Talukdar[1], Kangkon Saikia[2], Durlav N. Singha[1]

[1]Advanced Level Institutional Biotech Hub, Dakshin Kamrup College, Mirza, Assam, India.
[2]Department of Biotechnology, Gauhati University, Guwahati, Assam, India.

bhanitatalukdar13@gmail.com

Keywords: Antioxidants, HPLC-DAD, HPLC-ESI-MS/MS, nutritive value

Abstract

Moringa oleifera has been used widely as local food, in traditional medicine and in industrial applications. Leaves of *M. oleifera* are considered to be rich source of vitamins, minerals and exhibit strong antioxidant activity. In the present study, assessment of major nutritional parameters is carried out for protein, carbohydrates, minerals, lipid and dietary fibre; *in vitro* assessment of antioxidant potential was done by examining radical scavenging capability, ferric reducing antioxidant power, metal ion chelating ability and total antioxidant capacity. High-performance liquid chromatography (HPLC) coupled with electrospray ionization mass spectrometric (ESI-MS) detection in positive and negative ion mode has been used to identify the phenolic compounds. Photodiode-array detection (DAD) has been used for screening of different classes of phenolic compounds, whereas MS-MS fragmentation data were employed for their structural characterization. The *M. oleifera* leaves are found to be rich in protein and carbohydrate content with abundant composition of minerals like calcium (Ca), potassium (K), magnesium (Mg), phosphorous (P), sodium (Na), iron (Fe), zinc (Zn), copper (Cu). The high calcium content (2275 mg/100g) is essential as a part of human nutrition for growth and development. Among the nutraceutical parameters, *M. oleifera* was found to be rich in hydroxycinnamic acids. Major constituent being cholorogenic acid and p-coumaric acid, which is remarkable with total phenolics content of 14.89 mg GAE/g. Leaves are also found to be rich in flavonoid content (28.75 mg RE/g) where, majority of them were characterized as quercetin, kaempherol, apigenin and with some of their derivatives. The polyphenolic composition reflecting the antioxidant activity can be recognized as strongest among various conventional and non-conventional food plants.

Biochar for Soil Moisture Conservation

Kanku Deka[1], P. P. Hazarika,[2] and B. K. Medhi[3]

[1]Krishi Vigyan Kendra, Baksa, Assam, India
[2]Department of Soil Science, Assam Agricultural University, Jorhat, Assam, India
[3]AICRP on Irrigation Water Management, Assam Agricultural University, Jorhat, Assam, India

prarthana96780@gmail.com

Keywords: Biochar, Gravimetric water, pyrolysis

Abstract

"Biochar" is a relatively new term, yet it is not a new substance. Biochars, product of thermal decomposition or incomplete combustion of biomass or bio-wastes under limited oxygen supply, are fine-grained highly porous charcoal substances that are distinguished from other charcoals in its intended use as soil amendments. Recent estimate suggests that nearly 16, 12, 2.78, 58, and 188 lakh tons of rice straw, rice husk, *toria* stover, and bamboo leaves, respectively, remain unutilized annually and these farm wastes have the potential of further reutilization through production of biochar, which may effectively be used in sustainable production system. Characterization of biochar with respect to physico-chemical properties determines the suitability of biochar to conserve soil moisture, which is again regulated by kind and source of feed stock materials. Keeping these aspects in view, a study on characterization of biochars prepared from four different feed stocks, namely rice straw, rice husk, toria stover, and bamboo leaves was conducted at Assam Agricultural University during 2014–15 and 2015–16 to validate its efficiency. After determining the physicochemical properties of the four biochars, a set of pot culture experiment in poly house taking *toria* as test crop was conducted with four biochars. Four hundred gram of soil (preferably light textured) in 500 g capacity of plastic pot replicated thrice was designed statistically (*factorial* CRD) with four doses of biochars (0, 0.5, 1.0, and 1.5% wt/wt). Initially, a moisture level at field capacity was maintained and periodical volumetric soil moisture content (upto 70 days) was monitored to evaluate their efficiency. Gravimetric soil moisture content decreased significantly with the progress in days of experimentation irrespective of types of biochar used. However, increase in biochar doses increased the soil moisture content significantly over the one where no biochar was applied. Highest efficiency to conserve soil moisture over the days of study period was due to the application of bamboo leaves biochar.

In Vitro Antioxidant Activity of Some Non-Conventional Leafy Vegetables and Modelling of Isothermal Degradation of Dietary Antioxidants upon Cooking Time

Kangkon Saikia, Jharna Thakuria, Karabee Dutta, A. K. Handique

Department of Biotechnology, Gauhati University, Guwahati, Assam, India.

kangkonsaikia@gmail.com

Keywords: Antioxidant activity, dietary antioxidants, flavonoids, polyphenols, thermal degradation, HPLC-DAD.

Abstract

Use of non-conventional food plant particularly non-conventional leafy vegetables appears to be a relics of pre-historic age when human beings were hunter gatherer. Over millennium, the practice continued and is mostly common in tropical and subtropical countries in the world which are rich in biodiversity. Five different non-conventional leafy vegetables were selected as plant materials, undertaking major objectives of determination of antioxidant activity, estimation of dietary antioxidants and quantification of eight polyphenols namely, quercetin, naringin, rutin, chlorogenic acid, vanillic acid, sinapic acid, *p*-coumaric acid and gallic acid by high-performance liquid chromatography (HPLC) coupled with a photodiode-array detection (DAD) system. The degradation kinetics was studied for dietary antioxidants namely, total polyphenols and flavonoid content in correlation with different *in vitro* antioxidant assays namely, radical scavenging assay, ferric reducing antioxidant power, metal ion chelating ability and total antioxidant capacity. The kinetic model was designed to resemble the process of cooking (at 100 °C) within the time interval of 30 to 120 minutes. The fitting of experimental data in first order kinetic equation and the degradation rates were discussed. Significant variation in antioxidant content and antioxidant activity were observed among plant species. The kinetic model revealed that the total phenolics and flavonoid content along with their antioxidant activity increases within first 30 minutes and then reduces marginally after prolonged cooking (i.e. after 60 minutes) as a subject of degradation or loss of its antioxidant potential. However, marginal decrease in antioxidant capacity of some standard antioxidants were also observed within the same conditions but for some standards the decrease was insignificant. The primary conclusion drawn from the present study is that the heat treatment initially liberates some low molecular weight phenolic compounds depending on its intercellular occurrence which when subjected to prolonged heat treatment results in isothermal degradation and hence loss in antioxidant capacity.

Development of Broad-Spectrum Bacterial Blight Resistance in Traditional Basmati Rice Cultivar

Meenakshi Raina[1], Romesh Kumar Salgotra[1], and Deepak Kumar[2]

[1]School of Biotechnology, Sher-e-Kashmir University of Agricultural Sciences & Technology of Jammu, Jammu and Kashmir, India
[2]Department of Botany, Central University of Jammu, Rahya-Suchani (Bagla), Samba 181143 Jammu & Kashmir, India

raina.kesar592@gmail.com

Keywords: Bacterial blight, Basmati, marker-assisted backcross breeding, semi-dwarf

Abstract

The basmati rice variety "Ranbir Basmati" is very popular among farmers of Jammu region due to its palatable taste and short duration to fit in rice–wheat cropping system. However, the variety has recently succumbed to bacterial leaf blight (*Xoo*) caused by *Xanthomonas oryzae* pv. *oryzae* (*Xoo*) and prone to lodging. The severity and significance of damage caused by both the disease and lodging have necessitated the development of effective strategies for their management. Two major bacterial blight (BB) resistance genes and a semi-dwarf gene were introgressed into an Indian Basmati through marker-assisted backcross breeding. A high-yielding introgressed line PAU148 carrying *xa13*, *Xa21*, and *sd1* genes was used as a donor parent. Marker-assisted backcrossing was continued till BC2 generation wherein gene-specific markers specific for resistance genes were used for foreground selection, and a set of parental polymorphic microsatellite markers was used for the background selection at each stage of backcrossing. In BC_2F_2 population, 19 plants were found to be positive for all three genes, whereas the maximum genome recovery of Ranbir Basmati in BC2F2 was 86.9% in introgressed line SBTIL121. Introgressed lines carrying resistance genes were further evaluated for BB resistance. Genotypes carrying both resistance genes exhibited very high level of resistance against BB, whereas lines containing either *Xa21* or *xa13* gene alone showed moderate resistance. Pyramided lines were also analyzed for agro-morphological characters in randomized block design with two replications. All lines were found to be significant for all agro-morphological traits. Newly introgressed lines in the background of basmati will also be a unique genetic stock and a source for BB resistance genes along with semi-dwarfing gene. These lines can be used as semi-dwarf BB resistance donors for further basmati improvement program.

Production of Androgenic Haploid Plants Through *In Vitro* Anther Culture of *Camellia assamica* Ssp. *assamica* (Masters)

Vijay Kumar Mishra[1] and Rakhi Chaturvedi[2]

[1]Department of Bio-Sciences, Rajiv Gandhi University of Knowledge Technologies, Nuzvid, Krishna District 521202, Andhra Pradesh, India
[2]Department of Biosciences and Bioengineering, Indian Institute of Technology Guwahati, Guwahati 781039, Assam

vijay@rguktn.ac.in

Keywords: Androgenesis, anther culture, *Camellia assamica*, embryogenesis, haploids

Abstract

Tea, *Camellia assamica* ssp. *assamica* (Masters) (family Theaceae), is one of the most important commercial crop, which is consumed daily as a nonalcoholic beverage worldwide. Available *Camellia* species are genetically highly heterozygous due to cross-pollination nature with long gestation periods, which poses a hurdle in the production of elite clones or pure breeding/homozygous lines. In the present report, anthers cultures have been opted for the production of haploids. Androgenic haploid embryos were developed through callus formation from microspores during early-to-late uninucleate stages in anther cultures. Maximum callus induction (96%) was obtained on Murashige and Skoog's (MS) medium having 6% (w/v) glucose, supplemented with 5 µM 2,4-dichlorophenoxyacetic acid (2,4-D), 5 µM 6-furfurylaminopurine (kinetin), 800 mg/L L-glutamine, and 200 mg/L L-serine (callus-induction medium). Proliferation of callus occurred when glucose was replaced with 3% (w/v) sucrose in the callus-induction medium. Embryogenesis of nodulated callus was obtained in 85% of the androgenic callus cultures on MS medium containing 10 µM 6-benzylaminopurine (BAP), 3 µM gibberellic acid (GA_3), 800 mg/L L-glutamine and 200 mg/L L-serine (embryo-induction medium). Maturation of embryos occurred when the concentration of growth regulators and adjuvants present in the embryo-induction medium was reduced by 10-fold. Germination of embryos into the complete plantlets took place when the MS medium was supplemented with 10 µM BAP, 1 µM indole-3-butyric acid (IBA), 0.5 µM GA_3, 80 mg/L L-glutamine, and 20 mg/L L-serine. The chromosomal constitution of *in vitro* developed plantlets was confirmed as $2n = x = 15$ by cytological squash preparation of root tips. Flow cytometric analysis of leaves from these *in vitro* developed plantlets confirmed the ploidy status as haploid.

Unraveling the Role of Aldose Reductase Gene from a Resurrection Plant for Methylglyoxal Detoxification and Abiotic Stress Alleviation in Blackgram

Deepak Kumar,[1] Preeti Singh,[2] and Neera Bhalla Sarin[2]

[1]Department of Botany, Central University of Jammu, Rahya-Suchani (Bagla), Jammu, Jammu & Kashmir, India
[2]School of Life Sciences, Jawaharlal Nehru University, New Delhi, India

deepakkumar@cujammu.ac.in

Keywords: Aldose reductase, drought stress, methylglyoxal

Abstract

Drought and salinity exert osmotic stress on terrestrial plants causing water deficit, which consequently affects plant growth and development. Therefore, these stresses are recognized among the most serious challenges to crop production worldwide. Different environmental stresses imposed on plants may result in similar responses at the cellular and molecular levels. This is due to the fact that diverse environmental stresses often activate similar cell signaling pathways and cellular processes, such as the production of stress proteins, upregulation of antioxidants, and accumulation of compatible solutes. It has been demonstrated that the level of methylglyoxal (MG), a cytotoxic compound, increases upon exposure of plants to various abiotic stresses, in addition to Reactive oxygen species (ROS). Methylglyoxal (MG) is mainly catabolized by two major enzymatic pathways. The first is the ubiquitous detoxification pathway, the glyoxalase pathway. An alternate pathway involves aldose reductase that converts MG into acetol in a Nicotinamide adenine dinucleotide phosphate (NADPH)-dependent two-step reaction. It is this pathway that we have exploited in our study. Aldose reductase belongs to the aldo–keto reductase super family of enzymes and plays numerous roles in the metabolism of steroids, sugars, and other carbonyls in plants and animals. A detailed functional validation of aldose reductase homologue *ALDRXV4* was first carried out in a model plant, tobacco, and subsequently used for the transformation of a recalcitrant pulse crop, *Vigna mungo*. Studies with the model plant and crop plant revealed that overexpression of *ALDRXV4* in transgenics were more tolerant not only to osmotic stress but also to salinity stress. The increased aldose reductase activity, higher sorbitol content, and less accumulation of the toxic metabolite, MG,in the transgenic lines under nonstress and stress conditions resulted in increased protection through maintenance of better photosynthetic efficiency, higher relative water content, and less photo-oxidative damage. Together, these findings suggest the potential of engineering aldose reductase levels for better performance of agriculturally important crop plants growing under stress conditions in future.

Chrysanthemum morifolium Ramat. as Antigenotoxic Agent in *Allium cepa* Test System

Sakshi Bhushan[1], Jatinder Kaur Katnoria[2], and Saroj Arora[2]

[1]Department of Botany, University of Jammu, Jammu 180006, Jammu & Kashmir, India
[2]Department of Botanical and Environmental Sciences,Guru Nanak Dev University, Amritsar 143005, Punjab, India

bhushan.sakshi8@gmail.com

Keywords: *Allium cepa*, antigenotoxic, *Chrysanthemum*, secondary metabolites

Abstract

The demand for clear and safe environment has amplified over last few decades due to the accumulation of different pollutants such as pesticides, inorganic fertilizers, heavy metals, poly aromatic hydrocarbons, and so on. Among these pollutants, heavy metals have the tendency to persist in the environment and to cause several ill effects in human beings on exposure. Asteraceae, a family consisting of about 300 species, has a great importance in the field of medicine due to the presence of compounds that possess therapeutic properties such as anti-inflammatory, antimutagenicity, and anticarcinogenicity. Similarly, one of its species, *Chrysanthemum morifolium*, has been proved to be a rich source of secondary metabolites such as pyretheroids, sesquiterpenoids, flavonoids, coumarins, triterpenoids, steroids, phenolics, purines, lipids, aliphatic compounds, and so forth. Considering this, the present study was planned to explore the antigenotoxic potential of aqueous, butanol, and ethyl acetate extracts of *C. morifolium* Ramat. against nickel-induced genotoxicity in *Allium cepa* root chromosomal aberration assay. It was observed that all three extracts showed potential to reduce physiological as well as clastogenic aberrations in cells with significant percentage inhibition. Thus, ability of C. *morifolium* to act as antigenotoxic agent revealed its potential to act as an interesting candidate for future drug discovery using natural bioresources.

Floral Artifice in *Trichosanthes cucumerina* L.: Possible Implications and Effects

Mamata Jamwal and Namrata Sharma
Department of Botany, University of Jammu, Jammu 180006, Jammu & Kashmir, India

jamwal.mamta@gmail.com

Keywords: Anthesis, diurnal, fimbriae, monoecious, nocturnal

Abstract

Trichosanthes cucumerina L. is a herbaceous annual climbing vine valued for its medicinal importance. Plants of the species are monoecious and thus need a vector to transfer pollen from male to female flowers for successful fruit and seed set. Both male and female flowers in the species are small, tubular, and white in color with deeply fimbriate petals. Species is nocturnal, and flower opening is initiated between 20:00 and 20:30 hrs. Flowers of both the sexes remain open for full night and a major part of the next day, with their closing occurs around 18:30 hrs on this day. These flowers adopt interesting strategy during anthesis. Both the types show full expansion of fimbriate petals by 22:30 to 3:00 hrs, thus providing a suitable platform for the landing of nocturnal pollinators especially moths, which visit these flowers in ample numbers. After 3:00 hrs, the fimbriae of petals start retracting and retraction is complete by 5:00 hrs. Thereafter, flowers are seen visited by small diurnal visitors that include butterflies, ants, and *Ceratina* sp. Folding of petals give these pollinators better chance to probe flowers and help in pollen transfer. Dual, that is, both nocturnal and diurnal, pollination results in high fruit (64.2 ± 2.01) and seed (14.6 ± 0.29) set in the species. The presentation will elaborate in detail on this phenomenon.

Evaluation of Nutritional Value and Yield Characteristics of Different Species of *Pleurotus* species by Utilizing Various Agro-Forestry Wastes

Durlav Nr. Singha[1], Bhanita Talukdar[2], Manabendra Kalita[2], Nilakshi Gohain[2] and Gunajit Kalita[3]

[1]Department of Botany, Dakshin Kamrup College, Mirza, Assam, India
[2]Advanced Level Institutional Biotech Hub, Dakshin Kamrup College, Mirza, India
[3]Department of Botany, Nalbari College, Nalbari, Assam, India

durlavns@gmail.com

Keywords: Bagasse, bioconversion, ligno-cellulose, malnutrition, nutraceutical, spawn

Abstract

Mushrooms are organisms of fungal lineage whose popularity and nutraceutical properties have been realised in recent years and presently, identified as excellent food source to cope malnutrition in developing countries like India. Cultivation of edible mushroom through biotechnological intervention involves the bioconversion of various ligno-cellulosic rich agro-forestry wastes into proteinaceous foods. In the present study, cultivation of different *Pleurotus* spp. is carried out using low-cost eco-friendly technology wherein locally available agro-forestry wastes such as straw, saw dust, bagasse, husk and leaves. are utilized as raw material. Moreover, effectiveness of locally available substrates for spawn production were also examined and recorded for spawn running, pinhead formation, fruit body formation and mean yield. The experimental setup consist of complete randomized design with three replicates of each *Pleurotus* spp. and also substrate for spawn and mushroom production. The abiotic factors such as temperature, humidity and light and so forth, also plays an important role in production of mushrooms. Significant differences in yield characteristics was observed among different *Pleurotus* spp. growing in different substrates. Different substrates so chosen for cultivation of mushroom when supplemented with different carbon and nitrogen sources also shows enhanced nutrient content of mushroom. Mushroom cultivation, therefore, generates sufficient employment opportunities not only for unemployed youth but also provides adequate financial support to women folks and other weaker sections of the society.

Commercial Production of Micropropagated *Coccinia indica* (Ivy Gourd)—A Success Story

Daksha Bhatt, Prashant Bhatt, and Dipak Jagdale
Aranya Agri Biotech, Mogar, Gujarat, India

technical@aranyaagri.in

Keywords: *Coccinia indica*, commercialization, ivy gourd, medicinal value, micropropagation

Abstract

Coccinia indica, or ivy gourd, is a common vegetable, which belongs to the cucurbitaceae family. It is a perennial, vine-producing parthenocarpic fruit and propagated by vegetative cuttings during the rainy season. Apart from its use as a food, recent research highlights its excellent medicinal value. Considering the poor survival of vegetative propagation, followed by low yield, an alternative was sought to micropropagate selected mother plants of a locally cultivated variety. Here, we describe the first successful commercialization of micropropagated *C. indica*. Micropropagation was standardized using modified Murashige and Skoog (MS) medium fortified with different concentrations of N6-furfulryladenine (kinetin) and indole-3-acetic acid (IAA). After hardening, field planting was done. The first trials were taken in our own field in year 2009 when 100 plants were planted to monitor their performance. In subsequent years, large-scale plantations were done each year; till date, we have planted around four lac plants. On an average, the farmers harvested every third day between 1000 and 2500 kg vegetable per acre as compared with 150–250 kg from the conventional plantation. Moreover, the uniformity of fruit shape and size is 70% as compared with 30% in the conventional plantation. A typical plantation of 1 ha gives a total yield of 90–200 tons per year. Plants produced by micropropagation outperformed the conventionally propagated plants by a factor of 10. This method of production of micropropagated plants has now become a common practice for this crop, fulfilling demands not only in Gujarat but all over India and world too. On an average, we produce and sell about one lac tissue culture tindora plants per annum. Details on micro-propagated plants versus conventionally propagated plants are discussed.

Regulation of Growth, Photosynthetic Parameters and Sugar Metabolism in Rice (*Oryza sativa* L.) Seedlings by Arsenic and Their Possible Alteration by Silicon

Susmita Das and Asok K. Biswas

Plant Physiology and Biochemistry Laboratory, Centre of Advanced Studies, Department of Botany, University of Calcutta, Kolkata 700019, West Bengal, India

sdbot_rs@caluniv.ac.in

Keywords: Arsenic, arsenate reductase, carbohydrate metabolism, photosynthesis, rice, silicon

Abstract

Arsenic toxicity is a global concern owing to ever-increasing groundwater contamination, crops irrigation in many regions of the world including Bangladesh and West Bengal, India. Arsenic contaminated soil adversely affects economic development of Bengal due to its profound effects on growth and physiological activities of agriculturally important crops, and possess human health risk. Arsenate and arsenite are two important inorganic species of arsenic of which arsenite prevails in paddy soils. The effect of arsenate with or without silicate on growth, estimation of photosynthetic parameters by using LI-6400XT Portable Photosynthesis System and carbohydrate metabolism in rice (*Oryza sativa* L. cv. MTU-1010) seedlings were investigated. In the test cultivar, arsenic toxicity significantly decreased growth parameters while increased the level of oxidative stress markers. Rate of arsenate accumulation and its conversion to arsenite by arsenate reductase were significantly enhanced in all arsenate treated seedlings while in jointly treated seedlings with arsenate and silicate, arsenate accumulation and its conversion to arsenite decreased. Arsenate exposure hampered all chloroplast pigment content, namely, chlorophyll-a, chlorophyll-b, carotene and xanthophyll as well as photosynthetic parameters, namely intercellular-CO_2-concentration, net-photosynthesis, transpiration-rate and stomatal-conductance in rice seedlings. Arsenic toxicity increased the accumulation of sugar contents but decreased starch contents indicating major adaptive mechanisms of plants under arsenic stress that contributes to osmoregulation and provides protection of biomolecules. Activities of sucrose phosphate synthase, acid invertase, and starch phosphorylase were increased, sucrose phosphate synthase, activity was decreased. Co-application of silicate and arsenate showed significant alterations on all the examined parameters compared to arsenate treatment alone due to less accumulation of arsenic in tissue leading to better growth and productivity in rice seedlings. Such studies will help to develop a cost effective and farmer-friendly way to overcome this threat by the application of silicon-enriched fertilizers in arsenic-contaminated rice fields.

Development of Micropropagation Protocols of Apple Rootstocks for High-Density Plantation

Manju Modgil

Department of Biotechnology, Dr Y.S. Parmar University of Horticulture and Forestry Nauni, Solan 173230, Himachal Pradesh, India

manju_modgil@yahoo.com

Keywords: Apple, clonal rootstocks, commercial propagation, hardening, micropropagation, rooting

Abstract

As a result of collaborative efforts of University of Horticulture and Forestry and State Department of Horticulture, Shimla, dwarf and semi-dwarf clonal rootstocks of apple (M&MM series) have been recommended for commercial cultivation for high-density planting after testing. For large-scale production and to cater needs of growers, *in vitro* multiplication is of great use and application. Therefore, repeatable protocols have been developed for the micropropagation of recommended rootstocks using axillary buds/shoot meristems successfully for the production of quality planting material and their demonstration. Different-sized explants were excised and initiated to shoot proliferation on Murashige and Skoog (MS) medium supplemented with benzyl adenine (BA) and gibberellic acid (GA_3), with or without an auxin. Following establishment phase, small shoots emerged from clean explants were subcultured on multiplication medium. Shoot multiplication was influenced by cytokinin type, its concentration, and genotype. Of the cytokinins tested, BA was found superior to others. For rooting, vigorously growing shoots were either given a short pre-treatment in an auxin-containing medium or exposed to an auxin throughout the rooting phase. Rooted plantlets were hardened under controlled conditions and successfully transferred to field. Tissue culture (TC) raised plants were randomly tested for trueness to type and found genetically stable. Procedures were further refined for *in vitro* rooting and hardening of plants for their mass production. Nurseries were raised at different locations of the university where high survival, plant uniformity, and better growth have been observed and the plants were distributed to farmers. The methods standardized here are commercially viable and have provided the basis for rapid bulking up of plants. Therefore, *in vitro* shoot cultures /technologies were passed on to entrepreneurs/industry for further commercialization who are selling TC-raised rootstock plants to farmers of Himachal Pradesh and Jammu & Kashmir, and to Horticulture Department of Arunachal Pradesh.

Production and Characterization of Bio-Oil Derived from Banaba Seeds

Riturani Tamuly[1,2] and Prasenjit Saikia[1,2]

[1]CSIR North East Institute of Science and Technology, Jorhat 785006, Assam, India
[2]Academy of Scientific and Industrial Research, 110020, New Delhi, India

rituranitamuly2012@gmail.com

Keywords: Banaba seeds, bio-char, bio-oil, FTIR, pyrolysis

Abstract

One of the main thermochemical processes to develop useful and valuable bio-fuel such as bio-oil and bio-char is pyrolysis. This paper represents a study on the characterization of the bio-oil derived from banaba seeds (*Lagerstroemia speciosa*, a member of the family Lythraceae) through thermal pyrolysis method at various terminal temperatures from 350 to 650°C with a heating rate of 10°C/min. Maximum bio-oil yield (38%) was obtained at a pyrolysis temperature of 550°C. The product was characterized by Fourier Transform Infrared Spectroscopy (FTIR), Nuclear Magnetic Resonance (NMR; ^{1}H and ^{13}C), and Gas Chromatography–Mass Spectroscopy (GC-MS), which confirmed the presence of various oxygenated hydrocarbons and alcohols. Fuel properties were studied by measuring flash point, pour point, calorific value, and rheological properties.

Quantitative Estimation of Lysine, Cadaverine, Piperideine, and Piperine in *Piper longum* L. Using Reverse-Phase High-Performance Liquid Chromatography

Mrinalini Prasad and Prem Kumar Dantu

Department of Botany, Dayalbagh Educational Institute (Deemed University), Dayalbagh, Agra 282005, Uttar Pradesh, India

premdantu@gmail.com

Keywords: Cadaverine, lysine, piperine, *Piper longum*

Abstract

India seems to be the center of origin for about 110 species of family Piperaceae, and genus *Piper* in India has 50 species. Use of this genus in folk medicines is because of the presence of secondary metabolites. This study shows phytochemical analysis of *Piper longum* (pippali or long pepper). It is widely distributed in tropical and subtropical regions of the world. Its fruit is very important and used in spices after drying. Both fruit and roots of *P. longum* are rich in alkaloids and the key component is piperine. In Ayurveda, it is a good rejuvenator, stimulates the appetite, dispels gas from intestines, and cures respiratory diseases. The main purpose of this study is to emphasize on the recent pharmacology and pharmacognosy research on *P. longum*. The plant is an accepted source of drugs pippali and pippalimulam throughout India. We analyzed the spatio-temporal study of plants in different seasons of India by using chromatography. Main compounds of the study were lysine, cadaverine, piperideine, and piperine in different parts and seasons. All compounds were found to be higher in spikes followed by roots, petiole of the first leaf, and the first leaf. In case of the prespike stage, the maximum quantity of all compounds were found in roots and then in petiole of the first leaf. Interestingly, piperidine accumulation was high in all parts during all the three stages. Cadaverine and piperine contents were minimum in the second leaf and petiole, third leaf and petiole, lower leaf, and internode. While cadaverine was present at reasonably good quantity in nodes during all the three stages, piperine was perceptibly low in this part during all the three stages. Almost all the data show the same pattern in lysine, cadaverine, piperidine, and piperine in three different stages except nodes of the plant.

Characterization of *Areca catechu* Husk Char and its Utilization for Heavy Metal Adsorption

Silvia Saikia[1] and Prasenjit Saikia[2]

[1]Tezpur Central University, Tezpur 784028, Assam, India
[2]North East Institute of Science and Technology, Jorhat 785006, Assam, India

silviasaikia22@gmail.com

Keywords: Areca catechu, char, thermochemical conversion

Abstract

The current investigation deals with the evaluation of properties of char obtained from thermochemical conversion of *Areca catechu* (Ac) husk. Properties of *A. catechu* husk were determined in terms of physico-chemical, biochemical, ultimate, and thermogravimetric (TGA/DTG) and Fourier transform infrared spectroscopic (FTIR) analyses. The thermochemical conversion of *A. catechu* husk was performed using four terminal temperatures namely, 350°C, 450°C, 550°C, and 650°C, with a heating rate of 10°C/min. Characterization and properties of char obtained were evaluated by proximate/ultimate analysis, TGA/DTG, FTIR, X-ray diffraction (XRD), scanning electron microscopy (SEM), and so forth. The heavy metals' adsorption property of *A. catechu* is also reported here.

Effect of Synchronized Sound Waves in the Form of Indian Classical Instrumental Music (Strings and Closed Pipes) on Fruit Ripening Aspect of *Psidium guajava* And *Manilkara zapota*

Sanhita Padhi, P. P. Rout, and R. Swain

Department of Botany, Ravenshaw University, Cuttack, Odisha, India

san_puri9828@rediffmail.com

Keywords: Consonants, dissonant, harmonic frequencies, music therapy, octaves, sound wave technology

Abstract

Harmonic frequencies of octaves in Indian classical music influence the growth of plants starting from the germination to the fruit ripening. It can either enhance or restrict the growth of plants depending on the type of consonants or dissonant being played. Today, the sound waves technology has gained much popularity in this field. It has been studied that sound waves at different frequencies, amplitude, intensity/sound pressure levels; exposure periods; and distances from the sound source influence plants' growth. In recent times, music therapy has become more effective and popular. Soothing and rhythmic harmonic frequencies play a vital role on the basic physical and physiological processes of the living organisms such as plants and animals. Human-played harmonious melodic frequency of instrumental music is proved to exert extraordinary and magical influence on the fruit ripening and many biochemical processes of the experimental species of the present research work: grafted fruits-bearing plants, *Psidium guajava* (guava) and *Manilkara zapota* (sapodilla). In the present research work, experiments were conducted to study the effects of Indian classical *Ragas* (instrumental) having different harmonic frequencies of octaves on fruit-bearing plants. Observations inferred that plants exposed to "dose-dependent" and "time-dependent" soothing, harmonic frequencies of octaves of Indian instrumental classical music showed an earlier fruiting and fruit ripening. Moreover, other biochemical analyses of some primary metabolites in the ripened fruits treated with harmonic, melodious classical music showed astounding results of an increased concentration of metabolites such as reducing sugar, carbohydrates, and proteins inferring the certain development of fruit qualities.

Subcellular Bioengineering of *Artemisia annua* L. for Enhanced Biosynthesis and Accumulation of Artemisinin

M. Z. Abdin

Department of Biotechnology, School of Chemical and Life Sciences, Jamia Hamdard, Hamdard Nagar, 110062, New Delhi, India

mzabdin@jamiahamdard.ac.in

Keywords: Active pharmaceutical ingredient, *Artemisia annua* L., artemisnin biosynthesis, secondary plant metabolism

Abstract

Plants synthesize a vast array of secondary plant metabolites through a network of complex metabolic pathways regulated by endogenous and environmental factors. Although these compounds are believed to be nonessential for plants to live, they play an important role in interaction of plants with the environment ensuring their survival in the ecosystem. Some of these secondary metabolites are also of immense medicinal importance because of their therapeutic value. These are referred as active pharmaceutical ingredients (APIs). The concentrations of these compounds in medicinal plants are, however, very low, limiting their commercial exploitation. Artemisinin is one of these APIs isolated from aerial parts of *Artemisia annua* L. It is a potent antimalarial drug against drug-resistant malaria. In recent times, the demand for artemisinin is exponentially increasing with the increased incidence of drug-resistant malaria throughout the world, especially African and Asian continents. However, the commercial production of artemisinin-based combination therapies has limitation due to the presence of low concentration of artemisinin in plants. Therefore, we employed bioengineering approach to develop transgenic lines of *A. annua* L., overexpressing HMG-Co A reductase *(hmgr)*, amorpha-4, 11-diene synthase *(ads)*, and cytochrome P450 monooxygenase *(cyp71av1)* genes to enhance artemisinin content. The selected transgenic lines were found to accumulate 1.29% to 1.44% artemisinin. Thus, results obtained in these studies, clearly indicate that the synthesis of APIs in medicinal plants is tightly regulated, and bioengineering approach can be used in modulating plant metabolism to improve their biosynthesis, so that drugs manufactured from these APIs could be available at cheaper rates to the public.

A Case Study of Organic Farming in Sonapur, Assam

Kunal Swargiary

Department of Environmental science, Gauhati University, Guwahati, Assam, India

swargairykunal@gmail.com

Keywords: Fertilizer, OFAA, organic cultivation, pesticide

Abstract

Sonapur area with the richness of soil qualities, vegetation, and other suitable requirements is a future hub for organic cultivation. Current study was undertaken to investigate the existing organic farms, both private and government, including interaction with the local farmers with their views and opinions about welcoming the organic sector in their respective regions. The condition is favorable because of the support from the agriculture department toward farmers' efforts in cultivating organic crops. Farmers are in dire need of management skills and development programs. A recent study by Indian Council for Agriculture Research and Assam Agriculture University revealed that the soil is virgin and untouched by chemical fertilizers or pesticides, so can be categorized as organic by default. In 2006, Organic farmers and farms in Assam (OFAA), Sonapur belt was accorded "organic certification" by SKAl International of the Netherlands. The Indian government has also sanctioned a huge amount of money under organic farming schemes for the development of farms and research areas. Different schemes of the government are left unnoticed by the local people due to the lack of awareness. Awareness camps should be organized by government officials in selected areas. An emphasis on the both private and government organic farms could be the leading direction toward the future of organic farming in Assam.

Comparative Assessment of Morphological and Molecular Markers for Describing Genetic Relationships in Some Non-Commercial Banana Cultivars (*Musa* L.) of Assam, India

Debanjali Saikia, Kongkona Borborah, S. K. Borthakur, and Bhaben Tanti

Department of Botany, Gauhati University, Guwahati 781014, Assam, India

btanti@gauhati.ac.in

Keywords: Genetic variation, ISSR markers, morphometrics, non-commercial banana cultivars

Abstract

This study aimed to compare the genetic diversity of 14 non-commercial banana (*Musa* L.) cultivars of Assam, India, by using 48 morphological traits and inter simple sequence repeats (ISSR) markers. Results showed that the accessions of banana samples exhibited a significant amount of variation for their morphological studied parameters. The morphological data were presented in the form of hierarchical clusters and principal components. Furthermore, genetic variability among accessions using ISSR markers, showed the average percentage of polymorphism. Total seven primers were selected to evaluate the genetic diversity among the experimental banana samples. Each primer could generate polymorphism among the accessions, which may be due to a mutation at priming sites. A total of 62 bands were detected, of which 56 bands showed polymorphism. The highest number of bands was detected with the primer UBC-843, and the lowest number of bands was observed with UBC-848. The genetic similarity index was prepared using Jaccard's similarity coefficient, and the range of genetic similarity was from 0.28 to 0.77, with an average of 0.51. Dendrogram produced from the cluster analysis showed the clear division of the genotypes into two distinct clusters. Seven accessions namely *Bharatmoni*, *Assamiya-malbhog*, *Gobin tulashi*, *Bokmoni*, *Bangali-malbhog*, *Katiya-jahaji*, and *Abor-malbhog* were clustered in the first group. Second group contains the accessions namely *Jatikol*, *Adeel*, *Guwahatia-kol*, *Fessa-monohar*, *Athiya kol*, *Ximalu-monohar*, and *Bogi-monohar*. Relationships between morphological traits and ISSR markers variation were estimated using Mantel test. Morphological characters are good markers for overall genetic variation, whereas ISSR markers cannot resolve plant groups defined by visible traits up to interspecies level.

Isolation and Identification of Endophytic Bacterial Communities Associated with *Piper logum* L., An Important Medicinal Plant using Metagenomic Approach

Mubashar Nazir Mintoo and Prem Kumar Dantu

Plant Biotechnology Lab, Department of Botany, Dayalbagh Educational Institute, Dayalbagh, Agra 282005, Uttar Pradesh, India

premdantu@gmail.com

Keywords: Bioinformatics, endophytic bacteria, phenotypic characterization, *Piper longum* 16S rRNA

Abstract

Endophytic bacteria were isolated from nodes, internodes, petiole, and root segments of *Piper longum*. These bacteria were cultured on specific culture medium. The aim of this work was to isolate bacterial endophytes from *P. longum* and identify them by using both conventional and metagenomic approaches. These bacteria were first differentiated on the basis of morphological parameters. Six different colonies were isolated, purified, and selected for further analysis based upon morphological and colony characterization. These isolated bacteria were used to compare the phenological characterization with molecular identification on the basis of analysis using 16S rDNA sequences. 16S rDNA gene is highly diverse in different bacteria and serves as a phylogenetic marker for identification. In these bacteria, the conserved 16S rRNA was amplified using specific primers and amplicons sequenced. Sequences have been deposited with National Center for Biotechnology Information (NCBI), and accession numbers were obtained. Bioinformatics of the sequences identified bacteria as endophytic bacteria 135L-3, *Enterobacter* sp. SQ6-43, *Bacillus casamacensis* strain TN3, *Alishwanella* sp. JS-30, *Bacterium* B28, and *Enterobacter ludwigii* strain g45 belonging to two different bacterial groups, that is, γ-proterobacteria and firmicutes. These bacteria were present in different branches within a tree, suggesting that these clusters showed different phylotypes. These bacteria were known to perform beneficial roles in plant growth promotion and several other processes of plant metabolism. The present study is the first report about the endophytic bacterial population in *P. longum* using 16S rRNA technique through which six endophytic bacteria have been identified and characterized. The identified bacteria belong to γ-proterobacteria and firmicutes groups.

Purification of Biogas by Pressure Swing Adsorption Process with Mitigation of Methane Gas Emission

Goldy Shah,[1] Virendra Kumar Vijay,[1] and Kamal Kishore Pant[2]
[1]Centre for Rural Development and Technology, Indian Institute of Technology Delhi, Hauz Khas, New Delhi

[2]Department of Chemical Engineering, Indian Institute of Technology Delhi, Hauz Khas, New Delhi

goldyshah@gmail.com

Keywords: Biogas upgradation, methane loss, pressure swing adsorption

Abstract

Multi-fold increase in energy demand worldwide has led to a situation where quest for alternate energy sources has become unavoidable. Beside many other problems, the world is challenged by global warming, and the target is to reduce greenhouse gas (GHG) emissions. Moreover, biogas has potential as a renewable source of energy for rural areas because the majority of the rural population is agrarian, thereby generating substantial agricultural waste throughout the year. Interestingly, the biogas application at commercial scale may help in getting rid of biomass waste and reducing GHG emissions. The upgraded biogas can be utilized in many ways but local use is still the most common option and the most economically viable. However, compressed biogas is considered to be a potential alternative to compressed natural gas (CNG) because of its compositional similarity to CNG. Pressure-swing-adsorption (PSA) based biogas upgradation technology has the remarkable potential to produce bio-CNG as an alternative for natural gas. Over the decades, PSA performance has been described by different mathematical models, but there is limited work done in the field of process simulation, where adsorbent models can be incorporated with other unit operations using commercially available simulator. A major issue faced in the majority of the upgradation plant is that an off-gas stream with a significant methane (CH_4) content is produced and released directly into the atmosphere, which requires to be treated further to avoid emission into the environment. Therefore, this work aims to evaluate parameters and off-gas treatment to develop an efficient model for reducing CH_4 loss. Effects of the parameters such as flow rate, cycle time, and CH_4 concentration on CH_4 and carbon dioxide (CO_2) concentrations were optimized. The optimum conditions for the maximal CH_4 concentration (90.82%) and the lowest CO_2 concentration (7.5%) were 17.56 m^3/h flow rate, 51.89 min duration, and 53.04% CH_4 concentration in raw biogas.

Leaf-Waste-Based Biochar as a Promising Adsorbent for Effectual Hydrogen Sulfide Removal from Biogas

Shivali, Virendra Kumar Vijay, P. M. V. Subbarao, and Ram Chandra

Centre for Rural Development and Technology, Indian Institute of Technology Delhi, Hauz Khas, New Delhi

shivalisahota@gmail.com

Keywords: Adsorption, biogas, H_2S removal, leaf waste biochar

Abstract

Biogas is arguably a more versatile renewable energy source due to its determinate energy value and abundant biomass availability. The main hurdle for implementing the technology for vehicular application is that biogas contains certain concentration of impurity of hydrogen sulfide (H_2S), which leads to the corrosion of expensive metal parts inside plants and engines due to its high corrosiveness and toxicity. Previous literature studies have shown that activated carbon is an effective carbonaceous sorbent for H_2S removal due to its high surface area and porosity for sorption. But the high cost of activated carbon is still a competent challenge.To overcome such problems related to existing technologies, biochar has been found to be one of the best ways to remove H_2S at lower cost. Moreover, biochar-based adsorbents are eco-friendly and relatively 10 times cheaper than commercially available activated carbon. Especially, application of leaf-waste-based biochar will be beneficial for rural applications, where small, independent, robust and decentralized units for biogas production and subsequent upgradation can be installed. Thus, the present study is aimed to evaluate the suitability of leaf-waste-based biochar for H_2S removal from biogas for decentralized rural units. Leaf waste was carbonized at different temperatures to study its effectiveness for H_2S removal from biogas in an adsorption tower. Moreover, freshly prepared biochar and saturated biochar were characterized using attenuated total reflectance Fourier transform infrared spectroscopy, X-ray diffraction, scanning electron microscopy, and energy dispersive spectrometer to develop an insight into the adsorption mechanism. These observations show that leaf biochar can play two in one role as an H_2S absorbent and a nutrient-rich amendment for sulfur-deficient soils along with several factors affecting H_2S adsorption such as surface area of biochar, higher alkaline pH, carbonization temperature, and mineral elements present on the surface of biochar.

In Vitro Optimization Studies for the Production and Scale-Up of Artemisinin—An Important Antimalarial Drug

Rajendra Adak and Rakhi Chaturvedi

Department of Biosciences and Bioengineering, Indian Institute of Technology Guwahati, Guwahati 781039, Assam

rakhi_chaturvedi@iitg.ac.in

Keywords: Antimalarial, *Artemisia annua*, artemisinin, elicitors, secondary metabolites

Abstract

Artemisia annua L. is a diploid cross-pollinating species, which belongs to the family Asteraceae. It is also known as Qinghao, Wormwood, and Sweet Annie. This herb is a hub of several medicinally important secondary metabolites such as artemisinin, asarteannuin, absinthin, myricetin, quercetin, caffeic acid, and gallic acid. Although each one of these metabolites is important, artemisinin is in high demand owing to its wide therapeutic applications in malaria, cancer, and peptic ulcer. Additionally, the compound is reported to have potent antibacterial and anti-inflammatory properties. Essential oil obtained from *A. annua* is known to poses powerful insect-repelling properties. Despite being a high-value crop, extracting consistent amount of the metabolites from this plant becomes difficult due to occurrence of high variability within the existing genera.

In such a scenario, *in vitro* culture technique would serve as the best method for developing high-metabolite-producing cell lines independent of genetic and seasonal variations. The present review touches aspects for the best *in vitro* regeneration protocol to obtain elite clones in *A. annua* and further inspect into yield enhancement strategies using biotic (jasmonic acid and salicylic acid) and abiotic physical elicitors (UV-B, salinity, and temperature) to scrutinize and attain ideal parameters for scale-up studies of artemisinin production using a bioreactor.

Plant Regeneration Via Repetitive Secondary Embryogenesis from Androgenic Embryos in Suspension Cultures of *Camellia assamica* ssp. *assamica* and Clonal Fidelity Assessment using RAPD Marker

Ruchira Bajpai and Rakhi Chaturvedi

Department of Biosciences and Bioengineering, Indian Institute of Technology Guwahati, Guwahati 781039, Assam, India

rakhi_chaturvedi@iitg.ac.in

Keywords: FESEM, histology, hyperhydricity, RAPD, Recurrent embryogenesis, two-step protocol

Abstract

Rapid, recurrent embryogenesis was attained from *in vitro* developed androgenic haploid embryos of *Camellia assamica* ssp. *assamica* using a liquid medium. Primary globular embryos from an embryo multiplication medium, Murashige and Skoog (MS) medium supplemented with 10 µM 6-benzylaminopurine (BAP), 3 µM gibberellic acid (GA_3), 800 mg/L L-glutamine, and 200 mg/L L-serine, were subjected to pretreatment with 18 µM abscisic acid (ABA) either alone or in combination with either of the osmoticum, 25 g/L mannitol, 30 g/L polyethylene glycol (PEG-400) or 1 g/L glycine betaine. Higher multiplication with a fivefold increase in secondary embryogenesis was achieved after pretreatment of primary embryos on MS medium consisting of 18 µM ABA and 25 g/L mannitol for 30 days, followed by transfer to an embryo maturation medium (control), MS with 1 µM BAP, 0.3 µM GA_3, 80 mg/L L-glutamine, and 20 mg/L L-serine. Effect of physical state of the medium on embryo germination leading to shoot differentiation was also studied. An increased germination rate of 66.6% in secondary embryos was obtained when cultures were initially kept for 20 days in a liquid medium followed by transferred to a semisolid medium. In comparison, a liquid–solid medium favoured 44.4% embryo germination in 30 days, whereas a semisolid medium resulted in only 25% embryo germination in 45 days. The continuous immersion in a liquid medium led to hyperhydricity in cultures. Therefore, a two-step protocol, involving combination of liquid and semisolid medium transition was chosen for attaining large-scale multiplication of this elite clone in less time. Histology and field emission scanning electron microscopic (FESEM) analyses were performed to determine ontological stages of embryo development. Clonal fidelity of plants attained from liquid–semisolid medium transition was assessed using random amplified polymorphic DNA (RAPD) markers to rule out any somaclonal variations occurred during multiple transfer cycle.

International Plant Propagators' Society (IPPS)
174 Crestview Drive
Bellefonte, PA 16823-8516
USA

ISBN 978-1-7138-1369-9

9 781713 813699

CURRAN ASSOCIATES INC.
proceedings.com